Stochastic Mechanics
Random Media
Signal Processing and Image Synthesis
Mathematical Economics
Stochastic Optimization and Finance
Stochastic Control

Applications of Mathematics

Stochastic Modelling and Applied Probability

43

Springer-Science+Business Media, LLC

Applications of Mathematics

1 Fleming/Rishel, **Deterministic and Stochastic Optimal Control** (1975)
2 Marchuk, **Methods of Numerical Mathematics,** Second Ed. (1982)
3 Balakrishnan, **Applied Functional Analysis,** Second Ed. (1981)
4 Borovkov, **Stochastic Processes in Queueing Theory** (1976)
5 Liptser/Shiryayev, **Statistics of Random Processes I: General Theory** (1977)
6 Liptser/Shiryayev, **Statistics of Random Processes II: Applications** (1978)
7 Vorob'ev, **Game Theory: Lectures for Economists and Systems Scientists** (1977)
8 Shiryayev, **Optimal Stopping Rules** (1978)
9 Ibragimov/Rozanov, **Gaussian Random Processes** (1978)
10 Wonham, **Linear Multivariable Control: A Geometric Approach,** Third Ed. (1985)
11 Hida, **Brownian Motion** (1980)
12 Hestenes, **Conjugate Direction Methods in Optimization** (1980)
13 Kallianpur, **Stochastic Filtering Theory** (1980)
14 Krylov, **Controlled Diffusion Processes** (1980)
15 Prabhu, **Stochastic Storage Processes: Queues, Insurance Risk, Dams, and Data Communication,** Second Ed. (1998)
16 Ibragimov/Has'minskii, **Statistical Estimation: Asymptotic Theory** (1981)
17 Cesari, **Optimization: Theory and Applications** (1982)
18 Elliott, **Stochastic Calculus and Applications** (1982)
19 Marchuk/Shaidourov, **Difference Methods and Their Extrapolations** (1983)
20 Hijab, **Stabilization of Control Systems** (1986)
21 Protter, **Stochastic Integration and Differential Equations** (1990)
22 Benveniste/Métivier/Priouret, **Adaptive Algorithms and Stochastic Approximations** (1990)
23 Kloeden/Platen, **Numerical Solution of Stochastic Differential Equations** (1992)
24 Kushner/Dupuis, **Numerical Methods for Stochastic Control Problems in Continuous Time** (1992)
25 Fleming/Soner, **Controlled Markov Processes and Viscosity Solutions** (1993)
26 Baccelli/Brémaud, **Elements of Queueing Theory** (1994)
27 Winkler, **Image Analysis, Random Fields, and Dynamic Monte Carlo Methods: An Introduction to Mathematical Aspects** (1994)
28 Kalpazidou, **Cycle Representations of Markov Processes** (1995)
29 Elliott/Aggoun/Moore, **Hidden Markov Models: Estimation and Control** (1995)
30 Hernández-Lerma/Lasserre, **Discrete-Time Markov Control Processes: Basic Optimality Criteria** (1996)
31 Devroye/Györfi/Lugosi, **A Probabilistic Theory of Pattern Recognition** (1996)
32 Maitra/Sudderth, **Discrete Gambling and Stochastic Games** (1996)

Jiongmin Yong Xun Yu Zhou

Stochastic Controls

Hamiltonian Systems and HJB Equations

Springer

Jiongmin Yong
Department of Mathematics
Fudan University
Shanghai 200433, China

Xun Yu Zhou
Department of Systems Engineering
and Engineering Management
The Chinese University of Hong Kong
Shatin, Hong Kong

Managing Editors

I. Karatzas
Departments of Mathematics and Statistics
Columbia University
New York, NY 10027, USA

M. Yor
CNRS, Laboratoire de Probabilités
Université Pierre et Marie Curie
4, Place Jussieu, Tour 56
F-75252 Paris Cedex 05, France

Mathematics Subject Classification (1991): 93E20, 49K45, 49L20, 49L25, 60H30, 90A09, 70H20, 35C15

Library of Congress Cataloging-in-Publication Data
Yong, J. (Jiongmin), 1958–
Stochastic controls : Hamiltonian systems and HJB equations / Jiongmin Yong, Xun Yu Zhou.
p. cm. — (Applications of mathematics ; 43)
Includes bibliographical references (p. –) and index.
ISBN 978-1-4612-7154-3 ISBN 978-1-4612-1466-3 (eBook)
DOI 10.1007/978-1-4612-1466-3
1. Stochastic control theory. 2. Mathematical optimization. 3. Hamiltonian systems. 4. Hamilton–Jacobi equations. I. Zhou, Xun Yu. II. Title. III. Series.
QA402.37.Y66 1999
629.8′312—dc21 98-55411

Production managed by Allan Abrams; manufacturing supervised by Thomas King.
Photocomposed copy prepared from the authors' LaTeX files.

9 8 7 6 5 4 3 2 1

ISBN 978-1-4612-7154-3 SPIN 10707573

献给我们的父母：

雍文耀　陈湘霞

周庆基　沈采真

To Our Parents

Wenyao Yong and Xiangxia Chen

Qingji Zhou and Caizhen Shen

Preface

As is well known, Pontryagin's maximum principle and Bellman's dynamic programming are the two principal and most commonly used approaches in solving stochastic optimal control problems.* An interesting phenomenon one can observe from the literature is that these two approaches have been developed separately and independently. Since both methods are used to investigate the same problems, a natural question one will ask is the following:

(Q) *What is the relationship between the maximum principle and dynamic programming in stochastic optimal controls?*

There did exist some researches (prior to the 1980s) on the relationship between these two. Nevertheless, the results usually were stated in heuristic terms and proved under rather restrictive assumptions, which were not satisfied in most cases.

In the statement of a Pontryagin-type maximum principle there is an adjoint equation, which is an ordinary differential equation (ODE) in the (finite-dimensional) deterministic case and a stochastic differential equation (SDE) in the stochastic case. The system consisting of the adjoint equation, the original state equation, and the maximum condition is referred to as an *(extended) Hamiltonian system*. On the other hand, in Bellman's dynamic programming, there is a partial differential equation (PDE), of first order in the (finite-dimensional) deterministic case and of second order in the stochastic case. This is known as a *Hamilton–Jacobi–Bellman (HJB) equation*. This leads to the following question, which is essentially a rephrase of Question (Q):

(Q′) *What is the relationship between Hamiltonian systems and HJB equations?*

Or, even more generally,

(Q″) *What is the relationship between ODEs/SDEs and PDEs?*

Once the question is asked this way, one will immediately realize that similar questions have already been or are being addressed and studied in other fields. Let us briefly recall them below.

Analytic Mechanics. Using Hamilton's principle and Legendre's transformation, one can describe dynamics of a system of particles by a

---

* Here, by a stochastic optimal control problem we mean a completely observed control problem with a state equation of the Itô type and with a cost functional of the Bolza type; see Chapter 2 for details. When the diffusion coefficient is identically zero, and the controls are restricted to deterministic functions, the problem is reduced to a deterministic optimal control problem.

family of ODEs called *Hamilton's canonical system* or the *Hamiltonian system*. On the other hand, by introducing *Hamilton's principal function*, one can describe the particle system by a PDE called the *Hamilton–Jacobi (HJ) equation*. These two ways are in fact equivalent in the sense that the solutions of the canonical system can be represented by that of the HJ equation, and vice versa. One easily sees a strong analogy between optimal control and analytic mechanics. This is not surprising, however, since the classical calculus of variations, which is the foundation of analytic mechanics, is indeed the origin of optimal control theory.

Partial Differential Equations. There is a classical *method of characteristics* in solving PDEs. More specifically, for a first-order PDE, there is an associated family of ODEs for curves, called *characteristic strips*, by which the solutions to the PDE can be constructed. In the context of (deterministic) optimal controls, the Hamiltonian system involved in the maximum principle serves as the characteristics for the HJB equation involved in the dynamic programming.

Stochastic Analysis. The stochastic version of the method of characteristics is the *Feynman–Kac formula*, which represents the solutions to a *linear* second-order parabolic or elliptic PDE by those to some SDEs. On the other hand, a reversed representation has been recently developed via the so-called *four-step scheme*, which represents the solutions to a coupled forward–backward SDE by those to a PDE. A deterministic version of this is closely related to the so-called *invariant embedding*, which was studied by Bellman-Kalaba-Wing.

Economics. The key to understanding the economic interpretation of optimal control theory is the *shadow price* of a resource under consideration. The very definition of the shadow price originates from one of the relationships between the maximum principle and dynamic programming, namely, the shadow price (adjoint variable) is the rate of change of the performance measure (value function) with respect to the change of the resource (state variable).

Finance. The celebrated Black–Scholes formula indeed gives nothing but a way of representing the option price (which is the solution to a backward SDE) by the solution to the Black–Scholes equation (which is a parabolic PDE).

Interestingly enough, all the relationships described above can be captured by the following simple, generic mathematical formula:

$$y(t) = \theta(t, x(t)),$$

where $(x(t), y(t))$ satisfies some ODE/SDE and θ satisfies some PDE. For example, in the relationship between the maximum principle and dynamic programming, $(x(t), y(t))$ is the solution to the Hamiltonian system and $[-\theta]$ is the gradient in the spatial variable of the value function (which is the solution to the HJB equation). In the Black–Scholes model, $y(t)$ is the option price, $x(t)$ is the underlying stock price, and θ is the solution to the Black–Scholes PDE.

Before studying Question (Q), one has first to resolve the following two problems:

(P1) *What is a general stochastic maximum principle if the diffusion depends on the control and the control domain is not necessarily convex?*

This problem has been investigated since the 1960s. However, almost all the results prior to 1980 assume that the diffusion term does not depend on the control variable and/or the diffusion depends on the control but the control domain is convex. Under these assumptions, the statements of the maximum principle and their proofs are very much parallel to those of the deterministic case. One does not see much essential difference between stochastic and deterministic systems from those results. The stochastic maximum principle for systems with control-dependent diffusion coefficients and possibly nonconvex control domains had long been an outstanding open problem until 1988.

(P2) *How is one to deal with the inherent nonsmoothness when studying the relationship between the maximum principle and dynamic programming?*

The relationship unavoidably involves the derivatives of the value functions, which as is well known could be nonsmooth in even very simple cases.

During 1987–1989, a group led by Xunjing Li at the Institute of Mathematics, Fudan University, including Ying Hu, Jin Ma, Shige Peng, and the two authors of the present book, was studying those problems and related issues in their weekly seminars. They insisted on tackling the control-dependent diffusion cases, and this insistence was based on the following belief: Only when the controls/decisions could or would influence the scale of uncertainty (as is indeed the case in many practical systems, especially in the area of finance) do the stochastic problems differ from the deterministic ones. In the stimulating environment of the seminars, Problems (P1) and (P2) were solved almost at the same time in late 1988, based on the introduction of the so-called *second-order adjoint equation.* Specifically, Peng, then a postdoctoral fellow at Fudan, solved Problem (P1) by considering the quadratic terms in the Taylor expansion of the spike variation, via which he established a new form of maximum principle for stochastic optimal controls. On the other hand, Zhou (who was then a Ph.D. student at Fudan) found a powerful way for solving Problem (P2). By utilizing viscosity solution theory, he managed to disclose the relationship between the first-order (respectively, second-order) adjoint equations and the first-order (respectively, second-order) derivatives of the value functions. After 1989, members of the Fudan group went to different places in the world, but the research they carried out at Fudan formed the foundation of their further research. In particular, studies on nonlinear backward and forward–backward

SDEs by Pardoux–Peng, and Ma–Yong are natural extensions of those on the (linear) adjoint equations in the stochastic maximum principle. This fashionable theory soon became a notable topic among probabilists and control theorists, and found interesting applications in stochastic analysis, PDE theory, and mathematical finance. The remarkable work on the nonlinear Feynman–Kac formula (representing solutions to *nonlinear* PDEs by those to backward SDEs) by Peng and the four-step scheme (representing solutions to forward–backward SDEs by those to PDEs) by Ma–Protter–Yong once again remind us about their analogy in stochastic controls, namely, the relationship between stochastic Hamiltonian systems and HJB equations. On the other hand, (stochastic) verification theorems by Zhou and Zhou–Yong–Li by means of viscosity solutions are extensions of the relationship between the maximum principle and dynamic programming from open-loop controls to feedback controls. These verification theorems lead to optimal feedback synthesis without involving derivatives of the value functions. Finally, the recent work by Chen–Li–Zhou, Chen–Yong, and Chen–Zhou on stochastic linear quadratic (LQ) controls with *indefinite* control weighting matrices in costs demonstrates how fundamentally different it is when the control enters into the diffusion term. The LQ case also provides an important example where the maximum principle and dynamic programming are *equivalent* via the *stochastic Riccati equation.*

The purpose of this book is to give a systematic and self-contained presentation of the work done by the Fudan group and related work done by others, with the core being the study on Question (Q) or (Q′). In other words, the theme of the book is to unify the maximum principle and dynamic programming, and to demonstrate that viscosity solution theory provides a nice framework to unify them. While the main context is in stochastic optimal controls, we try whenever possible to disclose some intrinsic relationship among ODEs, SDEs, and PDEs that may go beyond control theory. When writing the book, we paid every attention to the coherence and consistency of the materials presented, so that all the chapters are closely related to each other to support the central theme. In some sense, the idea of the whole book may be boiled down to the single formula $y(t) = \theta(t, x(t))$, which was mentioned earlier. That said, we do not mean to trivialize things; rather we want to emphasize the common ground of seemingly different theories in different areas. In this perspective, the Black–Scholes formula, for instance, would not surprise a person who is familiar with mechanics or the Feynman–Kac formula.

Let us now sketch the main contents of each chapter of the book.

Chapter 1. Since the book is intended to be self-contained, some preliminary materials on stochastic calculus are presented. Specifically, this chapter collects notions and results in stochastic calculus scattered around in the literature that are related to stochastic controls. It also unifies terminology and notation (which may differ in different papers/books) that are to be used in later chapters. These materials are mainly for beginners (say, graduate students). They also serve as a quick reference for knowledgeable readers.

Chapter 2. The stochastic optimal control problem is formulated and some examples of real applications are given. However, the chapter starts with the deterministic case. This practice of beginning with deterministic problems is carried out in Chapters 3–6 as well. The reasons for doing so are not only that the deterministic case itself may contain important and interesting results, but also that readers can see the essential difference between the deterministic and stochastic systems. In the formulation of stochastic control problems we introduce strong and weak formulations and emphasize the difference between the two, which is not usually spelled out explicitly in the literature. Stochastic control models other than the "standard" one studied in this book are also briefly discussed. This chapter finally provides a very extensive literature review ranging from the very origin of optimal control problems to all the models and applied examples presented in this chapter.

Chapter 3. A stochastic Hamiltonian system is introduced that consists of two backward SDEs (adjoint equations) and one forward SDE (the original state equation) along with a maximum condition. The general stochastic maximum principle is then stated and proved. Cases with terminal state constraints and sufficiency of the maximum principle are discussed.

Chapter 4. First, a stochastic version of Bellman's principle of optimality is *proved* by virtue of the weak formulation, based on which HJB equations are derived. The viscosity solution is introduced as the tool to handle the inherent nonsmoothness of the value functions. Some properties of the value functions and viscosity solutions of the HJB equations are then studied. It is emphasized that the time variable here plays a special role due to the nonanticipativeness of the underlying system. Finally, a simplified proof (compared with the existing ones) of the uniqueness of the viscosity solutions is presented. Notice that the verification technique involved in the dynamic programming is deferred to Chapter 5.

Chapter 5. Classical Hamilton–Jacobi theory in mechanics is reviewed first to demonstrate the origin of the study of the relationship between the maximum principle and dynamic programming. The relationship for deterministic systems is investigated and is compared with the method of characteristics in PDE theory, the Feynman–Kac formula in probability theory, and the shadow price in economics. The relationship for stochastic systems is then studied. It starts with the case where the value function is smooth to give some insights, followed by a detailed analysis for the nonsmooth case. Finally, stochastic verification theorems workable for the nonsmooth situation are given, and the construction of optimal feedback controls is discussed.

Chapter 6. This chapter investigates a special case of optimal control problems, namely, the linear quadratic optimal control problems (LQ problems). They constitute an extremely important class of optimal control problems, and the solutions of LQ problems exhibit elegant properties due to their simple and nice structures. They also nicely exemplify the general theory developed in Chapters 3–5. In the chapter an LQ problem is first treated as an optimization problem in an infinite-dimensional space, and abstract results are obtained to give insights. Then linear optimal state feedback is established via the so-called stochastic Riccati equation. It is pointed out that both the maximum principle and dynamic programming can lead to the stochastic Riccati equation, by which one can see more clearly the relationship between the maximum principle and dynamic programming (actually, these two approaches are *equivalent* in the LQ case). We emphasize that the control weighting matrices in the cost are allowed to be *indefinite* in our formulation. Therefore, it is essentially different from the deterministic case. Stochastic Riccati equations are extensively studied for various cases. Finally, as an example, a mean–variance portfolio selection is solved by the LQ method developed.

Chapter 7. This chapter presents the latest development on backward and forward–backward SDEs, with an emphasis on the relationship between nonlinear SDEs and nonlinear PDEs. Although the topics in this chapter go beyond the scope of stochastic controls, they originate from stochastic controls as mentioned earlier. The chapter begins with the original argument of Bismut for studying linear backward SDEs by using the martingale representation theorem. Then the existence and uniqueness of solutions to nonlinear backward SDEs are investigated for two types of time durations, finite deterministic horizon and random horizon, by virtue of two different methods. Feynman–Kac-type formulae with respect to both forward and backward SDEs are presented. Next, a kind of inverse of the Feynman–Kac-type formulae, the so-called four-step scheme, which represents solutions to forward–backward SDEs by those to PDEs, is discussed. Solvability and nonsolvability of forward–backward SDEs are also analyzed. Finally, the Black–Scholes formula in option pricing is derived by the four-step scheme.

The idea of writing such a book was around in late 1994 when JY was visiting XYZ in Hong Kong. While discussing the stochastic verification theorems, they realized that the series of works done by the Fudan group were rich enough for a book, and there *should* be a book as a systematic account of these results. The plan became firm with encouragement from Wendell Fleming (Brown), Ioannis Karatzas (Columbia), and Xunjing Li (Fudan). The authors are greatly indebted to Robert Elliott (Alberta), Wendell Fleming (Brown), Ulrich Haussmann (British Columbia), Ioannis Karatzas (Columbia), Thomas Kurtz (Wisconsin-Madison), Mete Soner (Princeton), and Michael Taksar (SUNY-Stony Brook), who substantially reviewed some or all chapters, which led to a much improved version. Michael Kohlmann (Konstanz) and Andrew Lim (CUHK) read carefully large portions of the manuscript and offered numerous helpful suggestions. During various stages in the prolonged, four-year course of the project, many experts and friends have shown their concern and encouragement, conveyed their comments, or sent their research works for the book. Among them the following deserve special mention: Alain Bensoussan (CNES), Leonard Berkovitz (Purdue), Giuseppe Da Prato (Scuola Normale Superiore), Darrell Duffie (Stanford), Tyrone Duncan (Kansas), Fausto Gozzi (Scuola Normal Superiore), Suzanne Lenhart (Tennessee), Zhuangyi Liu (Minnesota-Duluth), John Moore (ANU), Makiko Nisio (Kobe), Bozenna Pasik-Duncan (Kansas), Thomas Seidman (Maryland-Baltimore County), Hiroshi Tanaka (Keio), Wing Wong (CUHK), Jia-an Yan (Academia Sinica), George Yin (Wayne State), and Qing Zhang (Georgia). Especially, both authors would like to express appreciation to their long-time teachers and/or colleagues in the Fudan group: Shuping Chen (Zhejiang), Ying Hu (Rennes), Xunjing Li (Fudan), Jin Ma (Purdue), Shige Peng (Shandong), and Shanjian Tang (Fudan), whose elegant research work provided a rich source for this book.

JY would like to acknowledge the partial support from the Natural Science Foundation of China, the Chinese Education Ministry Science Foundation, the National Outstanding Youth Foundation of China, and the Li Foundation at San Francisco, USA. In particular, with the financial support of a Research Fellowship of the Chinese University of Hong Kong (CUHK), JY visited XYZ at CUHK in the Spring of 1998 for half a year, which made it possible for the two authors to fully concentrate on finalizing the book.

In the career of XYZ, he has been influenced enormously by three scholars: Xunjing Li (Fudan), Makiko Nisio (Kobe), and Hiroshi Tanaka (Keio), whom he has the privilege of having worked with for a substantial period of time, and he would like to take this opportunity to pay them his highest respect. Also, he would like to acknowledge the support from the Research Grant Council and Additional Funding for Industry Support of the Hong Kong Government, and the Mainline Research Scheme of CUHK.

It has been a truly enjoyable experience to work with the staff at Springer-Verlag, especially the executive editor of statistics John Kimmel and the copyeditor David Kramer, whose helpful and professional services for the book are gratefully acknowledged.

Last, but not least, both authors would like to thank their families for their long-lasting support and love.

JY, Shanghai
XYZ, Hong Kong

November 1998

Contents

Notation

The following notation is frequently used in the book.

$\mathbb{R}^n$ — n-dimensional real Euclidean space.

$\mathbb{R}^{n\times m}$ — the set of all $(n\times m)$ real matrices.

$\mathcal{S}^n$ — the set of all $(n\times n)$ symmetric matrices.

$\mathcal{S}^n_+$ — the set of all $(n\times n)$ nonnegative definite matrices.

$\widehat{\mathcal{S}}^n_+$ — the set of all $(n\times n)$ positive definite matrices.

$\operatorname{tr}(A)$ — the trace of the square matrix A.

$x^\top$ — the transpose of the vector (or matrix) x.

$\langle\cdot\,,\cdot\rangle$ — inner product in some Hilbert space.

$\mathbf{Q}$ — the set of all rational numbers.

$\mathbb{N}$ — the set of natural numbers.

$|N|$ — Lebesgue measure of the set N.

$\stackrel{\Delta}{=}$ — Defined to be (see below).

I_A — the indicator function of the set A: $I_A(x)\stackrel{\Delta}{=}\begin{cases}1, & x\in A,\\ 0, & x\notin A.\end{cases}$

$\varphi^+\stackrel{\Delta}{=}\max\{\varphi,0\}, \qquad \varphi^-\stackrel{\Delta}{=}-\min\{\varphi,0\}.$

$a\vee b\stackrel{\Delta}{=}\max\{a,b\}, \qquad a\wedge b\stackrel{\Delta}{=}\min\{a,b\}.$

2^Ω — the set of all subsets of Ω.

ϕ — the empty set.

$C([0,T];\mathbb{R}^n)$ — the set of all continuous functions $\varphi:[0,T]\to\mathbb{R}^n$.

$C([0,\infty);\mathbb{R}^n)$ — the set of all continuous functions $\varphi:[0,\infty)\to\mathbb{R}^n$.

$C_b(U)$ — the set of all uniformly bounded, continuous functions on U.

$L^p(0,T;\mathbb{R}^n)$ — the set of Lebesgue measurable functions $\varphi:[0,T]\to\mathbb{R}^n$ such that $\int_0^T|\varphi(t)|^p dt<\infty$ $(p\in[1,\infty))$.

$L^\infty(0,T;\mathbb{R}^n)$ — the set of essentially bounded measurable functions $\varphi:[0,T]\to\mathbb{R}^n$.

$(\Omega,\mathcal{F},\mathbf{P})$ — probability space.

$\{\mathcal{F}_t\}_{t\geq 0}$ — filtration.

$(\Omega,\mathcal{F},\{\mathcal{F}_t\}_{t\geq 0},\mathbf{P})$ — filtered probability space.

$\mathcal{B}(U)$ — the Borel σ-field generated by all the open sets in U.

$\sigma(\xi)\stackrel{\Delta}{=}\xi^{-1}(\mathcal{F})$ — the σ-field generated by the random variable ξ.

$\sigma(\mathcal{A})$— the smallest σ-field containing the class $\mathcal{A}$,

$\bigvee_\alpha\mathcal{F}_\alpha\stackrel{\Delta}{=}\sigma(\bigcup_\alpha\mathcal{F}_\alpha), \qquad \bigwedge_\alpha\mathcal{F}_\alpha\stackrel{\Delta}{=}\bigcap_\alpha\mathcal{F}_\alpha.$

$\mathbf{P}_\xi = \mathbf{P}\circ\xi^{-1}$ — the probability measure induced by the random variable ξ.

EX — the expectation of the random variable X.

$\mathrm{Cov}(X,Y) \triangleq E[(X-EX)(Y-EY)^\top]$, $\quad$ $\mathrm{Var}\, X \triangleq \mathrm{Cov}(X,X)$.

$E(X|\mathcal{G})$ — conditional expectation of X given $\mathcal{G}$.

$L^p_{\mathcal{G}}(\Omega;\mathbb{R}^n)$ — the set of $\mathbb{R}^n$-valued $\mathcal{G}$-measurable random variables X such that $E|X|^p < \infty$ ($p \in [1,\infty)$).

$L^\infty_{\mathcal{G}}(\Omega;\mathbb{R}^n)$ — the set of bounded $\mathbb{R}^n$-valued $\mathcal{G}$-measurable random variables.

$L^p_{\mathcal{F}}(0,T;\mathbb{R}^n)$ — the set of all $\{\mathcal{F}\}_{t\geq 0}$-adapted $\mathbb{R}^n$-valued processes $X(\cdot)$ such that $E\int_0^T |X(t)|^p dt < \infty$.

$L^\infty_{\mathcal{F}}(0,T;\mathbb{R}^n)$ — the set of $\{\mathcal{F}\}_{t\geq 0}$-adapted $\mathbb{R}^n$-valued essentially bounded processes.

$L^p_{\mathcal{F}}(\Omega;C([0,T];\mathbb{R}^n))$ — the set of $\{\mathcal{F}_t\}_{t\geq 0}$-adapted $\mathbb{R}^n$-valued continuous processes $X(\cdot)$ such that $E\sup_{t\in[0,T]}|X(t)|^p < \infty$ ($p \in [1,\infty)$).

$\mathcal{M}^2[0,T]$ — the set of square-integrable martingales.

$\mathcal{M}^2_c[0,T]$ — the set of square-integrable continuous martingales.

$\mathcal{M}^{2,loc}[0,T]$ — the set of square-integrable local martingales.

$\mathcal{M}^{2,loc}_c[0,T]$ — the set of square-integrable continuous local martingales.

$\mathbf{W}^n[0,T] \triangleq C([0,T];\mathbb{R}^n)$, $\quad$ $\mathbf{W}^n \triangleq C([0,\infty);\mathbb{R}^n)$.

$\mathbf{C}_s$ — the set of Borel cylinders in $\mathbf{W}^n[0,s]$.

$\mathbf{C}$ — the set of Borel cylinders in $\mathbf{W}^n$.

$\mathbf{W}^n_t[0,T] \triangleq \{\zeta(\cdot\wedge t) \mid \zeta(\cdot) \in \mathbf{W}^n[0,T]\}$, $\quad t \in [0,T]$.

$\mathcal{B}_t(\mathbf{W}^n[0,T]) \triangleq \mathcal{B}(\mathbf{W}^n_t[0,T])$, $\quad t \in [0,T]$.

$\mathcal{B}_{t+}(\mathbf{W}^n[0,T]) \triangleq \bigcap_{s>t} \mathcal{B}_s(\mathbf{W}^n[0,T])$, $\quad t \in [0,T)$.

$\mathbf{W}^n_t \triangleq \{\zeta(\cdot\wedge t) \mid \zeta(\cdot) \in \mathbf{W}^n\}$, $\quad t \geq 0$.

$\mathcal{B}_t(\mathbf{W}^n) \triangleq \mathcal{B}(\mathbf{W}^n_t)$, $\quad t \geq 0$.

$\mathcal{B}_{t+}(\mathbf{W}^n) \triangleq \bigcap_{s>t} \mathcal{B}_s(\mathbf{W}^n)$, $\quad t \geq 0$.

$\mathcal{A}^n_T(U)$ — the set of all $\{\mathcal{B}_{t+}(\mathbf{W}^n[0,T])\}_{t\geq 0}$-progressively measurable processes $\eta : [0,T]\times\mathbf{W}^n[0,T] \to U$.

$\mathcal{A}^n(U)$ — the set of all $\{\mathcal{B}_{t+}(\mathbf{W}^n\}_{t\geq 0}$-progressively measurable processes $\eta : [0,\infty)\times\mathbf{W}^n \to U$.

$\mathcal{V}[0,T]$ — the set of all measurable functions $u : [0,T] \to U$.

$\mathcal{U}[0,T]$ — the set of all $\{\mathcal{F}_t\}_{t\geq 0}$-adapted processes $u : [0,T]\times\Omega \to U$.

$\mathcal{V}_{ad}[0,T]$ — the set of deterministic admissible controls.

$\mathcal{U}^s_{ad}[0,T]$ — the set of (stochastic) strong admissible controls.

$\mathcal{U}^w_{ad}[0,T]$ — the set of (stochastic) weak admissible controls.

Assumption Index

Problem Index

Chapter 1

Basic Stochastic Calculus

Stochastic calculus serves as a fundamental tool throughout this book. This chapter is meant to be a convenient "User's Guide" on stochastic calculus for use in the subsequent chapters. Specifically, it collects the definitions and results in stochastic calculus scattered around in the literature that are related to stochastic controls. It also unifies terminology and notation (which may differ in different papers/books) that are to be used in later chapters. Proofs of the results presented in this chapter are either given (which is the case when we think that the proof is important in understanding the subsequent material and/or when there is no immediate reference available) or else referred to standard and easily accessible books. Knowledgeable readers may skip this chapter or regard it as a quick reference.

1. Probability

In this section, some introductory probability will be briefly reviewed. The readers are assumed to have basic knowledge of real analysis.

1.1. Probability spaces

Definition 1.1. Let a set Ω be nonempty, and let $\mathcal{F} \subseteq 2^{\Omega}$ (2^{Ω} is the set of all subsets in Ω), called a *class*, be nonempty. We call $\mathcal{F}$

(i) a *π-system* if $A, B \in \mathcal{F} \quad \Rightarrow \quad A \cap B \in \mathcal{F}$,

(ii) a *λ-system* if

$$\begin{cases} \Omega \in \mathcal{F}; \\ A, B \in \mathcal{F},\ A \subseteq B \quad \Rightarrow \quad B \setminus A \in \mathcal{F}; \\ A_i \in \mathcal{F},\ A_i \uparrow A, i = 1, 2, \dots \quad \Rightarrow \quad A \in \mathcal{F}, \end{cases}$$

(iii) a *σ-field* if

$$\begin{cases} \Omega \in \mathcal{F}; \\ A, B \in \mathcal{F} \quad \Rightarrow \quad B \setminus A \in \mathcal{F}; \\ A_i \in \mathcal{F}\ , i = 1, 2, \dots \quad \Rightarrow \quad \bigcup_{i=1}^{\infty} A_i \in \mathcal{F}. \end{cases}$$

If $\mathcal{F}$ and $\mathcal{G}$ are both σ-fields on Ω and $\mathcal{G} \subseteq \mathcal{F}$, then $\mathcal{G}$ is called a *sub-σ-field* of $\mathcal{F}$. It is easy to see that $\mathcal{F}$ is a σ-field if and only if it is both a π-system and a λ-system. In what follows, for any class $\mathcal{A} \subseteq 2^{\Omega}$, we let $\sigma(\mathcal{A})$ be the smallest σ-field containing $\mathcal{A}$, called the *σ-field generated by $\mathcal{A}$*.

The following result is called a *monotone class theorem.*

Lemma 1.2. *Let $\mathcal{A} \subseteq \mathcal{F} \subseteq 2^{\Omega}$. Suppose $\mathcal{A}$ is a π-system and $\mathcal{F}$ is a λ-system. Then $\sigma(\mathcal{A}) \subseteq \mathcal{F}$.*

Proof. Let

$$\widehat{\mathcal{G}} \stackrel{\Delta}{=} \bigcap \{\mathcal{G} \supseteq \mathcal{A} \mid \mathcal{G} \text{ is a } \lambda\text{-system }\} \subseteq \mathcal{F}.$$

Then $\widehat{\mathcal{G}}$ is the smallest λ-system containing $\mathcal{A}$. Set

$$\widehat{\mathcal{F}} \stackrel{\Delta}{=} \{B \in \widehat{\mathcal{G}} \mid \forall A \in \mathcal{A},\ A \bigcap B \in \widehat{\mathcal{G}}\} \subseteq \widehat{\mathcal{G}}.$$

Clearly, $\widehat{\mathcal{F}}$ is also a λ-system, and since $\mathcal{A}$ is a π-system, $\mathcal{A} \subseteq \widehat{\mathcal{F}}$. Thus, we must have $\widehat{\mathcal{G}} = \widehat{\mathcal{F}}$. Now, let us define

$$\widetilde{\mathcal{F}} \stackrel{\Delta}{=} \{B \in \widehat{\mathcal{G}} \mid \forall A \in \widehat{\mathcal{G}},\ A \bigcap B \in \widehat{\mathcal{G}}\} \subseteq \widehat{\mathcal{G}}.$$

Again, $\widetilde{\mathcal{F}}$ is a λ-system, and by $\widehat{\mathcal{G}} = \widehat{\mathcal{F}}$, one has $\mathcal{A} \subseteq \widetilde{\mathcal{F}}$. Hence, $\widehat{\mathcal{G}} = \widetilde{\mathcal{F}}$. Consequently, for any $A, B \in \widehat{\mathcal{G}}$, regarding $B \in \widetilde{\mathcal{F}}$, we have $A \cap B \in \widehat{\mathcal{G}}$. This implies that $\widehat{\mathcal{G}}$ is also a π-system. Therefore, $\widehat{\mathcal{G}}$ is a σ-field (which contains $\mathcal{A}$). Then $\sigma(\mathcal{A}) \subseteq \widehat{\mathcal{G}} \subseteq \mathcal{F}$. □

Let $\{\mathcal{F}_\alpha\}$ be a family of σ-fields on Ω. Define

$$\bigvee_\alpha \mathcal{F}_\alpha \stackrel{\Delta}{=} \sigma\Big(\bigcup_\alpha \mathcal{F}_\alpha\Big) \tag{1.1}$$

and

$$\bigwedge_\alpha \mathcal{F}_\alpha \stackrel{\Delta}{=} \bigcap_\alpha \mathcal{F}_\alpha. \tag{1.2}$$

It is easy to show that $\vee_\alpha \mathcal{F}_\alpha$ and $\wedge_\alpha \mathcal{F}_\alpha$ are both σ-fields and that they are the smallest σ-field containing all $\mathcal{F}_\alpha$ and the largest σ-field contained in all $\mathcal{F}_\alpha$, respectively.

Let Ω be a nonempty set and $\mathcal{F}$ a σ-field on Ω. Then $(\Omega, \mathcal{F})$ is called a *measurable space.* A point $\omega \in \Omega$ is called a *sample.* A map $\mathbf{P} : \mathcal{F} \to [0, 1]$ is called a *probability measure* on $(\Omega, \mathcal{F})$ if

$$\begin{cases} \mathbf{P}(\phi) = 0, \qquad \mathbf{P}(\Omega) = 1; \\ A_i \in \mathcal{F},\ \ A_i \bigcap A_j = \phi,\ \ i, j = 1, 2, \ldots,\ i \neq j, \\ \qquad \Rightarrow\ \ \mathbf{P}\Big(\bigcup_{i=1}^{\infty} A_i\Big) = \sum_{i=1}^{\infty} \mathbf{P}(A_i). \end{cases} \tag{1.3}$$

The triple $(\Omega, \mathcal{F}, \mathbf{P})$ is called a *probability space.* Any $A \in \mathcal{F}$ is called an *event*, and $\mathbf{P}(A)$ represents the probability of the event A. Two events A and B are said to be *independent* if $\mathbf{P}(A \cap B) = \mathbf{P}(A)\mathbf{P}(B)$. An event A is said to be *independent* of a σ-field $\mathcal{F}$ if A is independent of any $B \in \mathcal{F}$.

Two σ-fields $\mathcal{F}$ and $\mathcal{G}$ are said to be *independent* if any event $A \in \mathcal{F}$ is independent of $\mathcal{G}$.

If an event A is such that $\mathbf{P}(A) = 1$, then we may alternatively denote this by

$$A \quad \text{holds}\ , \quad \mathbf{P}\text{-a.s.} \tag{1.4}$$

A set/event $A \in \mathcal{F}$ is called a $\mathbf{P}$-*null set/event* if $\mathbf{P}(A) = 0$. A probability space $(\Omega, \mathcal{F}, \mathbf{P})$ is said to be *complete* if for any $\mathbf{P}$-null set $A \in \mathcal{F}$, one has $B \in \mathcal{F}$ whenever $B \subseteq A$ (thus, it is necessary that B is also a $\mathbf{P}$-null set).

We present a simple example below.

Example 1.3. Let $\Omega = [0,1]$, $\mathcal{F}$ the set of all Lebesgue measurable sets in $[0,1]$, and $\mathbf{P}$ the Lebesgue measure on $[0,1]$. Then one can show that $(\Omega, \mathcal{F}, \mathbf{P})$ is complete. Let $x(\cdot) : \Omega \to \mathbb{R}$ be a continuous function and define the event $A = \{\omega \in \Omega \bigm| x(\omega) \geq 0\}$. The probability of this event is

$$\mathbf{P}(A) = \int_{\{\omega\in\Omega | x(\omega)\geq 0\}} d\mathbf{P}(\omega).$$

For any given probability space $(\Omega, \mathcal{F}, \mathbf{P})$, we define

$$\mathcal{N} = \{B \subseteq \Omega \bigm| \ \exists A \in \mathcal{F},\ \mathbf{P}(A) = 0,\ B \subseteq A\} \tag{1.5}$$

and $\widehat{\mathcal{F}} \triangleq \mathcal{F} \vee \mathcal{N}$. Then for any $\widehat{A} \in \widehat{\mathcal{F}}$, there exist $A, B \in \mathcal{F}$ such that $\mathbf{P}(B) = 0$ and $\widehat{A} \setminus A \subseteq B$. In such a case, we define $\mathbf{P}(\widehat{A}) = \mathbf{P}(A)$. This extends $\mathbf{P}$ to $\widehat{\mathcal{F}}$. Clearly, $(\Omega, \widehat{\mathcal{F}}, \mathbf{P})$ is a complete probability space. Any probability space can be made complete by the above *augmentation* procedure. In the sequel we shall assume that any probability space $(\Omega, \mathcal{F}, \mathbf{P})$ is complete.

Let $(\Omega, \mathcal{F})$ and $(\Omega', \mathcal{F}')$ be two measurable spaces and $f : \Omega \to \Omega'$ be a map. The map f is said to be $\mathcal{F}/\mathcal{F}'$-*measurable* or simply *measurable* if $f^{-1}(\mathcal{F}') \subseteq \mathcal{F}$. In particular, if $(\Omega', \mathcal{F}') = (\mathbb{R}^m, \mathcal{B}(\mathbb{R}^m))$, then f is said to be $\mathcal{F}$-*measurable* (see right below for the definition of $\mathcal{B}(\mathbb{R}^m)$).

Definition 1.4. We have the following definitions.

(i) A measurable space $(\Omega, \mathcal{F})$ is said to be *Borel isomorphic* to another measurable space $(\Omega', \mathcal{F}')$ if there exists a bijection $f : \Omega \to \Omega'$ such that both f and f^{-1} are measurable.

(ii) If Ω is a topological space, then the smallest σ-field containing all open sets of Ω is called the *Borel σ-field* of Ω, denoted by $\mathcal{B}(\Omega)$.

(iii) A separable complete metric space is called a *Polish space*.

(iv) A measurable space $(\Omega, \mathcal{F})$ is said to be *standard* if it is Borel isomorphic to one of the following: $(\langle 1,n \rangle, \mathcal{B}(\langle 1,n \rangle))$, $(\mathbb{N}, \mathcal{B}(\mathbb{N}))$, and $(\mathbb{M}, \mathcal{B}(\mathbb{M}))$, where $\langle 1,n \rangle \triangleq \{1,2,\ldots,n\}$, $\mathbb{N} \triangleq \{1,2,\ldots\}$, $\mathbb{M} \triangleq \{0,1\}^{\mathbb{N}} \equiv \{\omega = (\omega_1, \omega_2, \ldots) \bigm| \omega_i = 0, 1\}$ with the discrete topology on $\langle 1,n \rangle$ and $\mathbb{N}$, and the product topology on $\mathbb{M}$, respectively.

(v) A probability space $(\Omega, \mathcal{F}, \mathbf{P})$ is said to be *standard* if it is complete, and $(\Omega, \mathcal{F})$ is a standard measurable space.

(vi) A measurable space $(\Omega, \mathcal{F})$ is said to be *countably determined* (or simply, $\mathcal{F}$ is countably determined) if there exists a countable set $\mathcal{F}_0 \subseteq \mathcal{F}$ such that any two probability measures that agree on $\mathcal{F}_0$ must coincide (on $\mathcal{F}$).

Proposition 1.5. *Let Ω be a Polish space and $\mathcal{F} = \mathcal{B}(\Omega)$. Then:*

(i) *$(\Omega, \mathcal{F})$ is standard, and it is countably determined.*

(ii) *For any $\Omega' \in \mathcal{F}$, let $\mathcal{F}' = \Omega' \bigcap \mathcal{F} \triangleq \{\Omega' \bigcap A | A \in \mathcal{F}\}$. Then $(\Omega', \mathcal{F}')$ is standard.*

(iii) *$(\Omega, \mathcal{F})$ is Borel isomorphic to $(\mathbb{R}, \mathcal{B}(\mathbb{R}))$ if Ω is uncountable.*

See Parthasarathy [1, pp. 133–134] and Ikeda-Watanabe [1, p. 13] for a proof and descriptions.

1.2. Random variables

Let $(\Omega, \mathcal{F})$ and $(\Omega', \mathcal{F}')$ be two measurable spaces and $X : \Omega \to \Omega'$ an $\mathcal{F}/\mathcal{F}'$-measurable map. We call X an *$\mathcal{F}/\mathcal{F}'$-random variable*, or simply a *random variable* if there would be no confusion. It is called an *$\mathcal{F}$-random variable* when $(\Omega', \mathcal{F}') = (\mathbb{R}^m, \mathcal{B}(\mathbb{R}^m))$. Note that the notion of random variable is defined without probability measures. For a random variable $X : \Omega \to \Omega'$, $X^{-1}(\mathcal{F}')$ is a sub-σ-field of $\mathcal{F}$, which is called the *σ-field generated by X*, denoted by $\sigma(X)$. This is the smallest σ-field in Ω under which X is measurable. Also, if $\{X_\theta, \ \theta \in \Theta\}$ is a family of random variables from Ω to Ω', then we denote by

$$\sigma(X_\theta, \ \theta \in \Theta) \triangleq \bigvee_{\theta \in \Theta} X_\theta^{-1}(\mathcal{F}'),$$

the smallest sub-σ-field of $\mathcal{F}$ under which all X_θ $(\theta \in \Theta)$ are measurable.

Let $X, Y : \Omega \to \Omega'$ be two random variables and $\mathcal{G}$ a σ-field on Ω. Then X is said to be *independent* of $\mathcal{G}$ if $\sigma(X)$ is independent of $\mathcal{G}$, and X is said to be *independent* of Y if $\sigma(X)$ and $\sigma(Y)$ are independent.

Now we look at some measurability properties for random variables. The following result is closely related to Lemma 1.2.

Lemma 1.6. *Let $\mathcal{A}$ be a π-system on Ω. Let $\mathcal{H}$ be a linear space of functions from Ω to $\mathbb{R}$ such that*

$$\begin{cases} 1 \in \mathcal{H}; \qquad I_A \in \mathcal{H}, \quad \forall A \in \mathcal{A}; \\ \varphi_i \in \mathcal{H}, \ 0 \le \varphi_i \uparrow \varphi, \ \varphi \text{ is finite} \quad \Rightarrow \quad \varphi \in \mathcal{H}. \end{cases} \tag{1.6}$$

Then $\mathcal{H}$ contains all $\sigma(\mathcal{A})$-measurable functions from Ω to $\mathbb{R}$.

Proof. Set

$$\mathcal{F} = \{A \subseteq \Omega \mid I_A \in \mathcal{H}\}.$$

Then $\mathcal{F}$ is a λ-system containing $\mathcal{A}$. Thus, by Lemma 1.2, $\sigma(\mathcal{A}) \subseteq \mathcal{F}$.

Now, for any $\sigma(\mathcal{A})$-measurable function $\varphi : \Omega \to \mathbb{R}$, we set

$$\varphi_i = \sum_{j\geq 0} j2^{-i} I_{[j2^{-i}\leq\varphi(\omega)^+<(j+1)2^{-i}]}.$$

Clearly, $\varphi_i \in \mathcal{H}$ and $0 \leq \varphi_i \uparrow \varphi^+$. Hence, by our assumption, $\varphi^+ \in \mathcal{H}$. Similarly, $\varphi^- \in \mathcal{H}$. Thus, $\varphi \in \mathcal{H}$, proving our conclusion. □

The following result gives a representation of one random variable in terms of another, under some measurability conditions.

Theorem 1.7. *Let $(\Omega, \mathcal{F})$ and $(\Omega', \mathcal{F}')$ be two measurable spaces, and let (U, d) be a Polish space. Let $\xi : \Omega \to \Omega'$ and $\varphi : \Omega \to U$ be two random variables. Then φ is $\sigma(\xi)$-measurable, i.e.,*

$$\varphi^{-1}(\mathcal{B}(U)) \subseteq \xi^{-1}(\mathcal{F}'), \tag{1.7}$$

if and only if there exists a measurable map $\eta : \Omega' \to U$ such that

$$\varphi(\omega) = \eta(\xi(\omega)), \qquad \forall \omega \in \Omega. \tag{1.8}$$

Proof. We need only to prove the necessity. First, we assume that $U = \mathbb{R}$. For this case, set

$$\mathcal{H} \stackrel{\Delta}{=} \{\eta(\xi(\cdot)) \mid \eta : \Omega' \to U, \text{ measurable}\}.$$

Then $\mathcal{H}$ is a linear space, and $1 \in \mathcal{H}$. Also, if $A \in \sigma(\xi) \equiv \xi^{-1}(\mathcal{F}')$, then for some $B \in \mathcal{F}'$, $I_A(\cdot) = I_B(\xi(\cdot)) \in \mathcal{H}$. Now, suppose $\eta_i : \Omega' \to U$ is measurable and $\eta_i(\xi(\cdot)) \in \mathcal{H}$ such that $0 \leq \eta_i(\xi(\cdot)) \uparrow \zeta(\cdot)$, which is finite. Set

$$A = \{\omega' \in \Omega' \mid \sup_i \eta_i(\omega') < \infty\}.$$

Then $A \in \mathcal{F}'$ and $\xi(\Omega) \subseteq A$. Define

$$\eta(\omega') = \begin{cases} \sup_i \eta_i(\omega'), & r \in A, \\ 0, & \omega' \in \Omega' \setminus A. \end{cases}$$

Clearly, $\eta : \Omega' \to U$ is measurable and $\zeta(\cdot) = \eta(\xi(\cdot))$. Thus, $\zeta(\cdot) \in \mathcal{H}$. By Lemma 1.6, $\mathcal{H}$ contains all $\sigma(\xi)$-measurable random variables, in particular, $\varphi \in \mathcal{H}$, which leads to (1.8). This proves our conclusion for the case $U = \mathbb{R}$.

Now let (U, d) be an uncountable Polish space. By Proposition 1.5-(iii) and Definition 1.4, there exists a bijection $f : U \to \mathbb{R}$ such that $f(\mathcal{B}(U)) = \mathcal{B}(\mathbb{R})$. Consider the map $\widetilde{\varphi} = f \circ \varphi : \Omega \to \mathbb{R}$, which satisfies

$$\widetilde{\varphi}^{-1}(\mathcal{B}(\mathbb{R})) = \varphi^{-1} \circ f^{-1}(\mathcal{B}(\mathbb{R})) = \varphi^{-1}(\mathcal{B}(U)) \subseteq \xi^{-1}(\mathcal{F}').$$

Thus, there exists an $\widetilde{\eta} : \Omega' \to \mathbb{R}$ such that

$$\widetilde{\varphi}(\omega) = \widetilde{\eta}(\xi(\omega)), \qquad \forall \omega \in \Omega.$$

By taking $\eta = f^{-1} \circ \tilde{\eta}$, we obtain the desired result.

Finally, if (U, d) is countable or finite, we can prove the result replacing $\mathbb{R}$ in the above by $\mathbb{N}$ or $\langle 1, n \rangle$. □

Next, let $(\Omega, \mathcal{F}, \mathbf{P})$ be a probability space, $(\Omega', \mathcal{F}')$ a measurable space, and $X : \Omega \to \Omega'$ a random variable. Then X induces a probability measure $\mathbf{P}_X$ on $(\Omega', \mathcal{F}')$ as follows:

$$\begin{aligned} \mathbf{P}_X(A') &\stackrel{\Delta}{=} \mathbf{P} \circ X^{-1}(A') \equiv \mathbf{P}(X^{-1}(A')) \\ &= \mathbf{P}\{\omega \in \Omega \mid X(\omega) \in A'\} \stackrel{\Delta}{=} \mathbf{P}\{X \in A'\}, \quad \forall A' \in \mathcal{F}'. \end{aligned} \tag{1.9}$$

We call $\mathbf{P}_X$ the *distribution* of the random variable X. In the case where $(\Omega', \mathcal{F}') = (\mathbb{R}^m, \mathcal{B}(\mathbb{R}^m))$, $\mathbf{P}_X$ can be uniquely determined by the following function:

$$F(x) \equiv F(x_1, \ldots, x_m) \stackrel{\Delta}{=} \mathbf{P}\{\omega \in \Omega \mid X_i(\omega) \le x_i, \ 1 \le i \le m\}. \tag{1.10}$$

We call $F(x)$ the *distribution function* of X. Since $X = (X_1, \ldots, X_m)$, we also call $F(x)$ the *joint distribution function* of the (scalar-valued) random variables $X_1, \ldots, X_m$. Clearly, $F(x)$ is nonnegative, nondecreasing in each variable $x_i \in \mathbb{R}$, and

$$\lim_{\substack{x_i \to -\infty \\ \exists i}} F(x) = 0, \qquad \lim_{\substack{x_i \to \infty \\ \forall i}} F(x) = 1. \tag{1.11}$$

Further, if $\mathbf{P}_X$ is *absolutely continuous* with respect to the Lebesgue measure, i.e., for any $N \in \mathcal{B}(\mathbb{R}^m)$ with $|N| = 0$ ($|N|$ is the Lebesgue measure of N), one has $\mathbf{P}_X(N) = 0$, then by the Radon–Nikodým theorem, there exists a (nonnegative) function $f(\cdot) \in L^1(\mathbb{R}^m)$ such that

$$\mathbf{P}_X(A) = \int_A f(x)dx, \qquad \forall A \in \mathcal{B}(\mathbb{R}^m). \tag{1.12}$$

In particular,

$$F(x) = \int_{-\infty}^{x_1} \cdots \int_{-\infty}^{x_m} f(\xi_1, \ldots, \xi_m) d\xi_1 \cdots d\xi_m. \tag{1.13}$$

The function $f(x)$ is called the *density* of the random variable X. As a special case, if $f(x)$ is of the form

$$f(x) = \left[(2\pi)^m \det C\right]^{-1/2} e^{-\frac{1}{2}(x-\lambda)^\top C^{-1}(x-\lambda)}, \qquad x \in \mathbb{R}^m, \tag{1.14}$$

where $\lambda \in \mathbb{R}^m$, $C \in \mathbb{R}^{m \times m}$, and $C^\top = C > 0$, then we say that X has a *normal distribution* (denoted by $N(\lambda, C)$) with parameter (λ, C). (Later on we will see that λ is the *mean* and C is the *covariance matrix* of the normal distribution.) We also call $N(\lambda, C)$ the *Gaussian distribution* and X *Gaussian.*

We point out that to have a particular type of random variable, the underlying probability space $(\Omega, \mathcal{F})$ has to have certain particular properties. For example, if Ω is a finite or countable set, then $(\Omega, \mathcal{F})$ is not rich enough to accommodate a normal distribution.

A sequence of random variables $\{X_i\}$ is called *i.i.d.* if they are independent and identically distributed (i.e., their distribution functions coincide).

Now, let $X : (\Omega, \mathcal{F}, \mathbf{P}) \to \mathbb{R}^m$ be a random variable with the distribution function $F(x)$. Suppose the following integral exists:

$$\int_\Omega X(\omega) d\mathbf{P}(\omega) = \int_{\mathbb{R}^m} x dF(x) \triangleq EX. \tag{1.15}$$

Then we say that X has the *mean* EX. We also call EX the (*mathematical*) *expectation* of X. Note that a given random variable does not necessarily have a mean. We may also define the mean over a set $A \in \mathcal{F}$:

$$E(X; A) \triangleq \int_A X(\omega) d\mathbf{P}(\omega) = \int_{X(A)} x\, dF(x) \equiv E(XI_A). \tag{1.16}$$

Let $L^1_{\mathcal{F}}(\Omega; \mathbb{R}^m) \triangleq L^1(\Omega, \mathcal{F}, \mathbf{P}; \mathbb{R}^m)$ be the set of all random variables X, so that $|X|$ has means (over the whole space Ω). This is a Banach space with the norm

$$E|X| = \int_\Omega |X(\omega)| d\mathbf{P}(\omega) = \int_{\mathbb{R}^m} |x| dF(x). \tag{1.17}$$

Next, let $g : \mathbb{R}^m \to \mathbb{R}^k$ be a Borel measurable function. Then, for any random variable $X : \Omega \to \mathbb{R}^m$, $g(X) : \Omega \to \mathbb{R}^k$ is also a random variable. Further, if $g(X)$ has a mean, then

$$Eg(X) = \int_\Omega g(X(\omega)) d\mathbf{P}(\omega) = \int_{\mathbb{R}^m} g(x) dF(x). \tag{1.18}$$

The case $g(x) = |x|^p$ ($p \geq 1$) is of particular interest. If a random variable X is such that $|X|^p \in L^1_{\mathcal{F}}(\Omega; \mathbb{R})$, then

$$E|X|^p = \int_\Omega |X(\omega)|^p d\mathbf{P}(\omega). \tag{1.19}$$

Let $L^p_{\mathcal{F}}(\Omega; \mathbb{R}^m) \triangleq L^p(\Omega, \mathcal{F}, \mathbf{P}; \mathbb{R}^m)$ be the set of all random variables $X : \Omega \to \mathbb{R}^m$ with $|X|^p \in L^1_{\mathcal{F}}(\Omega; \mathbb{R})$. This is again a Banach space with the norm

$$\left(E|X|^p\right)^{1/p} = \left(\int_\Omega |X(\omega)|^p d\mathbf{P}(\omega)\right)^{1/p}. \tag{1.20}$$

In particular, if $p = 2$, then $L^2_{\mathcal{F}}(\Omega; \mathbb{R}^m)$ is a Hilbert space with the inner product

$$\langle X, Y \rangle_{L^2_{\mathcal{F}}(\Omega; \mathbb{R}^m)} \triangleq \int_\Omega \langle X(\omega), Y(\omega) \rangle \, d\mathbf{P}(\omega) \equiv E(X^\top Y). \tag{1.21}$$

Also, let $L^\infty_{\mathcal{F}}(\Omega; \mathbb{R}^m) \triangleq L^\infty(\Omega, \mathcal{F}, \mathbf{P}; \mathbb{R}^m)$ be the set of all essentially bounded $\mathcal{F}$-measurable random variables. Any $X \in L^p_{\mathcal{F}}(\Omega; \mathbb{R}^m)$ is called an *L^p-random variable*. From the Hölder inequality, we have $L^p_{\mathcal{F}}(\Omega; \mathbb{R}^m) \subseteq$

$L^q_{\mathcal{F}}(\Omega;\mathbb{R}^m)$ for any $p \geq q \geq 1$. In particular, $L^2_{\mathcal{F}}(\Omega;\mathbb{R}^m) \subseteq L^1_{\mathcal{F}}(\Omega;\mathbb{R}^m)$. Thus, for any $X, Y \in L^2_{\mathcal{F}}(\Omega;\mathbb{R}^m)$, EX and EY exist and the following is well-defined:

$$\text{Cov}\,(X,Y) \stackrel{\Delta}{=} E\big[(X - EX)(Y - EY)^\top\big], \tag{1.22}$$

which is called the *covariance* of the L^2-random variables X and Y. In particular,

$$\text{Var}\,X \stackrel{\Delta}{=} \text{Cov}\,(X,X) = E\big[(X - EX)(X - EX)^\top\big] \tag{1.23}$$

is called the *variance* of the L^2-random variable X. As an example, if X is Gaussian with the distribution $N(\lambda, C)$, then

$$EX = \lambda, \quad \text{Var}\,X = C.$$

Finally, for any $\mathbb{R}^m$-valued random variable X (which is not necessarily in $L^1_{\mathcal{F}}(\Omega;\mathbb{R}^m)$), the following is always well-defined:

$$\varphi_X(\xi) \stackrel{\Delta}{=} Ee^{i\xi^\top X} = \int_\Omega e^{i\xi^\top X(\omega)} d\mathbf{P}(\omega), \qquad \forall \xi \in \mathbb{R}^m. \tag{1.24}$$

We call $\varphi_X(\xi)$ the *characteristic function* of X. It is known that a random variable X is Gaussian with the distribution $N(\lambda, C)$ if and only if

$$\varphi_X(\xi) = e^{i\xi^\top \lambda - \frac{1}{2}\xi^\top C\xi}, \qquad \xi \in \mathbb{R}^m. \tag{1.25}$$

1.3. Conditional expectation

Let $X \in L^1_{\mathcal{F}}(\Omega;\mathbb{R}^m)$ and let $\mathcal{G}$ be a sub-σ-field of $\mathcal{F}$. Define a function $\mu : \mathcal{G} \to \mathbb{R}^m$ as follows:

$$\mu(A) \stackrel{\Delta}{=} E(X;A) = \int_A X(\omega) d\mathbf{P}(\omega), \qquad \forall A \in \mathcal{G}. \tag{1.26}$$

Then μ is a *vector-valued measure* on $\mathcal{G}$ with a bounded *total variation*

$$\|\mu\| \stackrel{\Delta}{=} \int_\Omega |X(\omega)| d\mathbf{P}(\omega) \equiv E|X|.$$

Moreover, μ is absolutely continuous with respect to $\mathbf{P}|_{\mathcal{G}}$, the restriction of $\mathbf{P}$ on $\mathcal{G}$. Thus, by the Radon Nikodým theorem, there exists a unique $f \in L^1_{\mathcal{G}}(\Omega;\mathbb{R}^m) \equiv L^1(\Omega,\mathcal{G},\mathbf{P}\big|_{\mathcal{G}};\mathbb{R}^m)$ (called the *Radon–Nikodým derivative* of μ with respect to $\mathbf{P}\big|_{\mathcal{G}}$) such that

$$\mu(A) = \int_A f(\omega) d\mathbf{P}|_{\mathcal{G}}(\omega) \equiv \int_A f(\omega) d\mathbf{P}(\omega), \qquad \forall A \in \mathcal{G}. \tag{1.27}$$

Here, note that $\mathbf{P}$ is an extension of $\mathbf{P}|_{\mathcal{G}}$. The function f is called the *conditional expectation* of X given $\mathcal{G}$, denoted by $E(X\big|\mathcal{G})$. Using this notation, we may rewrite (1.27) as follows:

$$\int_A X(\omega) d\mathbf{P}(\omega) = \int_A E(X\big|\mathcal{G})(\omega) d\mathbf{P}(\omega), \qquad \forall A \in \mathcal{G}, \tag{1.28}$$

or

$$E(X; A) = E(E(X|\mathcal{G}); A), \qquad \forall A \in \mathcal{G}. \tag{1.29}$$

Indeed, we can alternatively define $E(X|\mathcal{G})$ to be the unique $\mathcal{G}$-random variable satisfying (1.29).

Let us collect some basic properties of the conditional expectation.

Proposition 1.8. *Let $\mathcal{G}$ be a sub-σ-field of $\mathcal{F}$. Then*

(i) *Map $E(\cdot\,|\mathcal{G}) : L^1_{\mathcal{F}}(\Omega; \mathbb{R}^m) \to L^1_{\mathcal{G}}(\Omega; \mathbb{R}^m)$ is linear and bounded.*

(ii) $E(a|\mathcal{G}) = a$, $\quad$ **P**-a.s. , $\qquad \forall a \in \mathbb{R}$.

(iii) *If $X, Y \in L^1_{\mathcal{F}}(\Omega; \mathbb{R})$ with $X \geq Y$, then*

$$E(X|\mathcal{G}) \geq E(Y|\mathcal{G}), \quad \mathbf{P}\text{-a.s.} \tag{1.30}$$

In particular,

$$X \geq 0, \quad \mathbf{P}\text{-a.s.} \quad \Rightarrow \quad E(X|\mathcal{G}) \geq 0, \quad \mathbf{P}\text{-a.s.} \tag{1.31}$$

(iv) *Let $X \in L^1_{\mathcal{F}}(\Omega; \mathbb{R}^m)$, $Y \in L^1_{\mathcal{G}}(\Omega; \mathbb{R}^m)$ and $Z \in L^1_{\mathcal{F}}(\Omega; \mathbb{R}^k)$ with $XZ^\top, YZ^\top \in L^1_{\mathcal{F}}(\Omega; \mathbb{R}^{m\times k})$. Then*

$$E(YZ^\top|\mathcal{G}) = YE(Z|\mathcal{G})^\top, \quad \mathbf{P}\text{-a.s.} \tag{1.32}$$

In particular,

$$\begin{cases} E(E(X|\mathcal{G})Z^\top|\mathcal{G}) = E(X|\mathcal{G})E(Z|\mathcal{G})^\top, \\ E(Y|\mathcal{G}) = Y, \end{cases} \qquad \mathbf{P}\text{-a.s.} \tag{1.33}$$

(v) *A random variable X is independent of $\mathcal{G}$ if and only if for any Borel measurable function f such that $Ef(X)$ exists, it holds*

$$E\big(f(X)|\mathcal{G}\big) = Ef(X), \quad \mathbf{P}\text{-a.s.} \tag{1.34}$$

In particular, if X is independent of $\mathcal{G}$, then $E(X|\mathcal{G}) = EX$, $\quad$ **P**-a.s.

(vi) *Let $\mathcal{G}_1 \subseteq \mathcal{G}_2 \subseteq \mathcal{F}$. Then*

$$E\big(E(X|\mathcal{G}_2)|\mathcal{G}_1\big) = E\big(E(X|\mathcal{G}_1)|\mathcal{G}_2\big) = E(X|\mathcal{G}_1), \quad \mathbf{P}\text{-a.s.} \tag{1.35}$$

(vii) *(Jensen's inequality) Let $X \in L^1_{\mathcal{F}}(\Omega; \mathbb{R}^m)$ and $\varphi : \mathbb{R}^m \to \mathbb{R}$ be a convex function such that $\varphi(X) \in L^1_{\mathcal{F}}(\Omega; \mathbb{R})$. Then*

$$\varphi(E(X|\mathcal{G})) \leq E\big(\varphi(X)|\mathcal{G}\big), \quad \mathbf{P}\text{-a.s.} \tag{1.36}$$

In particular, for any $p \geq 1$, provided that $E|X|^p$ exists, we have

$$\big|E(X|\mathcal{G})\big|^p \leq E\big(|X|^p|\mathcal{G}\big), \quad \mathbf{P}\text{-a.s.} \tag{1.37}$$

Proofs of the above results are straightforward by the definitions.

Next, we consider the situation involving two random variables. Let $X \in L^1_{\mathcal{F}}(\Omega;\mathbb{R}^m)$ and $\xi : (\Omega,\mathcal{F},\mathbf{P}) \to (S,\mathcal{B})$ be two random variables. Then $\xi^{-1}(\mathcal{B})$ is a sub-σ-field of $\mathcal{F}$. Thus, we may define $E(X|\xi^{-1}(\mathcal{B}))$, the conditional expectation of X given $\xi^{-1}(\mathcal{B})$. On the other hand, we may treat this in a different way. Define (compare with (1.26))

$$\nu(B) = \int_{\xi^{-1}(B)} X(\omega)d\mathbf{P}(\omega), \qquad \forall B \in \mathcal{B}. \tag{1.38}$$

Then ν is a vector-valued measure on $(S,\mathcal{B})$ and it is absolutely continuous with respect to the measure $\mathbf{P}_\xi \stackrel{\Delta}{=} \mathbf{P} \circ \xi^{-1}$. We denote by $E(X|\xi = x)$ the Radon–Nikodým derivative of ν with respect to $\mathbf{P}_\xi$:

$$E(X|\xi = x) \stackrel{\Delta}{=} \frac{d\nu}{d\mathbf{P}_\xi}(x), \qquad \mathbf{P}_\xi\text{-a.s. } x \in S. \tag{1.39}$$

The above is called the *conditional expectation* of X given $\xi = x$. Thus, $E(X|\xi = x) : S \to \mathbb{R}^m$ is a function such that (noting (1.28) and (1.38)–(1.39))

$$\begin{aligned}
&\int_{\xi^{-1}(B)} E(X|\xi = \xi(\omega))d\mathbf{P}(\omega) \\
&\quad = \int_B E(X|\xi = x)d\mathbf{P}_\xi(x) \\
&\quad = \int_B d\nu(x) = \nu(B) = \int_{\xi^{-1}(B)} X(\omega)d\mathbf{P}(\omega) \\
&\quad = \int_{\xi^{-1}(B)} E(X|\xi^{-1}(\mathcal{B}))(\omega)d\mathbf{P}(\omega), \quad \forall B \in \mathcal{B}.
\end{aligned} \tag{1.40}$$

Consequently,

$$E(X|\xi^{-1}(\mathcal{B}))(\omega) = E(X|\xi = \xi(\omega)), \qquad \mathbf{P}|_{\xi^{-1}(\mathcal{B})}\text{-a.s. } \omega \in \Omega. \tag{1.41}$$

Next we turn to the conditional probability. For a sub-σ-field $\mathcal{G}$ of $\mathcal{F}$, define

$$\mathbf{P}(A|\mathcal{G}) \stackrel{\Delta}{=} E(I_A|\mathcal{G}), \quad \forall A \in \mathcal{F}, \tag{1.42}$$

which is called the *conditional probability* of (the event) A given (the condition) $\mathcal{G}$. Note that in general, we have

$$\begin{cases}
\mathbf{P}(\phi|\mathcal{G}) = 0, \qquad \mathbf{P}(\Omega|\mathcal{G}) = 1, \quad \mathbf{P}|_{\mathcal{G}}\text{-a.s.}, \\
\mathbf{P}\Big(\bigcup_{i=1}^\infty A_i|\mathcal{G}\Big) = \sum_{i=1}^\infty \mathbf{P}(A_i|\mathcal{G}), \quad \mathbf{P}|_{\mathcal{G}}\text{-a.s.}, \\
\qquad\qquad\qquad\qquad \forall A_i \in \mathcal{F}, \text{ mutually disjoint.}
\end{cases} \tag{1.43}$$

For any given $A \in \mathcal{F}$, $\mathbf{P}(A|\mathcal{G})(\omega)$ is defined only for $\mathbf{P}|_{\mathcal{G}}$-a.s. $\omega \in \Omega$. Thus for a given $\omega \in \Omega$, $\mathbf{P}(\cdot|\mathcal{G})(\omega)$ is not necessarily a probability measure on $\mathcal{F}$. However, we have the following result.

Proposition 1.9. *Let $(\Omega, \mathcal{F}, \mathbf{P})$ be a standard probability space and $\mathcal{G}$ be a sub-σ-field of $\mathcal{F}$. Then the following hold.*

(i) *There exists a map $p : \Omega \times \mathcal{F} \to [0,1]$, called a regular conditional probability given $\mathcal{G}$, such that $p(\omega, \cdot)$ is a probability measure on $(\Omega, \mathcal{F})$ for any $\omega \in \Omega$, $p(\cdot\,, A)$ is $\mathcal{G}$-measurable for any $A \in \mathcal{F}$, and*

$$(1.44)\quad E(I_A \big| \mathcal{G})(\omega) \equiv \mathbf{P}(A \big| \mathcal{G})(\omega) = p(\omega, A), \quad \mathbf{P}\big|_{\mathcal{G}}\text{-a.s. } \omega \in \Omega, \ \forall A \in \mathcal{F}.$$

Moreover, the above p is unique in the following sense: If p' is another regular conditional probability given $\mathcal{G}$, then there exists a $\mathbf{P}$-null set $N \in \mathcal{G}$ such that for any $\omega \notin N$,

$$(1.45)\qquad p(\omega, A) = p'(\omega, A), \quad \forall A \in \mathcal{F}.$$

(ii) *Let $\mathcal{H} \subseteq \mathcal{G}$ be a countably determined sub-σ-field and let $p(\cdot\,,\cdot)$ be the regular conditional probability given $\mathcal{G}$. Then there exists a $\mathbf{P}$-null set $N \in \mathcal{G}$ such that for any $\omega \notin N$,*

$$(1.46)\qquad p(\omega, A) = I_A(\omega), \qquad \forall A \in \mathcal{H}.$$

In particular, if $\xi : (\Omega, \mathcal{F}) \to (S, \mathcal{B})$ is a $\mathcal{G}/\mathcal{B}$-random variable and $\mathcal{B}$ is countably determined with $\{x\} \in \mathcal{B}$ for all $x \in S$, then

$$(1.47)\qquad p\big(\omega, \{\omega' : \xi(\omega') = \xi(\omega)\}\big) = 1, \qquad \mathbf{P}\text{-a.s. } \omega \in \Omega.$$

For the conditional expectation $E(X \big| \xi = x)$ of X given $\xi = x$, we have the following parallel result.

Proposition 1.10. *Let $(\Omega, \mathcal{F}, \mathbf{P})$ be a standard probability space and $(S, \mathcal{B})$ a measurable space. Let $\xi : (\Omega, \mathcal{F}) \to (S, \mathcal{B})$ be a random variable and $\mathbf{P}_\xi \equiv \mathbf{P} \circ \xi^{-1}$ the induced probability measure on $(S, \mathcal{B})$.*

(i) *Then there exists a map $\widetilde{p} : S \times \mathcal{F} \to \mathbb{R}$, called a regular conditional probability given $\xi = x$, such that $\widetilde{p}(x, \cdot)$ is a probability measure on $(\Omega, \mathcal{F})$ for any $x \in S$, $\widetilde{p}(\cdot\,, A)$ is $\mathcal{B}$-measurable for any $A \in \mathcal{F}$, and*

$$(1.48)\quad \mathbf{P}\big(A \textstyle\bigcap \xi^{-1}(B)\big) = \displaystyle\int_B \widetilde{p}(x, A) d\mathbf{P}_\xi(x), \qquad \forall A \in \mathcal{F}, \ B \in \mathcal{B}.$$

Consequently, for any $X \in L^1_{\mathcal{F}}(\Omega; \mathbb{R}^m)$,

$$(1.49)\qquad E(X \big| \xi = x) = \int_\Omega X(\omega) \widetilde{p}(x, d\omega), \qquad \mathbf{P}_\xi\text{-a.e. } x \in S,$$

and

$$(1.50)\quad \widetilde{p}(x, A) = E(I_A \big| \xi = x) \stackrel{\Delta}{=} \mathbf{P}(A \big| \xi = x), \ \ \mathbf{P}_\xi\text{-a.e. } x \in S, \ \forall A \in \mathcal{F}.$$

(ii) *Further, if $(S, \mathcal{B})$ is also standard, then there exists a $\mathbf{P}_\xi$-null set $N \in \mathcal{B}$ such that for any $x \notin N$,*

$$(1.51)\qquad \widetilde{p}(x, \xi^{-1}(B)) = I_B(x), \qquad \forall B \in \mathcal{B},$$

and in particular,

$$\widetilde{p}(x,\xi^{-1}(x))=1. \tag{1.52}$$

For proofs of Propositions 1.9 and 1.10, see Parthasarathy [1, pp. 145–150].

We now give a relation between $p(\cdot\,,\cdot)$ and $\widetilde{p}(\cdot\,,\cdot)$. Take $A\in\mathcal{F}$ and $B\in\mathcal{B}$. Using (1.44) (with $\mathcal{G}=\xi^{-1}(\mathcal{B})$) and (1.48), one has

$$\begin{aligned}\int_{\xi^{-1}(B)}p(\omega,A)d\mathbf{P}(\omega)&=\int_{\xi^{-1}(B)}E(I_A|\xi^{-1}(\mathcal{B}))(\omega)d\mathbf{P}(\omega)\\&=\int_{\xi^{-1}(B)}I_A(\omega)d\mathbf{P}(\omega)=\mathbf{P}(A\cap\xi^{-1}(B))\\&=\int_B\widetilde{p}(x,A)d(\mathbf{P}\circ\xi^{-1})(x)\\&=\int_{\xi^{-1}(B)}\widetilde{p}(\xi(\omega),A)d\mathbf{P}(\omega).\end{aligned}\tag{1.53}$$

Thus,

$$p(\omega,A)=\widetilde{p}(\xi(\omega),A),\qquad \mathbf{P}\big|_{\xi^{-1}(\mathcal{B})}\text{-a.s. }\omega\in\Omega,\ \forall A\in\mathcal{F}. \tag{1.54}$$

From now on in this book we will identify $\mathbf{P}(\cdot\,|\,\mathcal{G})(\cdot)$ with $p(\omega\,,\cdot)$ and $\mathbf{P}(\cdot\,|\,\xi=x)$ with $\widetilde{p}(x,\cdot)$. In other words, provided that $(\Omega,\mathcal{F},\mathbf{P})$ is standard, we always let $\mathbf{P}(\cdot\,|\,\mathcal{G})$ and $\mathbf{P}(\cdot\,|\,\xi=x)$ be the corresponding regular conditional probabilities. In particular, they are well-defined for each and every sample $\omega\in\Omega$.

To conclude this subsection, we supply a result for later use.

Lemma 1.11. *Let $\xi:(\Omega,\mathcal{F})\to(\mathbb{R}^\ell,\mathcal{B}(\mathbb{R}^\ell))$ be a random variable, and $X\in L^1_{\mathcal{F}}(\Omega;\mathbb{R}^m)$. If $E(g(\xi)X)=0$ for any $g\in C_b(\mathbb{R}^\ell)$, where $C_b(\mathbb{R}^\ell)$ is the set of all bounded continuous functions from $\mathbb{R}^\ell$ to $\mathbb{R}$, then we must have*

$$E(I_AX)=0,\quad \forall A\in\sigma(\xi). \tag{1.55}$$

Proof. Define

$$\mathcal{H}\stackrel{\Delta}{=}\{\varphi:(\Omega,\mathcal{F})\to(\mathbb{R},\mathcal{B}(\mathbb{R}))\mid\varphi\text{ is measurable, }E(\varphi X)=0\},$$

and

$$\begin{aligned}\mathcal{A}\stackrel{\Delta}{=}\{\xi^{-1}(B)|B=(a_1,b_1)\times(a_2,b_2)\times\cdots\times(a_\ell,b_\ell),\\ -\infty\le a_i\le b_i\le+\infty,\quad i=1,2,\ldots,\ell\}.\end{aligned}$$

It is clear that $\mathcal{A}$ is a π-system and $\mathcal{H}$ is a linear space. Moreover, $1\in\mathcal{H}$, and if $\varphi_i\in\mathcal{H}$ with $0\le\varphi_i\uparrow\varphi$, where φ is finite, then Lemma 1.6 yields $\varphi\in\mathcal{H}$. On the other hand, for any $A=\xi^{-1}(B)\in\mathcal{A}$ with $B=(a_1,b_1)\times(a_2,b_2)\times\cdots\times(a_\ell,b_\ell)$, a standard direct construction shows that there is a sequence $\{g_i\} \subseteq C_b(\mathbb{R}^\ell)$ such that $g_i(x) \uparrow I_B(x)$, $i \to \infty$, $\forall x \in \mathbb{R}^\ell$. Therefore,

$$\begin{aligned} E(I_A X) = E(I_{\xi^{-1}(B)} X) &= \int_\Omega I_B(\xi(\omega)) X(\omega) d\mathbf{P}(\omega) \\ &= \lim_{i\to\infty} \int_\Omega g_i(\xi(\omega)) X(\omega) d\mathbf{P}(\omega) = \lim_{i\to\infty} E(g_i(\xi)X) = 0. \end{aligned}$$

This shows that $I_A \in \mathcal{H}$. Then applying Lemma 1.6 yields that $\mathcal{H}$ contains all $\sigma(\mathcal{A}) \equiv \sigma(\xi)$-measurable functions from Ω to $\mathbb{R}$, proving (1.55). □

Proposition 1.12. *Let $\xi_i : (\Omega, \mathcal{F}) \to (\mathbb{R}^{m_i}, \mathcal{B}(\mathbb{R}^{m_i}))$ be a sequence of random variables ($i = 1, 2, \ldots$). Let $\mathcal{G} \triangleq \bigvee_i \sigma(\xi_i)$ and $X \in L^2_{\mathcal{F}}(\Omega; \mathbb{R}^m)$. Then $E(X|\mathcal{G}) = 0$ if and only if $E(g(\xi_1, \xi_2, \cdots, \xi_i)X) = 0$ for any i and any $g \in C_b(\mathbb{R}^N)$, where $N = \sum_{j=1}^i jm_j$.*

Proof. The "only if" part is clear. For the "if" part, define

$$\mathcal{F}_i \triangleq \sigma(\xi_1, \cdots, \xi_i), \quad \mathcal{A} \triangleq \bigcup_{i=1}^\infty \mathcal{F}_i.$$

Then it is clear that $\mathcal{G} = \sigma(\mathcal{A})$. Set

$$\mathcal{H} \triangleq \{\varphi : (\Omega, \mathcal{F}) \to (\mathbb{R}, \mathcal{B}(\mathbb{R})) | \varphi \text{ is measurable, } E(\varphi X) = 0\}.$$

For any $A \in \mathcal{A}$, we have $A \in \mathcal{F}_i$ for some i. By the assumption and Lemma 1.11, $I_A \in \mathcal{H}$. Moreover, if $\varphi_i \in \mathcal{H}$ with $0 \le \varphi_i \uparrow \varphi$, where φ is finite, then the Lévy lemma implies that $\varphi \in \mathcal{H}$. Hence by Lemma 1.6, $\mathcal{H}$ contains all $\sigma(\mathcal{A}) \equiv \mathcal{G}$-measurable functions, which completes the proof. □

1.4. Convergence of probabilities

To prove some existence results in stochastic analysis and stochastic controls, it inevitably involves convergence of certain probabilities. We summarize in this subsection some important results concerning the convergence of probability measures.

Let (U, d) be a separable metric space and $\mathcal{B}(U)$ the Borel σ-field. Denote by $\mathcal{P}(U)$ the set of all probability measures on the measurable space $(U, \mathcal{B}(U))$.

Definition 1.13. A sequence $\{\mathbf{P}_i\} \subseteq \mathcal{P}(U)$ is said to be *weakly convergent* to $\mathbf{P} \in \mathcal{P}(U)$ if for any $f \in C_b(U)$,

$$\lim_{i\to\infty} \int_U f(u) d\mathbf{P}_i(u) = \int_U f(u) d\mathbf{P}(u). \tag{1.56}$$

Proposition 1.14. *There is a metric ρ on $\mathcal{P}(U)$ such that $\mathbf{P}_i \to \mathbf{P}$ weakly is equivalent to $\rho(\mathbf{P}_i, \mathbf{P}) \to 0$ as $i \to \infty$.*

Definition 1.15. A set $\Lambda \subseteq \mathcal{P}(U)$ is said to be

(i) *relatively compact* if any sequence $\{\mathbf{P}_i\} \subseteq \Lambda$ contains a weakly convergent subsequence;

(ii) *compact* if Λ is relatively compact and closed;

(iii) *tight* if for any $\varepsilon > 0$ there is a compact set $K \subseteq U$ such that $\inf_{\mathbf{P}\in\Lambda} \mathbf{P}(K) \geq 1 - \varepsilon$.

Proposition 1.16. *Let $\Lambda \subseteq \mathcal{P}(U)$. Then:*

(i) *Λ is relatively compact if it is tight.*

(ii) *If (U, d) is complete (i.e., it is a polish space), then Λ is tight if it is relatively compact.*

See Ikeda–Watanabe [1, pp. 6–8] for proofs of Propositions 1.14 and 1.16.

Corollary 1.17. *If (U, d) is compact, then any $\Lambda \subseteq \mathcal{P}(U)$ is tight and relatively compact. In particular, $\mathcal{P}(U)$ is compact.*

The proof is straightforward by Definition 1.15 and Proposition 1.16. See also Parthasarathy [1, p. 45, Theorem 6.4].

Recall that if $X : (\Omega, \mathcal{F}, \mathbf{P}) \to (U, d)$ is a random variable, then $\mathbf{P}_X \in \mathcal{P}(U)$ denotes the probability induced by X (see (1.9)). We say that a family of random variables $X_\alpha : (\Omega, \mathcal{F}, \mathbf{P}) \to (U, d)$ is *tight* if $\{\mathbf{P}_{X_\alpha}\}$ is tight.

Definition 1.18. Let $X_i : (\Omega_i, \mathcal{F}_i, \mathbf{P}_i) \to (U, d)$, $i = 1, 2, \ldots$, and $X : (\Omega, \mathcal{F}, \mathbf{P}) \to (U, d)$ be random variables. We say that X_i *converges to X in law* if $\mathbf{P}_{X_i} \to \mathbf{P}_X$ weakly as $i \to \infty$.

Definition 1.19. Let $X_i, X : (\Omega, \mathcal{F}, \mathbf{P}) \to (U, d)$, $i = 1, 2, \ldots$, be random variables. We say that X_i *converges to X almost surely* if

$$\lim_{i\to\infty} d(X_i, X) = 0, \quad \mathbf{P}\text{-a.s.},$$

and that X_i *converges to X in probability* if for any $\varepsilon > 0$,

$$\lim_{i\to\infty} \mathbf{P}\{\omega | d(X_i(\omega), X(\omega)) > \varepsilon\} = 0.$$

It is not difficult to verify that for a sequence of random variables,

$$\begin{aligned} \text{almost surely convergence} \quad &\Rightarrow \quad \text{convergence in probability} \\ &\Rightarrow \quad \text{convergence in law.} \end{aligned} \tag{1.57}$$

The following result is a sort of reverse of the above implications.

Theorem 1.20. *Let (U, d) be a Polish space and $\{\mathbf{P}_i, i = 1, 2, \ldots, \mathbf{P}\} \subseteq \mathcal{P}(U)$ be such that $\mathbf{P}_i$ converges to $\mathbf{P}$ weakly. Then on some probability space $(\widehat{\Omega}, \widehat{\mathcal{F}}, \widehat{\mathbf{P}})$ there are random variables X_i, X: $(\widehat{\Omega}, \widehat{\mathcal{F}}) \to (U, \mathcal{B}(U))$, $i = 1, 2, \ldots$, such that*

(i) *$\mathbf{P}_i = \mathbf{P}_{X_i}$, $i = 1, 2, \ldots$, and $\mathbf{P} = \mathbf{P}_X$.*

(ii) *$X_i \to X$ as $i \to \infty$, $\widehat{\mathbf{P}}$-a.s.*

This theorem is due to Skorohod. See Billingsley [1] for a proof.

2. Stochastic Processes

In this section we recall some results on stochastic processes.

2.1. General considerations

Definition 2.1. Let $\mathcal{I}$ be a nonempty index set and $(\Omega, \mathcal{F}, \mathbf{P})$ a probability space. A family $\{X(t), t \in \mathcal{I}\}$ of random variables from $(\Omega, \mathcal{F}, \mathbf{P})$ to $\mathbb{R}^m$ is called a *stochastic process*. For any $\omega \in \Omega$, the map $t \mapsto X(t, \omega)$ is called a *sample path*.

In what follows, we let $\mathcal{I} = [0, T]$ with $T > 0$, or $\mathcal{I} = [0, \infty)$. We shall interchangeably use $\{X(t), t \in \mathcal{I}\}$, $X(\cdot)$, $X(t)$, or even X to denote a stochastic process.

For any given stochastic process $X(t), t \in \mathcal{I}$, we can define the following:

$$
\left\{
\begin{aligned}
&F_{t_1}(x_1) \stackrel{\Delta}{=} \mathbf{P}\{X(t_1) \le x_1\},\\
&F_{t_1,t_2}(x_1, x_2) \stackrel{\Delta}{=} \mathbf{P}\{X(t_1) \le x_1, X(t_2) \le x_2\},\\
&\cdots\cdots\\
&F_{t_1,\cdots,t_j}(x_1, \ldots, x_j) \stackrel{\Delta}{=} \mathbf{P}\{X(t_1) \le x_1, \ldots, X(t_j) \le x_j\},\\
&\cdots\cdots
\end{aligned}
\right.
\tag{2.1}
$$

Here, $t_i \in \mathcal{I}$, $x_i \in \mathbb{R}^m$, and $X(t_i) \le x_i$ stands for componentwise inequalities. The functions defined in (2.1) are called the *finite-dimensional distributions* of the process $X(t)$. This family of functions satisfies the following conditions:

(a) *Symmetry*: If $\{i_1, \ldots, i_j\}$ is a *permutation* of $\{1, \ldots, j\}$, then

$$
F_{t_{i_1},\ldots,t_{i_j}}(x_{i_1}, \ldots, x_{i_j}) = F_{t_1,\ldots,t_j}(x_1, \ldots, x_j). \tag{2.2}
$$

(b) *Compatibility*: For all $i < j$,

$$
F_{t_1,\ldots,t_i,t_{i+1},\ldots,t_j}(x_1, \ldots, x_i, \infty, \ldots, \infty) = F_{t_1,\ldots,t_i}(x_1, \ldots, x_i). \tag{2.3}
$$

Thus, any stochastic process admits symmetric and compatible finite-dimensional distributions. Conversely, given a family of functions, denoted by $\mathbb{F}$, consisting of functions $F_{t_1,\ldots,t_j}(x_1, \ldots, x_j)$, $j \ge 1$, and satisfying the above conditions (a) and (b), it is natural to ask whether one can find a stochastic process $\{X(t), t \in \mathcal{I}\}$ whose finite-dimensional distributions coincide with $\mathbb{F}$. The following result of Kolmogorov gives a positive answer to this question.

Theorem 2.2. *Let $\mathbb{F} = \{F_{t_1,\ldots,t_j}(x_1, \ldots, x_j),\ j \ge 1\}$ be a family of functions satisfying the symmetry and compatibility conditions. Then there exists a probability space $(\Omega, \mathcal{F}, \mathbf{P})$ and a stochastic process $X(t)$ whose finite-dimensional distributions coincide with $\mathbb{F}$.*

For a proof, see Parthasarathy [1, pp. 143–144]. In what follows, any stochastic process will be called simply a *process* if no ambiguity should arise, and the probability space $(\Omega, \mathcal{F}, \mathbf{P})$ and the time interval $[0,T]$ will be fixed unless otherwise stated.

Definition 2.3. Two processes $X(t)$ and $\overline{X}(t)$ are said to be *stochastically equivalent* if

$$X(t) = \overline{X}(t), \qquad \mathbf{P}\text{-a.s.}, \ \forall t \in [0,T]. \tag{2.4}$$

In this case, one is called a *modification* of the other.

If $X(t)$ and $\overline{X}(t)$ are stochastically equivalent, then for any $t \in [0,T]$, there exists a $\mathbf{P}$-null set $N_t \in \mathcal{F}$ such that

$$X(t,\omega) = \overline{X}(t,\omega), \qquad \forall \omega \in \Omega \setminus N_t. \tag{2.5}$$

Consequently, it follows from (2.1) that the finite-dimensional distributions of $X(t)$ and $\overline{X}(t)$ are the same. However, the $\mathbf{P}$-null set N_t depends on t. Therefore, the sample paths of $X(t)$ and $\overline{X}(t)$ can differ significantly. Here is a simple example.

Example 2.4. Let $\Omega = [0,1]$, $T \geq 1$, $\mathbf{P}$ the Lebesgue measure, $X(t,\omega) \equiv 0$, and

$$\overline{X}(t,\omega) = \begin{cases} 0, & \omega \neq t, \\ 1, & \omega = t. \end{cases} \tag{2.6}$$

Then $X(t)$ and $\overline{X}(t)$ are stochastically equivalent. But each sample path $X(\cdot\,,\omega)$ is continuous, and none of the sample paths $\overline{X}(\cdot\,,\omega)$ is continuous. In the present case, we actually have

$$\bigcup_{t\in[0,1]} N_t = [0,1] \equiv \Omega. \tag{2.7}$$

Definition 2.5. The process $X(t)$ is said to be *stochastically continuous* at $s \in [0,T]$ if for any $\varepsilon > 0$,

$$\lim_{t\to s} \mathbf{P}\{\omega \in \Omega \mid |X(t,\omega) - X(s,\omega)| > \varepsilon\} = 0. \tag{2.8}$$

Moreover, $X(t)$ is said to be *continuous* if there exists a $\mathbf{P}$-null set $N \in \mathcal{F}$ such that for any $\omega \in \Omega \setminus N$, the sample path $X(\cdot\,,\omega)$ is continuous.

Similarly, one can define the left and right (stochastic) continuity of processes. It is clear that continuity implies stochastic continuity.

Next, for a given measurable space $(\Omega, \mathcal{F})$, we introduce a monotone family of sub-σ-fields $\mathcal{F}_t \subseteq \mathcal{F}$, $t \in [0,T]$. Here, by monotonicity we mean

$$\mathcal{F}_{t_1} \subseteq \mathcal{F}_{t_2}, \qquad \forall 0 \leq t_1 \leq t_2 \leq T. \tag{2.9}$$

Such a family is called a *filtration*. Set $\mathcal{F}_{t+} \triangleq \bigcap_{s>t} \mathcal{F}_s$ for any $t \in [0,T)$, and $\mathcal{F}_{t-} \triangleq \bigcup_{s<t} \mathcal{F}_s$ for any $t \in (0,T]$. If $\mathcal{F}_{t+} = \mathcal{F}_t$ (resp. $\mathcal{F}_{t-} = \mathcal{F}_t$), we say that $\{\mathcal{F}_t\}_{t\geq 0}$ is *right* (resp. *left*) *continuous*. We call $(\Omega, \mathcal{F}, \{\mathcal{F}_t\}_{t\geq 0})$ a *filtered measurable space* and $(\Omega, \mathcal{F}, \{\mathcal{F}_t\}_{t\geq 0}, \mathbf{P})$ a *filtered probability space.*

Definition 2.6. We say that $(\Omega, \mathcal{F}, \{\mathcal{F}_t\}_{t\geq 0}, \mathbf{P})$ satisfies the *usual condition* if $(\Omega, \mathcal{F}, \mathbf{P})$ is complete, $\mathcal{F}_0$ contains all the $\mathbf{P}$-null sets in $\mathcal{F}$, and $\{\mathcal{F}_t\}_{t\geq 0}$ is right continuous.

Note that Definition 2.1 for stochastic processes can be extended naturally to those taking values in a metric space (U, d).

Definition 2.7. Let $(\Omega, \mathcal{F}, \{\mathcal{F}_t\}_{t\geq 0})$ be a filtered measurable space and $X(t)$ a process taking values in a metric space (U, d).

(i) The process $X(t)$ is said to be *measurable* if the map $(t, \omega) \mapsto X(t, \omega)$ is $(\mathcal{B}[0,T] \times \mathcal{F})/\mathcal{B}(U)$-measurable.

(ii) The process $X(t)$ is said to be $\{\mathcal{F}_t\}_{t\geq 0}$*-adapted* if for all $t \in [0,T]$, the map $\omega \mapsto X(t, \omega)$ is $\mathcal{F}_t/\mathcal{B}(U)$-measurable.

(iii) The process $X(t)$ is $\{\mathcal{F}_t\}_{t\geq 0}$*-progressively measurable* if for all $t \in [0,T]$, the map $(s, \omega) \mapsto X(s, \omega)$ is $\mathcal{B}[0,t] \times \mathcal{F}_t/\mathcal{B}(U)$-measurable.

It is clear that if $X(t)$ is $\{\mathcal{F}_t\}_{t\geq 0}$-progressive measurable, it must be measurable and $\{\mathcal{F}_t\}_{t\geq 0}$-adapted. Conversely, on a filtered probability space, we have the following result.

Proposition 2.8. *Let $(\Omega, \mathcal{F}, \{\mathcal{F}\}_{t\geq 0}, \mathbf{P})$ be a filtered probability space, and let $X(t)$ be measurable and $\{\mathcal{F}_t\}_{t\geq 0}$-adapted. Then there exists an $\{\mathcal{F}_t\}_{t\geq 0}$-progressively measurable process $\overline{X}(t)$ stochastically equivalent to $X(t)$. If in addition, $X(t)$ is left or right continuous, then $X(t)$ itself is $\{\mathcal{F}_t\}_{t\geq 0}$-progressively measurable.*

For a proof, see Meyer [1, p. 68]. Due to the above result, we make the following convention hereafter for the sake of simplicity, nevertheless without affecting any of the results.

Convention 2.9. Unless otherwise indicated, in a filtered probability space, by saying that a process $X(t)$ is $\{\mathcal{F}_t\}_{t\geq 0}$-adapted, we mean that $X(t)$ is $\{\mathcal{F}_t\}_{t\geq 0}$-progressively measurable.

Note that in the definition of stochastic equivalence, the probability measure $\mathbf{P}$ has to be involved. Thus, Proposition 2.8 and Convention 2.9 make sense only in filtered probability spaces (rather than merely filtered measurable spaces). In other words, in a filtered measurable space, we have to distinguish progressive measurability and adaptiveness.

We point out that all the results in this and the next subsections carry over naturally to the case where $[0,T]$ is replaced by $[s,T]$ ($s \in [0,T)$). The only modification is to replace 0 by s (in, say, $\{\mathcal{F}_t\}_{t\geq 0}$, $\mathcal{B}[0,T]$, etc.).

Let H be a Banach space with the norm $\|\cdot\|_H$ and let $(\Omega, \mathcal{F}, \{\mathcal{F}_t\}_{t\geq 0}, \mathbf{P})$ be a filtered probability space satisfying the usual condition. We will denote hereafter by $L^p_{\mathcal{F}}(0,T;H)$ the set of all H-valued, $\{\mathcal{F}_t\}_{t\geq 0}$-adapted processes $X(\cdot)$ such that $E\int_0^T \|X(t)\|^p_H dt < \infty$, by $L^\infty_{\mathcal{F}}(0,T;H)$ the set of all H-valued, $\{\mathcal{F}_t\}_{t\geq 0}$-adapted bounded processes, and by $L^p_{\mathcal{F}}(\Omega;C([0,T];H))$ the set of all H-valued, $\{\mathcal{F}_t\}_{t\geq 0}$-adapted continuous processes $X(\cdot)$ such that $E\{\sup_{t\in[0,T]}\|X(t)\|^p_H\} < \infty$.

Next, for $T>0$, we define $\mathbf{W}^m[0,T] \stackrel{\Delta}{=} C([0,T];\mathbb{R}^m)$ and define

$$
\left\{
\begin{array}{ll}
\mathbf{W}^m_t[0,T] \stackrel{\Delta}{=} \{\zeta(\cdot\wedge t) \mid \zeta(\cdot)\in \mathbf{W}^m[0,T]\}, & \forall t\in[0,T],\\
\mathcal{B}_t(\mathbf{W}^m[0,T]) \stackrel{\Delta}{=} \sigma(\mathcal{B}(\mathbf{W}^m_t[0,T])), & \forall t\in[0,T],\\
\mathcal{B}_{t+}(\mathbf{W}^m[0,T]) = \bigcap_{s>t}\mathcal{B}_s(\mathbf{W}^m[0,T]), & \forall t\in[0,T).
\end{array}
\right. \tag{2.10}
$$

We note that $\mathcal{B}_t(\mathbf{W}^m[0,T])$ is the σ-field (in $W^m[0,T]$) generated by $\mathcal{B}(\mathbf{W}^m_t[0,T])$, thus it contains $W^m[0,T]$. Clearly, both the following are filtered measurable spaces:

$$
\left\{
\begin{array}{l}
(\mathbf{W}^m[0,T], \mathcal{B}(\mathbf{W}^m[0,T]), \{\mathcal{B}_t(\mathbf{W}^m[0,T])\}_{t\geq 0}),\\
(\mathbf{W}^m[0,T], \mathcal{B}(\mathbf{W}^m[0,T]), \{\mathcal{B}_{t+}(\mathbf{W}^m[0,T])\}_{t\geq 0}).
\end{array}
\right.
$$

We point out that

$$
\mathcal{B}_{t+}(\mathbf{W}^m[0,T]) \neq \mathcal{B}_t(\mathbf{W}^m[0,T]), \quad \forall t\in[0,T]; \tag{2.11}
$$

see Karatzas–Shreve [3, p. 122], as well as some remarks in the next subsection. In what follows, for any Polish space U (in particular, for $U=\mathbb{R}^n, \mathbb{R}^{n\times m}$, etc.), we let $\mathcal{A}^m_T(U)$ be the set of all $\{\mathcal{B}_{t+}(\mathbf{W}^m[0,T])\}_{t\geq 0}$-progressively measurable processes $\eta:[0,T]\times\mathbf{W}^m[0,T]\to U$.

The following result is a further extension of Theorem 1.7 and will play an important role in appropriately formulating stochastic optimal control problems studied in the subsequent chapters.

Theorem 2.10. *Let $(\Omega,\mathcal{F},\mathbf{P})$ be a complete probability space and (U,d) a Polish space. Let $\xi:[0,T]\times\Omega\to\mathbb{R}^m$ be a continuous process and $\mathcal{F}^\xi_t=\sigma(\xi(s): 0\leq s\leq t)$. Then $\varphi:[0,T]\times\Omega\to U$ is $\{\mathcal{F}^\xi_t\}_{t\geq 0}$-adapted if and only if there exists an $\eta\in\mathcal{A}^m_T(U)$ such that*

$$
\varphi(t,\omega)=\eta(t,\xi(\cdot\wedge t,\omega)), \qquad \mathbf{P}\text{-a.s. } \omega\in\Omega, \ \forall t\in[0,T]. \tag{2.12}
$$

In order to prove the above result, we need two lemmas. A set $B\subseteq\mathbf{W}^m[0,T]$ is called a *Borel cylinder* if there exist $0\leq t_1<t_2<\cdots<t_j\leq T$ and $E\in\mathcal{B}(\mathbb{R}^{jm})$ such that

$$
B=\{\zeta\in\mathbf{W}^m[0,T] \mid (\zeta(t_1),\zeta(t_2),\ldots,\zeta(t_j))\in E\}. \tag{2.13}
$$

We let $\mathbf{C}_s$ be the set of all Borel cylinders in $\mathbf{W}^m_s[0,T]$ of the form (2.13) with $t_1,\ldots,t_j\in[0,s]$.

Lemma 2.11. *The σ-field $\sigma(\mathbf{C}_T)$ generated by $\mathbf{C}_T$ coincides with the Borel σ-field $\mathcal{B}(\mathbf{W}^m[0,T])$ of $\mathbf{W}^m[0,T]$.*

Proof. Let $0 \le t_1 < t_2 < \cdots < t_j \le T$ be given. We define a map $\mathcal{T} : \mathbf{W}^m[0,T] \to \mathbb{R}^{jm}$ as follows:

$$\mathcal{T}(\zeta) = (\zeta(t_1), \zeta(t_2), \ldots, \zeta(t_j)), \qquad \forall \zeta \in \mathbf{W}^m[0,T].$$

Clearly, $\mathcal{T}$ is continuous. Consequently, for any $E \in \mathcal{B}(\mathbb{R}^{jm})$, it follows that $\mathcal{T}^{-1}(E) \in \mathcal{B}(\mathbf{W}^m[0,T])$. This implies

$$\mathbf{C}_T \subseteq \mathcal{B}(\mathbf{W}^m[0,T]). \tag{2.14}$$

Next, for any $\zeta_0 \in \mathbf{W}^m[0,T]$ and $\varepsilon > 0$, we have

$$\begin{aligned} &\{\zeta \in \mathbf{W}^m[0,T] \mid |\zeta - \zeta_0|_{\mathbf{W}^m[0,T]} \le \varepsilon\} \\ &= \bigcap_{r \in \mathbf{Q}, r \in [0,T]} \{\zeta \in \mathbf{W}^m[0,T] \mid |\zeta(r) - \zeta_0(r)| \le \varepsilon\} \in \sigma(\mathbf{C}_T), \end{aligned} \tag{2.15}$$

since $\{\zeta \in \mathbf{W}^m[0,T] \mid |\zeta(r) - \zeta_0(r)| \le \varepsilon\}$ is a Borel cylinder, and $\mathbf{Q}$ is the set of all rational numbers (which is countable). Because the set of all sets in the form of the left-hand side of (2.15) is a basis of the open sets in $\mathbf{W}^m[0,T]$, we have

$$\mathcal{B}(\mathbf{W}^m[0,T]) \subseteq \sigma(\mathbf{C}_T). \tag{2.16}$$

Combining (2.14) and (2.16), we obtain our conclusion. □

Lemma 2.12. Let $(\Omega, \mathcal{F}, \mathbf{P})$ be a complete probability space and $\xi : [0,T] \times \Omega \to \mathbb{R}^m$ a continuous process. Then there exists an $\Omega_0 \in \mathcal{F}$ with $\mathbf{P}(\Omega_0) = 1$ such that $\xi : \Omega_0 \to \mathbf{W}^m[0,T]$ and for any $s \in [0,T]$,

$$\Omega_0 \bigcap \mathcal{F}_s^\xi = \Omega_0 \bigcap \xi^{-1}(\mathcal{B}_s(\mathbf{W}^m[0,T])). \tag{2.17}$$

Proof. Let $t \in [0,s]$ and $E \in \mathcal{B}(\mathbb{R}^m)$ be fixed. Then

$$B_t \triangleq \{\zeta \in \mathbf{W}^m[0,T] \mid \zeta(t) \in E\} \in \mathbf{C}_s,$$

and

$$\begin{aligned} \omega \in \xi^{-1}(B_t) \quad &\iff \quad \xi(\cdot\,, \omega) \in B_t \\ &\iff \quad \xi(t,\omega) \in E \quad \iff \quad \omega \in \xi(t,\cdot)^{-1}(E). \end{aligned}$$

Thus,

$$\xi(t,\cdot)^{-1}(E) = \xi^{-1}(B_t).$$

Then by Lemma 2.11, we obtain (2.17). □

Now we are ready to prove Theorem 2.10.

Proof of Theorem 2.10. We prove only the "only if" part. The "if" part is clear.

For any $s \in [0,T]$, we consider a mapping

$$\theta^s(t,\omega) \triangleq \big(t \wedge s, \xi(\cdot \wedge s, \omega)\big) : [0,T] \times \Omega \to [0,s] \times \mathbf{W}_s^m[0,T].$$

By Lemma 2.12, we have $\mathcal{B}[0,s]\times\mathcal{F}_s^\xi=\sigma(\theta^s)$. On the other hand, $(t,\omega)\mapsto\varphi(t\wedge s,\omega)$ is $(\mathcal{B}[0,s]\times\mathcal{F}_s^\xi)/\mathcal{B}(U)$-measurable. Thus, by Theorem 1.7, there exists a measurable map $\eta_s:([0,T]\times\mathbf{W}_s^m[0,T],\mathcal{B}[0,s]\times\mathcal{B}_s(\mathbf{W}^m[0,T]))\to U$ such that

$$\varphi(t\wedge s,\omega)=\eta_s(t\wedge s,\xi(\cdot\wedge s,\omega)),\quad\forall\omega\in\Omega,t\in[0,T]. \tag{2.18}$$

Now, for any $i\geq1$, let $0=t_0^i<t_1^i<\cdots$ be a partition of $[0,T]$ (with $\max_{j\geq1}[t_j^i-t_{j-1}^i]\to0$ as $i\to\infty$), and define

$$\begin{aligned}\eta^i(t,\zeta)=\eta_0(0,\zeta(\cdot\wedge0))I_{\{0\}}(t)+\sum_{j\geq1}\eta_{t_j^i}(t,\zeta(\cdot\wedge t_j^i))I_{(t_{j-1}^i,t_j^i]}(t),\\ \forall(t,\zeta)\in[0,T]\times\mathbf{W}^m[0,T].\end{aligned} \tag{2.19}$$

For any $t\in[0,T]$, there exists j such that $t_{j-1}^i<t\leq t_j^i$. Then

$$\eta^i(t,\xi(\cdot\wedge t_j^i,\omega))=\eta_{t_j^i}(t,\xi(\cdot\wedge t_j^i,\omega))=\varphi(t,\omega). \tag{2.20}$$

Now, in the case $U=\mathbb{R},\mathbb{N},\langle1,n\rangle$, we may define

$$\eta(t,\zeta)=\overline{\lim_{i\to\infty}}\,\eta^i(t,\zeta) \tag{2.21}$$

to get the desired result. In the case where U is a general Polish space, we need to amend the proof in the same fashion as in that of Theorem 1.7. □

Note that the map $\eta(t,\zeta)$ in (2.21) is only $\{\mathcal{B}_{t+}(\mathbf{W}^m[0,T])\}_{t\geq0}$-progressively measurable, not necessarily $\{\mathcal{B}_t(\mathbf{W}^m[0,T])\}_{t\geq0}$-progressively measurable.

Proposition 2.13. *Let $(\Omega,\mathcal{F},\{\mathcal{F}_t\}_{t\geq0},\mathbf{P})$ be a filtered probability space satisfying the usual condition, and let ξ be an $\{\mathcal{F}_t\}_{t\geq0}$-adapted process. Then for any $s\in[0,T]$,*

$$\begin{aligned}\mathbf{P}\big(\{\omega'\mid\xi(t,\omega')=\xi(t,\omega)\}\mid\mathcal{F}_s\big)(\omega)=1,\\ \mathbf{P}\text{-a.s. }\omega\in\Omega,\ \forall t\in[0,s].\end{aligned} \tag{2.22}$$

If in addition $\xi(t)$ is continuous, then there exists a $\mathbf{P}$-null set $N\in\mathcal{F}$ such that for any $s\in[0,T]$ and $\omega\notin N$,

$$\xi(t,\omega')=\xi(t,\omega),\qquad\forall t\in[0,s],\ \mathbf{P}(\cdot\mid\mathcal{F}_s)(\omega)\text{-a.s. }\omega'\in\Omega. \tag{2.23}$$

Proof. By definition, for any $t\in[0,s]$,

$$\begin{aligned}&\mathbf{P}\big(\{\omega'\mid\xi(t,\omega')=\xi(t,\omega)\}\mid\mathcal{F}_s\big)(\omega)\\&=E\big(I_{\{\omega'|\xi(t,\omega')=\xi(t,\omega)\}}|\mathcal{F}_s\big)(\omega)\\&=I_{\{\omega'|\xi(t,\omega')=\xi(t,\omega)\}}(\omega)=1,\quad\mathbf{P}\text{-a.s. }\omega\in\Omega,\end{aligned}$$

proving (2.22). By the continuity of ξ, we can get (2.23) easily. □

The above result implies that under the probability $\mathbf{P}(\cdot \,|\, \mathcal{F}_s)(\omega)$, where ω is fixed, $\xi(t)$ is almost surely a deterministic constant $\xi(t,\omega)$ for any $t \leq s$. This fact will have important applications in stochastic dynamic programming studied in later chapters.

Finally, we discuss a convergence property of a sequence of continuous stochastic processes, based on the Skorohod theorem (Theorem 1.20).

Theorem 2.14. *Let $X_i = \{X_i(\cdot)\}$, $i = 1,2,\ldots$, be a sequence of m-dimensional continuous processes over $[0,T]$ on a probability space $(\Omega, \mathcal{F}, \mathbf{P})$ satisfying the following:*

$$\begin{cases} \sup_{i\geq 1} E|X_i(0)|^\gamma < +\infty, \\ \sup_{i\geq 1} E|X_i(t) - X_i(s)|^\alpha \leq K|t-s|^{1+\beta}, \quad \forall t,s \in [0,T], \end{cases}$$

for some constants $\alpha, \beta, \gamma > 0$. Then $\{X_i(\cdot)\}$ is tight as $\mathbf{W}^m[0,T]$-valued random variables. As a consequence, there exists a subsequence $\{i_j\}$, m-dimensional continuous processes $\widehat{X}_{i_j} = \{\widehat{X}_{i_j}(\cdot)\}$ $(j = 1,2,\ldots)$, and $\widehat{X}$ defined on a probability space $(\widehat{\Omega}, \widehat{\mathcal{F}}, \widehat{\mathbf{P}})$ such that

$$\begin{cases} \mathbf{P}(X_{i_j} \in A) = \widehat{\mathbf{P}}(\widehat{X}_{i_j} \in A), \quad \forall A \in \mathcal{B}(\mathbf{W}^m[0,T]), \\ \lim_{j\to\infty} \widehat{X}_{i_j} \to \widehat{X} \quad \textit{in } \mathbf{W}^m[0,T], \ \widehat{\mathbf{P}}\text{-a.s.} \end{cases}$$

Corollary 2.15. *Let $X(\cdot)$ be an m-dimensional stochastic process over $[0,T]$ such that*

$$E|X(t) - X(s)|^\alpha \leq K|t-s|^{1+\beta}, \quad \forall t,s \in [0,T],$$

for some constants $\alpha, \beta > 0$. Then there exists an m-dimensional continuous process that is stochastically equivalent to $X(\cdot)$.

See Ikeda–Watanabe [1, pp. 17–20] for proofs of Theorem 2.14 and Corollary 2.15.

2.2. Brownian motions

Let us now introduce an extremely important example of stochastic processes, called *Brownian motion*.

Definition 2.16. Let $(\Omega, \mathcal{F}, \{\mathcal{F}_t\}_{t\geq 0}, \mathbf{P})$ be a filtered probability space. An $\{\mathcal{F}_t\}_{t\geq 0}$-adapted $\mathbb{R}^m$-valued process $X(\cdot)$ is called an m-dimensional $\{\mathcal{F}_t\}_{t\geq 0}$-*Brownian motion* over $[0,\infty)$ if for all $0 \leq s < t$, $X(t) - X(s)$ is independent of $\mathcal{F}_s$ and is normally distributed with mean 0 and covariance $(t-s)I$. Namely, for any $0 \leq s \leq t$,

$$\begin{cases} E\big(X(t) - X(s) | \mathcal{F}_s\big) = 0, \quad \mathbf{P}\text{-a.s.}, \\ E\big((X(t) - X(s))(X(t) - X(s))^\top | \mathcal{F}_s\big) = (t-s)I, \quad \mathbf{P}\text{-a.s.} \end{cases} \tag{2.24}$$

In addition, if $\mathbf{P}(X(0) = 0) = 1$, then $X(\cdot)$ is called an m-dimensional *standard* $\{\mathcal{F}_t\}_{t\geq 0}$*-Brownian motion* over $[0, \infty)$.

Note that one can also define a Brownian motion $W(\cdot)$ naturally over any time interval $[a, b]$ or $[a, b)$ for any $0 \leq a < b \leq +\infty$. In particular, $W(\cdot)$ is said to be *standard* over $[a, b]$ if $W(a) = 0$.

It is easy to show that if $W(t), t \geq 0$, is a standard $\{\mathcal{F}_t\}_{t\geq 0}$-Brownian motion, so are $W(t + a) - W(a)$ $(a > 0)$ and $\lambda^{-1}W(\lambda^2 t)$ $(\lambda \neq 0)$.

We now construct a Brownian motion. Define $\mathbf{W}^m \stackrel{\Delta}{=} C([0, \infty); \mathbb{R}^m)$, and let

$$\widehat{\rho}(w, \widehat{w}) = \sum_{j\geq 1} 2^{-j} \big[|w - \widehat{w}|_{C([0,j];\mathbb{R}^m)} \wedge 1 \big], \quad \forall w, \widehat{w} \in \mathbf{W}^m.$$

Clearly, $\widehat{\rho}$ is a metric under which $\mathbf{W}^m$ is a Polish space. We call a set (of the form similar to (2.13))

$$B = \{\zeta \in \mathbf{W}^m \mid (\zeta(t_1), \zeta(t_2), \cdots, \zeta(t_j)) \in E\},$$

a Borel cylinder in $\mathbf{W}^m$, where $0 \leq t_1 < t_2 < \cdots < t_j < \infty$ and $E \in \mathcal{B}(\mathbb{R}^{jm})$. We let $\mathbf{C}$ be the set of all such Borel cylinders in $\mathbf{W}^m$. The following result is similar to Lemma 2.11.

Lemma 2.17. *The σ-field $\sigma(\mathbf{C})$ generated by $\mathbf{C}$ coincides with the Borel σ-field $\mathcal{B}(\mathbf{W}^m)$ of $\mathbf{W}^m$.*

For any $t \geq 0$, let

$$\begin{cases} \mathbf{W}_t^m \stackrel{\Delta}{=} \{\zeta(\cdot \wedge t) \mid \zeta(\cdot) \in \mathbf{W}^m\}, \\ \mathcal{B}_t(\mathbf{W}^m) \stackrel{\Delta}{=} \sigma(\mathcal{B}(\mathbf{W}_t^m)), \\ \mathcal{B}_{t+}(\mathbf{W}^m) = \bigcap_{s>t} \mathcal{B}_s(\mathbf{W}^m), \end{cases} \tag{2.25}$$

which is similar to (2.10). Let μ be a probability measure on $(\mathbb{R}^m, \mathcal{B}(\mathbb{R}^m))$ and let

$$f(x, t) = (2\pi t)^{-m/2} e^{-\frac{|x|^2}{2t}}, \qquad t > 0, \ x \in \mathbb{R}^m, \tag{2.26}$$

be the density of normal distribution $N(0, tI)$ (see (1.14)). We introduce $\mathbf{P}_\mu : \mathbf{C} \to [0, 1]$ as follows: For any $0 < t_1 < t_2 < \cdots < t_j \leq T$ and any $E_i \in \mathcal{B}(\mathbb{R}^m)$ $(1 \leq i \leq j)$,

$$\begin{aligned} &\mathbf{P}_\mu\{w \in \mathbf{W}^m \mid w(t_1) \in E_1, \ldots, w(t_j) \in E_j\} \\ &\stackrel{\Delta}{=} \int_{\mathbb{R}^m} \mu(dx_0) \int_{E_1} f(t_1, x_1 - x_0) dx_1 \int_{E_2} f(t_2 - t_1, x_2 - x_1) dx_2 \\ &\qquad \cdots \int_{E_j} f(t_j - t_{j-1}, x_j - x_{j-1}) dx_j. \end{aligned} \tag{2.27}$$

One can show that $\mathbf{P}_\mu$ is a probability measure on $(\mathbf{C}, \sigma(\mathbf{C}))$. Then, by Lemma 2.17, we may extend $\mathbf{P}_\mu$ to a probability measure on the space

$(\mathbf{W}^m, \mathcal{B}(\mathbf{W}^m))$, which is called the *Wiener measure* with the initial distribution μ. For given μ, such a Wiener measure is unique. Next, on the filtered probability space $(\mathbf{W}^m, \mathcal{B}(\mathbf{W}^m), \{\mathcal{B}_t(\mathbf{W}^m)\}_{t\geq 0}, \mathbf{P}_\mu)$ we define

$$X(t,w) = w(t), \qquad t \geq 0,\ w \in \mathbf{W}^m. \tag{2.28}$$

It can be shown that $X(t)$ is an m-dimensional Brownian motion with the initial distribution μ on $(\mathbf{W}^m, \mathcal{B}(\mathbf{W}^m), \{\mathcal{B}_t(\mathbf{W}^m)\}_{t\geq 0}, \mathbf{P}_\mu)$. In addition, if μ is a probability measure on $(\mathbb{R}^m, \mathcal{B}(\mathbb{R}^m))$ concentrated at 0, the above gives an m-dimensional standard Brownian motion.

We call the process $X(t)$ defined by (2.28) the *canonical realization* of an m-dimensional Brownian motion. For detailed proofs and comments, see Ikeda–Watanabe [1, pp. 40–41] or Karatzas–Shreve [3, pp. 49–56].

In general, if $X(\cdot)$ is a Brownian motion defined on some filtered probability space $(\Omega, \mathcal{F}, \{\mathcal{F}_t\}_{t\geq 0}, \mathbf{P})$, we may define

$$\mathcal{F}_t^X \stackrel{\Delta}{=} \sigma\{X(s), 0 \leq s \leq t\} \subseteq \mathcal{F}_t, \qquad \forall t \geq 0. \tag{2.29}$$

In particular, if $X(t)$ is given by (2.28), then (noting Lemma 2.17)

$$\begin{aligned} \mathcal{F}_t^X &= \sigma\Big\{\{w \in \mathbf{W}^m \mid w(s) \in B\} \Bigm| 0 \leq s \leq t,\ B \in \mathcal{B}(\mathbb{R}^m)\Big\} \\ &\equiv \mathcal{B}_t(\mathbf{W}^m). \end{aligned} \tag{2.30}$$

The filtration $\{\mathcal{F}_t^X\}_{t\geq 0}$ is left continuous, but not necessarily right continuous due to a fact similar to that stipulated in (2.11). On the other hand, the augmentation $\{\widehat{\mathcal{F}}_t^X\}_{t\geq 0}$ of $\{\mathcal{F}_t^X\}_{t\geq 0}$ by all the $\mathbf{P}$-null sets is continuous, and $X(t)$ is still a Brownian motion on the (augmented) filtered probability space $(\Omega, \mathcal{F}, \{\widehat{\mathcal{F}}_t^X\}_{t\geq 0}, \mathbf{P})$ (see Karatzas–Shreve [3, p. 89 and p. 122] for detailed discussions). In what follows, by saying that $\{\mathcal{F}_t\}_{t\geq 0}$ is the *natural filtration* generated by the Brownian motion X, we mean that $\{\mathcal{F}_t\}_{t\geq 0}$ is generated as (2.29) augmented by all the $\mathbf{P}$-null sets in $\mathcal{F}$ (thus $\{\mathcal{F}_t\}_{t\geq 0}$ is continuous).

3. Stopping Times

In this section we discuss a special class of random variables, which plays an interesting role in stochastic analysis.

Let $(\Omega, \mathcal{F}, \{\mathcal{F}_t\}_{t\geq 0}, \mathbf{P})$ be a filtered probability space satisfying the usual condition (see Definition 2.6).

Definition 3.1. A mapping $\tau : \Omega \to [0, \infty]$ is called an $\{\mathcal{F}_t\}_{t\geq 0}$-*stopping time* if

$$(\tau \leq t) \stackrel{\Delta}{=} \{\omega \in \Omega \mid \tau(\omega) \leq t\} \in \mathcal{F}_t, \quad \forall t \geq 0. \tag{3.1}$$

For any stopping time τ, define

$$\mathcal{F}_\tau \stackrel{\Delta}{=} \{A \in \mathcal{F} \mid A \cap (\tau \leq t) \in \mathcal{F}_t,\ \forall t \geq 0\}. \tag{3.2}$$

It is clear that $\mathcal{F}_\tau$ is a sub-σ-field of $\mathcal{F}$.

Proposition 3.2. *Stopping times have the following properties:*

(i) *A map $\tau : \Omega \to [0, \infty]$ is a stopping time if and only if*

$$(\tau < t) \in \mathcal{F}_t, \qquad \forall t > 0. \tag{3.3}$$

(ii) *If τ is a stopping time, then $A \in \mathcal{F}_\tau$ if and only if*

$$A \cap (\tau < t) \in \mathcal{F}_t, \qquad \forall t > 0. \tag{3.4}$$

Proof. We first prove (ii). If $A \in \mathcal{F}_\tau$, then for any $t > 0$,

$$A \cap (\tau < t) = \bigcup_{n \geq 1} \Big\{ A \cap \Big(\tau \leq t - \frac{1}{n} \Big) \Big\} \in \mathcal{F}_t. \tag{3.5}$$

Conversely, if (3.4) holds, then for any $t \geq 0$, by the right continuity of $\mathcal{F}_t$,

$$A \cap (\tau \leq t) = \bigcap_{n \geq 1} \Big\{ A \cap \Big(\tau < t + \frac{1}{n} \Big) \Big\} \in \mathcal{F}_{t+} = \mathcal{F}_t. \tag{3.6}$$

This proves (ii). Taking $A = \Omega$, we obtain (i). □

Now we give an example that will be useful later.

Example 3.3. Let $X(t)$ be $\{\mathcal{F}_t\}_{t \geq 0}$-adapted and continuous. Let $E \subseteq \mathbb{R}^m$ be an open set. Then the *first hitting time* of the process $X(t)$ to E,

$$\sigma_E(\omega) \triangleq \inf\{t \geq 0 \mid X(t, \omega) \in E\}, \tag{3.7}$$

and the *first exit time* of the process $X(t)$ from E,

$$\tau_E(\omega) \triangleq \inf\{t \geq 0 \mid X(t, \omega) \notin E\}, \tag{3.8}$$

are both $\{\mathcal{F}_t\}_{t \geq 0}$-stopping times. (Here, $\inf \phi \triangleq +\infty$.) Let us prove these two facts. First of all, for any $s > 0$, we claim that

$$(\sigma_E < s) = \bigcup_{r \in \mathbf{Q}, r < s} (X(r) \in E) \in \mathcal{F}_s. \tag{3.9}$$

To prove the equality, let ω be such that $\sigma_E(\omega) < s$. Then by the continuity of X, we have some rational number $r < s$, such that $X(r, \omega) \in E$. Hence ω belongs to the right-hand side of (3.9). Conversely, let ω be in the right-hand side of (3.9). Then, for some rational number $r < s$, $X(r, \omega) \in E$. This implies that $\sigma_E(\omega) < s$. Hence, the equality in (3.9) holds. Consequently, by Proposition 3.2, σ_E is a stopping time. Next, for any $t \geq 0$, we have

$$(\tau_E \geq t) = \bigcap_{r \in \mathbf{Q}, r < t} (X_r \in E) \in \mathcal{F}_t. \tag{3.10}$$

The proof of (3.10) is similar to that of (3.9). Hence

$$(\tau_E < t) = (\tau_E \geq t)^c \in \mathcal{F}_t. \tag{3.11}$$

Then, by Proposition 3.2, τ_E is a stopping time.

Proposition 3.4. *Let σ, τ, and σ_i be stopping times. Then*

(i) *The following are also stopping times:*

$$\sigma + \tau, \ \sup_i \sigma_i, \ \inf_i \sigma_i, \ \overline{\lim_i} \, \sigma_i, \ \underline{\lim_i} \, \sigma_i.$$

(ii) *The map $\tau : (\Omega, \mathcal{F}_\tau) \to ([0,\infty], \mathcal{B}[0,\infty])$ is measurable and the process $Y(t) \stackrel{\Delta}{=} \tau \wedge t$ is $\{\mathcal{F}_t\}_{t\geq 0}$-progressively measurable.*

(iii) *The following holds:*

$$(\sigma > \tau), (\sigma \geq \tau), (\sigma = \tau) \in \mathcal{F}_{\sigma\wedge\tau}. \tag{3.12}$$

Moreover, for any $A \in \mathcal{F}_\sigma$,

$$A \bigcap (\sigma \leq \tau) \in \mathcal{F}_\tau. \tag{3.13}$$

In particular,

$$\sigma \leq \tau \quad \text{on } \Omega \quad \Rightarrow \quad \mathcal{F}_\sigma \subseteq \mathcal{F}_\tau. \tag{3.14}$$

In the case where $(\Omega, \mathcal{F}, \{\mathcal{F}_t\}_{t\geq 0}, \mathbf{P})$ is complete, (3.14) can be replaced by

$$\sigma \leq \tau, \quad \mathbf{P}\text{-a.s.} \quad \Rightarrow \quad \mathcal{F}_\sigma \subseteq \mathcal{F}_\tau. \tag{3.15}$$

(iv) *Let $\bar\sigma = \inf_{i\geq 1} \sigma_i$. Then*

$$\bigcap_{i\geq 1} \mathcal{F}_{\sigma_i} = \mathcal{F}_{\bar\sigma}. \tag{3.16}$$

For proofs of the above results, see Karatzas–Shreve [3, pp. 6–10].

Proposition 3.5. *Let $(\Omega, \mathcal{F}, \{\mathcal{F}_t\}_{t\geq 0}, \mathbf{P})$ be a filtered probability space satisfying the usual condition. Let $X(t)$ be an $\{\mathcal{F}_t\}_{t\geq 0}$-progressively measurable process and τ an $\{\mathcal{F}_t\}_{t\geq 0}$-stopping time. Then the random variable $X(\tau)$ is $\mathcal{F}_\tau$-measurable, and the process $X(\tau \wedge t)$ is $\{\mathcal{F}_t\}_{t\geq 0}$-progressively measurable.*

Proof. We first prove that $X(\tau\wedge t)$ is $\{\mathcal{F}_t\}_{t\geq 0}$-progressively measurable. To this end, by Proposition 3.4, the process $\tau \wedge t$ is $\{\mathcal{F}_t\}_{t\geq 0}$-progressively measurable. Thus, for each $t \geq 0$, the map $(s,\omega) \mapsto (\tau(\omega) \wedge s, \omega)$ is measurable from $([0,t] \times \Omega, \mathcal{B}[0,t] \times \mathcal{F}_t)$ into itself. On the other hand, by the progressive measurability of $X(t)$, the map $(s,\omega) \mapsto X(s,\omega)$ is measurable from $([0,t] \times \Omega, \mathcal{B}[0,t] \times \mathcal{F}_t)$ to $(\mathbb{R}^m, \mathcal{B}(\mathbb{R}^m))$. Hence, the map $(s,\omega) \mapsto X(\tau(\omega) \wedge s, \omega)$ is measurable from $([0,t] \times \Omega, \mathcal{B}[0,t] \times \mathcal{F}_t)$ to $(\mathbb{R}^m, \mathcal{B}(\mathbb{R}^m))$. Hence the process $X(\tau \wedge t)$ is $\{\mathcal{F}_t\}_{t\geq 0}$-progressively measurable. In particular, $X(\tau \wedge t)$ is $\mathcal{F}_t$-measurable for each fixed $t \geq 0$. Now, for any $B \in \mathcal{B}(\mathbb{R}^m)$,

$$(X(\tau) \in B) \cap (\tau \leq t) = (X(\tau \wedge t) \in B) \cap (\tau \leq t) \in \mathcal{F}_t, \quad \forall t \geq 0. \tag{3.17}$$

Thus, $(X(\tau) \in B) \in \mathcal{F}_\tau$, for all $B \in \mathcal{B}(\mathbb{R}^m)$, proving the measurability of the map $X(\tau) : (\Omega, \mathcal{F}_\tau) \to (\mathbb{R}^m, \mathcal{B}(\mathbb{R}^m))$. □

Proposition 3.6. *Let τ be a stopping time and ξ a random variable. Then ξ is $\mathcal{F}_\tau$-measurable if and only if for all $t \geq 0$, $\xi I_{(\tau \leq t)}$ is $\mathcal{F}_t$-measurable.*

Proof. If ξ is $\mathcal{F}_\tau$ measurable, then there exists a sequence of $\mathcal{F}_\tau$-measurable simple functions

$$\xi_j \equiv \sum_{i \geq 1} \xi_j^i I_{A_j^i} \to \xi, \text{ as } j \to \infty, \quad \mathbf{P}\text{-a.s.}, \tag{3.18}$$

where $\xi_j^i \in \mathbb{R}$ and $A_j^i \in \mathcal{F}_\tau$. Hence,

$$\xi_j I_{(\tau \leq t)} = \sum_{i \geq 1} \xi_j^i I_{A_j^i \cap (\tau \leq t)} \tag{3.19}$$

is $\mathcal{F}_t$-measurable. Letting $j \to \infty$, we see that $\xi I_{(\tau \leq t)}$ is $\mathcal{F}_t$-measurable. Conversely, if for all $t \geq 0$, $\xi I_{(\tau \leq t)}$ is $\mathcal{F}_t$-measurable, then for any $\alpha < 0$, we have

$$(\xi \leq \alpha) \bigcap (\tau \leq t) = (\xi I_{(\tau \leq t)} \leq \alpha) \in \mathcal{F}_t, \tag{3.20}$$

and for any $\alpha \geq 0$,

$$(\xi > \alpha) \bigcap (\tau \leq t) = (\xi I_{(\tau \leq t)} > \alpha) = (\xi I_{(\tau \leq t)} \leq \alpha)^c \in \mathcal{F}_t. \tag{3.21}$$

This implies

$$\begin{cases} (\xi \leq \alpha) \in \mathcal{F}_\tau, & \alpha < 0, \\ (\xi \leq \alpha) = (\xi > \alpha)^c \in \mathcal{F}_\tau, & \alpha \geq 0. \end{cases} \tag{3.22}$$

Consequently, ξ is $\mathcal{F}_\tau$-measurable. □

Proposition 3.7. *Let σ and τ be stopping times and X an integrable random variable. Then*

$$\begin{cases} I_{(\sigma > \tau)} E(X \big| \mathcal{F}_\tau) = E\big(I_{(\sigma > \tau)} X \big| \mathcal{F}_\tau\big) = I_{(\sigma > \tau)} E\big(X \big| \mathcal{F}_{\sigma \wedge \tau}\big), \\ I_{(\sigma \geq \tau)} E(X \big| \mathcal{F}_\tau) = E\big(I_{(\sigma \geq \tau)} X \big| \mathcal{F}_\tau\big) = I_{(\sigma \geq \tau)} E\big(X \big| \mathcal{F}_{\sigma \wedge \tau}\big), \\ E\big(E(X | \mathcal{F}_\tau) \big| \mathcal{F}_\sigma\big) = E\big(X \big| \mathcal{F}_{\sigma \wedge \tau}\big). \end{cases} \tag{3.23}$$

Proof. By Proposition 3.4-(iii), we have the first equalities in the first two assertions. Now, since

$$I_{(\sigma > \tau)} E(X | \mathcal{F}_\tau) I_{(\sigma \wedge \tau \leq t)} = E(X | \mathcal{F}_\tau) I_{(\tau \leq t)} I_{(\sigma > \tau, \sigma \wedge \tau \leq t)}, \tag{3.24}$$

$E(X|\mathcal{F}_\tau)$ is $\mathcal{F}_\tau$-measurable, and $(\sigma > \tau) \in \mathcal{F}_{\sigma \wedge \tau}$ (see (3.12)), it follows that the right-hand side of (3.24) is $\mathcal{F}_t$-measurable by Proposition 3.6. Hence, by Proposition 3.6 again, $I_{(\sigma > \tau)} E(X \big| \mathcal{F}_\tau)$ is $\mathcal{F}_{\sigma \wedge \tau}$-measurable. Then, by the basic property of the conditional expectation, we have

$$\begin{aligned} E\big(I_{(\sigma > \tau)} X \big| \mathcal{F}_\tau\big) &= I_{(\sigma > \tau)} E(X | \mathcal{F}_\tau) = E\big(I_{(\sigma > \tau)} E(X | \mathcal{F}_\tau) \big| \mathcal{F}_{\sigma \wedge \tau}\big) \\ &= I_{(\sigma > \tau)} E\big(E(X | \mathcal{F}_\tau) \big| \mathcal{F}_{\sigma \wedge \tau}\big) = I_{(\sigma > \tau)} E(X | \mathcal{F}_{\sigma \wedge \tau}). \end{aligned} \tag{3.25}$$

This gives the first relation in (3.23). The second relation can be proved similarly. Finally,

$$
\begin{aligned}
&E\big(E(X|\mathcal{F}_\tau)\big|\mathcal{F}_\sigma\big)\\
&\quad= E\big(I_{(\sigma>\tau)}E(X|\mathcal{F}_\tau)\big|\mathcal{F}_\sigma\big)+E\big(I_{(\sigma\le\tau)}E(X|\mathcal{F}_\tau)\big|\mathcal{F}_\sigma\big)\\
&\quad= E\big(I_{(\sigma>\tau)}E(X|\mathcal{F}_{\sigma\wedge\tau})\big|\mathcal{F}_\sigma\big)+E\big(I_{(\tau\ge\sigma)}E(X|\mathcal{F}_{\sigma\wedge\tau})\big|\mathcal{F}_\sigma\big)\\
&\quad= I_{(\sigma>\tau)}E(X|\mathcal{F}_{\sigma\wedge\tau})+I_{(\tau\ge\sigma)}E(X|\mathcal{F}_{\sigma\wedge\tau})=E(X|\mathcal{F}_{\sigma\wedge\tau}).
\end{aligned}
\tag{3.26}
$$

This proves the third relation. □

4. Martingales

In this section we will briefly recall some results on martingales, which form a special class of stochastic processes.

Let $(\Omega,\mathcal{F},\{\mathcal{F}_t\}_{t\ge0},\mathbf{P})$ be a filtered probability space satisfying the usual condition.

Definition 4.1. A real-valued process $X(t)$ is called a (continuous) $\{\mathcal{F}_t\}_{t\ge0}$-*martingale* (resp. *submartingale, supermartingale*) if it is $\{\mathcal{F}_t\}_{t\ge0}$-adapted and for any $t\ge0$, $EX(t)$ exists with the property that

$$
E(X(t)\big|\mathcal{F}_s)=X(s)\quad(\text{resp. }\ge,\ \le),\quad \mathbf{P}\text{-a.s.},\ \forall 0\le s\le t. \tag{4.1}
$$

It is clear that any martingale must be both a sub- and a supermartingale. As an example, any Brownian motion $X(t)$ is a martingale by the first relation in (2.24).

Proposition 4.2. *Let $\{\mathcal{F}_t\}_{t\ge0}$ and $\{\mathcal{G}_t\}_{t\ge0}$ be two families of sub-σ-fields of $\mathcal{F}$ with $\mathcal{G}_t\subseteq\mathcal{F}_t$, $\forall t\ge0$. If $X(t)$ is an $\{\mathcal{F}_t\}_{t\ge0}$-martingale (resp. submartingale, supermartingale), then $Y(t)\stackrel{\Delta}{=}E(X(t)|\mathcal{G}_t)$ is a $\{\mathcal{G}_t\}_{t\ge0}$-martingale (resp. submartingale, supermartingale). In particular, if $X(t)$ is $\{\mathcal{G}_t\}_{t\ge0}$-adapted, then $X(t)$ itself is a $\{\mathcal{G}_t\}_{t\ge0}$-martingale (resp. submartingale, supermartingale).*

Proof. The first assertion follows easily from Proposition 1.8-(vi) and the definition of a martingale, while the second one follows from the first along with Proposition 1.8-(iv). □

Roughly speaking, this proposition implies that a martingale with respect to a "larger" filtration is also a martingale with respect to a "smaller" filtration.

Proposition 4.3. *Let $X(t)$ be a submartingale and $\varphi:\mathbb{R}\to\mathbb{R}$ a nondecreasing convex function such that $E\varphi(X(t))$ exists for all $t\ge0$. Then $\varphi(X(t))$ is a submartingale. In particular, when $X(t)$ is a nonnegative submartingale and $E|X(t)|^p$ exists for some $p\ge1$, then $|X(t)|^p$ is a submartingale.*

Proof. By the monotonicity of $\varphi(\cdot)$ and Jensen's inequality (see (1.36)), we have

$$
\varphi(X(s))\le\varphi(E(X(t)\big|\mathcal{F}_s))\le E(\varphi(X(t))\big|\mathcal{F}_s),\quad \mathbf{P}\text{-a.s.},\ s\le t. \tag{4.2}
$$

This proves our assertion. □

As we know, a stochastic process is not necessarily left or right continuous. However, for submartingales (in particular, martingales), we have the following result.

Theorem 4.4. *Let $X(t)$ be an $\{\mathcal{F}_t\}_{t\geq 0}$-submartingale. Then there exists a* $\mathbf{P}$*-null set N such that for any $\omega \in \Omega \setminus N$, the following limits exist:*

$$\lim_{r\in\mathbf{Q}, r\downarrow t} X(r), \qquad \lim_{r\in\mathbf{Q}, r\uparrow t} X(r), \quad \forall t \in [0,\infty). \tag{4.3}$$

If we define

$$\widehat{X}(t) = \lim_{r\in\mathbf{Q}, r\downarrow t} X(r), \qquad \mathbf{P}\text{-a.s.}, \ \forall t \geq 0, \tag{4.4}$$

and define $\widehat{X}(t)$ as an arbitrary constant on N, then $\widehat{X}(t)$ is a submartingale that is right continuous with left limits, and for any sequence $\varepsilon_j \downarrow 0$,

$$X(t+\varepsilon_j) \to \widehat{X}(t), \qquad \forall t \geq 0, \ \mathbf{P}\text{-a.s.}, \tag{4.5}$$

$$X(t) \leq \widehat{X}(t), \qquad \forall t \geq 0, \ \mathbf{P}\text{-a.s.} \tag{4.6}$$

Moreover, $\mathbf{P}(X(t) = \widehat{X}(t)) = 1$ for all $t \geq 0$ if and only if $t \mapsto EX(t)$ is right continuous.

The process $\widehat{X}(t)$ in the above result is called the *right-continuous modification* of $X(t)$. By Theorem 4.4, if $EX(\cdot)$ is right continuous (which is the case when $X(\cdot)$ is a martingale), we may assume that $X(\cdot)$ itself is right continuous.

Theorem 4.5. *Let $p \geq 1$ and $X(t)$ be a right-continuous $\{\mathcal{F}_t\}_{t\geq 0}$-martingale such that for all $t \geq 0$, $E|X(t)|^p < \infty$. Then, for any $T > 0$,*

$$\mathbf{P}\Big(\sup_{t\in[0,T]} |X(t)| > \lambda\Big) \leq \frac{E|X(T)|^p}{\lambda^p}, \tag{4.7}$$

and for $p > 1$,

$$E\Big(\sup_{t\in[0,T]} |X(t)|^p\Big) \leq \Big(\frac{p}{p-1}\Big)^p E|X(T)|^p. \tag{4.8}$$

Theorem 4.6. (Optional sampling theorem) *Let $(\Omega, \mathcal{F}, \{\mathcal{F}_t\}_{t\geq 0}, \mathbf{P})$ be a filtered probability space satisfying the usual condition. Let $X(t)$ be a right-continuous $\{\mathcal{F}_t\}_{t\geq 0}$-martingale (resp. submartingale, supermartingale). Let $\{\sigma_t\}_{t\geq 0}$ be a family of bounded $\{\mathcal{F}_t\}_{t\geq 0}$-stopping times such that*

$$\mathbf{P}(\sigma_s \leq \sigma_t) = 1, \qquad s < t. \tag{4.9}$$

Let $\widetilde{X}(t) = X(\sigma_t)$ and $\widetilde{\mathcal{F}}_t = \mathcal{F}_{\sigma_t}$. Then $\{\widetilde{X}(t)\}_{t\geq 0}$ is an $\{\widetilde{\mathcal{F}}_t\}_{t\geq 0}$-martingale (resp. submartingale, supermartingale).

See Ikeda–Watanabe [1, pp. 32–34] for proofs of Theorems 4.4–4.6.

Corollary 4.7. *Let $(\Omega, \mathcal{F}, \{\mathcal{F}_t\}_{t\geq 0}, \mathbf{P})$ be the same as in Theorem 4.6. Let $\sigma \leq \tau$ be two bounded stopping times. Then, for any $\{\mathcal{F}_t\}_{t\geq 0}$-martingale (resp. submartingale, supermartingale) $X(t)$,*

$$E(X(\tau)|\mathcal{F}_\sigma) = X(\sigma) \quad (\text{resp. } \geq, \leq), \quad \mathbf{P}\text{-a.s.} \tag{4.10}$$

Proof. Let

$$\sigma_t = \sigma I_{[0,1]}(t) + \tau I_{(1,\infty)}(t), \qquad t \geq 0. \tag{4.11}$$

Then we have (4.9). Appealing to Theorem 4.6 by taking $t > 1$ and $s \leq 1$, one obtains (4.10). □

Corollary 4.8. *Let $X(t)$ be an $\{\mathcal{F}_t\}_{t\geq 0}$-martingale, and let $\sigma \leq \tau$ be two stopping times. Then*

$$E\big[X(t\wedge\tau) - X(t\wedge\sigma)\big|\mathcal{F}_\sigma\big] = 0, \qquad \forall t \geq 0. \tag{4.12}$$

Proof. In view of Propositions 3.4 and 3.5, $t \wedge \tau$ and $t \wedge \sigma$ are stopping times with $t\wedge\tau \geq t\wedge\sigma$, $\mathbf{P}$-a.s., and $X(t\wedge\tau)$ is $\mathcal{F}_t$-measurable. Thus, by Corollary 4.7 and Proposition 3.7, we obtain

$$\begin{aligned} X(t\wedge\sigma) &= E\big[X(t\wedge\tau)\big|\mathcal{F}_{t\wedge\sigma}\big] \\ &= E\big\{E\big[X(t\wedge\tau)\big|\mathcal{F}_t\big]\big|\mathcal{F}_\sigma\big\} = E\big[X(t\wedge\tau)\big|\mathcal{F}_\sigma\big]. \end{aligned} \tag{4.13}$$

This yields (4.12). □

Sometimes, we need to generalize the notion of martingale.

Definition 4.9. A real-valued process $X(t)$ is called a *local* $\{\mathcal{F}_t\}_{t\geq 0}$-*martingale* if there exists a nondecreasing sequence of stopping times $\{\sigma_j\}_{j\geq 1}$ with

$$\mathbf{P}\Big(\lim_{j\to\infty} \sigma_j = \infty\Big) = 1, \tag{4.14}$$

so that $X(t\wedge\sigma_j)$ is an $\{\mathcal{F}_t\}_{t\geq 0}$-martingale for each j.

We point out that every martingale is a local martingale, but it is not necessarily true the other way around (see Karatzas–Shreve [3, p. 168] for an example).

We will mainly restrict ourselves to the finite interval $[0, T]$ in this book. A local martingale $X(t)$ restricted to $[0, T]$ becomes a local martingale on $[0, T]$. We may also define a local martingale on $[0, T]$ directly. In fact, we may simply copy Definition 4.9 with (4.14) replaced by the following:

$$\mathbf{P}\Big(\lim_{j\to\infty} \sigma_j \geq T\Big) = 1. \tag{4.15}$$

This observation will be useful later.

5. Itô's Integral

In this section we are going to define the integral of type

$$\int_0^T f(t)dW(t), \tag{5.1}$$

where f is some stochastic process and $W(t)$ is a Brownian motion. Such an integral will play an essential role in the rest of this book. Note that if for $\omega \in \Omega$, the map $t \mapsto W(t,\omega)$ was of *bounded variation*, then a natural definition of (5.1) would be a Lebesgue–Stieltjes-type integral, regarding ω as a parameter. Unfortunately, we will see below that the map $t \mapsto W(t,\omega)$ is *not* of bounded variation for almost all $\omega \in \Omega$. Thus one needs to define (5.1) in a different way. A proper definition for such an integral is due to Kiyoshi Itô [1, 2]. We will introduce it after the following subsection.

5.1. Nondifferentiability of Brownian motion

In this subsection we prove the almost sure nondifferentiability of Brownian motion. Actually, we will prove a little bit more than that. Again, let $(\Omega, \mathcal{F}, \{\mathcal{F}_t\}_{t\geq 0}, \mathbf{P})$ be a filtered probability space satisfying the usual conditions.

Proposition 5.1. *Let $W(t)$ be a (one-dimensional) Brownian motion. Let $\gamma > \frac{1}{2}$ and set*

$$\begin{aligned} S_\gamma \stackrel{\Delta}{=} \{\omega \in \Omega \mid\, & W(\cdot\,,\omega) \text{ is Hölder continuous with} \\ & \text{exponent } \gamma \text{ at some } t \in [0,\infty)\}. \end{aligned} \tag{5.2}$$

Then

$$\mathbf{P}(S_\gamma) = 0. \tag{5.3}$$

In particular, for almost all $\omega \in \Omega$, the map $t \mapsto W(t,\omega)$ is nowhere differentiable.

Proof. Since $\gamma > \frac{1}{2}$, there exists an integer μ such that $\mu(\gamma - \frac{1}{2}) > 1$. For any $j, k \geq 1$, define

$$\begin{aligned} A_{jk} = \big\{\omega \in \Omega \mid\, & |W(t+h,\omega) - W(t,\omega)| \leq jh^\gamma, \ \forall h \in [0, \tfrac{1}{k}] \\ & \text{for some } t \in [0,1]\big\}. \end{aligned} \tag{5.4}$$

Also, for $m = 0, 1, 2, \ldots$, set

$$\begin{aligned} S_\gamma[m, m+1] \stackrel{\Delta}{=} \{\omega \in \Omega \mid\, & W(\cdot\,,\omega) \text{ is Hölder continuous with} \\ & \text{exponent } \gamma \text{ at some } t \in [m, m+1]\}. \end{aligned}$$

Clearly, one has

$$S_\gamma[0,1] = \bigcup_{j,k\geq 1} A_{jk}. \tag{5.5}$$

We will show that $\mathbf{P}(A_{jk}) = 0$ for all $j, k \geq 1$. To this end, let us fix $j, k \geq 1$. For any $\omega \in A_{jk}$, by definition we can find a $t \in [0,1]$ such that

$$|W(t+h,\omega) - W(t,\omega)| \leq jh^\gamma, \qquad \forall h \in [0, \tfrac{1}{k}]. \tag{5.6}$$

Now let $n \geq \mu k$. Then, for the above t, there exists an i, $1 \leq i \leq n$, such that $\frac{i-1}{n} \leq t < \frac{i}{n}$. Since $n \geq \mu k$, we have

$$\frac{i+\nu}{n} - t \leq \frac{\nu+1}{n} \leq \frac{1}{k}, \qquad 1 \leq \nu \leq \mu - 1. \tag{5.7}$$

Observe that for $1 \leq \nu \leq \mu - 1$,

$$\begin{aligned} &\big|W(\tfrac{i+\nu}{n},\omega) - W(\tfrac{i+\nu-1}{n},\omega)\big| \\ &\quad\leq \big|W(\tfrac{i+\nu}{n},\omega) - W(t,\omega)\big| + \big|W(\tfrac{i+\nu-1}{n},\omega) - W(t,\omega)\big| \\ &\quad\leq j\Big(\frac{i+\nu}{n} - t\Big)^\gamma + j\Big(\frac{i+\nu-1}{n} - t\Big)^\gamma \leq j\Big(\frac{(\nu+1)^\gamma + \nu^\gamma}{n^\gamma}\Big). \end{aligned} \tag{5.8}$$

This means that any $\omega \in A_{jk}$ must satisfy (5.8). Set

$$C_i^{(n)} = \bigcap_{\nu=1}^{\mu} \Big\{\omega \in \Omega \mid |W(\tfrac{i+\nu}{n},\omega) - W(\tfrac{i+\nu-1}{n},\omega)| \leq j(\tfrac{(\nu+1)^\gamma+\nu^\gamma}{n^\gamma})\Big\}. \tag{5.9}$$

Clearly,

$$A_{jk} \subseteq \bigcup_{i=1}^{n} C_i^{(n)}, \qquad \forall n \geq \mu k. \tag{5.10}$$

Noting that $W(\frac{i+\nu}{n}) - W(\frac{i+\nu-1}{n}) \sim N(0, \frac{1}{n})$, we obtain

$$Z_\nu \stackrel{\Delta}{=} \sqrt{n}\big[W(\tfrac{i+\nu}{n}) - W(\tfrac{i+\nu-1}{n})\big] \sim N(0,1). \tag{5.11}$$

Since $W(\cdot)$ is a Brownian motion, then Z_ν, $1 \leq \nu \leq \mu$, are independent. On the other hand, by (5.11), for any $\varepsilon > 0$,

$$\mathbf{P}\Big(|Z_\nu| \leq \varepsilon\Big) = \frac{1}{\sqrt{2\pi}} \int_{-\varepsilon}^{\varepsilon} e^{-\frac{|z|^2}{2}} dz \leq \sqrt{\frac{2}{\pi}}\varepsilon. \tag{5.12}$$

Therefore,

$$\begin{aligned} \mathbf{P}(C_i^{(n)}) &= \mathbf{P}\Big(|Z_\nu| \leq j\frac{(\nu+1)^\gamma + \nu^\gamma}{n^{\gamma-1/2}}, \ 1 \leq \nu \leq \mu\Big) \\ &\leq \Big(\sqrt{\frac{2}{\pi}}\frac{j}{n^{\gamma-1/2}}\Big)^\mu \prod_{\nu=1}^{\mu} \big((\nu+1)^\gamma + \nu^\gamma\big) \equiv K n^{-\mu(\gamma-1/2)}, \end{aligned} \tag{5.13}$$

where $K = K(j,\gamma,\mu)$ is a constant independent of $n \geq \mu k$. Hence (noting that $\mu(\gamma - \frac{1}{2}) > 1$),

$$\mathbf{P}\Big(\bigcup_{i=1}^{n} C_i^{(n)}\Big) \leq \sum_{i=1}^{n} \mathbf{P}(C_i^{(n)}) \leq K n^{1-\mu(\gamma-1/2)} \to 0, \qquad n \to \infty. \tag{5.14}$$

This shows that A_{jk} is a subset of a $\mathbf{P}$-null set. By the completeness of the probability space, we have $A_{jk} \in \mathcal{F}$ with $\mathbf{P}(A_{jk}) = 0$. Hence, $\mathbf{P}(S_\gamma[0,1]) = 0$. Similarly, for any $m \geq 1$, $\mathbf{P}(S_\gamma[m, m+1]) = 0$. Then (5.3) follows. Consequently, it is clear that for almost all $\omega \in \Omega$, the map $t \mapsto W(t, \omega)$ is nowhere differentiable. $\square$

5.2. Definition of Itô's integral and basic properties

In this section we give the definition of the *Itô integral* as well as some basic properties of such an integral. We shall describe the basic idea of defining the Itô integral without giving proofs. Those who are interested in the details should consult standard books such as Ikeda–Watanabe [1, pp. 45–51] and Karatzas–Shreve [3, pp. 129–141].

We first introduce the function space consisting of all possible integrands. Let $(\Omega, \mathcal{F}, \{\mathcal{F}_t\}_{t \geq 0}, \mathbf{P})$ be a fixed filtered probability space satisfying the usual condition. Let $T > 0$ and recall that $L^2_{\mathcal{F}}(0, T; \mathbb{R})$ is the set of all measurable processes $f(t, \omega)$ adapted to $\{\mathcal{F}_t\}_{t \geq 0}$ such that

$$\|f\|_T^2 \stackrel{\Delta}{=} E\Big\{ \int_0^T f(t,\omega)^2 dt \Big\} < \infty. \tag{5.15}$$

It is seen that $L^2_{\mathcal{F}}(0, T; \mathbb{R})$ is a Hilbert space.

Next, we introduce the following sets, which are related to the integrals we are going to define:

$$\begin{cases} \mathcal{M}^2[0,T] = \{X \in L^2_{\mathcal{F}}(0,T;\mathbb{R}) \mid X \text{ is a right-continuous} \\ \qquad\qquad\qquad \{\mathcal{F}_t\}_{t\geq 0}\text{-martingale with } X(0) = 0, \ \mathbf{P}\text{-a.s.}\}, \\ \mathcal{M}^2_c[0,T] = \{X \in \mathcal{M}^2[0,T] \mid t \mapsto X(t) \text{ is continuous, } \mathbf{P}\text{-a.s.}\}. \end{cases} \tag{5.16}$$

We identify $X, Y \in \mathcal{M}^2[0,T]$ if there exists a set $N \in \mathcal{F}$ with $\mathbf{P}(N) = 0$ such that $X(t,\omega) = Y(t,\omega)$, for all $t \geq 0$ and $\omega \notin N$. Define

$$|X|_T = \big(EX(T)^2\big)^{1/2}, \qquad \forall X \in \mathcal{M}^2[0,T]. \tag{5.17}$$

We can show by the martingale property that (5.17) is a norm under which $\mathcal{M}^2[0,T]$ is a Hilbert space. Moreover, $\mathcal{M}^2_c[0,T]$ is a closed subspace of $\mathcal{M}^2[0,T]$. We should distinguish the norms $\|\cdot\|_T$ (defined by (5.15)) and $|\cdot|_T$ (defined by (5.17)). It is important to note that any Brownian motion $W(\cdot)$ is in $\mathcal{M}^2_c[0,T]$ with $|W|_T^2 = T$ (see (2.24)).

Now we are ready to itemize the steps in defining the Itô integral for a given one-dimensional Brownian motion $W(t)$ defined on the filtered probability space $(\Omega, \mathcal{F}, \{\mathcal{F}_t\}_{t\geq 0}, \mathbf{P})$.

Step 1. Consider a subset $\mathcal{L}_0[0,T] \subseteq L^2_{\mathcal{F}}(0,T;\mathbb{R})$ consisting of all real processes $f(t,\omega)$ of the following form (called *simple processes*):

$$f(t,\omega) = f_0(\omega) I_{\{t=0\}}(t) + \sum_{i \geq 0} f_i(\omega) I_{(t_i, t_{i+1}]}(t), \qquad t \in [0,T], \tag{5.18}$$

where $0 = t_0 < t_1 < \cdots$, $t_i \le T$ and $f_i(\omega)$ is $\mathcal{F}_{t_i}$-measurable with $\sup_i \sup_\omega |f_i(\omega)| < \infty$. One can show that the set $\mathcal{L}_0[0,T]$ is dense in $L^2_{\mathcal{F}}(0,T;\mathbb{R})$.

Step 2. Define an integral for any simple process $f \in \mathcal{L}_0[0,T]$ of the form (5.18): For any $t \in [t_j, t_{j+1}]$ $(j \ge 0)$, let

$$\begin{aligned}\widehat{I}(f)(t,\omega) \stackrel{\Delta}{=} \sum_{i=0}^{j-1} f_i(\omega)\big[W(t_{i+1},\omega) - W(t_i,\omega)\big]\\ + f_j(\omega)\big[W(t,\omega) - W(t_j,\omega)\big].\end{aligned} \tag{5.19}$$

Equivalently, we have the following:

$$\widehat{I}(f)(t,\omega) = \sum_{i\ge 0} f_i(\omega)\big[W(t\wedge t_{i+1},\omega) - W(t\wedge t_i,\omega)\big], \quad t \in [0,T]. \tag{5.20}$$

It is seen that $\widehat{I}$ is a linear operator from $\mathcal{L}_0[0,T]$ to $\mathcal{M}^2_c[0,T]$. Moreover, $\widehat{I}$ has the property that

$$|\widehat{I}(f)|^2_T = \|f\|^2_T, \quad \forall f \in \mathcal{L}_0[0,T]. \tag{5.21}$$

Step 3. For any $f \in L^2_{\mathcal{F}}(0,T;\mathbb{R})$, by *Step 1* there are $f_j \in \mathcal{L}_0[0,T]$ such that $\|f - f_j\|_T \to 0$ as $j \to \infty$. From (5.21), $\{\widehat{I}(f_j)\}$ is Cauchy in $\mathcal{M}^2_c[0,T]$. Thus, it has a unique limit in $\mathcal{M}^2_c[0,T]$, denoted by $\widehat{I}(f)$. It is seen from (5.21) that this limit depends only on f and is independent of the choice of the sequence f_j. Hence $\widehat{I}(f)$ is well-defined on $L^2_{\mathcal{F}}(0,T;\mathbb{R})$ and is called the *Itô integral*, denoted by

$$\int_0^t f(s)dW(s) \stackrel{\Delta}{=} \widehat{I}(f)(t), \qquad 0 \le t \le T. \tag{5.22}$$

Further, for any $f \in L^2_{\mathcal{F}}(0,T;\mathbb{R})$ and any two stopping times σ and τ with $0 \le \sigma \le \tau \le T$, $\mathbf{P}$-a.s. , we define

$$\int_\sigma^\tau f(s)dW(s) \stackrel{\Delta}{=} \widehat{I}(f)(\tau) - \widehat{I}(f)(\sigma). \tag{5.23}$$

Now let us collect some fundamental properties of the Itô integral.

Proposition 5.2. *The Itô integral has the following properties:*

(i) *For any $f, g \in L^2_{\mathcal{F}}(0,T;\mathbb{R})$ and stopping times σ and τ with $\sigma \le \tau$ ($\mathbf{P}$-a.s.),*

$$E\Big\{\int_{t\wedge\sigma}^{t\wedge\tau} f(r)dW(r)\big|\mathcal{F}_\sigma\Big\} = 0, \quad \mathbf{P}\text{-a.s.}, \tag{5.24}$$

$$\begin{aligned}&E\Big\{\Big[\int_{t\wedge\sigma}^{t\wedge\tau} f(r)dW(r)\Big]\Big[\int_{t\wedge\sigma}^{t\wedge\tau} g(r)dW(r)\Big]\Big|\mathcal{F}_\sigma\Big\}\\ &\quad = E\Big\{\int_{t\wedge\sigma}^{t\wedge\tau} f(r)g(r)dr\big|\mathcal{F}_\sigma\Big\}, \qquad \mathbf{P}\text{-a.s.}\end{aligned} \tag{5.25}$$

In particular, for any $0 \le s \le t \le T$,

$$E\Big\{\int_s^t f(r)dW(r)\big|\mathcal{F}_s\Big\} = 0, \quad \mathbf{P}\text{-a.s.}, \tag{5.26}$$

$$\begin{aligned} &E\Big\{\Big[\int_s^t f(r)dW(r)\Big]\Big[\int_s^t g(r)dW(r)\Big]\big|\mathcal{F}_s\Big\} \\ &\quad = E\Big\{\int_s^t f(r)g(r)dr\big|\mathcal{F}_s\Big\}, \qquad \mathbf{P}\text{-a.s.} \end{aligned} \tag{5.27}$$

(ii) *For any stopping time* σ *and* $f \in L^2_{\mathcal{F}}(0,T;\mathbb{R})$, *let* $\widetilde{f}(t,\omega) = f(t,\omega)I_{(\sigma(\omega)\ge t)}$. *Then*

$$\int_0^{t\wedge\sigma} f(s)dW(s) = \int_0^t \widetilde{f}(s)dW(s). \tag{5.28}$$

See Ikeda–Watanabe [1, pp. 49–51] for a proof.

We now extend the Itô integral to a bigger class of integrands than $L^2_{\mathcal{F}}(0,T;\mathbb{R})$. To this end, we introduce

$$\begin{aligned} L^{2,loc}_{\mathcal{F}}(0,T;\mathbb{R}) = \{X : [0,T]\times\Omega \to \mathbb{R} \mid X(\cdot) \text{ is } \{\mathcal{F}_t\}_{t\ge 0}\text{-adapted} \\ \text{and } \int_0^T |X(t)|^2 dt < \infty, \quad \mathbf{P}\text{-a.s.}\}, \end{aligned} \tag{5.29}$$

and

$$\begin{cases} \mathcal{M}^{2,loc}[0,T] = \{X : [0,T]\times\Omega \to \mathbb{R} \mid \exists \text{ nondecreasing} \\ \qquad \text{stopping times } \sigma_j \text{ with } \mathbf{P}(\lim\limits_{j\to\infty}\sigma_j \ge T) = 1, \\ \qquad \text{and } X(\cdot\wedge\sigma_j) \in \mathcal{M}^2[0,T], \quad \forall j = 1,2,\ldots\}, \\ \mathcal{M}^{2,loc}_c[0,T] = \{X \in \mathcal{M}^{2,loc}[0,T] \mid t\mapsto X(t) \text{ continuous}, \mathbf{P}\text{-a.s.}\}. \end{cases} \tag{5.30}$$

For any $f(\cdot) \in L^{2,loc}_{\mathcal{F}}(0,T;\mathbb{R})$, define

$$\sigma_j(\omega) \triangleq \inf\{t \in [0,T] \mid \int_0^t |f(s)|^2 ds \ge j\}, \quad j = 1,2,\ldots. \tag{5.31}$$

In the above, we define $\inf\phi \triangleq T$. Clearly, $\{\sigma_j\}_{j\ge 1}$ is a sequence of nondecreasing stopping times satisfying (4.15). Set $f_j(t) \triangleq f(t)I_{\{\sigma_j \ge t\}}$. Then

$$\int_0^T |f_j(s)|^2 ds = \int_0^{\sigma_j} |f(s)|^2 ds \le j, \tag{5.32}$$

which implies $f_j(\cdot) \in L^2_{\mathcal{F}}(0,T;\mathbb{R})$. By (5.28), we have

$$\begin{aligned} &\int_0^{t\wedge\sigma_i} f_j(s)dW(s) = \int_0^t f_j(s)I_{\{\sigma_i\ge s\}}dW(s) \\ &= \int_0^t f(s)I_{\{\sigma_j\ge s\}}I_{\{\sigma_i\ge s\}}dW(s) = \int_0^t f_i(s)dW(s), \quad \forall i \le j. \end{aligned} \tag{5.33}$$

Hence, the following is well-defined:

$$\int_0^t f(s)dW(s) \triangleq \int_0^t f_j(s)dW(s), \qquad \forall t \in [0,\sigma_j],\ j=1,2,\ldots \tag{5.34}$$

This is called the *Itô integral* of $f(\cdot) \in L^{2,loc}_{\mathcal{F}}(0,T;\mathbb{R})$. It is easy to see that $\int_0^t f(s)dW(s) \in \mathcal{M}^{2,loc}_c[0,T]$for any $f(\cdot) \in L^{2,loc}_{\mathcal{F}}(0,T;\mathbb{R})$.

We point out that for $f(\cdot) \in L^{2,loc}_{\mathcal{F}}(0,T;\mathbb{R})$, (5.24)–(5.27) do not hold in general, but (5.28) remains true.

Next, we briefly discuss the higher-dimensional case. Let $W(t) = (W^1(t),\ldots,W^m(t))$ be an m-dimensional Brownian motion and $f = (f_1,\ldots,f_m) \in L^2_{\mathcal{F}}(0,T;\mathbb{R}^m)$. Then, for $j=1,2,\ldots,m$, $\int_0^t f_j(s)dW^j(s)$ is well-defined. We define

$$\int_0^t \langle f(s), dW(s)\rangle \triangleq \sum_{j=1}^m \int_0^t f_j(s)dW^j(s), \qquad t \geq 0. \tag{5.35}$$

It is easy to see that the above defines an element in $\mathcal{M}^2_c[0,T]$. Similarly, for any $\sigma = (\sigma_{ij}) \in L^2_{\mathcal{F}}(0,T;\mathbb{R}^{n\times m})$, we may define an element in $(\mathcal{M}^2_c[0,T])^n$ as follows:

$$\int_0^t \sigma(s)dW(s) \triangleq \begin{pmatrix} \sum_{j=1}^m \int_0^t \sigma_{1j}(s)dW^j(s) \\ \vdots \\ \sum_{j=1}^m \int_0^t \sigma_{nj}(s)dW^j(s) \end{pmatrix}, \qquad t \geq 0. \tag{5.36}$$

Due to (2.24), $W^i(t)$ and $W^j(t)$ $(i \neq j)$ are independent. Thus, we have the following result.

Proposition 5.3. *Let $f \in L^2_{\mathcal{F}}(0,T;\mathbb{R}^m)$ and $\sigma, \widehat{\sigma} \in L^2_{\mathcal{F}}(0,T;\mathbb{R}^{n\times m})$. Then, for all $0 \leq s \leq t \leq T$,*

$$\begin{aligned} &E\Big\{\int_s^t f_i(r)dW^i(r)\int_s^t f_j(r)dW^j(r)\big|\mathcal{F}_s\Big\} \\ &\quad= \delta_{ij}E\Big\{\int_s^t f_i(r)f_j(r)dr\big|\mathcal{F}_s\Big\}, \qquad 1 \leq i,j \leq m, \end{aligned} \tag{5.37}$$

where

$$\delta_{ij} \triangleq \begin{cases} 1, & i=j, \\ 0, & i \neq j, \end{cases}$$

$$\begin{aligned} &E\Big\{\langle\int_s^t \sigma(r)dW(r), \int_s^t \widehat{\sigma}(r)dW(r)\rangle\,\big|\mathcal{F}_s\Big\} \\ &\qquad= E\Big\{\int_s^t \mathrm{tr}\,[\widehat{\sigma}(r)^\top\sigma(r)]dr\big|\mathcal{F}_s\Big\}. \end{aligned} \tag{5.38}$$

The proof is straightforward.

We may define $L_{\mathcal{F}}^{2,loc}(0,T;\mathbb{R}^m)$, $\mathcal{M}^{2,loc}[0,T]^m$ and $\mathcal{M}_c^{2,loc}[0,T]^m$ similarly to (5.29)–(5.30). The Itô integrals for integrands in $L_{\mathcal{F}}^{2,loc}(0,T;\mathbb{R}^m)$ can be defined accordingly. In addition, we may also define $L_{\mathcal{F}}^{p,loc}(0,T;\mathbb{R}^m)$, $p \geq 1$, naturally. The details are left to the reader.

To conclude this subsection, we present an important result called the *Burkholder–Davis–Gundy inequality.*

Theorem 5.4. *Let $(\Omega, \mathcal{F}, \{\mathcal{F}_t\}_{t\geq 0}, \mathbf{P})$ be given as before and let $W(t)$ be an m-dimensional standard Brownian motion. Let $\sigma \in L_{\mathcal{F}}^{2,loc}(0,T;\mathbb{R}^{n\times m})$. Then, for any $r > 0$, there exists a constant $K_r > 0$ such that for any stopping time τ,*

$$
\begin{aligned}
\frac{1}{K_r}E\Big\{\int_0^\tau |\sigma(s)|^2 ds\Big\}^r &\leq E\Big\{\sup_{0\leq t\leq \tau}\Big|\int_0^t \sigma(s)dW(s)\Big|^{2r}\Big\} \\
&\leq K_r E\Big\{\int_0^\tau |\sigma(s)|^2 ds\Big\}^r .
\end{aligned}
\tag{5.39}
$$

See Karatzas–Shreve [3, p. 166] for a proof of Theorem 5.4.

5.3. Itô's formula

In this subsection we present a stochastic version of the *chain rule*, or *change-of-variable formula*, called *Itô's formula/lemma/rule*, which plays one of the most important roles in stochastic calculus.

Let us first recall the chain rule for deterministic functions. Suppose

$$
X(t) = X(0) + \int_0^t b(s)ds, \qquad t \in [0,T],
\tag{5.40}
$$

for some $b(\cdot) \in L^1(0,T;\mathbb{R}^n)$. Then for any $F \in C^1([0,T]\times\mathbb{R}^n)$,

$$
\begin{aligned}
F(t,X(t)) =& F(0,X(0)) \\
&+ \int_0^t \{F_t(s,X(s)) + \langle F_x(s,X(s)), b(s)\rangle\}ds, \ t \in [0,T].
\end{aligned}
\tag{5.41}
$$

For the stochastic case, we have a similar formula, albeit significantly different from (5.41). Consider the following:

$$
X(t) = X(0) + \int_0^t b(s)ds + \int_0^t \sigma(s)dW(s), \qquad t \in [0,T],
\tag{5.42}
$$

with $b(\cdot) \in L_{\mathcal{F}}^{1,loc}(0,T;\mathbb{R}^n)$ and $\sigma(\cdot) \in L_{\mathcal{F}}^{2,loc}(0,T;\mathbb{R}^{n\times m})$. Any process $X(\cdot)$ of form (5.42) is called an *Itô process*.

Theorem 5.5. (Itô's formula) *Let $(\Omega, \mathcal{F}, \{\mathcal{F}_t\}_{t\geq 0}, \mathbf{P})$ be a filtered probability space satisfying the usual condition, $W(t)$ an m-dimensional $\{\mathcal{F}_t\}_{t\geq 0}$-Brownian motion, $b(\cdot) \in L_{\mathcal{F}}^{1,loc}(0,T;\mathbb{R}^n)$, $\sigma \in L_{\mathcal{F}}^{2,loc}(0,T;\mathbb{R}^{n\times m})$, and let*

$X(\cdot)$ be given by (5.42). Let $F(t,x) : [0,T] \times \mathbb{R}^n \to \mathbb{R}$ be C^1 in t and C^2 in x with F_t, F_x, and F_{xx} being continuous such that

$$
\left\{
\begin{aligned}
& F_t(\cdot\,, X(\cdot)), \langle F_x(\cdot\,, X(\cdot)), b(\cdot) \rangle \in L^{1,loc}_{\mathcal{F}}(0,T;\mathbb{R}), \\
& \operatorname{tr}\{\sigma(\cdot)^\top F_{xx}(\cdot\,, X(\cdot))\sigma(\cdot)\} \in L^{1,loc}_{\mathcal{F}}(0,T;\mathbb{R}), \\
& \sigma(\cdot)^\top F_x(\cdot\,, X(\cdot)) \in L^{2,loc}_{\mathcal{F}}(0,T;\mathbb{R}^m).
\end{aligned}
\right.
\tag{5.43}
$$

Then

$$
\begin{aligned}
F(t, X(t)) = & F(0, X(0)) + \int_0^t \{F_s(s, X(s)) + \langle F_x(s, X(s)), b(s) \rangle \\
& + \frac{1}{2}\operatorname{tr}\left[\sigma(s)^\top F_{xx}(s, X(s))\sigma(s)\right]\}ds \\
& + \int_0^t \langle F_x(s, X(s)), \sigma(s)dW(s) \rangle, \quad \forall t \in [0,T], \mathbf{P}\text{-a.s.}
\end{aligned}
\tag{5.44}
$$

Note that for fixed $\omega \in \Omega$ and $t \geq 0$, $X(s,\omega)$ is bounded on $[0,t]$. Thus, the first integral exists. The second integral is defined as in the previous subsection.

Let us make an observation. Let $(\Omega, \mathcal{F}, \{\mathcal{F}_t\}_{t\geq 0}, \mathbf{P})$ be a filtered probability space satisfying the usual condition and let $W(\cdot)$ be a standard Brownian motion. Take $\sigma(\cdot) \in L^2_{\mathcal{F}}(0,T;\mathbb{R})$ and consider

$$
X(t) = \int_0^t \sigma(s)dW(s), \quad t \in [0,T]. \tag{5.45}
$$

Then $X(\cdot) \in \mathcal{M}^2_c[0,T]$. By Itô's formula above, we have

$$
X(t)^2 = \int_0^t \sigma(s)^2 ds + 2\int_0^t X(s)\sigma(s)dW(s), \quad t \in [0,T]. \tag{5.46}
$$

Note that $X(\cdot)\sigma(\cdot)$ is in $L^{2,loc}_{\mathcal{F}}(0,T;\mathbb{R})$, but not necessarily in $L^2_{\mathcal{F}}(0,T;\mathbb{R})$. Hence, the introduction of Itô's integral for integrands in $L^{2,loc}_{\mathcal{F}}(0,T;\mathbb{R})$ is not just some routine generalization. It is really necessary even for as simple a calculus as the above.

For a proof of the Itô formula, see Ikeda–Watanabe [1, pp. 66–73] (for a more general case).

As an application of the Itô formula, we present the following result, which will be frequently used in this book.

Corollary 5.6. *Let Z and $\widehat{Z}$ be $\mathbb{R}^n$-valued continuous processes satisfying*

$$
\left\{
\begin{aligned}
& dZ(t) = b(t)dt + \sigma(t)dW(t), \\
& d\widehat{Z}(t) = \widehat{b}(t)dt + \widehat{\sigma}(t)dW(t),
\end{aligned}
\right.
\tag{5.47}
$$

where $b, \widehat{b}, \sigma, \widehat{\sigma}$ are $\{\mathcal{F}_t\}_{t\geq 0}$-adapted measurable processes taking values in $\mathbb{R}^n$, and $W(t)$ is a one-dimensional standard Brownian motion. Then

$$
\begin{aligned}
\langle Z(t), \widehat{Z}(t) \rangle &= \langle Z(0), \widehat{Z}(0) \rangle \\
&\quad + \int_0^t \{\langle Z(s), \widehat{b}(s)\rangle + \langle b(s), \widehat{Z}(s)\rangle + \langle \sigma(s), \widehat{\sigma}(s)\rangle\} dt \\
&\quad + \int_0^t \{ \langle \sigma(s), \widehat{Z}(s) \rangle + \langle Z(s), \widehat{\sigma}(s) \rangle \} dW(s).
\end{aligned}
\tag{5.48}
$$

This lemma can be easily proved by Itô's formula with $F(x,y) = \langle x, y \rangle$ for $(x,y) \in \mathbb{R}^{2n}$.

5.4. Martingale representation theorems

In view of (5.26), the Itô integral

$$
M(t) \triangleq \int_0^t f(s) dW(s) \tag{5.49}
$$

is a martingale (or local martingale). It is natural to ask whether a martingale (or local martingale) can be represented as an Itô integral of the form (5.49). Results on such a problem are called *martingale representation theorems.* They play very important roles in stochastic calculus itself as well as in stochastic control theory.

There are two components on the right-hand side of (5.49), namely, the integrand f and the Brownian motion $W(t)$. The first martingale representation theorem is concerned with how to represent a martingale (local martingale) with a fixed Brownian motion.

Theorem 5.7. *Let $(\Omega, \mathcal{F}, \{\mathcal{F}_t\}_{t\geq 0}, \mathbf{P})$ be a filtered probability space satisfying the usual condition. Assume that $\{\mathcal{F}_t\}_{t\geq 0}$ is the natural filtration generated by an m-dimensional standard Brownian motion $W(t)$. Let $X \in \mathcal{M}^2[0,T]$ (resp. $\mathcal{M}^{2,loc}[0,T]$). Then there exists a unique $\varphi(\cdot) \in L^2_{\mathcal{F}}(0,T;\mathbb{R}^m)$ (resp. $L^{2,loc}_{\mathcal{F}}(0,T;\mathbb{R}^m)$) such that*

$$
X(t) = \int_0^t \langle \varphi(s), dW(s) \rangle, \qquad \forall t \in [0,T], \ \mathbf{P}\text{-a.s.} \tag{5.50}
$$

A proof can be found in Ikeda–Watanabe [1, pp. 80–83].

Recall that Theorem 2.10 shows that if $(\Omega, \mathcal{F}, \{\mathcal{F}_t\}_{t\geq 0}, \mathbf{P})$ and $W(\cdot)$ are as in Theorem 5.7, and $X(t)$ is a process adapted to $\{\mathcal{F}_t\}_{t\geq 0}$, then $X(\cdot)$ can be represented as a functional η of $W(\cdot)$. Theorem 5.7 says that if in addition, $X(t)$ is a square integrable $\{\mathcal{F}_t\}_{t\geq 0}$-martingale, then we have a more explicit functional of the Brownian motion $W(\cdot)$, namely,

$$
\eta(t, W(t\wedge\cdot)) = X(0) + \int_0^t \langle \varphi(s), dW(s) \rangle, \tag{5.51}
$$

for some $\varphi(\cdot) \in L^2_{\mathcal{F}}(0,T;\mathbb{R}^m)$.

We emphasize that Theorem 5.7 works only for the case where $\{\mathcal{F}_t\}_{t\geq 0}$ is generated by the (given) Brownian motion. On the other hand, it implies that if $\{\mathcal{F}_t\}_{t\geq 0}$ *is* generated by a standard Brownian motion, then any square-integrable $\{\mathcal{F}_t\}_{t\geq 0}$-martingale must be continuous, i.e., $\mathcal{M}^2[0,T] = \mathcal{M}^2_c[0,T]$.

Let us now look at an interesting consequence of Theorem 5.7. Let $(\Omega, \mathcal{F}, \{\mathcal{F}_t\}_{t\geq 0}, \mathbf{P})$ and $W(\cdot)$ be the same as in the above. Let $\xi \in L^2_{\mathcal{F}_T}(\Omega;\mathbb{R})$. Then $E(\xi|\mathcal{F}_t)$ is an $\{\mathcal{F}_t\}_{t\geq 0}$-martingale. Thus, by Theorem 5.7, there exists a $z(\cdot) \in L^2_{\mathcal{F}}(0,T;\mathbb{R}^m)$ such that

$$E(\xi|\mathcal{F}_t) = E\xi + \int_0^t \langle z(s), dW(s) \rangle, \qquad t \in [0,T]. \tag{5.52}$$

In particular,

$$\xi = E\xi + \int_0^T \langle z(s), dW(s) \rangle . \tag{5.53}$$

This shows that

$$L^2_{\mathcal{F}_T}(\Omega;\mathbb{R}) = \mathbb{R} + \Big\{ \int_0^T \langle z(s), dW(s) \rangle \mid z(\cdot) \in L^2_{\mathcal{F}}(0,T;\mathbb{R}^m) \Big\}. \tag{5.54}$$

Next we turn to the question of how to represent a martingale (local martingale) with a fixed integrand. To do this, we first need to define several notions.

Let $(\Omega, \mathcal{F}, \{\mathcal{F}_t\}_{t\geq 0}, \mathbf{P})$ be a filtered probability space satisfying the usual condition. Further, let $W(t)$ be an m-dimensional standard $\{\mathcal{F}_t\}_{t\geq 0}$-Brownian motion and $f \in L^2_{\mathcal{F}}(0,T;\mathbb{R}^{n\times m})$ (resp. $f \in L^{2,loc}_{\mathcal{F}}(0,T;\mathbb{R}^{n\times m})$). Define

$$M(t) \stackrel{\Delta}{=} \int_0^t f(s) dW(s). \tag{5.55}$$

By Itô's formula, $M(t)M(t)^\top - \int_0^t f(s)f(s)^\top ds$ is an $\{\mathcal{F}_t\}_{t\geq 0}$-martingale (resp. local martingale). Inspired by this observation, we have the following definition.

Definition 5.8. Let $M \in \mathcal{M}^2[0,T]^n$ (resp. $\mathcal{M}^{2,loc}[0,T]^n$). An increasing process $A(t)$ is called the *quadratic variation (process)* of $M(t)$ if $M(t)M(t)^\top - A(t)$ is an $\{\mathcal{F}_t\}_{t\geq 0}$-martingale (resp. local martingale). We denote it by $\langle M \rangle(t) \stackrel{\Delta}{=} A(t)$.

The above definition makes sense, as the above $A(t)$ is uniquely determined by Doob–Meyer's decomposition theorem. See Ikeda–Watanabe [1, p. 53] for details.

According to Definition 5.8, for M defined by (5.55) we have

$$\langle M \rangle(t) = \int_0^t f(s)f(s)^\top ds. \tag{5.56}$$

Definition 5.9. Let $(\Omega, \mathcal{F}, \{\mathcal{F}_t\}_{t\geq 0}, \mathbf{P})$ and $(\widehat{\Omega}, \widehat{\mathcal{F}}, \{\widehat{\mathcal{F}}_t\}_{t\geq 0}, \widehat{\mathbf{P}})$ be two filtered probability spaces satisfying the usual condition. The latter is called an *extension* of the former if there exists a random variable $\theta : (\widehat{\Omega}, \widehat{\mathcal{F}}) \to (\Omega, \mathcal{F})$ such that:

(i) $\theta^{-1}(\mathcal{F}_t) \subseteq \widehat{\mathcal{F}}_t, \quad \forall t.$

(ii) $\mathbf{P} = \widehat{\mathbf{P}} \circ \theta^{-1} \equiv \widehat{\mathbf{P}}_\theta$ (see (1.9)).

(iii) for any $X \in L^\infty(\Omega, \mathcal{F}, \mathbf{P})$,

$$\widehat{E}(\widehat{X}(\widehat{\omega})|\widehat{\mathcal{F}}_t)(\widehat{\omega}) = E(X|\mathcal{F}_t)(\theta\widehat{\omega}), \quad \widehat{\mathbf{P}}\text{-a.s. } \widehat{\omega} \in \widehat{\Omega},$$

where we define $\widehat{X}(\widehat{\omega}) \stackrel{\Delta}{=} X(\theta\widehat{\omega}), \forall \widehat{\omega} \in \widehat{\Omega}$.

Let $(\widehat{\Omega}, \widehat{\mathcal{F}}, \{\widehat{\mathcal{F}}_t\}_{t\geq 0}, \widehat{\mathbf{P}})$ be an extension of $(\Omega, \mathcal{F}, \{\mathcal{F}_t\}_{t\geq 0}, \mathbf{P})$. If M is a square-integrable martingale with respect to $(\Omega, \mathcal{F}, \{\mathcal{F}_t\}_{t\geq 0}, \mathbf{P})$, then $\widehat{M}$ where $\widehat{M}(t, \widehat{\omega}) \stackrel{\Delta}{=} M(t, \theta\widehat{\omega})$ is a square-integrable martingale with respect to $(\widehat{\Omega}, \widehat{\mathcal{F}}, \{\widehat{\mathcal{F}}_t\}_{t\geq 0}, \widehat{\mathbf{P}})$. Therefore, the space $\mathcal{M}^2[0,T]^n$ ($\mathcal{M}^{2,loc}[0,T]^n$, etc.) is naturally embedded into the corresponding spaces $\widehat{\mathcal{M}}^2[0,T]^n$ ($\widehat{\mathcal{M}}^{2,loc}[0,T]^n$, etc.) in the extension space. For this reason, we shall not distinguish M and $\widehat{M}$ in what follows.

Theorem 5.10. *Let $(\Omega, \mathcal{F}, \{\mathcal{F}_t\}_{t\geq 0}, \mathbf{P})$ be a filtered probability space satisfying the usual condition. Let $M \in \mathcal{M}^2[0,T]^n$ (resp. $\mathcal{M}^{2,loc}[0,T]^n$), and $\sigma \in L^2_{\mathcal{F}}(0,T;\mathbb{R}^{n\times m})$ (resp. $L^{2,loc}_{\mathcal{F}}(0,T;\mathbb{R}^{n\times m})$) with $\sigma\sigma^\top \in L^{1,loc}_{\mathcal{F}}(0,T;\mathbb{R}^{n\times n})$ (resp. $L^{1,loc}_{\mathcal{F}}(0,T;\mathbb{R}^{n\times n})$). If*

$$\langle M \rangle(t) = \int_0^t \sigma(s)\sigma(s)^\top ds, \tag{5.57}$$

then there exists an extension $(\widehat{\Omega}, \widehat{\mathcal{F}}, \{\widehat{\mathcal{F}}_t\}_{t\geq 0}, \widehat{\mathbf{P}})$ of $(\Omega, \mathcal{F}, \{\mathcal{F}_t\}_{t\geq 0}, \mathbf{P})$ on which there lives an m-dimensional $\{\widehat{\mathcal{F}}_t\}_{t\geq 0}$-Brownian motion $W(t)$ such that

$$M(t) = \int_0^t \sigma(s)dW(s). \tag{5.58}$$

See Ikeda–Watanabe [1, pp. 90–92] for a proof.

Note that in Theorem 5.10 the filtration is not required to be generated by the Brownian motion. On the other hand, if $n = m$ and the matrix $\sigma(s)$ is invertible $\mathbf{P}$-a.s., for any s, then the representation (5.58) holds on the *original* (rather than an extension of) $(\Omega, \mathcal{F}, \{\mathcal{F}_t\}_{t\geq 0}, \mathbf{P})$.

6. Stochastic Differential Equations

In this section we are going to study stochastic differential equations (SDEs, for short), which can be regarded as a generalization of ordinary differential equations (ODEs, for short). Since the Itô integral will be involved, the situation is much more complicated than that of ODEs, and the corresponding theory is very rich.

Let us first recall the space $\mathbf{W}^n \equiv C([0,\infty);\mathbb{R}^n)$ and its metric $\widehat{\rho}$ defined in §2.2. Let U be a Polish space and $\mathcal{A}^n(U)$ the set of all $\{\mathcal{B}_{t+}(\mathbf{W}^n)\}_{t\geq 0}$-progressively measurable processes $\eta : [0,\infty) \times \mathbf{W}^n \to U$.

Lemma 6.1. *Let $b \in \mathcal{A}^n(\mathbb{R}^n)$. Let $(\Omega,\mathcal{F},\{\mathcal{F}_t\}_{t\geq 0},\mathbf{P})$ be given satisfying the usual condition and let X be a continuous $\mathbb{R}^n$-valued $\{\mathcal{F}_t\}_{t\geq 0}$-adapted process. Then the process $t \mapsto b(t, X(\cdot,\omega))$ is $\{\mathcal{F}_t\}_{t\geq 0}$-adapted.*

Proof. For any $t \in [0,\infty)$, let $\Phi : [0,t] \times \Omega \to [0,t] \times \mathbf{W}^n$ be the map $(s,\omega) \mapsto (s, X(\cdot,\omega))$. Then

$$b(s, X(\cdot,\omega)) = (b \circ \Phi)(s,\omega), \qquad (s,\omega) \in [0,t] \times \Omega. \tag{6.1}$$

By assumption, $b^{-1}(\mathcal{B}(\mathbb{R}^n)) \stackrel{\Delta}{=} \{b^{-1}(B) | B \in \mathcal{B}(\mathbb{R}^n)\} \subseteq \mathcal{B}[0,t] \times \mathcal{B}_{t+}(\mathbf{W}^n)$. Moreover, $\Phi^{-1}(\mathcal{B}[0,t] \times \mathcal{B}_{t+}(\mathbf{W}^n)) \subseteq \mathcal{B}[0,t] \times \mathcal{F}_{t+} = \mathcal{B}[0,t] \times \mathcal{F}_t$. This implies that $b(s, X(\cdot,\omega))$ is $\mathcal{B}[0,t] \times \mathcal{F}_t/\mathcal{B}(\mathbb{R}^n)$-measurable, proving the desired result. □

Next, let $b \in \mathcal{A}^n(\mathbb{R}^n)$ and $\sigma \in \mathcal{A}^n(\mathbb{R}^{n\times m})$. Consider the following equation:

$$\begin{cases} dX(t) = b(t,X)dt + \sigma(t,X)dW(t), \\ X(0) = \xi. \end{cases} \tag{6.2}$$

In the above equation, X is the unknown. Such an equation is called a *stochastic differential equation*. There are different notions of solutions to (6.2) depending on different roles that the underlying filtered probability space $(\Omega,\mathcal{F},\{\mathcal{F}_t\}_{t\geq 0},\mathbf{P})$ and the Brownian motion $W(\cdot)$ are playing. Let us introduce them in the following subsections.

6.1. Strong solutions

Definition 6.2. Let $(\Omega,\mathcal{F},\{\mathcal{F}_t\}_{t\geq 0},\mathbf{P})$ be given, $W(t)$ be a given m-dimensional standard $\{\mathcal{F}_t\}_{t\geq 0}$-Brownian motion, and ξ $\mathcal{F}_0$-measurable. An $\{\mathcal{F}_t\}_{t\geq 0}$-adapted continuous process $X(t), t \geq 0$, is called a *strong solution* of (6.2) if

$$X(0) = \xi, \qquad \mathbf{P}\text{-a.s.}, \tag{6.3}$$

$$\int_0^t \{|b(s,X)| + |\sigma(s,X)|^2\} ds < \infty, \qquad \forall t \geq 0,\ \mathbf{P}\text{-a.s.}, \tag{6.4}$$

$$X(t) = X(0) + \int_0^t b(s,X)ds + \int_0^t \sigma(s,X)dW(s), \qquad t \geq 0,\ \mathbf{P}\text{-a.s.} \tag{6.5}$$

If for any two strong solutions X and Y of (6.2) defined on any given $(\Omega,\mathcal{F},\{\mathcal{F}_t\}_{t\geq 0},\mathbf{P})$ along with any given standard $\{\mathcal{F}_t\}_{t\geq 0}$-Brownian motion, we have

$$\mathbf{P}(X(t) = Y(t),\ 0 \leq t < \infty) = 1, \tag{6.6}$$

then we say that the strong solution is *unique* or that *strong uniqueness* holds.

In the above (6.5), the first integral on the right is a usual Lebesgue integral (regarding $\omega \in \Omega$ as a parameter), and the second is the Itô integral defined in the previous section. If (6.4) holds, then these two integrals are well-defined. We refer to $\int_0^t b(s, X)ds$ as the *drift* term and $\int_0^t \sigma(s, X)dW(s)$ as the *diffusion* term.

One should pay particular attention to the notion of strong uniqueness. It requires (6.6) to hold for any two solutions X, Y associated with *every* given filtered probability space $(\Omega, \mathcal{F}, \{\mathcal{F}_t\}_{t\geq 0}, \mathbf{P})$ and m-dimensional standard $\{\mathcal{F}_t\}_{t\geq 0}$-Brownian motion, rather than particular ones. So it may be more appropriate to talk about strong uniqueness for the pair (b, σ), which are the coefficients of (6.2). See Karatzas–Shreve [3, p. 286] for a discussion on this point.

Next we give conditions that ensure the existence and uniqueness of strong solutions.

Let $(\Omega, \mathcal{F}, \{\mathcal{F}_t\}_{t\geq 0}, \mathbf{P})$ be a filtered probability space satisfying the usual condition, $W(t)$ an m-dimensional standard $\{\mathcal{F}_t\}_{t\geq 0}$-Brownian motion, and ξ an $\mathcal{F}_0$-measurable random variable. We make the following assumption for the coefficients of (6.2).

(H) $b \in \mathcal{A}^n(\mathbb{R}^n)$, $\sigma \in \mathcal{A}^n(\mathbb{R}^{n\times m})$, and there exists an $L > 0$ such that for all $t \in [0, \infty)$, $x(\cdot), y(\cdot) \in \mathbf{W}^n$,

$$
\left\{
\begin{aligned}
&|b(t, x(\cdot)) - b(t, y(\cdot))| \leq L|x(\cdot) - y(\cdot)|_{\mathbf{W}^n},\\
&|\sigma(t, x(\cdot)) - \sigma(t, y(\cdot))| \leq L|x(\cdot) - y(\cdot)|_{\mathbf{W}^n},\\
&|b(\cdot\,, 0)| + |\sigma(\cdot\,, 0)| \in L^2(0, T; \mathbb{R}), \quad \forall T > 0.
\end{aligned}
\right.
\tag{6.7}
$$

We say that $b(t, x(\cdot))$ and $\sigma(t, x(\cdot))$ are *Lipschitz continuous* in $x(\cdot)$ if the first two inequalities in (6.7) hold.

Theorem 6.3. *Let (H) hold. Then, for any $\xi \in L^\ell_{\mathcal{F}_0}(\Omega; \mathbb{R}^n)$ $(\ell \geq 1)$, equation (6.2) admits a unique strong solution X such that for any $T > 0$,*

$$
\left\{
\begin{aligned}
&E \sup_{0\leq s\leq T} |X(s)|^\ell \leq K_T(1 + E|\xi|^\ell),\\
&E|X(t) - X(s)|^\ell \leq K_T(1 + E|\xi|^\ell)|t - s|^{\ell/2}, \qquad \forall s, t \in [0, T].
\end{aligned}
\right.
\tag{6.8}
$$

Moreover, if $\widehat{\xi} \in L^\ell_{\mathcal{F}_0}(\Omega; \mathbb{R}^n)$ is another random variable and $\widehat{X}(t)$ is the corresponding strong solution of (6.2), then for any $T > 0$, there exists a $K_T > 0$ such that

$$
E \sup_{0\leq s\leq T} |X(s) - \widehat{X}(s)|^\ell \leq K_T E|\xi - \widehat{\xi}|^\ell.
\tag{6.9}
$$

The above theorem can be proved by a routine *successive approximation* argument and the Borel–Cantelli lemma; see Ikeda–Watanabe [1, pp. 166-168] or Karatzas–Shreve [3, pp. 289-290]. However, below we will give another proof, which is based on the *contraction mapping theorem* (Zeidler [1, p. 17]).

Proof of Theorem 6.3. Let $T > 0$. We are going to prove that (6.2) admits a strong solution on $[0, T]$. For any $0 \le \tau \le T$, set (see Section 2)

$$(6.10)\qquad \begin{aligned} \mathcal{X}_\ell[0,\tau] &\stackrel{\Delta}{=} L^\ell_{\mathcal{F}}(\Omega; C([0,\tau]; \mathbb{R}^n)) \\ &\equiv \{x : [0,\tau]\times\Omega \to \mathbb{R}^n \mid x(\cdot) \text{ is } \{\mathcal{F}_t\}_{t\ge 0}\text{-adapted}, \\ &\qquad \text{continuous, and } E \sup_{0\le t\le\tau} |x(t)|^\ell < \infty\}. \end{aligned}$$

Clearly, $\mathcal{X}_\ell[0,\tau]$ is a Banach space with the norm

$$|x(\cdot)|_{\mathcal{X}_\ell[0,\tau]} \stackrel{\Delta}{=} \Big\{ E\Big[\sup_{0\le t\le\tau} |x(t)|^\ell\Big]\Big\}^{\frac{1}{\ell}}.$$

For any $x(\cdot), y(\cdot) \in \mathcal{X}_\ell[0,\tau]$, define

$$(6.11)\qquad \begin{cases} X(t) = \xi + \displaystyle\int_0^t b(s,x)ds + \int_0^t \sigma(s,x)dW(s), \\ Y(t) = \xi + \displaystyle\int_0^t b(s,y)ds + \int_0^t \sigma(s,y)dW(s), \end{cases} \qquad t \in [0,\tau].$$

By (6.7) and the Burkholder–Davis–Gundy inequality (Theorem 5.4), we have

$$(6.12)\qquad \begin{cases} |X(\cdot)|^\ell_{\mathcal{X}_\ell[0,\tau]} \le K, \\ |X(\cdot) - Y(\cdot)|^\ell_{\mathcal{X}_\ell[0,\tau]} \le K\Big\{\tau^{\frac{\ell}{2}} |x(\cdot) - y(\cdot)|^\ell_{\mathcal{X}_\ell[0,\tau]}\Big\}. \end{cases}$$

Here, the constant K is independent of $\tau, \xi, x(\cdot), y(\cdot)$.

We let $\tau \in [0, T]$ be given such that $K\tau^{\frac{\ell}{2}} < 1$, where K is in (6.12). From (6.12), it follows that for any $\xi \in L^\ell_{\mathcal{F}_0}(\Omega; \mathbb{R}^n)$, the map $x(\cdot) \mapsto X(\cdot)$ defined via (6.11) is from $\mathcal{X}_\ell[0,\tau]$ to itself and is contractive. Thus, there exists a unique fixed point, which gives a strong solution $X(\cdot)$ to (6.2) on $[0,\tau]$. Next, repeating the procedure on $[\tau, 2\tau]$, etc., we are able to get the unique strong solution on $[0, T]$. Since $T > 0$ is arbitrary, we obtain the strong solution on $[0,\infty)$. The proof of the remaining conclusions follow easily from the Burkholder–Davis–Gundy inequality. □

Next, we introduce a special case of SDEs. Let $b : [0,\infty)\times\mathbb{R}^n \to \mathbb{R}^n$ and $\sigma : [0,\infty)\times\mathbb{R}^n \to \mathbb{R}^{n\times m}$. Then the maps $(t, w) \mapsto b(t, w(t))$ and $(t, w) \mapsto \sigma(t, w(t))$ are progressively measurable when regarded as maps from $[0,\infty)\times\mathbf{W}^n$ to $\mathbb{R}^n$ and $\mathbb{R}^{n\times m}$, respectively. In this case, (6.2) becomes

$$(6.13)\qquad \begin{cases} dX(t) = b(t, X(t))dt + \sigma(t, X(t))dW(t), \\ X(0) = \xi. \end{cases}$$

Such an SDE is said to be of *Markovian type.* If in addition, b and σ are time-invariant, then (6.13) is said to be of *time homogeneous Markovian type.* Note that in the case $\sigma \equiv 0$, (6.2) is reduced to a functional differential equation and (6.13) is reduced to an ordinary differential equation.

Now let us present an existence and uniqueness result for (6.13). First we introduce the following assumption:

(H)′ The maps $b : [0,\infty) \times \mathbb{R}^n \to \mathbb{R}^n$ and $\sigma : [0,\infty) \times \mathbb{R}^n \to \mathbb{R}^{n\times m}$ are measurable in $t \in [0,\infty)$ and there exists a constant $L > 0$, such that

$$(6.14) \qquad \begin{cases} |b(t,x) - b(t,\widehat{x})| + |\sigma(t,x) - \sigma(t,\widehat{x})| \le L|x - \widehat{x}|, \\ \qquad\qquad\qquad\qquad t \in [0,\infty),\ x,\widehat{x} \in \mathbb{R}^n, \\ |b(\cdot\,,0)| + |\sigma(\cdot\,,0)| \in L^2(0,T;\mathbb{R}), \quad \forall T > 0. \end{cases}$$

Clearly, by regarding b and σ as maps from $[0,\infty) \times \mathbf{W}^n$ to $\mathbb{R}^n$ and $\mathbb{R}^{n\times m}$, respectively, we see that (H)′ implies (H). Hence, the following result is clear.

Corollary 6.4. *Let (H)′ hold. Then (6.13) admits a unique strong solution.*

6.2. Weak solutions

Definition 6.5. A 6-tuple $(\Omega, \mathcal{F}, \{\mathcal{F}_t\}_{t\ge 0}, \mathbf{P}, W, X)$ is called a *weak solution* of (6.2) if

(i) $(\Omega, \mathcal{F}, \{\mathcal{F}_t\}_{t\ge 0}, \mathbf{P})$ is a filtered probability space satisfying the usual condition;

(ii) W is an m-dimensional standard $\{\mathcal{F}_t\}_{t\ge 0}$-Brownian motion and X is $\{\mathcal{F}_t\}_{t\ge 0}$-adapted and continuous;

(iii) $X(0)$ and ξ have the same distribution;

(iv) (6.4)–(6.5) hold.

The essential difference between the strong and weak solutions is the following: For the former, the filtered probability space $(\Omega, \mathcal{F}, \{\mathcal{F}_t\}_{t\ge 0}, \mathbf{P})$ and the $\{\mathcal{F}_t\}_{t\ge 0}$-Brownian motion W on it are fixed a priori, while for the latter $(\Omega, \mathcal{F}, \{\mathcal{F}_t\}_{t\ge 0}, \mathbf{P})$ and W are *parts* of the solution.

Definition 6.6. If for any two weak solutions $(\Omega, \mathcal{F}, \{\mathcal{F}_t\}_{t\ge 0}, \mathbf{P}, W, X)$ and $(\widetilde{\Omega}, \widetilde{\mathcal{F}}, \{\widetilde{\mathcal{F}}_t\}_{t\ge 0}, \widetilde{\mathbf{P}}, \widetilde{W}, \widetilde{X})$ of (6.2) with

$$(6.15) \qquad \mathbf{P}(X(0) \in B) = \widetilde{\mathbf{P}}(\widetilde{X}(0) \in B), \qquad \forall B \in \mathcal{B}(\mathbb{R}^n),$$

we have

$$(6.16) \qquad \mathbf{P}(X \in A) = \widetilde{\mathbf{P}}(\widetilde{X} \in A), \qquad \forall A \in \mathcal{B}(\mathbf{W}^n),$$

then we say that the weak solution of (6.2) is *unique (in the sense of probability law)*, or that *weak uniqueness* holds.

Definition 6.7. If

$$(6.17) \qquad \mathbf{P}(X(t) = \widetilde{X}(t),\ 0 \le t < \infty) = 1$$

for any two weak solutions $(\Omega, \mathcal{F}, \{\mathcal{F}_t\}_{t\geq 0}, \mathbf{P}, W, X)$ and $(\Omega, \mathcal{F}, \{\widetilde{\mathcal{F}}_t\}_{t\geq 0}, \mathbf{P}, W, \widetilde{X})$ of (6.2) with

$$\mathbf{P}(X(0) = \widetilde{X}(0)) = 1, \tag{6.18}$$

then we say that the weak solutions have *pathwise uniqueness.*

Note that in the definition of pathwise uniqueness, $\Omega, \mathcal{F}, \mathbf{P}$, and W are the same for the two solutions under comparison.

Existence of weak solutions does not imply that of strong solutions, and weak uniqueness does not imply pathwise uniqueness nor strong uniqueness. See Karatzas–Shreve [3, pp. 301–302] for an example. Relations between the strong and weak solutions are presented in the following two theorems.

Theorem 6.8. *Let $b \in \mathcal{A}^n(\mathbb{R}^n)$ and $\sigma \in \mathcal{A}^n(\mathbb{R}^{n\times m})$. Then (6.2) admits a unique strong solution if and only if for any probability measure μ on $(\mathbb{R}^n, \mathcal{B}(\mathbb{R}^n))$, (6.2) admits a weak solution with the initial distribution μ and pathwise uniqueness holds for (6.2).*

By and large, Theorem 6.8 tells that strong existence and uniqueness is equivalent to weak existence plus pathwise uniqueness.

Theorem 6.9. *Pathwise uniqueness implies weak uniqueness.*

See Karatzas–Shreve [3, pp. 308–311] for proofs of Theorems 6.8–6.9.

The following is a general existence result of weak solutions for equations with only continuous (not necessarily Lipschitz continuous) coefficients.

Theorem 6.10. *Let $b \in \mathcal{A}^n(\mathbb{R}^n)$ and $\sigma \in \mathcal{A}^n(\mathbb{R}^{n\times m})$ be bounded and continuous. Then there exists a weak solution of (6.2).*

The proof can be found in Ikeda–Watanabe [1, pp. 155–158].

By the well-known *Girsanov transformation*, one can obtain the existence and uniqueness of weak solutions for equations with merely *measurable* drift coefficients. Let us now introduce this approach.

Given $\sigma \in \mathcal{A}^n(\mathbb{R}^{n\times n})$ with the matrix $\sigma(t,x)$ invertible for any (t,x), suppose $(\Omega, \mathcal{F}, \{\mathcal{F}_t\}_{t\geq 0}, \mathbf{P}, W, X)$ is a weak solution of

$$\begin{cases} dX(t) = \sigma(t,X)dW(t), \\ X(0) = \xi. \end{cases} \tag{6.19}$$

Let $b \in \mathcal{A}^n(\mathbb{R}^n)$ such that

$$Ee^{\frac{1}{2}\int_0^t |(\sigma^{-1}b)(s,X)|^2 ds} < \infty, \quad \forall t \in [0,\infty). \tag{6.20}$$

Then the Girsanov theorem says that

$$M(t) \triangleq e^{\int_0^t (\sigma^{-1}b)(s,X)dW(s) - \frac{1}{2}\int_0^t |(\sigma^{-1}b)(s,X)|^2 ds} \tag{6.21}$$

is an $\{\mathcal{F}_t\}_{t\geq 0}$-martingale, and for each fixed $T \in [0,\infty)$,

$$\widehat{W}(t) \triangleq W(t) - \int_0^t (\sigma^{-1}b)(s,X)ds \tag{6.22}$$

is a standard $\{\mathcal{F}_t\}_{t\geq 0}$-Brownian motion on $(\Omega, \mathcal{F}_T, \widehat{\mathbf{P}}_T)$, where the new probability measure $\widehat{\mathbf{P}}_T$ is defined on $\mathcal{F}_T$ as

$$d\widehat{\mathbf{P}} \triangleq M(T)d\mathbf{P}. \tag{6.23}$$

The family of probability measures $\{\widehat{\mathbf{P}}_T\}$ is consistent in the sense that $\widehat{\mathbf{P}}_T(A) = \widehat{\mathbf{P}}_{T'}(A)$ for any $A \in \mathcal{F}_T$ whenever $0 \leq T \leq T'$. See Karatzas–Shreve [3, pp. 190–196] for all the proofs of the above claims. By (6.19) and (6.22), it is easily seen that $(\Omega, \mathcal{F}, \{\mathcal{F}_t\}_{t\geq 0}, \widehat{\mathbf{P}}, \widehat{W}, X)$ is a weak solution of

$$\begin{cases} dX(t) = b(t,X)dt + \sigma(t,X)d\widehat{W}(t), \\ X(0) = \xi. \end{cases} \tag{6.24}$$

Theorem 6.11. *Let $\sigma \in \mathcal{A}^n(\mathbb{R}^{n\times n})$ be bounded and continuous, and let $b \in \mathcal{A}^n(\mathbb{R}^n)$ be bounded. Moreover, assume that $\sigma(t,x)^{-1}$ exists for all $(t,x) \in [0,\infty) \times \mathbf{W}^n$ with $|\sigma(t,x)^{-1}|$ uniformly bounded. Then (6.2) has a weak solution.*

Proof. Since $\sigma \in \mathcal{A}^n(\mathbb{R}^n)$ is bounded and continuous, by Theorem 6.10, equation (6.19) has a weak solution $(\Omega, \mathcal{F}, \{\mathcal{F}_t\}_{t\geq 0}, \mathbf{P}, W, X)$. By the assumptions, it is easily verified that (6.20) is satisfied. Therefore, the weak existence of solutions to (6.2) follows from (6.21)–(6.24). □

In Ikeda–Watanabe [1, pp. 178–180], it is proved that under the assumptions of Theorem 6.11, the weak uniqueness of (6.2) is equivalent to that of (6.19), by a sort of "reverse transformation" of (6.21). As a consequence, we have the following uniqueness result.

Theorem 6.12. *Under the same assumptions of Theorem 6.11, if in addition, $\sigma(t,x)$ is Lipschitz in x, uniformly in $t \in [0,T]$, then weak uniqueness holds for solutions of (6.2).*

Proof. Since $\sigma(t,x)$ is Lipschitz in x, uniformly in $t \in [0,T]$, (6.19) has a unique strong solution by Theorem 6.3. This implies weak uniqueness by Theorems 6.8 and 6.9. Therefore, the conclusion follows from the above observation. □

For Markovian-type equations, we have the following elegant weak existence result for merely measurable coefficients, due to Krylov.

Theorem 6.13. *Let $b : [0,\infty) \times \mathbb{R}^n \to \mathbb{R}^n$ and $\sigma : [0,\infty) \times \mathbb{R}^n \to \mathcal{S}^n$ be measurable and bounded. Moreover, assume that σ is uniformly positive definite, namely, there is $\delta > 0$ such that $\langle \sigma(t,x)\lambda, \lambda\rangle \geq \delta|\lambda|^2$, $\forall \lambda \in \mathbb{R}^n$. Then (6.13) has a weak solution.*

For a proof see Krylov [2, pp. 87–91].

6.3. Linear SDEs

In this subsection we consider the linear SDE

$$
\begin{cases} dX(t) = [A(t)X(t) + b(t)]dt + [C(t)X(t) + \sigma(t)]dW(t), \\ X(0) = \xi, \end{cases} \tag{6.25}
$$

where $W(\cdot)$ is a one-dimensional standard Brownian motion and

$$
\begin{cases} A(\cdot), C(\cdot) \in L^\infty(0, T; \mathbb{R}^{n\times n}), \\ b(\cdot), \sigma(\cdot) \in L^2(0, T; \mathbb{R}^n). \end{cases} \tag{6.26}
$$

From Theorem 6.3, we know that under (6.26), SDE (6.25) admits a unique strong solution. Our main result of this subsection is the following *variation of constants formula:*

Theorem 6.14. *For any $\xi \in L^2_{\mathcal{F}_0}(\Omega; \mathbb{R}^n)$, equation (6.25) admits a unique strong solution $X(\cdot)$, which is represented by the following:*

$$
\begin{aligned} X(t) = \Phi(t)\xi + \Phi(t)\int_0^t \Phi(s)^{-1}\big[b(s) - C(s)\sigma(s)\big]ds \\ + \Phi(t)\int_0^t \Phi(s)^{-1}\sigma(s)dW(s), \qquad t \in [0, T], \end{aligned} \tag{6.27}
$$

where $\Phi(\cdot)$ is the unique solution of the following matrix-valued SDE:

$$
\begin{cases} d\Phi(t) = A(t)\Phi(t)dt + C(t)\Phi(t)dW(t), \\ \Phi(0) = I, \end{cases} \tag{6.28}
$$

and $\Phi(t)^{-1} \equiv \Psi(t)$ exists, satisfying

$$
\begin{cases} d\Psi(t) = \Psi(t)\big[-A(t) + C(t)^2\big]dt - \Psi(t)C(t)dW(t), \\ \Psi(0) = I. \end{cases} \tag{6.29}
$$

Proof. By Theorem 6.3, we see that (6.28) admits a unique solution $\Phi(\cdot)$. To show that $\Phi(t)^{-1}$ exists, let $\Psi(\cdot)$ be the unique strong solution of (6.29), which exists by, once again, Theorem 6.3. Applying Itô's formula to $\Phi(t)\Psi(t)$, one easily obtains that $d\big[\Phi(t)\Psi(t)\big] = 0$. Hence $\Phi(t)\Psi(t) \equiv I$, leading to $\Psi(t) = \Phi(t)^{-1}$. Next, applying Itô's formula to $\Psi(t)X(t)$, where $X(t)$ is the unique strong solution of (6.25), we obtain

$$
d\big[\Psi(t)X(t)\big] = \Psi(t)(b(t) - C(t)\sigma(t))dt + \Psi(t)\sigma(t)dW(t).
$$

This yields (6.27), where we note the fact that $\Psi(t) = \Phi(t)^{-1}$. □

Let us now briefly discuss the case where $W(\cdot)$ is multidimensional. In this case, we can write the linear equation as

$$
\begin{cases} dX(t) = \big[A(t)X(t) + b(t)\big]dt + \sum_{j=1}^m \big[C_j(t)X(t) + \sigma_j(t)\big]dW^j(t), \\ X(0) = \xi. \end{cases} \tag{6.30}
$$

Let $\Phi(t)$ be the solution of the following:

$$
\begin{cases} d\Phi(t) = A(t)\Phi(t)dt + \sum_{j=1}^{m} C_j(t)\Phi(t)dW^j(t), \\ \Phi(0) = I. \end{cases} \tag{6.31}
$$

We can similarly prove that $\Phi(t)^{-1}$ exists, which is the solution of

$$
\begin{cases} d(\Phi(t)^{-1}) = \Phi(t)^{-1}\big[- A(t) + \sum_{j=1}^{m} C_j(t))^2\big]dt \\ \qquad\qquad - \sum_{j=1}^{m} \Phi(t)^{-1}C_j(t)dW^j(t), \\ \Phi(0)^{-1} = I. \end{cases} \tag{6.32}
$$

The strong solution X of (6.30) can be represented as

$$
\begin{aligned} X(t) = \Phi(t)\xi + \Phi(t) \int_0^t \Phi(s)^{-1}\big[b(s) - \sum_{j=1}^{m} C_j\sigma_j(s)\big]ds \\ + \sum_{j=1}^{m} \Phi(t) \int_0^t \Phi(s)^{-1}\sigma_j(s)dW^j(s). \end{aligned} \tag{6.33}
$$

6.4. Other Types of SDEs

In this subsection we discuss two types of SDEs that are encountered in stochastic control problems. One of them does not fit into the form of (6.2), while the other one does after a certain transformation.

We note that in equation (6.2) the coefficients b and σ do not depend on the sample ω *explicitly* (they do implicitly through the solution $X(\cdot)$). In the case where these coefficients are dependent on ω explicitly, we have *SDEs with random coefficients*:

$$
\begin{cases} dX(t,\omega) = b(t, X, \omega)dt + \sigma(t, X, \omega)dW(t, \omega), \\ X(0, \omega) = \xi(\omega). \end{cases} \tag{6.34}
$$

Since the coefficients b and σ should be given a priori, equation (6.34) should be defined on a *given* probability space $(\Omega, \mathcal{F}, \mathbf{P})$, where b and σ are defined. In other words, for an SDE with random coefficients, it is not sensible to have the notion of weak solutions.

Definition 6.15. Let the maps $b : [0, \infty) \times \mathbf{W}^n \times \Omega \to \mathbb{R}^n$ and $\sigma : [0, \infty) \times \mathbf{W}^n \times \Omega \to \mathbb{R}^{n\times m}$ be given on a given filtered probability space $(\Omega, \mathcal{F}, \{\mathcal{F}_t\}_{t\ge 0}, \mathbf{P})$ and an m-dimensional standard $\{\mathcal{F}_t\}_{t\ge 0}$-Brownian motion $W(t)$. Let ξ be $\mathcal{F}_0$-measurable. An $\{\mathcal{F}_t\}_{t\ge 0}$-adapted continuous process $X(t), t \ge 0$, is called a *solution* of (6.34) if

$$
X(0) = \xi, \qquad \mathbf{P}\text{-a.s.}, \tag{6.35}
$$

$$\int_0^t \{|b(s,X(\omega),\omega)| + |\sigma(s,X(\omega),\omega)|^2\}ds < \infty, \quad t \geq 0, \ \mathbf{P}\text{-a.s.}\, \omega \in \Omega, \tag{6.36}$$

$$X(t,\omega) = \xi(\omega) + \int_0^t b(s,X(\omega),\omega)ds + \int_0^t \sigma(s,X(\omega),\omega)dW(s,\omega), \quad t \geq 0, \ \mathbf{P}\text{-a.s.}\, \omega \in \Omega. \tag{6.37}$$

If

$$\mathbf{P}(X(t) = Y(t), \ 0 \leq t < \infty) = 1, \tag{6.38}$$

holds for any two solutions X and Y of (6.34) defined on the given probability space $(\Omega, \mathcal{F}, \{\mathcal{F}_t\}_{t\geq 0}, \mathbf{P})$ along with the given $\{\mathcal{F}_t\}_{t\geq 0}$-Brownian motion $W(t)$, then we say that the solution is *unique*.

Note the differences between Definitions 6.2 and 6.15. First of all, one does not need to distinguish between the strong and weak solutions for (6.34), so we simply term them as "solutions". Second, the uniqueness in Definition 6.15 is based only on the *given* $(\Omega, \mathcal{F}, \{\mathcal{F}_t\}_{t\geq 0}, \mathbf{P})$ and $W(t)$ rather than on *any given* $(\Omega, \mathcal{F}, \{\mathcal{F}_t\}_{t\geq 0}, \mathbf{P})$ and $W(t)$ as in Definition 6.2.

Let us now introduce the following assumption.

(RC) For any $\omega \in \Omega$, $b(\cdot\,,\cdot\,,\omega) \in \mathcal{A}^n(\mathbb{R}^n)$ and $\sigma(\cdot\,,\cdot\,,\omega) \in \mathcal{A}^n(\mathbb{R}^{n\times m})$ and for any $x \in \mathbf{W}^n$, $b(\cdot\,,x,\cdot)$ and $\sigma(\cdot\,,x,\cdot)$ are both $\{\mathcal{F}_t\}_{t\geq 0}$-adapted processes. Moreover, there exists an $L > 0$ such that for all $t \in [0,\infty)$, $x(\cdot), y(\cdot) \in \mathbf{W}^n$, and $\omega \in \Omega$,

$$\begin{cases} |b(t,x(\cdot),\omega) - b(t,y(\cdot),\omega)| \leq L|x(\cdot) - y(\cdot)|_{\mathbf{W}^n}, \\ |\sigma(t,x(\cdot),\omega) - \sigma(t,y(\cdot),\omega)| \leq L|x(\cdot) - y(\cdot)|_{\mathbf{W}^n}, \\ |b(\cdot\,,0,\cdot)| + |\sigma(\cdot\,,0,\cdot)| \in L^2_{\mathcal{F}}(0,T;\mathbb{R}), \quad \forall T > 0. \end{cases} \tag{6.39}$$

Theorem 6.16. *Let (RC) hold. Then, for any $\xi \in L^\ell_{\mathcal{F}_0}(\Omega;\mathbb{R}^n)$ $(\ell \geq 1)$, (6.34) admits a unique solution X such that for any $T > 0$,*

$$E \max_{0\leq s\leq T} |X(s)|^\ell \leq K_T(1 + E|\xi|^\ell) \tag{6.40}$$

and

$$E|X(t) - X(s)|^\ell \leq K_T(1 + E|\xi|^\ell)|t-s|^{\ell/2}, \quad \forall s,t \in [0,T]. \tag{6.41}$$

Moreover, if $\widehat{\xi} \in L^\ell_{\mathcal{F}_0}(\Omega;\mathbb{R}^n)$ is another random variable and $\widehat{X}(t)$ is the corresponding solution of (6.34), then for any $T > 0$, there exists a $K_T > 0$ such that

$$E \max_{0\leq s\leq T} |X(s) - \widehat{X}(s)|^\ell \leq K_T E|\xi - \widehat{\xi}|^\ell. \tag{6.42}$$

The proof of this theorem is the same as that of Theorem 6.3.

Next, we discuss another type of equations that will play a vital role in establishing an appropriate formulation for optimal stochastic control problems.

Consider the following SDE:

$$\begin{cases} dX(t) = b(t, X, W)dt + \sigma(t, X, W)dW(t), \\ X(0) = \xi, \end{cases} \tag{6.43}$$

where b and σ are defined on $[0, \infty] \times \mathbf{W}^n \times \mathbf{W}^m$. This equation is not readily seen to be in the form of (6.2). However, if we define $Y(t) \equiv W(t)$, then the above equation is equivalent to

$$\begin{cases} dX(t) = b(t, X, Y)dt + \sigma(t, X, Y)dW(t), \\ dY(t) = dW(t), \\ X(0) = \xi, \quad Y(0) = 0. \end{cases} \tag{6.44}$$

This is a special case of (6.2). Thus all the notions of strong and weak solutions, their uniqueness and pathwise uniqueness, apply to (6.43).

Let us make the following assumption.

(W) $b \in \mathcal{A}^{n+m}(\mathbb{R}^n)$, $\sigma \in \mathcal{A}^{n+m}(\mathbb{R}^{n\times m})$, and there exists an $L > 0$ such that for all $t \in [0, \infty)$, $x(\cdot), y(\cdot) \in \mathbf{W}^n$ and $w(\cdot) \in \mathbf{W}^m$,

$$\begin{cases} |b(t, x(\cdot), w(\cdot)) - b(t, y(\cdot), w(\cdot))| \leq L|x(\cdot) - y(\cdot)|_{\mathbf{W}^n}, \\ |\sigma(t, x(\cdot), w(\cdot)) - \sigma(t, y(\cdot), w(\cdot))| \leq L|x(\cdot) - y(\cdot)|_{\mathbf{W}^n}, \\ |b(t, x(\cdot), w(\cdot))| + |\sigma(t, x(\cdot), w(\cdot))| \leq K(1 + |x(\cdot)|_{\mathbf{W}^n}). \end{cases} \tag{6.45}$$

Theorem 6.17. *Under (W), equation (6.43) has a unique strong solution. As a consequence, both pathwise uniqueness and weak existence and uniqueness hold.*

Proof. Given any filtered probability space $(\Omega, \mathcal{F}, \{\mathcal{F}_t\}_{t\geq 0}, \mathbf{P})$ along with an m-dimensional $\{\mathcal{F}_t\}_{t\geq 0}$-Brownian motion $W(\cdot)$, $Y(t) \equiv W(t)$ will then be fixed. Then by (W) and the same arguments employed in proving Theorem 6.3, we obtain the existence and uniqueness of $X(\cdot)$, which, together with $Y(\cdot)$, solves (6.44) or, equivalently, (6.43). The second assertion follows then from Theorems 6.8 and 6.9. $\square$

We emphasize that in (W), no continuity is assumed for the coefficients in the variable $w(\cdot)$. Strong uniqueness still holds due to the particular role of $W(\cdot)$.

Chapter 2

Stochastic Optimal Control Problems

1. Introduction

Uncertainty is inherent in most real-world systems. It places many disadvantages (and sometimes, surprisingly, advantages) on humankind's efforts, which are usually associated with the quest for optimal results. The systems mainly studied in this book are *dynamic*, namely, they evolve over time. Moreover, they are described by Itô's stochastic differential equations and are sometimes called *diffusion models*. The basic source of uncertainty in diffusion models is *white noise*, which represents the joint effects of a large number of *independent* random forces acting on the systems. Since the systems are dynamic, the relevant *decisions* (*controls*), which are made based on the most updated information available to the *decision makers* (*controllers*), must also change over time. The decision makers must select an *optimal* decision among all possible ones to achieve the best expected result related to their goals. Such optimization problems are called *stochastic optimal control problems*. The range of stochastic optimal control problems covers a variety of physical, biological, economic, and management systems, just to mention a few. In this chapter we shall set up a rigorous mathematical framework for stochastic optimal control problems.

The remainder of this chapter will be organized as follows. In Section 2, the formulation of deterministic optimal control problems will be recalled. This practice of beginning with the deterministic case will be carried out in later chapters whenever appropriate. The reason for doing so is not only that the deterministic situation itself may contain interesting results, but also that readers can see the essential differences between the deterministic and stochastic problems. In Section 3, several practical examples arising in manufacturing, finance, insurance, and industry that can be formulated as stochastic optimal control models are given. Section 4 presents strong and weak formulations of stochastic optimal control problems that will be mainly dealt with throughout this book. Section 5 is concerned with the existence of stochastic optimal controls for both strong and weak formulations. In Section 6 we demonstrate that there are fundamental differences between the stochastic and deterministic cases in terms of *reachable sets*, which are closely related to the existence of optimal controls as well as controllability. Section 7 introduces some stochastic optimal control models other than the one formulated in Section 3. These models, including optimal stopping, singular controls, risk-sensitive controls, and partially observed models, have been widely studied in the literature. Although they are not to be investigated in this book, the two approaches, namely Pontryagin's maximum principle and Bellman's dynamic programming, are also the principal ones to solve those problems. Finally, Section 8 gives historical remarks.

2. The Deterministic Cases Revisited

We recall deterministic optimal control problems in this section. Let us start with a production planning problem. A machine is producing one type of product. The raw materials are processed by the machine, and the finished products are stored in a *buffer*. Suppose at time t the *production rate* is $u(t)$ and the *inventory level* in the buffer is $x(t)$. If the *demand rate* for this product is a known function $z(t)$ and the inventory is x_0 at time $t = 0$, then the relationship among these quantities can be described by

$$\begin{cases} \dot{x}(t) = u(t) - z(t), & t \geq 0, \\ x(0) = x_0. \end{cases} \tag{2.1}$$

Here $x(t)$ may take either positive or negative values. The product has a *surplus* when $x(t) > 0$, and a *backlog* when $x(t) < 0$. Suppose the cost of having the inventory x and production rate u per unit time is $h(x, u)$. A typical example of h is

$$h(x, u) = c^+ x^+ + c^- x^- + pu, \tag{2.2}$$

where $x^+ \stackrel{\Delta}{=} \max\{x, 0\}, x^- \stackrel{\Delta}{=} \max\{-x, 0\}$, $c^+, c^- \geq 0$ are the marginal cost/penalty for surplus and backlog, respectively, and p is the unit cost of production. The production management wants to choose a $u(\cdot)$ so as to minimize the total *discounted* cost over the *planning horizon* $[0, T]$. Namely the following functional is to be minimized:

$$J(u(\cdot)) = \int_0^T e^{-\gamma t} h(x(t), u(t)) dt, \tag{2.3}$$

where $\gamma > 0$ is the *discount rate*. Note here that the decision $u(\cdot)$ is a *function* on $[0, T]$, which is called a *production plan*. If the machine has a maximum production rate k (called the *production capacity*), then any production plan must be subject to

$$0 \leq u(t) \leq k, \quad \forall t \in [0, T]. \tag{2.4}$$

On the other hand, if the buffer size is $b > 0$, then the inventory level $x(t)$ must satisfy the *constraint*

$$x(t) \leq b. \tag{2.5}$$

Any production plan that satisfies (2.4)–(2.5) (via (2.1)) is called an *admissible plan*. The problem is to minimize the cost (2.3) over all admissible plans.

The above problem is perhaps one of the simplest examples of an optimal control problem. It is deterministic because there is no uncertainty in the system dynamics (2.1) and the constraints (2.4)–(2.5).

Now we present the general formulation of (deterministic) optimal control problems. Let T, $0 < T \leq +\infty$, $x_0 \in \mathbb{R}^n$, and a metric space Γ

be given. The *dynamics* of a system under consideration are given by the following ordinary differential equation:

$$
(2.6) \qquad \begin{cases} \dot{x}(t) = b(t, x(t), u(t)), & \text{a.e. } t \in [0, T], \\ x(0) = x_0, \end{cases}
$$

where $b : [0, T] \times \mathbb{R}^n \times \Gamma \to \mathbb{R}^n$ is a given map. When $T = \infty$, $[0, T]$ above should be replaced by $[0, \infty)$. A measurable map $u(\cdot) : [0, T] \to \Gamma$ is called a *control*, x_0 is called the *initial state*, and $x(\cdot)$, a solution of (2.6), is called a *state trajectory* corresponding to $u(\cdot)$. In the production planning example, the dynamics are (2.1), the control is the production rate, and the state is the inventory level.

Hereafter, we assume that for any $x_0 \in \mathbb{R}^n$ and any control $u(\cdot)$, there exists a unique solution $x(\cdot) \equiv x(\cdot\,; u(\cdot))$ to (2.6). In such a case, we obtain an *input–output relation* with *input* $u(\cdot)$ and *output* $x(\cdot)$. We also call (2.6) a *controlled*, or *control, system*.

A particular case of (2.6) is the *linear* one, where the controlled system is of the form

$$
(2.7) \qquad \begin{cases} \dot{x}(t) = A(t)x(t) + B(t)u(t), & \text{a.e. } t \in [0, T], \\ x(0) = x_0, \end{cases}
$$

with $A : [0, T] \to \mathbb{R}^{n \times n}$ and $B : [0, T] \to \mathbb{R}^{n \times k}$ being measurable maps. Sometimes we call (2.7) a *time-varying linear system*, since $A(\cdot)$ and $B(\cdot)$ are time-dependent. Further, the following system is called a *time-invariant linear system*:

$$
(2.8) \qquad \begin{cases} \dot{x}(t) = Ax(t) + Bu(t), & \text{a.e. } t \in [0, T], \\ x(0) = x_0, \end{cases}
$$

where $A \in \mathbb{R}^{n \times n}$ and $B \in \mathbb{R}^{n \times k}$.

As we have seen in the production planning example, there may be various constraints involved. For example, the *state constraint* could be given by

$$
(2.9) \qquad x(t) \in S(t), \quad \forall t \in [0, T],
$$

and the *control constraint* given by

$$
(2.10) \qquad u(t) \in U(t), \quad \text{a.e. } t \in [0, T],
$$

where $S(t) : [0, T] \to 2^{\mathbb{R}^n}$ and $U(t) : [0, T] \to 2^{\Gamma}$ are given *multifunctions* (i.e., for each $t \in [0, T]$, $S(t) \subseteq \mathbb{R}^n$ and $U(t) \subseteq \Gamma$). Some other types of constraints (e.g., constraints in an integral form) are also possible.

In this book we treat only the case where the control constraint $U(t)$ is time invariant, namely, $U(t) \equiv U$. Remember that U itself can be regarded as a metric space. Thus, from now on, we replace Γ by U. Define

$$
\mathcal{V}[0, T] \triangleq \{u : [0, T] \to U \mid u(\cdot) \text{ is measurable}\}.
$$

Any $u(\cdot) \in \mathcal{V}[0,T]$ is called a *feasible control.*

Next, we are given a *cost functional* measuring the performance of the controls:

$$J(u(\cdot)) = \int_0^T f(t,x(t),u(t))dt + h(x(T)) \tag{2.11}$$

for given maps $f : [0,T] \times \mathbb{R}^n \times U \to \mathbb{R}$ and $h : \mathbb{R}^n \to \mathbb{R}$. The first and second terms on the right-hand side of (2.11) are called the *running cost* and the *terminal cost*, respectively.

Definition 2.1. A control $u(\cdot)$ is called an *admissible* control, and $(x(\cdot),u(\cdot))$ called an *admissible pair*, if:

(i) $u(\cdot) \in \mathcal{V}[0,T]$;
(ii) $x(\cdot)$ is the unique solution of equation (2.6) under $u(\cdot)$;
(iii) the state constraint (2.9) is satisfied;
(iv) $t \to f(t,x(t),u(t)) \in L^1[0,T]$.

The set of all admissible controls is denoted by $\mathcal{V}_{ad}[0,T]$. Our deterministic optimal control problem can be stated as follows.

Problem (D). Minimize (2.11) over $\mathcal{V}_{ad}[0,T]$.

Problem (D) is said to be *finite* if (2.11) has a finite lower bound, and is said to be *(uniquely) solvable* if there is a (unique) $\bar{u}(\cdot) \in \mathcal{V}_{ad}[0,T]$ satisfying

$$J(\bar{u}(\cdot)) = \inf_{u(\cdot)\in\mathcal{V}_{ad}[0,T]} J(u(\cdot)). \tag{2.12}$$

Any $\bar{u}(\cdot) \in \mathcal{V}_{ad}[0,T]$ satisfying (2.12) is called an *optimal control*, which the corresponding state trajectory $\bar{x}(\cdot) \equiv x(\cdot\,;\bar{u}(\cdot))$ and $(\bar{x}(\cdot),\bar{u}(\cdot))$ are called an *optimal state trajectory* and an *optimal pair*, respectively.

A particularly interesting and important case is that in which the controlled system is given by (2.7), the state constraint (2.9) is absent (i.e., there is no state constraint, or $S(t) \equiv \mathbb{R}^n$), the control constraint U is $\mathbb{R}^k$, and the cost functional is of the form

$$\begin{aligned} J(u) =& \frac{1}{2}\int_0^T \big[\,\langle\, Q(t)x(t),x(t)\,\rangle + \langle\, R(t)u(t),u(t)\,\rangle\,\big]dt \\ &+ \frac{1}{2}\,\langle\, Gx(T),x(T)\,\rangle, \end{aligned} \tag{2.13}$$

for some suitable symmetric matrix G and symmetric matrix-valued functions $Q(\cdot)$ and $R(\cdot)$. The corresponding Problem (D) is called a *linear quadratic optimal control problem* (LQ problem, for short).

Finally, one may note that in the production planning example the terminal cost h is equal to 0. In general, an optimal control problem with $h = 0$ is called the *Lagrange problem*, that with $f = 0$ called the *Mayer problem*, and that with $f \neq 0, h \neq 0$ called the *Bolza problem.* It is well known (and easy to show) that these three problems are mathematically equivalent.

3. Examples of Stochastic Control Problems

This section is devoted to a presentation of various practical problems that can later be categorized into one model, namely, the *stochastic optimal control model* under consideration throughout this book.

3.1. Production planning

Consider the optimal production planning problem with the presence of uncertainties in demand. The uncertainties are described by a filtered probability space $(\Omega, \mathcal{F}, \{\mathcal{F}_t\}_{t\geq 0}, \mathbf{P})$ satisfying the usual condition, on which a one-dimensional standard Brownian motion $W(t)$ is defined. The demand process $z(t)$ is no longer deterministic. Rather, it is given by the following:

$$z(t) = z_0 + \int_0^t \xi(s)ds + \int_0^t \sigma(s)dW(s), \quad t \in [0, T]. \tag{3.1}$$

Here, $\xi(t)$ represents the expected *demand rate* at time t, and the term $\int_0^t \sigma(s)dW(s)$ represents the fluctuation of the demand due to the uncertainty of the environment. Note that the notion of "demand" is in a rather broad sense: Any withdrawal from or addition to the inventory can be regarded as an action of the demand. Therefore, it may cover those processes such as *sales return* and *inventory spoilage*, which are in general quite random. We assume that both $\xi(\cdot)$ and $\sigma(\cdot)$ are $\{\mathcal{F}_t\}_{t\geq 0}$-adapted, and the integrals in (3.1) are well-defined. To meet the demand, the factory has to adjust its production rate all the time to accommodate changing situations. Suppose the machine is reliable (i.e., it will never break down) and, as in the deterministic case, at time t the production rate is $u(t)$ and the inventory level in the buffer is $x(t)$. If at time $t = 0$ the inventory is x_0, then the system can be modeled as

$$\begin{cases} dx(t) = (u(t) - z(t))\, dt, & x(0) = x_0, \\ dz(t) = \xi(t)dt + \sigma(t)dW(t), & z(0) = z_0. \end{cases} \tag{3.2}$$

Once again, the control or the production rate $u(\cdot)$ is subject to the production capacity, namely,

$$0 \leq u(t) \leq k, \qquad \text{a.e. } t \in [0, T], \quad \mathbf{P}\text{-a.s.} \tag{3.3}$$

There is another implicit constraint on the production policy $u(\cdot)$ due to the stochastic environment of the problem, namely, at any time the production management makes a decision based only on past information rather than any future information. Therefore, the control or decision must be *nonanticipative*. The precise mathematical translation of this constraint will be given in Section 4.

Here, in order to avoid an infinite backlog, the maximum production capacity should be large enough overall to meet the demand. So the following minimum condition should be imposed for the problem to be meaningful:

$$kT > E \int_0^T z(t)dt \equiv z_0 T + E \int_0^T \int_0^t \xi(s)ds\, dt. \tag{3.4}$$

On the other hand, the inventory state should not exceed the buffer size b:

$$x(t) \leq b, \qquad \forall t \in [0,T], \quad \mathbf{P}\text{-a.s.} \tag{3.5}$$

The expected total (discounted) cost is the following:

$$J(u(\cdot)) = E\Big\{ \int_0^T e^{-\gamma t} f(x(t),u(t))dt + e^{-\gamma T} h(x(T)) \Big\}, \tag{3.6}$$

where the first term represents the total running cost for inventory and production, the second term is a penalty of the inventory left over at the end of the production horizon (e.g., disposal cost), and $\gamma > 0$ is the discount rate.

The objective of the production management is to choose a suitable production plan $u(\cdot)$ so that (3.2), (3.3), and (3.5) are satisfied and $J(u(\cdot))$ is minimized.

3.2. Investment vs. consumption

Suppose there is a market in which $n+1$ *assets* (or *securities*) are traded continuously. One of the assets is called *bond*, whose *price process* $P_0(t)$ is subject to the following (deterministic) ordinary differential equation:

$$\begin{cases} dP_0(t) = r(t)P_0(t)dt, & \text{a.e. } t \in [0,T], \\ P_0(0) = p_0 > 0, \end{cases} \tag{3.7}$$

where $r(t) > 0$ is called the *interest rate* (of the bond). Clearly $P_0(t)$ will grow steadily for sure as time goes by, and the bond is therefore called a *riskless* asset. The other n assets are called *stocks*, whose price processes $P_1(t), \ldots, P_m(t)$ satisfy the following stochastic differential equation:

$$\begin{cases} dP_i(t) = P_i(t)\{b_i(t)dt + \langle \sigma_i(t), dW(t) \rangle\}, & t \in [0,T], \\ P_i(0) = p_i > 0, \end{cases} \tag{3.8}$$

where $b_i : [0,T] \times \Omega \to \mathbb{R}^1$ with $b_i(t) > 0$ (almost surely) is called the *appreciation rate*, and $\sigma_i : [0,T] \times \Omega \to \mathbb{R}^m$ is called the *volatility* or the *dispersion* of the stocks. All these processes are assumed to be $\{\mathcal{F}_t\}_{t\geq 0}$-adapted. Here, $W(t)$ is an m-dimensional standard Brownian motion defined on some fixed filtered complete probability space $(\Omega, \mathcal{F}, \{\mathcal{F}_t\}_{t\geq 0}, \mathbf{P})$. The diffusion term in (3.8) reflects the fluctuation of the stock prices, and the stocks are therefore called *risky* assets. Usually, one has

$$Eb_i(t) > r(t) > 0, \qquad \forall t \in [0,T], \quad 1 \leq i \leq n. \tag{3.9}$$

This is a very natural assumption, since otherwise nobody is willing to invest in the risky stocks. However, we point out that the following discussion does not depend on this assumption.

We now consider an investor whose total wealth at time $t \geq 0$ is denoted by $x(t)$. Suppose he/she decides to hold $N_i(t)$ shares of the ith asset ($i =$

$0, 1, \ldots, n)$ at time t. Then

$$x(t) = \sum_{i=0}^{n} N_i(t) P_i(t), \qquad t \geq 0. \tag{3.10}$$

Suppose that the trading of shares and the payment of *dividends* (at the rate $\mu_i(t)$ per unit time and per unit of money invested in the ith stock) takes place continuously. Further, we let $c(t)$ be the rate of withdrawal for *consumption*. Then one has

$$\begin{aligned} x(t+\Delta t) - x(t) &= \sum_{i=0}^{n} N_i(t)\big[P_i(t+\Delta t) - P_i(t)\big] \\ &\quad + \sum_{i=1}^{n} \mu_i(t) N_i(t) P_i(t) \Delta t - c(t)\Delta t. \end{aligned} \tag{3.11}$$

Letting $\Delta t \to 0$, we obtain

$$\begin{aligned} dx(t) &= \sum_{i=0}^{n} N_i(t) dP_i(t) + \sum_{i=1}^{n} \mu_i(t) N_i(t) P_i(t) dt - c(t) dt \\ &= \Big\{ r(t) N_0(t) P_0(t) + \sum_{i=1}^{n} \big[b_i(t) + \mu_i(t)\big] N_i(t) P_i(t) - c(t) \Big\} dt \\ &\quad + \sum_{i=1}^{n} N_i(t) P_i(t) \langle \sigma_i(t), dW(t) \rangle \\ &= \Big\{ r(t) x(t) + \sum_{i=1}^{n} \big[b_i(t) + \mu_i(t) - r(t)\big] u_i(t) - c(t) \Big\} dt \\ &\quad + \langle \sum_{i=1}^{n} \sigma_i(t) u_i(t), dW(t) \rangle, \end{aligned} \tag{3.12}$$

where

$$u_i(t) \stackrel{\Delta}{=} N_i(t) P_i(t), \qquad i = 0, 1, 2, \ldots, n, \tag{3.13}$$

denotes the market value of the investor's wealth in the ith bond/stock. When $u_i(t) < 0$ $(i = 1, 2, \ldots, n)$, the investor is *short-selling* the ith stock. When $u_0(t) < 0$, the investor is borrowing the amount $|u_0(t)|$ with an interest rate $r(t)$. It is clear that by changing $u_i(t)$, the investor changes the "allocation" of his/her wealth in these $n + 1$ assets. We call $u(t) \stackrel{\Delta}{=} (u_1(t), \ldots, u_n(t))$ a *portfolio* of the investor. Note that we do not include the allocation to the bond in the portfolio, as it will be uniquely determined by the allocation to the stocks, given the total wealth. As in the stochastic production planning case, any admissible portfolio is required to be nonanticipative. Now, for $x(0) = x_0 > 0$, the investor wants to choose the investment portfolio $u(\cdot)$ and the consumption plan $c(\cdot)$ such that

$$x(t) \geq 0, \qquad \forall t \in [0, T], \quad \mathbf{P}\text{-a.s.}, \tag{3.14}$$

and such that the stream of discounted *utility*,

$$J(u(\cdot),c(\cdot)) = E\Big\{\int_0^T e^{-\gamma t}\varphi(c(t))dt + e^{-\gamma T}h(x(T))\Big\}, \tag{3.15}$$

is maximized, where $\gamma > 0$ is the discount rate, $\varphi(c)$ is the instantaneous utility from consumption c, and $e^{-\gamma T}h(x(T))$ is the (discounted) utility that is derived from *bequests*.

We may impose some further constraints for the above problem. For example, one constraint may be that short-selling has to be bounded. Then we will have the constraints

$$u_i(t) \geq -L_i, \qquad \forall t \in [0,T], \quad \mathbf{P}\text{-a.s.}\,, \ i = 0,1,\ldots,n,$$

for some $L_i \geq 0$. That $L_i = 0$ means that short-selling is prohibited.

3.3. Reinsurance and dividend management

We consider a model of an insurance company, which can choose a *reinsurance* policy to manage risks. In addition, there is a choice of the amount of *dividends* to be paid out to shareholders. Notwithstanding any policy decision, there is a constant payment of a *corporate debt*, such as bond liability or loan amortization. It is a classical approach to assess the value of a company based on the total dividend payouts. Therefore, the objective is to find the reinsurance and dividend policy that maximize the expected total discounted dividend payouts over a time period.

To formulate the model, consider the company's *cash reserve*. There are three different components that affect it. The first two are deterministic – the payment of *premiums* by customers and debt repayment by the company, both happening at a constant rate. The third is payments on *claims*. If we denote the number of claims received up to time t by $A(t)$ and the size of the ith claim by η_i, then $R(t)$, the company's reserve at time t, is given by

$$R(t) = R(0) + pt - \delta t - \sum_{i=1}^{A(t)} \eta_i. \tag{3.16}$$

Here p is the amount of premiums received per unit time, and δ is the rate of debt repayment. If $A(t)$ is a *Poisson process* with *intensity* λ and all claims are i.i.d, then $R(t)$ can be approximated by a Brownian motion with a drift $p - \delta - \lambda E\eta$ and a diffusion coefficient $\sigma = (\lambda E\eta^2)^{1/2}$.

Reinsurance allows the company to divert a proportion $1 - a$ of all premiums to another company, and consequently, the fraction $1 - a$ of each claim is paid by the other company. In this case, in equation (3.16) describing the reserve, p is replaced by ap and η_i by $a\eta_i$. In the limiting diffusion model the quantity $p - \lambda E\eta$ in the drift term is changed into $a(p - \lambda E\eta)$, while σ is changed into $a\sigma$. With reinsurance one reduces the risk incurred by the company at the expense of reducing the potential profit. This explains why reinsurance can be considered in the framework of risk management.

With the above motivation we describe our model as follows. Denote the value of the liquid assets of the company at time t by $x(t)$. Let $c(t)$ be the dividend rate paid out to the shareholders at time t. Then the dynamics of $x(t)$ are given by

$$\begin{cases} dx(t) = [a(t)\mu - \delta - c(t)]dt + a(t)\sigma dw(t), \\ x(0) = x_0. \end{cases} \tag{3.17}$$

Here μ is the difference between the premium rate and the expected payments on claims per unit time (called the *safety loading*), $1 - a(t)$ is the reinsurance fraction at time t, and x_0 is the initial value of the liquid assets of the company. By the nature of the problem, the following constraints must be satisfied:

$$x(t) \geq 0, \qquad 0 \leq a(t) \leq 1, \qquad \forall t \in [0, T]. \tag{3.18}$$

The objective of the management of the company is to choose the dividend payout scheme $c(\cdot)$ and the risk management policy $a(\cdot)$, both being nonanticipative, such that (3.18) is satisfied and the following expected total dividend payouts over $[0, T]$ is maximized:

$$E \int_0^T e^{-\gamma t} c(t) dt, \tag{3.19}$$

where $\gamma > 0$ is the discount rate.

3.4. Technology diffusion

There are three phases in the life cycle of a new technology: research and development (R&D), transfer and commercialization, and operation and regeneration. *Technology diffusion* refers to the transition of technology's economic value during the transfer and operation phases of a technology life cycle. The modeling of technology diffusion must address two aspects: *regularity* due to the mean *depletion rate* of the technology's economic value, and *uncertainty* due to disturbances occurring in technological evolution and innovation. Therefore, stochastic differential equations will be a good framework to model technology diffusion problems.

Suppose the process of adopting a new technology starts at time 0 with an initial economic value x_0, and the *value depletion process* $x(\cdot)$ of the technology satisfies the following Itô's stochastic differential equation on a filtered probability space $(\Omega, \mathcal{F}, \{\mathcal{F}_t\}_{t\geq 0}, \mathbf{P})$:

$$\begin{cases} dx(t) = b(t, x(t), u(t))dt + \sigma(t, x(t), u(t))dW(t), \\ x(0) = x_0. \end{cases} \tag{3.20}$$

Here b is the mean depletion rate, and σ represents exogenous disturbances. The initial value x_0 is assumed to be deterministic and positive, which signifies that the underlying technology must have a definite and positive value measured according to the best information available at the initial time. By the nature of depletion, the drift b must be decreasing in time (i.e., $\frac{\partial b(t,x,u)}{\partial t} \leq 0$). On the other hand, $u(\cdot)$ represents the company's policies in influencing the depletion rate and disturbances in the process of technology diffusion, such as technology utilization degree, learning and training, budgeting, and resource allocation. A typical drift will be

$$b(t,x,u_1,u_2) = -\frac{u_1 x}{1+u_2 x}, \tag{3.21}$$

where u_1 is the natural depletion rate per unit of technology utilized (which reflects the degree of technology utilization) and u_2 is the compensation of depletion by the learning and training effect in terms of knowledge, skill, and new developments. As for the disturbance $\sigma(t,x,u)$, it is reasonable to assume that it diminishes as the economic value of the technology goes down. Namely, we require

$$\sigma(t,0,u) \approx 0, \qquad \frac{\partial \sigma(t,x,u)}{\partial x} > 0.$$

The exact form of σ depends on different situations. If the uncertainty goes down at a roughly constant rate related to the value of the technology, then a function such as

$$\sigma(t,x,u) = \sqrt{ax} \tag{3.22}$$

would be a suitable one. If the uncertainty does not vary much at the initial stage and falls significantly at some later stage, then it is more reasonable to use a function such as

$$\sigma(t,x,u) = a\sqrt{1-\eta e^{-\lambda u x}}, \tag{3.23}$$

where a, λ are positive constants and η is constant smaller than but close to one.

The objective of the model is to choose an appropriate nonanticipative policy $u(\cdot)$ so as to maximize the following expected total discounted profit in a time period $[0,T]$:

$$J(u(\cdot)) = E\left\{\int_0^T e^{-\gamma t} f(t,x(t),u(t))dt + e^{-\gamma T}h(x(T))\right\}, \tag{3.24}$$

where $\gamma > 0$ is the discount rate, f is the profit rate, and $e^{-\gamma T}h(x(T))$ is the *salvage value* of the technology at the end time.

3.5. Queueing systems in heavy traffic

A queueing system in *heavy traffic* (namely, the average customer arrival rate is close to the average server service rate) can be approximated by a (nonstandard) Brownian motion using so-called *diffusion approximation*. Therefore, a controlled queueing system can be approximately formulated as a stochastic control problem, and sometimes it is advantageous from both the analytical and computational points of view. Let us consider a particular single-server queueing system, a manufacturing system with a single *unreliable* machine (*server*) producing a single type of product to meet incoming orders (*customers*). Let $\{\alpha_i\}$ and $\{\beta_i\}$, $i = 1, 2, \ldots$, be two sequences of i.i.d.random variables representing successive up and down periods of the machine, and let $\{\xi_i\}$ and $\{\eta_i\}$, $i = 1, 2, \ldots$, be two sequences of i.i.d.random variables representing successive interarrival times of demand batches and their sizes, respectively. Suppose

$$
(3.25)\qquad \begin{cases} E\alpha_i = \dfrac{1}{a}, & E\beta_i = \dfrac{1}{b}, \quad E\xi_i = \dfrac{1}{c}, \quad E\eta_i = \dfrac{1}{d}, \\ \text{Var } \alpha_i = \sigma^2, & \text{Var } \beta_i = \lambda^2, \quad \text{Var } \xi_i = \mu^2, \quad \text{Var } \eta_i = \delta^2. \end{cases}
$$

Let $X(t)$ be the inventory of the product and $U(t)$ the production rate at time t. Suppose that $A(t)$ is the cumulative number of arrivals of the demand batches up to time t, and

$$
G(t) = \begin{cases} 1, & \text{if the machine is up,} \\ 0, & \text{if the machine is down.} \end{cases}
$$

Then the inventory process satisfies

$$
(3.26)\qquad X(t) = x_0 + \int_0^t G(s)U(s)ds - \sum_{i=1}^{A(t)} \eta_i.
$$

A production plan $U(\cdot)$ must satisfy the constraint $0 \le U(t) \le k$, where k is the maximum production capacity. The objective of this controlled queueing system problem is to select a nonanticipative production plan $U(\cdot)$ so as to minimize the expected total discounted costs of inventory and production:

$$
(3.27)\qquad J(U(\cdot)) = E\int_0^T e^{-\gamma t} h(X(t), U(t))dt,
$$

where $\gamma > 0$ is the discount rate. This actually is a very hard problem, and a closed-form solution seems impossible to obtain even for simple forms of the cost function h. Now let us approximate this problem by another one involving a diffusion process for which there are more convenient tools to apply. To this end, we first note that the expected long-run production capacity is $\frac{b}{a+b}k$, and the expected long-run demand rate is $\frac{c}{d}$. Therefore, the *traffic intensity* of the queueing system is

$$
(3.28)\qquad \rho = \left(\frac{c}{d}\right)^{-1} \left(\frac{bk}{a+b}\right).
$$

Usually, ρ is assumed to be greater than 1; otherwise, the backlog will grow to infinity in the long run. The queueing system is said to be in *heavy traffic* if ρ is close to 1, namely,

$$0 < \varepsilon \equiv \frac{bk}{a+b} - \frac{c}{d} \approx 0. \tag{3.29}$$

When the original system is in heavy traffic, its optimal control problem can be approximated by the following problem:

$$\begin{cases} \text{minimize } J(u(\cdot)) = E \displaystyle\int_0^T e^{-\gamma t} h(t, x(t), u(t)) dt, \\ \text{subject to } \begin{cases} dx(t) = (1 - u(t))dt + g dW(t), \\ x(0) = \varepsilon x_0, \end{cases} \end{cases} \tag{3.30}$$

where the diffusion coefficient is

$$g = \sqrt{k^2(a+b)^{-3}(a^3 b\sigma^2 + ab^3\lambda^2) + c^3 d^{-2}\mu^2 + c\delta^2}. \tag{3.31}$$

4. Formulations of Stochastic Optimal Control Problems

The examples presented in the previous section have some common features: There is a *diffusion system*, which is described by Itô's stochastic differential equation; there are many alternative *decisions* that can affect the dynamics of the system; there are some *constraints* that the decisions and/or the state are subject to; and there is a *criterion* that measures the performance of the decisions. The goal is to *optimize* (maximize or minimize) the criterion by selecting a *nonanticipative* decision among the ones satisfying all the constraints. Such problems are called *stochastic optimal control problems.*

Let us now present two mathematical formulations (*strong* and *weak* formulations) of stochastic optimal control problems in the following two subsections, respectively.

4.1. Strong formulation

Given a filtered probability space $(\Omega, \mathcal{F}, \{\mathcal{F}_t\}_{t\geq 0}, \mathbf{P})$ satisfying the usual condition (see Chapter 1, Definition 2.6) on which an m-dimensional standard Brownian motion $W(\cdot)$ is defined, consider the following *controlled* stochastic differential equation:

$$\begin{cases} dx(t) = b(t, x(t), u(t))dt + \sigma(t, x(t), u(t))dW(t), \\ x(0) = x_0 \in \mathbb{R}^n, \end{cases} \tag{4.1}$$

where $b : [0,T] \times \mathbb{R}^n \times U \to \mathbb{R}^n$, $\sigma : [0,T] \times \mathbb{R}^n \times U \to \mathbb{R}^{n\times m}$, with U being a given separable metric space, and $T \in (0, \infty)$ being fixed. The function $u(\cdot)$ is called the *control* representing the action, decision, or policy of the decision-makers (*controllers*). At any time instant the controller is knowledgeable about *some* information (as specified by the information field $\{\mathcal{F}_t\}_{t\geq 0}$) of what has happened up to that moment, but *not* able to foretell what is going to happen afterwards due to the uncertainty of the system (as a consequence, for any t the controller cannot exercise his/her decision $u(t)$ before the time t really comes). This nonanticipative restriction in mathematical terms can be represented as "$u(\cdot)$ is $\{\mathcal{F}_t\}_{t\geq 0}$-adapted" (recall Convention 2.9 in Chapter 1). Namely, the control $u(\cdot)$ is taken from the set

$$\mathcal{U}[0,T] \triangleq \big\{u : [0,T] \times \Omega \to U \;\big|\; u(\cdot) \text{ is } \{\mathcal{F}_t\}_{t\geq 0}\text{-adapted}\big\}.$$

Any $u(\cdot) \in \mathcal{U}[0,T]$ is called a *feasible control.* In addition, we may have some state constraints. For any $u(\cdot) \in \mathcal{U}[0,T]$, equation (4.1) is one with random coefficients (see Chapter 1, Section 6.4). Let $S(t) : [0,T] \to 2^{\mathbb{R}^n}$ be a given multifunction. The *state constraint* may be given by

$$x(t) \in S(t), \quad \forall t \in [0,T], \quad \mathbf{P}\text{-a.s.} \tag{4.2}$$

Note that some other types of state constraints are also possible.

Next, we introduce the *cost functional* as follows:

$$J(u(\cdot)) = E\Big\{\int_0^T f(t, x(t), u(t))dt + h(x(T))\Big\}. \tag{4.3}$$

Definition 4.1. Let $(\Omega, \mathcal{F}, \{\mathcal{F}_t\}_{t\geq 0}, \mathbf{P})$ be given satisfying the usual conditions (see Chapter 1, Definition 2.6) and let $W(t)$ be a given m-dimensional standard $\{\mathcal{F}_t\}_{t\geq 0}$-Brownian motion. A control $u(\cdot)$ is called an *s-admissible* control, and $(x(\cdot), u(\cdot))$ an *s-admissible pair*, if

(i) $u(\cdot) \in \mathcal{U}[0,T]$;

(ii) $x(\cdot)$ is the unique solution of equation (4.1) in the sense of Chapter 1, Definition 6.15;

(iii) some prescribed state constraints (for example, (4.2)) are satisfied;

(iv) $f(\cdot\,, x(\cdot), u(\cdot)) \in L^1_{\mathcal{F}}(0,T;\mathbb{R})$ and $h(x(T)) \in L^1_{\mathcal{F}_T}(\Omega;\mathbb{R})$.

The set of all s-admissible controls is denoted by $\mathcal{U}^s_{ad}[0,T]$. Our stochastic optimal control problem under strong formulation can be stated as follows:

Problem (SS). Minimize (4.3) over $\mathcal{U}^s_{ad}[0,T]$.

The goal is to find $\bar{u}(\cdot) \in \mathcal{U}^s_{ad}[0,T]$ (if it ever exists), such that

$$J(\bar{u}(\cdot)) = \inf_{u(\cdot)\in\mathcal{U}^s_{ad}[0,T]} J(u(\cdot)). \tag{4.4}$$

Problem (SS) is said to be *s-finite* if the right-hand side of (4.4) is finite, and it is said to be (uniquely) *s-solvable* if there exists a (unique) $\bar{u}(\cdot) \in \mathcal{U}^s_{ad}[0,T]$ such that (4.4) holds. Any $\bar{u}(\cdot) \in \mathcal{U}^s_{ad}[0,T]$ satisfying (4.4) is called an *s-optimal control.* The corresponding state process $\bar{x}(\cdot)$ and the state-control pair $(\bar{x}(\cdot), \bar{u}(\cdot))$ are called an *s-optimal state process* and an *s-optimal pair*, respectively.

4.2. Weak formulation

We note that in the strong formulation, the underlying filtered probability space $(\Omega, \mathcal{F}, \{\mathcal{F}_t\}_{t\geq 0}, \mathbf{P})$ along with the Brownian motion $W(\cdot)$ are all fixed. However, in certain situations it will be convenient or necessary to vary $(\Omega, \mathcal{F}, \{\mathcal{F}_t\}_{t\geq 0}, \mathbf{P})$ as well as $W(\cdot)$ and to consider them as *parts* of the control. This is the case, for example, when one applies the dynamic programming approach to solve a stochastic optimal control problem originally under the strong formulation (see Chapter 4 for details). Therefore, we need to have another formulation of the problem.

Definition 4.2. A 6-tuple $\pi = (\Omega, \mathcal{F}, \{\mathcal{F}_t\}_{t\geq 0}, \mathbf{P}, W(\cdot), u(\cdot))$ is called a *w-admissible* control, and $(x(\cdot), u(\cdot))$ a *w-admissible pair*, if

(i) $(\Omega, \mathcal{F}, \{\mathcal{F}_t\}_{t\geq 0}, \mathbf{P})$ is a filtered probability space satisfying the usual conditions;

(ii) $W(\cdot)$ is an m-dimensional standard Brownian motion defined on $(\Omega, \mathcal{F}, \{\mathcal{F}_t\}_{t\geq 0}, \mathbf{P})$;

(iii) $u(\cdot)$ is an $\{\mathcal{F}_t\}_{t\geq 0}$-adapted process on $(\Omega, \mathcal{F}, \mathbf{P})$ taking values in U;

(iv) $x(\cdot)$ is the unique solution (in the sense of Chapter 1, Definition 6.15) of equation (4.1) on $(\Omega, \mathcal{F}, \{\mathcal{F}_t\}_{t\geq 0}, \mathbf{P})$ under $u(\cdot)$;

(v) some prescribed state constraints (for example, (4.2)) are satisfied;

(vi) $f(\cdot\,, x(\cdot), u(\cdot)) \in L^1_{\mathcal{F}}(0,T;\mathbb{R})$ and $h(x(T)) \in L^1_{\mathcal{F}_T}(\Omega;\mathbb{R})$. Here, the spaces $L^1_{\mathcal{F}}(0,T;\mathbb{R})$ and $L^1_{\mathcal{F}_T}(\Omega;\mathbb{R})$ are defined on the given filtered probability space $(\Omega, \mathcal{F}, \{\mathcal{F}_t\}_{t\geq 0}, \mathbf{P})$ associated with the 6-tuple π.

The set of all w-admissible controls is denoted by $\mathcal{U}^w_{ad}[0,T]$. Sometimes, we might write $u(\cdot) \in \mathcal{U}^w_{ad}[0,T]$ instead of $(\Omega, \mathcal{F}, \{\mathcal{F}_t\}_{t\geq 0}, \mathbf{P}, W(\cdot), u(\cdot)) \in \mathcal{U}^w_{ad}[0,T]$ if it is clear from the context that weak formulation is under consideration. Our stochastic optimal control problem under the weak formulation can be stated as follows:

Problem (WS). Minimize (4.3) over $\mathcal{U}^w_{ad}[0,T]$.

Namely, one seeks $\bar{\pi} \in \mathcal{U}^w_{ad}[0,T]$ (if it exists) such that

$$J(\bar{\pi}) = \inf_{\pi \in \mathcal{U}^w_{ad}[0,T]} J(\pi). \tag{4.5}$$

As in the strong formulation, Problem (WS) is said to be *w-finite* if the right-hand side of (4.5) is finite. One may also define the *w-solvability*, the *w-optimal control*, the *w-optimal state process*, and the *w-optimal pair* accordingly.

We emphasize here that the strong formulation is the one that stems from the practical world, whereas the weak formulation sometimes serves as an auxiliary but effective mathematical model aiming at ultimately solving problems with the strong formulation. A main reason why this might work is that the objective of a stochastic control problem is to minimize the *mathematical expectation* of a certain random variable that depends only on the *distribution* of the processes involved. Therefore, if the solutions of (4.1) in different probability spaces have the same probability distribution, then one has more freedom in choosing a convenient probability space to work with. A good example is when one tries to employ the dynamic programming principle to solve the problem (see Chapter 4). It should also be noted, however, that the weak formulation fails if any of the given coefficients b, σ, f, and h are also random (namely, they depend on ω *explicitly*), because in this case the probability space has to be specified and fixed a priori.

As in the deterministic case, the following type of the state equation is of particular interest:

$$\begin{cases} dx(t) = \big[A(t)x(t) + B(t)u(t)\big]dt \\ \qquad\qquad + \displaystyle\sum_{j=1}^{m} \big[C_j(t)x(t) + D_j(t)u(t)\big]dW^j(t), \\ x(0) = x_0, \end{cases} \tag{4.6}$$

where $A(\cdot)$, $B(\cdot)$, $C_j(\cdot)$, and $D_j(\cdot)$ are matrix-valued functions of suitable sizes. The associated cost functional is of a quadratic form:

$$\begin{aligned} J(u(\cdot)) = E\Big\{\frac{1}{2}\int_0^T \big[\langle Q(t)x(t), x(t)\rangle + \langle R(t)u(t), u(t)\rangle\big]dt \\ + \frac{1}{2}\langle Gx(T), x(T)\rangle\Big\}. \end{aligned} \tag{4.7}$$

Problem (SS) or (WS) with state equation (4.6) and cost functional (4.7) is called a *stochastic linear quadratic optimal control problem* (stochastic LQ problem, for short). We will study such problems in great detail in Chapter 6.

We note that in the above formulation, $T > 0$ is a fixed constant. Thus, Problem (SS) or (WS) is also referred to as a *fixed duration problem*. Sometimes, we may face problems with nonfixed durations. We will discuss those problems as well as some others in Section 7.

Finally, throughout this book when the type of formulation (strong or weak) being considered is clear from the context, we shall simply refer to a control as an "admissible control" and the set of admissible controls as $\mathcal{U}_{ad}[0,T]$, etc., omitting the superscript/prefix "s" or "w".

5. Existence of Optimal Controls

In this section we are going to discuss the existence of optimal controls. The essence of the theory is the following: A lower semicontinuous function defined on some compact metric space attains it minimum.

5.1. A deterministic result

Let us first consider Problem (D) formulated in Section 2. We make the following assumptions.

(DE1) (U, d) is a Polish space and $T > 0$.

(DE2) The maps $b : [0,T] \times \mathbb{R}^n \times U \to \mathbb{R}^n$, $f : [0,T] \times \mathbb{R}^n \times U \to \mathbb{R}$, and $h : \mathbb{R}^n \to \mathbb{R}$ are measurable and there exist a constant $L > 0$ and a modulus of continuity $\bar{\omega} : [0,\infty) \to [0,\infty)$ such that for $\varphi(t,x,u) = b(t,x,u), f(t,x,u), h(x)$,

$$(5.1)\qquad \begin{cases} |\varphi(t,x,u) - \varphi(t,\widehat{x},\widehat{u})| \leq L|x - \widehat{x}| + \bar{\omega}(d(u,\widehat{u})), \\ \qquad\qquad\qquad \forall t \in [0,T], x, \widehat{x} \in \mathbb{R}^n,\ u, \widehat{u} \in U, \\ |\varphi(t,0,u)| \leq L, \qquad \forall (t,u) \in [0,T] \times U. \end{cases}$$

(DE3) For every $(t,x) \in [0,T] \times \mathbb{R}^n$, the set

$$(b,f)(t,x,U) \stackrel{\Delta}{=} \{(b_i(t,x,u), f(t,x,u)) \mid u \in U,\ i = 1,2,\ldots,n\}$$

is convex and closed in $\mathbb{R}^{n+1}$.

(DE4) $S(t) \equiv \mathbb{R}^n$ (i.e., there is no state constraint).

Theorem 5.1. *Under (DE1)–(DE4), if Problem (D) is finite, then it admits an optimal control.*

Proof. Let $(x_j(\cdot), u_j(\cdot))$ be a *minimizing sequence* (i.e., $J(u_j(\cdot)) \to \inf_{\mathcal{V}[0,T]} J(u(\cdot))$ as $j \to \infty$). By (DE2), we see immediately that the sequences $x_j(\cdot)$, $b(\cdot\,, x_j(\cdot), u_j(\cdot))$, and $f(\cdot\,, x_j(\cdot), u_j(\cdot))$ are uniformly bounded in j. Hence, $x_j(\cdot)$ is equicontinuous (by the state equation). Then we may assume that (one might extract subsequences, if necessary) as $j \to \infty$,

$$(5.2)\qquad \begin{cases} x_j(\cdot) \to \bar{x}(\cdot) \quad \text{in } C([0,T];\mathbb{R}^n), \\ b(\cdot\,, x_j(\cdot), u_j(\cdot)) \to \bar{b}(\cdot) \quad \text{weakly in } L^2(0,T;\mathbb{R}^n), \\ f(\cdot\,, x_j(\cdot), u_j(\cdot)) \to \bar{f}(\cdot) \quad \text{weakly in } L^2(0,T;\mathbb{R}), \\ h(x_j(T)) \to h(\bar{x}(T)) \quad \text{in } \mathbb{R}. \end{cases}$$

By the first convergence in (5.2) and Mazur's theorem (see Yosida [1, p. 120, Theorem 2]), we have some $\alpha_{ij} \geq 0$ with $\sum_{i\geq 1} \alpha_{ij} = 1$, such that as $j \to \infty$,

$$(5.3)\qquad \begin{cases} \displaystyle\sum_{i\geq 1} \alpha_{ij} b(\cdot\,, \bar{x}(\cdot), u_{i+j}(\cdot)) \to \bar{b}(\cdot) \quad \text{strongly in } L^2(0,T;\mathbb{R}^n), \\ \displaystyle\sum_{i\geq 1} \alpha_{ij} f(\cdot\,, \bar{x}(\cdot), u_{i+j}(\cdot)) \to \bar{f}(\cdot) \quad \text{strongly in } L^2(0,T;\mathbb{R}). \end{cases}$$

Hence, by assumption (DE3),

$$(5.4)\qquad (\bar{b}(t), \bar{f}(t)) \in (b,f)(t, \bar{x}(t), U).$$

Consequently, by Filippov's lemma (see Li–Yong [1, p. 102, Corollary 2.26]), there exists a $\bar{u}(\cdot) \in \mathcal{V}[0,T]$ such that

$$(5.5)\qquad \bar{b}(t) = b(t, \bar{x}(t), \bar{u}(t)), \quad \bar{f}(t) = f(t, \bar{x}(t), \bar{u}(t)), \ \text{a.e. } t \in [0,T].$$

Thus, $(\overline{x}(\cdot), \overline{u}(\cdot))$ is an admissible pair of Problem (D). Further, by Fatou's lemma (Yosida [1, p. 17]), we obtain

$$
\begin{aligned}
J(\overline{u}(\cdot)) &= \int_0^T f(t, \overline{x}(t), \overline{u}(t))dt + h(\overline{x}(T)) \\
&\leq \varliminf_{j\to\infty} \sum_{i\geq 1} \alpha_{ij} \Big\{ \int_0^T f(t, x_{i+j}(t), u_{i+j}(t))dt + h(x_{i+j}(T)) \Big\} \qquad (5.6) \\
&= \varliminf_{j\to\infty} \sum_{i\geq 1} \alpha_{ij} J(u_{i+j}(\cdot)) = \inf_{u(\cdot)\in\mathcal{V}[0,T]} J(u(\cdot)).
\end{aligned}
$$

Thus, $\overline{u}(\cdot) \in \mathcal{V}[0,T]$ is an optimal control. □

There are two essential issues in the above proof: The space $\mathbb{R}^n$ where the state $x(t)$ takes values is *locally compact* (i.e., any bounded closed set is compact), which ensures that the sequence $\{x_j(\cdot)\}$ is relatively compact (i.e., it contains a convergence subsequence) under condition (DE2). The set $(b,f)(t,x,U)$ is convex and closed (called *Roxin's condition*), which guarantees that Mazur's theorem and Filippov's measurable selection theorem are applicable to obtain an optimal control.

We also point out that condition (DE4) can be relaxed. A certain kind of continuity for $S(t)$ is enough. We leave the details to the interested reader.

5.2. Existence under strong formulation

We now consider the stochastic case. Let us first look at the existence under strong formulation.

Let $(\Omega, \mathcal{F}, \{\mathcal{F}_t\}_{t\geq 0}, \mathbf{P})$ be given as before and let $W(\cdot)$ be a given one-dimensional Brownian motion. Consider the following stochastic linear controlled system:

$$
(5.7) \qquad \begin{cases} dx(t) = \big[Ax(t) + Bu(t)\big]dt + \big[Cx(t) + Du(t)\big]dW(t), & t \in [0,T], \\ x(0) = x_0, \end{cases}
$$

where A, B, C, D are matrices of suitable sizes. The state $x(\cdot)$ takes values in $\mathbb{R}^n$, and the control $u(\cdot)$ is in

$$
(5.8) \qquad \mathcal{U}^L[0,T] \stackrel{\Delta}{=} \{u(\cdot) \in L^2_{\mathcal{F}}(0,T;\mathbb{R}^k) \mid u(t) \in U, \ \text{a.e.}\ t \in [0,T], \ \mathbf{P}\text{-a.s.}\},
$$

with $U \subseteq \mathbb{R}^k$. Note that we have an additional constraint that a control must be square-integrable just to ensure the existence of solutions of (5.7) under $u(\cdot)$. If U is bounded, then this restriction is satisfied automatically. The cost functional is

$$
(5.9) \qquad J(u(\cdot)) = E\Big\{ \int_0^T f(x(t), u(t))dt + h(x(T)) \Big\},
$$

with $f : \mathbb{R}^n \times U \to \mathbb{R}$ and $h : \mathbb{R}^n \to \mathbb{R}$. The optimal control problem can be stated as follows.

Problem (SL). Minimize (5.6) subject to (5.7) over $\mathcal{U}^L[0,T]$.

Now we introduce the following assumptions:

(H1) The set $U \subseteq \mathbb{R}^k$ is convex and closed, and the functions f and h are convex and for some $\delta, K > 0$,

$$f(x,u) \geq \delta|u|^2 - K, \quad h(x) \geq -K, \quad \forall (x,u) \in \mathbb{R}^n \times U. \tag{5.10}$$

(H2) The set $U \subseteq \mathbb{R}^k$ is convex and compact, and the functions f and h are convex.

When either (H1) or (H2) is assumed, Problem (SL) is called a *stochastic linear-convex optimal control problem*, since the controlled system is linear and the cost functional is convex. A special case is that in which both f and h are convex quadratic functions. Problem (SL) is then reduced to a stochastic linear quadratic problem.

Theorem 5.2. *Under either (H1) or (H2), if Problem (SL) is finite, then it admits an optimal control.*

Proof. First suppose (H1) holds. Let $(x_j(\cdot), u_j(\cdot))$ be a minimizing sequence. By (5.10), we have

$$E\int_0^T |u_j(t)|^2 dt \leq K, \qquad \forall j \geq 1, \tag{5.11}$$

for some constant $K > 0$. Thus, there is a subsequence, which is still labeled by $u_j(\cdot)$, such that

$$u_j(\cdot) \to \bar{u}(\cdot), \quad \text{weakly in } L^2_{\mathcal{F}}(0,T;\mathbb{R}^k). \tag{5.12}$$

By Mazur's theorem, we have a sequence of convex combinations

$$\widetilde{u}_j(\cdot) \stackrel{\Delta}{=} \sum_{i\geq 1} \alpha_{ij} u_{i+j}(\cdot), \qquad \text{with } \alpha_{ij} \geq 0, \ \sum_{i\geq 1} \alpha_{ij} = 1,$$

such that

$$\widetilde{u}_j(\cdot) \to \bar{u}(\cdot), \quad \text{strongly in } L^2_{\mathcal{F}}(0,T;\mathbb{R}^k). \tag{5.13}$$

Since the set $U \subseteq \mathbb{R}^k$ is convex and closed, it follows that $\bar{u}(\cdot) \in \mathcal{U}^L[0,T]$. On the other hand, if $\widetilde{x}_j(\cdot)$ is the state under the control $\widetilde{u}_j(\cdot)$, then we have the convergence

$$\widetilde{x}_j(\cdot) \to \bar{x}(\cdot), \quad \text{strongly in } C_{\mathcal{F}}([0,T],\mathbb{R}^n). \tag{5.14}$$

Clearly, $(\bar{x}(\cdot), \bar{u}(\cdot))$ is admissible, and the convexity of f and h implies

$$\begin{aligned} J(\bar{u}(\cdot)) &= \lim_{j\to\infty} J(\widetilde{u}_j(\cdot)) \leq \lim_{j\to\infty} \sum_{i\geq 1} \alpha_{ij} J(u_{i+j}(\cdot)) \\ &= \inf_{u(\cdot)\in\mathcal{U}^L[0,T]} J(u(\cdot)). \end{aligned} \tag{5.15}$$

Hence, $(\bar{x}(\cdot), \bar{u}(\cdot))$ is optimal.

In the case where (H2) holds, we automatically have (5.11). The above proof then applies. □

We have seen that the above proof seems very similar to that of Theorem 5.1. However, there is a crucial difference. In Theorem 5.1, we do not have (weak) convergence of $u_j(\cdot)$, and we obtained the convergence of $x_j(\cdot)$ directly by the equicontinuity of the sequence (due to the local compactness of $\mathbb{R}^n$). On the other hand, in Theorem 5.2, the uniform convergence of $x_j(\cdot)$ in (5.14) was shown *via* the convergence of $u_j(\cdot)$. The linearity plays an essential role here. For the general Problem (SS), we do not have the convergence of $u_j(\cdot)$, and the convergence of $x_j(\cdot)$ is not necessarily valid in the given probability space due to the lack of local compactness of infinite-dimensional spaces such as $L^2_{\mathcal{F}}(\Omega; \mathbb{R}^n)$. Thus, the same arguments of the deterministic case do not apply for the strong formulation of general nonlinear stochastic optimal control problems.

5.3. Existence under weak formulation

In this subsection we examine the existence of optimal controls under the weak formulation.

Let us make the following assumptions.

(SE1) (U, d) is a compact metric space and $T > 0$.

(SE2) The maps b, σ, f, and h are all continuous, and there exists a constant $L > 0$ such that for $\varphi(t,x,u) = b(t,x,u), \sigma(t,x,u), f(t,x,u), h(x)$,

$$\begin{cases} |\varphi(t,x,u) - \varphi(t,\widehat{x},u)| \leq L|x - \widehat{x}|, \\ \qquad\qquad\qquad\qquad \forall t \in [0,T], \quad x, \widehat{x} \in \mathbb{R}^n, \ u \in U, \\ |\varphi(t,0,u)| \leq L, \qquad \forall (t,u) \in [0,T] \times U. \end{cases} \tag{5.16}$$

(SE3) For every $(t,x) \in [0,T] \times \mathbb{R}^n$, the set

$$\begin{aligned}(b, \sigma\sigma^\top, f)(t,x,U) \stackrel{\Delta}{=} \{(b_i(t,x,u), (\sigma\sigma^\top)^{ij}(t,x,u), f(t,x,u)) \mid \\ u \in U, \ i = 1, \ldots, n, \ j = 1, \ldots, m\}\end{aligned}$$

is convex in $\mathbb{R}^{n+nm+1}$.

(SE4) $S(t) \equiv \mathbb{R}^n$.

In proving the existence of optimal controls one typically seeks a certain compactness structure. The weak formulation enables us to find the compactness of the *image measure* of some stochastic processes involved on a certain functional space. However, since the control $u(\cdot)$ is measurable only in t and there is no convenient compactness property on the space of merely measurable functions, we need to *embed* it in a larger space with proper compactness.

To this end, let us denote by Λ the set of all (nonnegative) measures λ on $[0,T] \times U$ such that

$$\lambda([0,s] \times U) = s, \qquad \forall s \in [0,T]. \tag{5.17}$$

Since U is compact, Λ is tight when endowed with the weak convergence topology, which is metrizable; see Chapter 1, Proposition 1.14 and Corollary 1.17 (note that the discussions in Chapter 1, Section 1.4, for probability measures can easily be adapted to the present case). On the other hand, by (5.17), λ can be represented as $\lambda(dt,du) = \lambda'(t,du)dt$, where $\lambda'(t,\cdot)$ is a probability measure on U for almost all t and is determined uniquely except on a t-null set. Under this framework, any U-valued measurable (deterministic) function $u(\cdot)$ may be *embedded* into the space Λ in the sense that $u(\cdot)$ corresponds to the Dirac measure $\lambda_{u(\cdot)}(dt,du) = \lambda'_{u(\cdot)}(t,du)dt$ with the following property: For any bounded and uniformly continuous function $\rho(t,x,u)$ defined on $[0,T] \times \mathbb{R}^n \times U$,

$$\rho(t,x,u(t)) = \int_U \rho(t,x,u)\lambda'_{u(\cdot)}(t,du) \stackrel{\Delta}{=} \tilde{\rho}(t,x,\lambda_{u(\cdot)}). \tag{5.18}$$

Now we introduce a suitable filtration on Λ. First of all, each $\lambda \in \Lambda$ can be identified as a linear functional on $C([0,T] \times U)$ in the following way:

$$\lambda(f) \stackrel{\Delta}{=} \int_0^T \int_U f(t,u)\lambda(dt,du), \quad \forall f \in C([0,T] \times U).$$

For any $f \in C([0,T] \times U)$ and $t \in [0,T]$, define $f^t \in C([0,T] \times U)$ by

$$f^t(s,u) \stackrel{\Delta}{=} f(s \wedge t, u).$$

Since $C([0,T] \times U)$ is separable, we may let $\{f_j\}_{j\geq 1}$ be a countable dense subset (with respect to the uniform norm). Clearly, for any $t \in [0,T]$, $\{f_j^t\}_{j\geq 1}$ is dense in the set $\{f^t \mid f \in C([0,T] \times U)\}$. Define

$$\mathcal{B}_t(\Lambda) \stackrel{\Delta}{=} \sigma\Big\{\{\lambda \in \Lambda | \lambda(f^s) \in B\} : s \in [0,t],\ B \in \mathcal{B}(\mathbb{R})\Big\}. \tag{5.19}$$

Then, as in Chapter 1, Lemma 2.17, we can show that $\mathcal{B}_t(\Lambda)$ can be generated by cylinders of the following form:

$$\begin{aligned}\mathcal{B}_t(\Lambda) = \sigma\Big\{&\{\lambda \in \Lambda | \lambda(f_j^s) \in (a,b)\} : t \geq s \in \mathbf{Q},\\ &j = 1,2,\ldots,\ a,b \in \mathbf{Q}\Big\}.\end{aligned} \tag{5.20}$$

Next, μ is called an $\{\mathcal{F}_t\}_{t\geq 0}$-adapted Λ-valued random variable (Λ-r.v. for short) on a given filtered probability space $(\Omega, \mathcal{F}, \{\mathcal{F}_t\}_{t\geq 0}, \mathbf{P})$ if $\mu(B_1 \times B_2)$ is $\{\mathcal{F}_t\}_{t\geq 0}$-measurable for any $B_1 \in \mathcal{B}([0,t])$ and $B_2 \in \mathcal{B}(U)$. It is clear that if $u(t,\omega)$ is an $\{\mathcal{F}_t\}_{t\geq 0}$-adapted process, then its embedding $\lambda_{u(\cdot)}$ is an $\{\mathcal{F}_t\}_{t\geq 0}$-adapted Λ-r.v. and vice versa.

We now state and prove the existence of optimal controls in the weak formulation.

Theorem 5.3. *Under (SE1)–(SE4), if Problem (WS) is finite, then it admits an optimal control.*

Proof. Let $\pi_k \equiv (\Omega_k, \mathcal{F}_k, \{\mathcal{F}_{kt}\}_{t\ge 0}, \mathbf{P}_k, W_k(\cdot), u_k(\cdot)) \in \mathcal{U}_{ad}^w[0,T]$ be a minimizing sequence, namely,

$$\lim_{k\to\infty} J(\pi_k) = \inf_{\pi\in\mathcal{U}_{ad}^w[0,T]} J(\pi). \tag{5.21}$$

Let $x_k(\cdot)$ be the state trajectory corresponding to π_k. Define

$$X_k(\cdot) \stackrel{\Delta}{=} (x_k(\cdot), B_k(\cdot), \Sigma_k(\cdot), F_k(\cdot), W_k(\cdot)), \tag{5.22}$$

where

$$\begin{cases} B_k(t) \stackrel{\Delta}{=} \displaystyle\int_0^t b(s, x_k(s), u_k(s))ds, \\ \Sigma_k(t) \stackrel{\Delta}{=} \displaystyle\int_0^t \sigma(s, x_k(s), u_k(s))dW_k(s), \\ F_k(t) \stackrel{\Delta}{=} \displaystyle\int_0^t f(s, x_k(s), u_k(s))ds. \end{cases} \tag{5.23}$$

By (SE2), it is routine to show that there is a constant $K > 0$ such that

$$E_k|X_k(t) - X_k(s)|^4 \le K|t-s|^2, \qquad \forall t, s \in [0,T], \ \forall k,$$

where E_k is the expectation under $\mathbf{P}_k$. Noting the compactness of Λ, $\{(X_k(\cdot), \lambda_{u_k(\cdot)})\}$ is tight as a sequence of $C([0,T]; \mathbb{R}^{3n+m+1}) \times \Lambda$-random variables (cf. Chapter 1, Theorem 2.14 and Corollary 1.17). By Skorohod's theorem (Chapter 1, Theorem 1.20), one can choose a subsequence (still labeled as $\{k\}$) and have

$$\begin{cases} \{(\overline{X}_k(\cdot), \overline{\lambda}_k)\} \equiv \{(\overline{x}_k(\cdot), \overline{B}_k(\cdot), \overline{\Sigma}_k(\cdot), \overline{F}_k(\cdot), \overline{W}_k(\cdot), \overline{\lambda}_k)\}, \\ (\overline{X}(\cdot), \overline{\lambda}) \equiv (\overline{x}(\cdot), \overline{B}(\cdot), \overline{\Sigma}(\cdot), \overline{F}(\cdot), \overline{W}(\cdot), \overline{\lambda}), \end{cases}$$

on a suitable probability space $(\overline{\Omega}, \overline{\mathcal{F}}, \overline{\mathbf{P}})$ such that

$$\text{law of } (\overline{X}_k(\cdot), \overline{\lambda}_k) = \text{law of } (X_k(\cdot), \lambda_{u_k(\cdot)}), \quad \forall k \ge 1, \tag{5.24}$$

and $\overline{\mathbf{P}}$-a.s.,

$$\overline{X}_k(t) \to \overline{X}(t) \qquad \text{uniformly on } t \in [0,T], \tag{5.25}$$

and

$$\overline{\lambda}_k \to \overline{\lambda} \quad \text{weakly on } \Lambda. \tag{5.26}$$

Set

$$
(5.27)\qquad \begin{cases} \overline{\mathcal{F}}_{kt} \stackrel{\Delta}{=} \Big(\sigma\{\overline{W}_k(s), \overline{x}_k(s) : s \le t\} \vee \overline{\lambda}_k^{-1}(\mathcal{B}_t(\Lambda))\Big), \\ \overline{\mathcal{F}}_t \stackrel{\Delta}{=} \Big(\sigma\{\overline{W}(s), \overline{x}(s) : s \le t\} \vee \overline{\lambda}^{-1}(\mathcal{B}_t(\Lambda))\Big). \end{cases}
$$

By (5.20) and Chapter 1, Lemma 2.17, $\overline{\mathcal{F}}_{kt}$ is the σ-field generated by $\overline{W}_k(t_1), \ldots, \overline{W}_k(t_\ell), \overline{x}_k(t_1), \ldots, \overline{x}_k(t_\ell), \overline{\lambda}_k(f_j^{t_1}), \ldots, \overline{\lambda}_k(f_j^{t_\ell})$, $0 \le t_1 \le t_2 \le \cdots \le t_\ell \le t$ and $j, \ell = 1, 2, \ldots$. A similar statement can be made for $\overline{\mathcal{F}}_t$. Now we need to show that $\overline{W}_k(\cdot)$ is an $\{\overline{\mathcal{F}}_{kt}\}_{t \ge 0}$-Brownian motion. To this end, note that $W_k(\cdot)$ is a $\sigma\{W_k(s), x_k(s); s \le t\} \vee \lambda_{u_k(\cdot)}^{-1}(\mathcal{B}_t(\Lambda))$-Brownian motion in view of Chapter 1, Proposition 4.2. Thus, by Chapter 1, Proposition 1.12, for any $0 \le s \le t \le T$ and any bounded continuous function g on $\mathbb{R}^{(m+n+\beta)\ell}$, we have

$$
E_k\{g(Y_k)(W_k(t) - W_k(s))\} = 0,
$$

where

$$
(5.28)\qquad \begin{aligned} Y_k &\stackrel{\Delta}{=} \{W_k(t_i), x_k(t_i), \lambda_k(f_{j_\alpha}^{t_i})\}, \\ &\qquad 0 \le t_1 \le t_2 \le \cdots \le t_\ell \le s, \quad \alpha = 1, 2, \ldots, \beta. \end{aligned}
$$

In view of (5.24),

$$
\overline{E}_k\{g(\overline{Y}_k)(\overline{W}_k(t) - \overline{W}_k(s))\} = 0,
$$

where

$$
(5.29)\qquad \begin{aligned} \overline{Y}_k &\stackrel{\Delta}{=} \{\overline{W}_k(t_i), \overline{x}_k(t_i), \overline{\lambda}_k(f_{j_\alpha}^{t_i})\}, \\ &\qquad 0 \le t_1 \le t_2 \le \cdots \le t_\ell \le s, \quad \alpha = 1, 2, \ldots, \beta. \end{aligned}
$$

By Chapter 1, Proposition 1.12, again, we get the first equality of (2.24) in Chapter 1. Similarly, one can prove the second equality of (2.24) in Chapter 1. This shows that $\overline{W}_k(\cdot)$ is an $\{\overline{\mathcal{F}}_{kt}\}_{t \ge 0}$-Brownian motion.

By (5.24), we have the following SDE on $(\overline{\Omega}, \overline{\mathcal{F}}, \{\overline{\mathcal{F}}_{kt}\}_{t \ge 0}, \overline{\mathbf{P}})$:

$$
(5.30)\qquad \begin{aligned} \overline{x}_k(t) &= x_0 + \overline{B}_k(t) + \overline{\Sigma}_k(t) \\ &\equiv x_0 + \int_0^t \int_U b(s, \overline{x}_k(s), u) \overline{\lambda}_k'(s, du) ds \\ &\quad + \int_0^t \int_U \sigma(s, \overline{x}_k(s), u) \overline{\lambda}_k'(s, du) d\overline{W}_k(s) \\ &\stackrel{\Delta}{=} x_0 + \int_0^t \tilde{b}(s, \overline{x}_k(s), \overline{\lambda}_k) ds + \int_0^t \tilde{\sigma}(s, \overline{x}_k(s), \overline{\lambda}_k) d\overline{W}_k(s). \end{aligned}
$$

Note that all the integrals in (5.30) are well-defined due to the fact that

$\overline{W}_k(\cdot)$ is an $\{\overline{\mathcal{F}}_{kt}\}_{t\geq 0}$-Brownian motion. Moreover,

$$
\begin{aligned}
\overline{E}\,\overline{F}_k(T) &\equiv \overline{E}\left\{\int_0^T \int_U f(s,\overline{x}_k(s),u)\overline{\lambda}_k'(s,du)ds + h(\overline{x}_k(T))\right\} \\
&\stackrel{\Delta}{=} \overline{E}\left\{\int_0^T \tilde{f}(s,\overline{x}_k(s),\overline{\lambda}_k)ds + h(\overline{x}_k(T))\right\} \qquad (5.31)\\
&= J(\pi_k) \to \inf_{\pi\in\mathcal{U}_{ad}^w[0,T]} J(\pi), \text{ as } k\to\infty,
\end{aligned}
$$

where $\overline{E}$ is the expectation under $\overline{\mathbf{P}}$. Letting $k\to\infty$ in (5.30) and (5.31), and noting (5.25), we get

$$
(5.32)\qquad \begin{cases} \overline{x}(t) = x_0 + \overline{B}(t) + \overline{\Sigma}(t), \quad \forall t\in[0,T], \ \overline{\mathbf{P}}\text{-a.s.}, \\ \overline{E}\,\overline{F}(T) = \inf\limits_{\pi\in\mathcal{U}_{ad}^w[0,T]} J(\pi). \end{cases}
$$

Next let us consider the sequence

$$
a_k(s) \stackrel{\Delta}{=} \tilde{\sigma}\tilde{\sigma}^\top(s,\overline{x}_k(s),\overline{\lambda}_k), \quad s\in[0,T].
$$

By (SE2), $\sup_k \overline{E}\int_0^T |a_k(s)|^2 ds < +\infty$, and hence $\{a_k\}$ is weakly relatively compact in the space $L^2([0,T]\times\overline{\Omega};\mathcal{S}^n)$. We can then find a subsequence (still labeled by $\{k\}$) and a function $a\in L^2([0,T]\times\overline{\Omega};\mathcal{S}^n)$ such that

$$
(5.33)\qquad a_k\to a, \qquad \text{weakly on } L^2([0,T]\times\overline{\Omega};\mathcal{S}^n).
$$

Denoting by a^{ij} the ijth element of a matrix a, we claim that for almost all (s,ω),

$$
(5.34)\qquad \varliminf_{k\to\infty} a_k^{ij}(s,\omega) \leq a^{ij}(s,\omega) \leq \varlimsup_{k\to\infty} a_k^{ij}(s,\omega), \quad i,j=1,\ldots,n.
$$

Indeed, if (5.34) is not true and on a set $A\subseteq[0,T]\times\overline{\Omega}$ of positive measure,

$$
\varliminf_{k\to\infty} a_k^{ij}(s,\omega) > a^{ij}(s,\omega),
$$

then we have by Fatou's lemma that

$$
\varliminf_{k\to\infty} \int_A a_k^{ij}(s,\omega)dsd\overline{\mathbf{P}}(\omega) > \int_A a^{ij}(s,\omega)dsd\overline{\mathbf{P}}(\omega),
$$

which is a contradiction to (5.33). The same can be said for the $\varlimsup$, which proves (5.34). Moreover, by (SE2) and (5.25), for almost all (s,ω),

$$
(5.35)\qquad \begin{cases} \varliminf\limits_{k\to\infty} a_k^{ij}(s,\omega) = \varliminf\limits_{k\to\infty} \tilde{\sigma}\tilde{\sigma}^\top(s,\overline{x}(s),\overline{\lambda}_k), \\ \varlimsup\limits_{k\to\infty} a_k^{ij}(s,\omega) = \varlimsup\limits_{k\to\infty} \tilde{\sigma}\tilde{\sigma}^\top(s,\overline{x}(s),\overline{\lambda}_k). \end{cases}
$$

Then combining (5.34), (5.35), and (SE3) gives

$$
(5.36)\quad a^{ij}(s,\omega)\in(\sigma\sigma^\top)^{ij}(s,\overline{x}(s,\omega),U), \quad \text{a.e. } (s,\omega), \quad i,j=1,2,\ldots,n.
$$

Modify a^{ij} on a null set, if necessary, so that (5.36) holds for all $(s,\omega) \in [0,T] \times \overline{\Omega}$. Similarly, setting

$$\begin{cases} b_k^i(s) \triangleq \tilde{b}^i(s, \overline{x}_k(s), \overline{\lambda}_k), \ i = 1, \ldots, n, \\ f_k(s) \triangleq \tilde{f}(s, \overline{x}_k(s), \overline{\lambda}_k), \end{cases}$$

one can prove that there are $b^i, f \in L^2([0,T] \times \overline{\Omega}; \mathbb{R})$ such that

$$b_k^i \to b^i \ (i = 1, \ldots, n), \quad f_k \to f, \quad \text{weakly on } L^2([0,T] \times \overline{\Omega}; \mathbb{R}), \tag{5.37}$$

and

$$\begin{aligned} &b^i(s,\omega) \in b^i(s, \overline{x}(s,\omega), U), \quad f(s,\omega) \in f(s, \overline{x}(s,\omega), U), \\ &\qquad \forall (s,\omega) \in [0,T] \times \overline{\Omega}, \quad i = 1, 2, \ldots, n. \end{aligned} \tag{5.38}$$

By (5.36), (5.38), (SE3), and a measurable selection theorem (see Li–Yong [1, p. 102, Corollary 2.26]), there is a U-valued, $\{\overline{\mathcal{F}}_t\}_{t \geq 0}$-adapted process $\overline{u}(\cdot)$ such that

$$\begin{aligned} &(b, \sigma\sigma^\top, f)(s,\omega) = (b, \sigma\sigma^\top, f)(s, \overline{x}(s,\omega), \overline{u}(s,\omega)), \\ &\qquad \forall (s,\omega) \in [0,T] \times \overline{\Omega}. \end{aligned} \tag{5.39}$$

Next we claim that $\overline{\Sigma}(t)$ is an $\{\overline{\mathcal{F}}_t\}_{t \geq 0}$-martingale. To see this, once again let $0 \leq s \leq t \leq T$, and define $\overline{Y}_k$ as (5.29) and

$$\begin{aligned} \overline{Y} \triangleq &\{\overline{W}(t_i), \overline{x}(t_i), \overline{\lambda}(f_{j_\alpha}^{t_i})\}, \\ &0 \leq t_1 \leq t_2 \leq \cdots \leq t_\ell \leq s, \quad \alpha = 1, 2, \ldots, \beta. \end{aligned}$$

Since $\overline{\Sigma}_k(t)$ is an $\overline{\mathcal{F}}_{kt}$-martingale, for any bounded continuous function g on $\mathbb{R}^{(m+n+\beta)\ell}$, we have

$$0 = \overline{E}\big[g(\overline{Y}_k)(\overline{\Sigma}_k(t) - \overline{\Sigma}_k(s))\big] \to \overline{E}\big[g(\overline{Y})(\overline{\Sigma}(t) - \overline{\Sigma}(s))\big], \tag{5.40}$$

by (5.25), (5.26), and the dominated convergence theorem. This proves that $\overline{\Sigma}(t)$ is an $\{\overline{\mathcal{F}}_t\}_{t \geq 0}$-martingale in view of Chapter 1, Proposition 1.12. Furthermore, from (5.30), it follows that

$$\langle \overline{\Sigma}_k \rangle(t) = \int_0^t \tilde{\sigma}\tilde{\sigma}^\top(s, \overline{x}_k(s), \overline{\lambda}_k) ds \equiv \int_0^t a_k(s) ds,$$

where $\langle \overline{\Sigma}_k \rangle$ is the quadratic variation of $\overline{\Sigma}_k$ (see Chapter 1, Definition 5.8). Hence $\overline{\Sigma}_k \overline{\Sigma}_k^T(t) - \int_0^t a_k(s) ds$ is an $\{\overline{\mathcal{F}}_{kt}\}_{t \geq 0}$-martingale. Recalling $a_k(s) \to a(s) \equiv \sigma\sigma^\top(s, \overline{x}(s), \overline{u}(s))$ weakly on $L^2([0,T] \times \overline{\Omega})$, we have for any $s, t \in [0,T]$,

$$\int_s^t a_k(r) dr \to \int_s^t \sigma\sigma^\top(r, \overline{x}(r), \overline{u}(r)) dr, \quad \text{weakly on } L^2(\overline{\Omega}).$$

On the other hand, by the dominated convergence theorem,

$$g(\overline{Y}_k) \to g(\overline{Y}), \qquad \text{strongly on } L^2(\Omega).$$

Thus,

$$\overline{E}\left(g(\overline{Y}_k)\int_s^t a_k(r)dr\right) \to \overline{E}\left(g(\overline{Y})\int_s^t \sigma\sigma^\top(r,\overline{x}(r),\overline{u}(r))dr\right).$$

Therefore, using an argument similar to the above, we obtain that $\overline{\Sigma\Sigma}^\top(t) - \int_0^t \sigma\sigma^\top(s,\overline{x}(s),\overline{u}(s))ds$ is an $\{\overline{\mathcal{F}}_t\}_{t\geq 0}$-martingale. This implies

$$\langle\overline{\Sigma}\rangle(t) = \int_0^t \sigma\sigma^\top(s,\overline{x}(s),\overline{u}(s))ds. \tag{5.41}$$

By a martingale representation theorem (Chapter 1, Theorem 5.10), there is an extension $(\widehat{\Omega},\widehat{\mathcal{F}},\{\widehat{\mathcal{F}}_t\}_{t\geq 0},\widehat{\mathbf{P}})$ of $(\overline{\Omega},\overline{\mathcal{F}},\{\overline{\mathcal{F}}_t\}_{t\geq 0},\overline{\mathbf{P}})$ on which lives an m-dimensional $\{\widehat{\mathcal{F}}_t\}_{t\geq 0}$-Brownian motion $\widehat{W}(t)$ such that

$$\overline{\Sigma}(t) = \int_0^t \sigma(s,\overline{x}(s),\overline{u}(s))d\widehat{W}(s). \tag{5.42}$$

Similarly (in fact more easily), one can show that

$$\begin{aligned}\overline{B}(t) &= \int_0^t b(s,\overline{x}(s),\overline{u}(s))ds,\\ \overline{F}(t) &= \int_0^t f(s,\overline{x}(s),\overline{u}(s))ds.\end{aligned} \tag{5.43}$$

Putting this into (5.32) and noting Definition 5.9 of Chapter 1, we arrive at the conclusion that $\overline{\pi} \stackrel{\Delta}{=} (\widehat{\Omega},\widehat{\mathcal{F}},\{\widehat{\mathcal{F}}_t\}_{t\geq 0},\widehat{\mathbf{P}},\widehat{W}(\cdot),\overline{u}(\cdot)) \in \mathcal{U}_{ad}^w[0,T]$ is an optimal control. $\square$

The basic idea behind the above proof is the so-called *relaxed control*, which is needed in order to provide some compact structure. To be more precise, the set of all admissible controls can be regarded in general as a subset of a much larger set of Λ-valued random variables (relaxed controls), whereas Λ is compact under the metric induced by the weak* topology. This approach works, since under Assumption (SE3) (Roxin's condition) along with (SE1), the set of admissible controls coincides with that of relaxed controls.

6. Reachable Sets of Stochastic Control Systems

Stochastic control systems are quite different from their deterministic counterparts. In this section we present several unexpected properties of stochastic control systems in terms of the reachable sets, which will help us understand that many results for deterministic control systems do not necessarily extend to stochastic ones. For the sake of simplicity, we present the results only for linear systems. It is expected that the situation for nonlinear systems will be more complicated.

6.1. Nonconvexity of the reachable sets

Consider the following deterministic controlled system:

$$(6.1)\qquad \begin{cases} \dot{x}(t) = Ax(t) + Bu(t). & t \in [0, T], \\ x(0) = x_0, \end{cases}$$

where $A \in \mathbb{R}^{n\times n}$ and $B \in \mathbb{R}^{n\times k}$. Define

$$(6.2)\qquad \mathcal{R}_D(T) = \{x(T; u(\cdot)) \mid u(\cdot) \in \mathcal{V}[0, T]\} \subseteq \mathbb{R}^n,$$

which is called the *reachable set* of the system (6.1). Clearly, the set $\mathcal{R}_D(T)$ changes as T varies. Investigating properties of $\mathcal{R}_D(T)$ is very important for the study of *controllability* and *time optimal control* problems. Let us elaborate these a bit more.

Let $M \subseteq \mathbb{R}^n$ be a nonempty set. We would like to find a control that steers the initial state $x_0 \in \mathbb{R}^n$ to the *target* M; if it is possible, we further would like to reach M as fast as possible. To formulate the problems rigorously, we define

$$(6.3)\qquad \mathcal{V}_M \stackrel{\Delta}{=} \{u(\cdot) \in \mathcal{V}[0, \infty) \mid x(T; u(\cdot)) \in M, \text{ for some } T \geq 0\}$$

and

$$(6.4)\qquad J(u(\cdot)) = \inf\{T \geq 0 \mid x(T; u(\cdot)) \in M\}, \quad \forall u(\cdot) \in \mathcal{V}_M,$$

with the convention $\inf \phi \stackrel{\Delta}{=} \infty$. Then we pose the following problems:

Problem (DC). Find conditions under which $\mathcal{V}_M$ is nonempty.

Problem (DT). Minimize (6.4) subject to (6.1) over $\mathcal{V}_M$ when $\mathcal{V}_M \neq \phi$.

We call Problems (DC) and (DT) the *controllability* problem and *time optimal control problem*, respectively.

Let us now assume $\mathcal{V}_M \neq \phi$, which is equivalent to

$$(6.5)\qquad \Big[\bigcup_{T\geq 0} \mathcal{R}_D(T)\Big] \bigcap M \neq \phi.$$

Under this assumption, the goal of Problem (DT) is to find the first moment $\overline{T}$ when $\mathcal{R}_D(\overline{T})$ intersects M and a control $\overline{u}(\cdot) \in \mathcal{V}[0, \infty)$ such that

$$(6.6)\qquad x(\overline{T}; \overline{u}(\cdot)) \in \mathcal{R}_D(\overline{T}) \bigcap M.$$

When they exist, we call $\overline{T}$, $\overline{u}(\cdot)$, $\overline{x}(\cdot) \equiv x(\cdot\,; \overline{u}(\cdot))$, and $(\overline{x}(\cdot), \overline{u}(\cdot))$ the *minimum time*, a (*time*) *optimal control*, *optimal trajectory*, and *optimal pair*, respectively.

Now assume that M is convex and closed. If we also have the convexity of $\mathcal{R}_D(T)$ for any $T \geq 0$, then it is possible to use some powerful techniques in convex analysis to study (6.6) by which the minimum time $\overline{T}$ and the time optimal pair $(\overline{x}(\cdot), \overline{u}(\cdot))$ could be characterized. Hence, let us study when $\mathcal{R}_D(T)$ is convex.

First of all, it is clear that if $U \subseteq \mathbb{R}^k$ is convex, then $\mathcal{R}_D(T)$ is convex for any $T \geq 0$. Surprisingly, we have the following result without the convexity of U!

Proposition 6.1. *Let $U \subseteq \mathbb{R}^k$ be compact. Then for any $T > 0$, $\mathcal{R}_D(T)$ is convex and compact in $\mathbb{R}^n$, and it coincides with $\widehat{\mathcal{R}}_D(T)$, the reachable set with U replaced by $\overline{\text{co}}\, U$.*

A consequence of Proposition 6.1 is the following existence result for Problem (DT).

Corollary 6.2. *Let $U \subseteq \mathbb{R}^k$ be compact and suppose that (6.5) holds with M closed. Then Problem (DT) admits an optimal control.*

Proof. Using the same argument as in the proof of Theorem 5.1, one can show that Problem (DT) with U replaced by $\overline{\text{co}}\, U$ admits a time optimal control, say, $\widehat{u}(\cdot) \in \widehat{\mathcal{V}}[0, \infty) \equiv \{u : [0, \infty) \to \overline{\text{co}}\, U \mid u(\cdot) \text{ measurable }\}$ with the optimal trajectory $\widehat{x}(\cdot)$, and the minimum time $\widehat{T}$. Thus,

$$\begin{cases} \widehat{\mathcal{R}}_D(T) \cap M = \phi, & \forall T \in [0, \widehat{T}\,), \\ \widehat{x}(T) \in \widehat{\mathcal{R}}_D(\widehat{T}) \cap M. \end{cases}$$

By Proposition 6.1, $\mathcal{R}_D(T) = \widehat{\mathcal{R}}_D(T)$ for all $T \geq 0$. Thus, there exists a $\overline{u}(\cdot) \in \mathcal{V}[0, \infty)$ such that $x(\widehat{T}; \overline{u}(\cdot)) = \widehat{x}(\widehat{T})$. Hence, $\overline{u}(\cdot)$ is a time optimal control of Problem (DT), with the minimum time $\overline{T} = \widehat{T}$. □

The proof of Proposition 6.1 is based on a very deep theorem due to Lyapunov, which we now state.

Theorem 6.3. (Lyapunov's theorem) *Suppose $f \in L^1(0, T; \mathbb{R}^n)$. Then the set*

$$\mathcal{R} \stackrel{\Delta}{=} \Big\{ \int_S f(t)dt \mid S \in \mathcal{B}[0, T] \Big\} \tag{6.7}$$

is convex, where $\mathcal{B}[0, T]$ is the Borel σ-field of $[0, T]$.

We refer the reader to Hermes–LaSalle [1] for proofs of Proposition 6.1 and Theorem 6.3 (see Diestel–Uhl [1] for a more detailed discussion related to Theorem 6.3).

It is seen that Proposition 6.1 and Corollary 6.2 give an elegant existence result of time optimal control for the deterministic linear system (6.1). One can actually further characterize the time optimal control and the minimum time by using the convexity of the reachable set $\mathcal{R}_D(T)$. A natural question is then whether we have a similar convexity property for the reachable set of a linear *stochastic* system.

Let $(\Omega, \mathcal{F}, \mathbf{P})$ be a given complete probability space on which a standard Brownian motion $W(\cdot)$ is defined with $\{\mathcal{F}_t\}_{t\geq 0}$ being the natural filtration of $W(\cdot)$, augmented by all the $\mathbf{P}$-null sets in $\mathcal{F}$. For system (5.7), we define

$$\mathcal{R}_S(T) = \{x(T; u(\cdot)) \mid u(\cdot) \in \mathcal{U}^L[0,T]\} \subseteq L^2_{\mathcal{F}_T}(\Omega; \mathbb{R}^n).$$

We call $\mathcal{R}_S(T)$ the *reachable set* of system (5.7) at T. Let us present a simple example below showing that $\mathcal{R}_S(T)$ may no longer be convex.

Example 6.4. Consider the following one-dimensional control system:

$$\begin{cases} dx(t) = u(t)dW(t), \\ x(0) = 0. \end{cases} \tag{6.8}$$

The control set U is $\{-1, 1\}$. Thus, we have a linear controlled system with a nonconvex control set (which is, however, compact). By taking $u(t) \equiv 1$ and $u(t) \equiv -1$, we see that

$$\mathcal{R}_S(t) \supseteq \{W(t), -W(t)\}, \qquad \forall t \in [0,T].$$

Thus,

$$0 \in \operatorname{co} \mathcal{R}_S(t), \qquad t \in [0,T].$$

However, for any $u(\cdot) \in \mathcal{U}^L[0,T]$,

$$E|x(t)|^2 = E\Big| \int_0^t u(s)dW(s)\Big|^2 = t.$$

Hence,

$$0 \in \operatorname{co} \mathcal{R}_S(t) \setminus \overline{\mathcal{R}_S(t)}, \quad \forall t \in (0,T], \tag{6.9}$$

showing that for any $t > 0$, neither $\mathcal{R}_S(t)$ nor $\overline{\mathcal{R}_S(t)}$ is convex.

The essence behind the above example is the nonconvexity of the range of vector measures defined through the Itô integral. Let us make this more precise. Let

$$\mathcal{S}[0,T] \stackrel{\Delta}{=} \{S \subseteq [0,T] \times \Omega \mid I_S \text{ is } \{\mathcal{F}_t\}_{t\geq 0}\text{-adapted}\}. \tag{6.10}$$

Given $f \in L^2_{\mathcal{F}}(0,T;\mathbb{R}^n)$, define

$$\mu(S) \stackrel{\Delta}{=} \int_0^T I_S(t) f(t) dW(t), \quad \forall S \in \mathcal{S}[0,T]. \tag{6.11}$$

Then $\mu : \mathcal{S}[0,T] \to L^2_{\mathcal{F}_T}(\Omega; \mathbb{R}^n)$ is a *vector-valued measure*. Define the *range* of μ as $\mu(\mathcal{S}[0,T]) \stackrel{\Delta}{=} \{\mu(S) | S \in \mathcal{S}[0,T]\}$. We have the following result.

Theorem 6.5. *Suppose* $f \in L^2_{\mathcal{F}}(0,T;\mathbb{R}^n)$ *such that*

$$E\int_0^T |f(t)|^2 dt > 0. \tag{6.12}$$

Then there exist $S_1, S_2 \in \mathcal{S}[0,T]$ *with*

$$|S_i| \triangleq \int_{S_i} dt\, d\mathbf{P} > 0, \quad S_1 \cap S_2 = \phi, \tag{6.13}$$

such that for any $\lambda \in (0,1)$,

$$\begin{aligned} E|\lambda\mu(S_1) + (1-\lambda)\mu(S_2) - \mu(S)|^2 \geq \delta(S_1, S_2, \lambda) > 0, \\ \forall S \in \mathcal{S}[0,T]. \end{aligned} \tag{6.14}$$

In particular, the range $\mu(\mathcal{S}[0,T])$ *of the vector-valued measure* μ *as well as its closure* $\overline{\mu(\mathcal{S}[0,T])}$ *in* $L^2_{\mathcal{F}_T}(\Omega;\mathbb{R})$ *are nonconvex.*

Proof. For any $\varepsilon > 0$, define

$$S^\varepsilon = \{(t,\omega) \in [0,T] \mid |f(t,\omega)| \geq \varepsilon\} \in \mathcal{S}[0,T]. \tag{6.15}$$

Let

$$h^\varepsilon(t) = \int_0^t \mathbf{P}\big(S^\varepsilon \cap ([0,s] \times \Omega)\big) ds, \quad t \in [0,T]. \tag{6.16}$$

Clearly, $h^\varepsilon(t)$ is continuous, and by condition (6.12), $h^\varepsilon(T) > 0$ for some $\varepsilon > 0$. Thus, one can find $t_0 \in [0,T)$ such that

$$0 < h^\varepsilon(t_0) < h^\varepsilon(T). \tag{6.17}$$

Next, set

$$S_1 = S^\varepsilon \cap \big([0,t_0] \times \Omega)\big), \quad S_2 = S^\varepsilon \cap \big((t_0,T] \times \Omega)\big). \tag{6.18}$$

From (6.17), we see that

$$\begin{cases} S_1, S_2 \in \mathcal{S}[0,T], \quad S_1 \cap S_2 = \phi, \quad |S_1|, |S_2| > 0, \\ |f(t,\omega)| \geq \varepsilon, \qquad \forall (t,\omega) \in S_1 \cup S_2. \end{cases} \tag{6.19}$$

Now, for any $\lambda \in (0,1)$, we obtain

$$\begin{aligned} &E|\lambda\mu(S_1) + (1-\lambda)\mu(S_2) - \mu(S)|^2 \\ &= E\Big|\int_0^T \{\lambda I_{S_1}(t) + (1-\lambda)I_{S_2}(t) - I_S(t)\} f(t) dW(t)\Big|^2 \\ &= \int_0^T E\big\{|\lambda I_{S_1}(t) + (1-\lambda)I_{S_2}(t) - I_S(t)|^2 |f(t)|^2\big\} dt \\ &= \int_{S_1} |\lambda I_{S_1}(t,\omega) - I_S(t,\omega)|^2 |f(t,\omega)|^2 dt\, d\mathbf{P}(\omega) \\ &\quad + \int_{S_2} |(1-\lambda)I_{S_2}(t,\omega) - I_S(t,\omega)|^2 |f(t,\omega)|^2 dt\, d\mathbf{P}(\omega) \\ &\geq \varepsilon^2[\lambda^2 \wedge (1-\lambda)^2]\big(|S_1| + |S_2|\big) > 0. \end{aligned}$$

This proves (6.14). ∎

The above result implies that the vector measure defined through the Itô integral has a nonconvex range, provided that (6.12) holds. The following is a simple consequence of the above result, which generalizes Example 6.4.

Proposition 6.6. *Suppose in (5.7)–(5.8) that*

$$B = CD, \tag{6.20}$$

and $U = \{u_0, u_1\}$ *with*

$$D(u_1 - u_0) \neq 0. \tag{6.21}$$

Then, for any $x_0 \in \mathbb{R}^n$ *and* $T > 0$, $\overline{\mathcal{R}_S(T)}$ *is nonconvex.*

Proof. Without loss of generality, we may assume $U = \{0, \bar{u}\}$ with $D\bar{u} \neq 0$ and $x_0 = 0$. Let $\Phi(\cdot)$ be an $\mathbb{R}^{n\times n}$-valued process satisfying

$$\begin{cases} d\Phi(t) = A\Phi(t)dt + C\Phi(t)dW(t), \\ \Phi(0) = I. \end{cases} \tag{6.22}$$

Then $\Phi(t)^{-1}$ exists, and for any $u(\cdot) \in \mathcal{U}^L[0,T]$, noting the condition (6.20) and $x_0 = 0$, we have (see Chapter 1, Theorem 6.14)

$$x(t; u(\cdot)) = \Phi(t) \int_0^t \Phi(s)^{-1} Du(s)dW(s), \quad t \in [0,T]. \tag{6.23}$$

Since $U = \{0, \bar{u}\}$, any $u(\cdot) \in \mathcal{U}^L[0,T]$ corresponds to an $S \in \mathcal{S}[0,T]$ such that

$$u(t,\omega) = I_S(t,\omega)\bar{u}, \qquad (t,\omega) \in [0,T] \times \Omega. \tag{6.24}$$

Then (6.23) implies

$$\Phi(T)^{-1}x(T) = \int I_S(s)\Phi(s)^{-1} D\bar{u}\, dW(s).$$

Hence, applying Theorem 6.5 to the function $f(t) = \Phi(t)^{-1}D\bar{u}$, we obtain the nonconvexity of $\overline{\Phi(T)^{-1}\mathcal{R}_S(T)}$. Our assertion follows then, since $\Phi(T)^{-1}$ is an invertible linear map. $\square$

Let us make a final observation. When condition (6.20) does not hold, instead of (6.23), we would have

$$\begin{aligned} x(t; u(\cdot)) &= \Phi(t) \int_0^t \Phi(s)^{-1}(B - CD)u(s)ds \\ &\quad + \Phi(t) \int_0^t \Phi(s)^{-1} Du(s)dW(s), \quad t \in [0,T]. \end{aligned} \tag{6.25}$$

Thus, there would be some interaction between the diffusion and drift. The situation would become much more complicated and interesting. We encourage the reader to explore this situation and to obtain results that are not yet known to us.

6.2. Noncloseness of the reachable sets

Proposition 6.1 tells us that the reachable set $\mathcal{R}_D(T)$ of the deterministic system (6.1) is convex and compact if $U \subseteq \mathbb{R}^m$ is compact. For the reachable set $\mathcal{R}_S(T)$ of the stochastic system (5.7), the results of the previous subsection imply that it is not convex in general even when U is compact. We now investigate whether $\mathcal{R}_S(T)$ is closed in $L^2_{\mathcal{F}_T}(\Omega;\mathbb{R}^n)$.

Our first result is the following.

Theorem 6.7. *Let*

$$\begin{cases} \mathcal{R}_1 = \Big\{ \int_0^T h(s)ds \mid h(\cdot) \in L^2_{\mathcal{F}}(0,T;\mathbb{R}) \Big\}, \\ \mathcal{R}_2 = \Big\{ \int_0^T f(s)dW(s) \mid f(\cdot) \in C_{\mathcal{F}}([0,T];L^2(\Omega;\mathbb{R})) \Big\}. \end{cases} \tag{6.26}$$

Then

$$\begin{cases} \overline{\mathcal{R}}_1 = \overline{\mathcal{R}}_2 + \mathbb{R} = L^2_{\mathcal{F}_T}(\Omega;\mathbb{R}), \\ \text{Int}\,(\mathcal{R}_1 + \mathcal{R}_2) = \phi. \end{cases} \tag{6.27}$$

In particular, both $\mathcal{R}_1$ and $\mathcal{R}_2$ are not closed.

Proof. First of all, it is clear that

$$\overline{\mathcal{R}}_1 \subseteq L^2_{\mathcal{F}_T}(\Omega;\mathbb{R}), \quad \overline{\mathcal{R}}_2 + \mathbb{R} \subseteq L^2_{\mathcal{F}_T}(\Omega;\mathbb{R}). \tag{6.28}$$

Thus, we need only to prove the other direction of the inclusions. For any $\eta \in L^2_{\mathcal{F}_T}(\Omega;\mathbb{R})$, applying the martingale representation theorem (see Chapter 1, Theorem 5.7) to $E(\eta|\mathcal{F}_t)$, we have $Z(\cdot) \in L^2_{\mathcal{F}}(0,T;\mathbb{R})$ such that

$$E(\eta|\mathcal{F}_t) = E\eta + \int_0^t Z(s)dW(s).$$

In particular,

$$\eta = E\eta + \int_0^T Z(s)dW(s). \tag{6.29}$$

For any $\varepsilon > 0$, we can find a $Z_\varepsilon(\cdot) \in C_{\mathcal{F}}([0,T];L^2(\Omega;\mathbb{R}))$ such that

$$E\int_0^T |Z(s) - Z_\varepsilon(s)|^2 ds < \varepsilon.$$

Then

$$\begin{cases} \eta_\varepsilon \triangleq \int_0^T Z_\varepsilon(s)dW(s) \in \mathcal{R}_2, \\ E|\eta - [E\eta + \eta_\varepsilon]|^2 < \varepsilon, \end{cases}$$

proving $\overline{\mathcal{R}}_2+\mathbb{R}=L^2_{\mathcal{F}_T}(\Omega;\mathbb{R})$. Next, for any $\eta\in L^2_{\mathcal{F}_T}(\Omega;\mathbb{R})$, again we have (6.29) for some $Z(\cdot)\in L^2_{\mathcal{F}}(0,T;\mathbb{R})$. For any $\varepsilon>0$, let

$$h_\varepsilon(s)=\frac{1}{\varepsilon}E\big(\eta\big|\mathcal{F}_{T-\varepsilon}\big)I_{[T-\varepsilon,T]}(s),\quad s\in[0,T]. \tag{6.30}$$

Clearly, $h_\varepsilon(\cdot)\in L^2_{\mathcal{F}}(0,T;\mathbb{R})$, and

$$\mathcal{R}_1\ni\eta_\varepsilon\overset{\Delta}{=}\int_0^T h_\varepsilon(s)ds=E\big(\eta\big|\mathcal{F}_{T-\varepsilon}\big)=E\eta+\int_0^{T-\varepsilon}Z(s)dW(s).$$

Thus,

$$E|\eta_\varepsilon-\eta|^2=\int_{T-\varepsilon}^T E|Z(s)|^2ds\to 0,\qquad \text{as } \varepsilon\to 0,$$

which proves $\overline{\mathcal{R}}_1=L^2_{\mathcal{F}_T}(\Omega;\mathbb{R})$.

Next, we prove the second relation in (6.27). To this end, we construct a deterministic Lebesgue measurable function β satisfying the following:

$$\begin{cases}\beta(s)=\pm1, & \forall s\in[0,T],\\ |\{s\in[T_i,T]\mid\beta(s)=1\}|=|\{s\in[T_i,T]\mid\beta(s)=-1\}| & \\ \qquad\qquad=\dfrac{T-T_i}{2}, & i\ge 1,\end{cases} \tag{6.31}$$

for a sequence $T_i\uparrow T$, where $|A|$ stands for the Lebesgue measure of a measurable set A. Such a function exists by some elementary construction. We claim that

$$\int_0^T\beta(s)dW(s)\notin\mathcal{R}_1+\mathcal{R}_2. \tag{6.32}$$

Suppose there exist $h(\cdot)\in L^2_{\mathcal{F}}(0,T;\mathbb{R})$ and $f(\cdot)\in C_{\mathcal{F}}([0,T];L^2(\Omega;\mathbb{R}))$ such that

$$\int_0^T\beta(s)dW(s)=\int_0^T h(s)ds+\int_0^T f(s)dW(s). \tag{6.33}$$

Define

$$\zeta(t)=\int_0^t h(s)ds+\int_0^t[f(s)-\beta(s)]dW(s),\quad t\in[0,T]. \tag{6.34}$$

Then, by Itô's formula, we have

$$\begin{aligned}E|\zeta(t)|^2&+E\int_t^T|f(s)-\beta(s)|^2ds=-2E\int_t^T h(s)\zeta(s)ds\\&=2E\int_t^T h(s)\Big[\int_s^T h(r)dr+\int_s^T(f(r)-\beta(r))dW(r)\Big]ds\\&=2E\int_t^T h(s)\int_s^T h(r)drds\\&=E\Big|\int_t^T h(s)ds\Big|^2\le(T-t)\int_t^T Eh(s)^2ds.\end{aligned} \tag{6.35}$$

On the other hand, since $h(\cdot) \in L^2_{\mathcal{F}}(0,T;\mathbb{R})$ and $f(\cdot) \in C_{\mathcal{F}}([0,T];L^2(\Omega;\mathbb{R}))$, we have

$$
\begin{aligned}
&E\int_t^T |f(T)-\beta(s)|^2 ds \\
&\le 2E\int_t^T |f(s)-\beta(s)|^2 ds + 2E\int_t^T |f(T)-f(s)|^2 ds \\
&\le 2(T-t)\int_t^T Eh(s)^2 ds + 2E\int_t^T |f(T)-f(s)|^2 ds \\
&= o(T-t).
\end{aligned} \tag{6.36}
$$

By (6.31), the definition of $\beta(\cdot)$, we have

$$
\begin{aligned}
E\int_{T_i}^T |f(T)-\beta(s)|^2 ds &= \frac{T-T_i}{2}\Big(E|f(T)-1|^2 + E|f(T)+1|^2\Big) \\
&\ge T-T_i, \quad \forall i \ge 2.
\end{aligned} \tag{6.37}
$$

Combining (6.36) and (6.37), we obtain a contradiction. Thus, (6.32) holds. Since $\mathcal{R}_1+\mathcal{R}_2$ is a subspace, the second relation in (6.27) follows from the fact that for any $\eta \in \mathcal{R}_1+\mathcal{R}_2$ and any $\varepsilon > 0$,

$$
\eta + \varepsilon\int_0^T \beta(s)dW(s) \notin \mathcal{R}_1+\mathcal{R}_2.
$$

This proves our result. □

Let us discuss some simple consequences of the above result. The first is related to the problem of controllability.

Corollary 6.8. *Let a control system be given by (5.7) with $A, C \in \mathbb{R}^{n\times n}$ and $B, D \in \mathbb{R}^{n\times k}$. Let*

$$
\operatorname{rank} D < n. \tag{6.38}
$$

Then

$$
\mathcal{R}_S(T) \ne L^2_{\mathcal{F}_T}(\Omega;\mathbb{R}^n), \quad \forall T > 0. \tag{6.39}
$$

Proof. Suppose (6.38) holds. Then we can find some $\eta \in \mathbb{R}^n$, $|\eta| = 1$, such that

$$
\eta^\top D = 0. \tag{6.40}
$$

Multiplying the integral version of (5.7) by $\eta^\top$, we obtain

$$
\begin{aligned}
\eta^\top x(T) = \eta^\top x_0 &+ \int_0^T \eta^\top [Ax(t)+Bu(t)]dt \\
&+ \int_0^T \eta^\top Cx(t)dW(t).
\end{aligned} \tag{6.41}
$$

By (6.27), we see that

$$\text{Int}\,\big[\eta^\top \mathcal{R}_S(T)\big] \subseteq \text{Int}\,(\mathcal{R}_1+\mathcal{R}_2)=\phi. \tag{6.42}$$

Hence, (6.39) holds. □

We know that for the deterministic system (6.1), provided that *Kalman's rank condition*

$$\text{rank}\,(B, AB, \ldots, A^{n-1}B)=n \tag{6.43}$$

holds, for any $x_0, x_1 \in \mathbb{R}^n$ and $T>0$, there exists a $u(\cdot)\in\mathcal{V}[0,T]$ such that $x(T;x_0,u(\cdot))=x_1$. In other words, under condition (6.40), the system (6.1) is *completely controllable*. However, Corollary 6.8 tells that if D has rank smaller than n, in particular if $k<n$, then there is some $\xi\in L^2_{\mathcal{F}_T}(\Omega;\mathbb{R}^n)$ that will not be hit by the state $x(T)$.

Another consequence of Theorem 6.7 is the noncloseness of reachable sets and the nonexistence of optimal controls for some stochastic systems. Let us look at such an example.

Example 6.9. Consider the controlled system

$$\begin{cases} dx(t)=u(t)dt+\beta(t)dW(t), \quad x(0)=0,\\ dy(t)=2\beta(t)dW(t), \quad y(0)=0,\end{cases} \tag{6.44}$$

where $\beta(\cdot)\in L^2_{\mathcal{F}}(0,T;\mathbb{R})$ satisfying

$$\int_0^T \beta(s)dW(s)\in L^2_{\mathcal{F}_T}(\Omega;\mathbb{R})\setminus\mathcal{R}_1. \tag{6.45}$$

Such a $\beta(\cdot)$ can be constructed as in (6.31). We let the control set be $\mathcal{U}[0,T]\triangleq L^2_{\mathcal{F}}(0,T;\mathbb{R})$ and the cost functional be

$$J(u(\cdot))=E|x(T)-y(T)|^2. \tag{6.46}$$

We now show that this (very simple) optimal control problem does not admit an optimal control. In fact,

$$J(u(\cdot))=E\Big|\int_0^T u(s)ds-\int_0^T\beta(s)dW(s)\Big|^2. \tag{6.47}$$

By (6.27), $L^2_{\mathcal{F}_T}(\Omega;\mathbb{R})=\overline{\mathcal{R}}_1$. Thus, $\inf_{u(\cdot)\in L^2_{\mathcal{F}}(0,T;\mathbb{R})}J(u(\cdot))=0$. However, the infimum cannot be achieved, due to the fact (6.45). Thus, an optimal control does not exist.

The above argument actually shows that there is a sequence $\{u_k(\cdot)\}\subseteq\mathcal{U}[0,T]$ such that the corresponding state $(x(\cdot\,;u_k(\cdot)),y(\cdot\,;u_k(\cdot)))$ satisfies

$$E\big|x(T;u_k(\cdot))-y(T;u_k(\cdot))\big|^2\to 0.$$

However, no $u(\cdot)\in\mathcal{U}[0,T]$ exists such that

$$x(T;u(\cdot))=y(T;u(\cdot)).$$

This means that the "diagonal line"

$$\{(\xi, \xi) \mid \xi \in L^2_{\mathcal{F}_T}(\Omega; \mathbb{R})\} \subseteq \overline{\mathcal{R}_S(T)} \setminus \mathcal{R}_S(T).$$

Thus, $\mathcal{R}_S(T)$ is not closed in $L^2_{\mathcal{F}_T}(\Omega; \mathbb{R}^2)$. This is indeed the very reason why the optimal control does not exist in this example.

7. Other Stochastic Control Models

In Section 4 we have formulated optimal control problems in both strong and weak forms where the system is governed by an Itô stochastic differential equation, the system is completely observable (meaning that the controller is able to observe the system state completely), and the cost is of an integral form calculated over a fixed, deterministic time duration. This type of stochastic control problems will be the subject of study throughout this book. The reason for doing so is that we would like to capture some fundamental structure and properties of optimal stochastic control in a clean and clear way via this standard and relatively simple model, without being blurred by technicalities caused by more complicated models. Certainly, there are many other types of stochastic control problems that are of significant theoretical and practical importance. In this section we shall introduce and formulate several representative models that have been commonly encountered in practice and investigated in theory. It should be noted that a real-life problem could involve a combination of some or all of those models and therefore be very complicated. On the other hand, since this section is rather introductory, we shall omit all the proofs when we state relevant results, and the interested readers may find more details in the references surveyed in Section 8. Furthermore, only strong formulations will be discussed, and weak formulations are completely parallel.

7.1. Random duration

In Problem (SS) formulated in Section 4, the time duration under consideration (called a *control horizon*) is a fixed, deterministic finite interval $[0, T]$. In many real applications the control horizon may be a *random* duration $[0, \tau]$ where the terminal time τ is a random variable at which the state of the controlled system changes critically and the control beyond τ may no longer be meaningful or necessary. For example, in the insurance company model (Section 3.3) the control after the time when the company is bankrupt becomes unnecessary. Therefore, it makes better sense to consider the problem in $[0, \tau]$ where τ is the bankruptcy time defined as $\tau = \inf\{t \geq 0 : x(t) = 0\}$.

To formulate a general model that incorporates a random control horizon, consider the following stochastic differential equation:

$$\begin{cases} dx(t) = b(t, x(t), u(t))dt + \sigma(t, x(t), u(t))dW(t), \\ x(0) = x_0 \in \mathcal{O}, \end{cases} \tag{7.1}$$

on a complete filtered probability space $(\Omega, \mathcal{F}, \{\mathcal{F}_t\}_{t\geq 0}, \mathbf{P})$ on which an m-dimensional Brownian motion $W(\cdot)$ is defined. Here $\mathcal{O} \subseteq \mathbb{R}^n$ is a given open set. An admissible control $u(\cdot)$ is defined similarly as in Section 4.1. Define the first *exit time*

$$\tau \stackrel{\Delta}{=} \inf\{t \geq 0 : x(t) \notin \mathcal{O}\}.$$

Note that τ implicitly depends on the control taken. Given the cost functional

$$J(u(\cdot)) = E\Big\{\int_0^\tau f(t, x(t), u(t))dt + h(x(\tau))\Big\}, \tag{7.2}$$

the problem is to minimize (7.2) over the set of admissible controls.

An interesting case is that in which $f \equiv 1$ and $h \equiv 0$, which corresponds to a stochastic time optimal control problem with the target set $M = \mathcal{O}^c$. Another special model has $\mathcal{O} = \mathbb{R}^n$, in which case $\tau = \infty$. It is called an *infinite horizon* problem if the integral in (7.2) remains finite (as is the case when, for example, the function f is of the form $e^{-\gamma t}g(t, x, u)$ with g bounded and $\gamma > 0$).

7.2. Optimal stopping

In the random duration model introduced in the previous subsection, the control influences the exit time only *indirectly*. One has a different model if the controller controls the ending time *directly*, which is called an *optimal stopping problem*. In such a situation, the stopping time itself becomes a part of the control. For instance, in the technology transfer example (Section 3.4), while one can consider the *natural ending time* of the technology (namely, the time when the economic value of the technology reaches zero), it is practically more important to study the *early phase-out* decision, that is, to decide when to terminate the use of the technology *before* its natural ending time.

In the formulation of such models, an admissible control/stopping time is a *pair* $(u(\cdot), \tau)$ defined on a filtered probability space $(\Omega, \mathcal{F}, \{\mathcal{F}_t\}_{t\geq 0}, \mathbf{P})$ along with an m-dimensional Brownian motion $W(\cdot)$, where $u(\cdot)$ is the control satisfying the usual conditions and τ is an $\{\mathcal{F}_t\}_{t\geq 0}$-stopping time. The optimal control/stopping problem is to minimize

$$J(u(\cdot), \tau) = E\Big\{\int_0^\tau f(t, x(t), u(t))dt + h(x(\tau))\Big\} \tag{7.3}$$

subject to (4.1) over the set of admissible controls and stopping times.

7.3. Singular and impulse controls

In Problem (SS) presented in Section 4.1, the state changes continuously in time t in response to the efforts of the control. However, in many real situations, the state may be displaced drastically due to possible "singular behavior" of a control at a certain time instant. For example, in the insurance company model in Section 3.3, rather than paying out the dividends continuously, the company is more likely to distribute them once or twice a year. As a result, on a dividend payday the liquid assets reserve of the company changes drastically. The investment vs. consumption model presented in Section 3.2 is also just an approximation of what is really happening where changes of portfolios and withdrawals for consumption occur only at discrete time instants. A *singular control* can better model such situations.

To begin, let us first introduce the following function space: $D \equiv D([0,T];\mathbb{R}^n)$ is the space of all functions $\zeta : [0,T] \to \mathbb{R}^n$ that are right continuous with left limits (*càdlàg* for short). For $\zeta \in D$, we define the total variation of ζ on $[0,T]$ by $\int_{[0,T]} |d\zeta(s)| \equiv |\zeta|_{[0,T]} = \sum_{i=1}^n |\zeta^i|_{[0,T]}$, where $|\zeta^i|_{[0,T]}$ is the total variation of the ith component of ζ on $[0,T]$ in the usual sense. We define $|\zeta|_t = |\zeta|_{[0,t]}$, $t > 0$, for simplicity. For $\zeta \in D$, we define $\Delta\zeta(s) \stackrel{\Delta}{=} \zeta(s) - \zeta(s-)$ and $S_\zeta \stackrel{\Delta}{=} \{s \in [0,T] \mid \Delta\zeta(s) \neq 0\}$. Further, we define

$$BV([0,T];\mathbb{R}^n) = \{\zeta \in D \bigm| |\zeta|_T < \infty\}.$$

For any $\zeta \in BV([0,T];\mathbb{R}^n)$ we define the *pure jump* part of ζ by $\zeta^{jp}(t) \stackrel{\Delta}{=} \sum_{0 \le s < t} \Delta\zeta(s)$, and the *continuous part* of ζ by $\zeta^c(t) = \zeta(t) - \zeta^{jp}(t)$. Since ζ^c is still of bounded variation, it is differentiable almost everywhere, and we have $\zeta^c(t) = \zeta^{ac}(t) + \zeta^{sc}(t)$, $t \in [0,T]$, where $\zeta^{ac}(t) \stackrel{\Delta}{=} \int_0^t \dot{\zeta}^c(s)ds$. We call ζ^{ac} the *absolutely continuous part* of ζ, and ζ^{sc} the *singularly continuous part* of ζ. Thus, we obtain the so-called *Lebesgue decomposition* for $\zeta \in BV([0,T];\mathbb{R}^n)$:

$$\zeta(t) = \zeta^{ac}(t) + \zeta^{sc}(t) + \zeta^{jp}(t), \qquad t \in [0,T]. \tag{7.4}$$

It is clear that such a decomposition is unique.

Now let $(\Omega, \mathcal{F}, \{\mathcal{F}_t\}_{t\ge0}, \mathbf{P})$ be a complete filtered probability space on which is defined an m-dimensional standard Brownian motion $W(\cdot)$. Let $BV_{\mathcal{F}}([0,T];\mathbb{R}^n)$ be the set of all $\{\mathcal{F}_t\}_{t\ge0}$-adapted processes $\xi(\cdot)$ such that

$$\begin{cases} \xi(\cdot\,,\omega) \in BV([0,T];\mathbb{R}^n), \\ E|\xi|_T < \infty. \end{cases}$$

For any $x_0 \in \mathbb{R}^n$ and $\xi \in BV_{\mathcal{F}}([0,T];\mathbb{R}^n)$, we consider a controlled stochastic process

$$x(t) = x_0 + \int_0^t b(s,x(s))ds + \int_0^t \sigma(s,x(s))dW(s) + \xi(t), \; 0 \le t \le T, \tag{7.5}$$

and introduce the cost functional

$$\begin{aligned} J(\xi(\cdot)) = E\Big\{ \int_0^T f(t,x(t))dt + \int_0^T f^a(t)\|\dot{\xi}^{ac}(t)\|_1 dt \\ + \int_0^T f^s(t)|d\xi^{sc}(t)| + \sum_{t \in S_\xi[0,T]} \ell(t,\Delta\xi(t)) + h(x(T))\Big\}. \end{aligned} \tag{7.6}$$

Here, f, f^a, f^s, ℓ, and h are given functions, $\xi(t) = \xi^{ac}(t) + \xi^{sc}(t) + \xi^{jp}(t)$ is its Lebesgue decomposition, and $\|\dot{\xi}^{ac}(t)\|_1 dt$ and $|d\xi^{sc}(t)|$ are the measures generated by the total variations of ξ^{ac} and ξ^{sc}, respectively, with $\|\cdot\|_1$ denoting the L^1-norm in $\mathbb{R}^n$. Our goal is to minimize the cost functional (7.6) over $BV_{\mathcal{F}}([0,T];\mathbb{R}^n)$. This is called an *optimal singular control problem.*

A special case is that in which $\xi^{sc} \equiv \xi^{jp} \equiv 0$. In this case $\xi^{ac}(t) \equiv \int_0^t \dot{\xi}^{ac}(s)ds$, and the problem reduces to a standard stochastic control problem as formulated in Section 4 if we regard $\dot{\xi}^{ac}(\cdot)$ as a new control variable.

Another interesting case is that in which ξ takes a special form of a pure jump process, i.e., $\xi^{ac} \equiv \xi^{sc} \equiv 0$, in which case we get the so-called *impulse (impulsive) control problem.*

7.4. Risk-sensitive controls

In the *risk-sensitive control* model, both the system dynamics and an admissible control are defined in the same way as those of Problem (SS) in Section 4.1. The only difference is the way of defining the performance criterion. Instead of minimizing the direct cost as in (4.3), the controller aims to minimize some *disutility* of the cost in a risk-sensitive control problem. To be more precise, given a monotonically increasing function $\varphi : \mathbb{R} \to \mathbb{R}$, the objective is to find an admissible control $u(\cdot)$ so as to minimize

$$J(u(\cdot)) = E\varphi\Big(\int_0^T f(t,x(t),u(t))dt + h(x(T))\Big) \tag{7.7}$$

subject to the system (4.1).

In economic theory the disutility function φ is usually taken to be either convex or concave, corresponding respectively to the *risk-averse* or *risk-seeking* attitude of the decision-maker. That is why the model is called a risk-sensitive one. Let us elaborate this point by an intuitive argument. Define $X \equiv \int_0^T f(t,x(t),u(t))dt + h(x(T))$ and suppose φ is differentiable at EX. Then we have the Taylor expansion

$$\varphi(X) \approx \varphi(EX) + \varphi'(EX)(X - EX) + \frac{1}{2}\varphi''(EX)(X - EX)^2.$$

If φ is strictly convex near EX, then $\varphi''(EX) > 0$, which introduces a penalty to the variance term $(X - EX)^2$ in the overall cost. This means that the controller tries to avoid large a deviation of X from its mean EX, that is, the controller is risk-averse. Conversely, if φ is strictly concave near EX, then $\varphi''(EX) < 0$, which reduces the overall cost with a large $|X - EX|$. This is the so-called risk-seeking case. Finally, if $\varphi''(EX)$ is close or equal to 0, then $E\varphi(x) \approx \varphi(EX)$, in which case the risk-sensitive model reduces to the standard one in Section 4 (recall that φ is monotonically increasing). This corresponds to the *risk-neutral* situation.

A commonly used and extensively studied disutility function is the following exponential function:

$$\varphi^\theta(x) = \theta\exp(\theta x),$$

where $\theta \in \mathbb{R}$ is a parameter representing the risk sensitivity degree of the criterion. In particular, φ^θ is convex (respectively, concave) when $\theta > 0$ (respectively, $\theta < 0$). When θ is sufficiently small, the problem can be well approximated by the standard stochastic control problem introduced in Section 4.

Another frequently used disutility function is the so-called *HARA utility*:

$$\varphi^\gamma(x) = \begin{cases} \frac{1}{\gamma} x^\gamma, \text{ if } \gamma \neq 0, \\ \ln x, \text{ if } \gamma = 0, \end{cases} \qquad x > 0,$$

in which $\gamma < 1$, $\gamma > 1$, and $\gamma = 1$ correspond to risk-seeking, risk-averse, and risk-neutral situations, respectively.

7.5. Ergodic controls

Many stochastic systems, though randomly fluctuating in nature, do exhibit certain regular patterns over a sufficiently long time period. This regular pattern can be characterized by the so-called *steady-state distribution* or *invariant measure* or *ergodic distribution*. The ergodic distribution, if it does exist, is independent of the initial state of the system and may be obtained by calculating the average of the states over a long time (called a *long-run average*) under some conditions. Similarly, in an optimal stochastic control problem over a long time period, minimizing a *long-run average cost* sometimes may better reflect the controller's desire to improve the performance on a long-term and average basis. Such problems are called *ergodic control* problems.

The system under this model is the same as (4.1), but now over the infinite time horizon $[0, +\infty)$. The admissible control is defined similarly. One typically seeks to minimize the expected long-run average cost

$$J(u(\cdot)) = \overline{\lim_{T\to+\infty}} \frac{1}{T} E \int_0^T f(t, x(t), u(t))dt, \tag{7.8}$$

or almost surely minimize the long-run average cost

$$J(u(\cdot)) = \overline{\lim_{T\to+\infty}} \frac{1}{T} \int_0^T f(t, x(t), u(t))dt. \tag{7.9}$$

Cost (7.8) is called a *mean average cost*, while (7.9) is called a *pathwise average cost*.

7.6. Partially observable systems

In all the stochastic optimal control problems introduced so far, including Problem (SS), it is assumed that the controller is able to completely observe the system state. However, in many practical situations, it often happens that the state can be only partially observed via other variables, and there could be noise existing in the observation systems. For example, in the technology diffusion example (Section 3.4) the true economic value of a technology may not be completely observable by the management; rather, they can be estimated only by other directly measurable entities such as quality level of the products or improvement degree of efficiency due to the adoption of the technology. Optimal control of such partially observed systems is called a *partially observable (observed) stochastic control model.*

Let W and $\tilde{W}$ be two independent standard Brownian motions on a filtered probability space $(\Omega, \mathcal{F}, \{\mathcal{F}_t\}_{t\geq 0}, \mathbf{P})$, with values in $\mathbb{R}^m$ and $\mathbb{R}^{\tilde{m}}$, respectively. Consider the following controlled *signal process*:

$$\tag{7.10} \begin{cases} dx(t) = b(t, x(t), y(t), u(t))dt + \sigma(t, x(t), y(t), u(t))dW(t) \\ \qquad\qquad + \alpha(t, x(t), y(t), u(t))d\tilde{W}(t), \\ x(0) = x_0, \end{cases}$$

with an *observation process*

$$\tag{7.11} \begin{cases} dy(t) = \beta(t, x(t), y(t), u(t))dt + d\tilde{W}(t), \\ y(0) = y_0. \end{cases}$$

Here α is called the *correlation* between the signal noise and observation noise. Let U be a given set in a metric space and set

$$\mathcal{Y}_t \stackrel{\Delta}{=} \sigma\{y(s) : 0 \leq s \leq t\}.$$

A control $u(\cdot)$ is called *admissible* if it takes values in U and is $\{\mathcal{Y}_t\}_{t\geq 0}$-adapted. The objective of the problem is to choose an admissible control such that the following cost functional is minimized:

$$\tag{7.12} J(u(\cdot)) = E\left\{\int_0^T f(t, x(t), y(t), u(t))dt + h(x(T), y(T))\right\}.$$

This is a substantially difficult problem compared with the completely observable case. One way of analyzing the problem is to formulate it as a completely observable stochastic control problem for a *stochastic partial differential equation* (SPDE for short). To do this, first we change the probability measure as follows:

$$d\tilde{\mathbf{P}} \stackrel{\Delta}{=} \rho^{-1}(T)d\mathbf{P},$$

where

$$\begin{aligned} \rho(t) \stackrel{\Delta}{=} \exp\Big\{ &\int_0^t \beta(s, x(s), y(s), u(s))dy(s) \\ &- \frac{1}{2}\int_0^t |\beta(s, x(s), y(s), u(s))|^2 ds\Big\}. \end{aligned}$$

It can be shown that under the new probability $\tilde{\mathbf{P}}$, W and y become independent Brownian motions. Now consider the *unnormalized conditional probability* $p(t)(\psi)$ and its density $p(t,x)$ determined by

$$p(t)(\psi) \stackrel{\Delta}{=} \widetilde{E}\left(\psi(x(t))\rho(t)\Big|\mathcal{Y}_t\right) \equiv \int_{\mathbb{R}^n} \psi(x)p(t,x)dx, \tag{7.13}$$

where ψ is any bounded Borel function on $\mathbb{R}^n$ (called a *test function*). Then $p(t) \equiv p(t,\cdot)$ satisfies the following SPDE (the *Duncan–Mortensen–Zakai equation*):

$$\begin{cases} dp(t) = A(t,y(t),u(t))p(t)dt + \sum_{k=1}^{\tilde{m}} M^k(t,y(t),u(t))p(t)dy_k(t), \\ p(0) = p_0. \end{cases} \tag{7.14}$$

Here y_k, $k = 1,2,\ldots,\tilde{m}$, are the components of y, p_0 is the density function of x_0, and the differential operators A and M^k are given by

$$\begin{cases} A(t,y,u)\psi(x) \stackrel{\Delta}{=} \sum_{i,j=1}^{n} \frac{\partial}{\partial x_i}\left(a^{ij}(t,x,y,u)\frac{\partial\psi(x)}{\partial x_j}\right) \\ \qquad\qquad - \sum_{i=1}^{n} \frac{\partial}{\partial x_i}\left(a^i(t,x,y,u)\psi(x)\right), \\ M^k(t,y,u)\psi(x) \stackrel{\Delta}{=} -\sum_{i=1}^{n} \alpha^{ik}(t,x,y,u)\frac{\partial\psi(x)}{\partial x_i} + \beta^k(t,x,y,u)\psi(x), \end{cases} \tag{7.15}$$

with

$$\begin{cases} (a^{ij}(t,x,y,u))_{ij} \equiv a(t,x,y,u) \stackrel{\Delta}{=} \frac{1}{2}\left(\sigma\sigma^T + \alpha\alpha^T\right)(t,x,y,u), \\ a^i(t,x,y,u) \stackrel{\Delta}{=} b^i(t,x,y,u) - \sum_{j=1}^{n} \frac{\partial a^{ij}(t,x,y,u)}{\partial x_j}, \\ h^k(t,x,y,u) \stackrel{\Delta}{=} \beta^k(t,x,y,u) - \sum_{i=1}^{n} \frac{\partial \alpha^{ik}(t,x,y,u)}{\partial x_i}, \\ \qquad i,j = 1,2,\ldots,n, \qquad k = 1,2,\ldots,\tilde{m}. \end{cases}$$

To properly define the solution to (7.14), consider the *Gelfand triple* $H^1 \hookrightarrow H^0 \hookrightarrow H^{-1}$, where $H^0 \stackrel{\Delta}{=} L^2(\mathbb{R}^n)$, H^1 is the Sobolev space

$$H^1 \stackrel{\Delta}{=} \left\{\varphi \in H^0 \Big| \frac{\partial\varphi}{\partial x_i} \in H^0, \quad i = 1,2,\ldots,n\right\}$$

with the norm

$$\|\partial\varphi\|_1 \stackrel{\Delta}{=} \left\{\int_{\mathbb{R}^n}\left(|\varphi(x)|^2 + \sum_{i=1}^{n} \left|\frac{\partial\varphi(x)}{\partial x_i}\right|^2\right)dx\right\}^{1/2},$$

and H^{-1} is the dual of H^1 given that H^0 is identified to its dual. Denote by $(\cdot\,,\cdot)$ the inner product on H^0 and by $\langle\,\cdot\,,\cdot\rangle$ the duality pairing between

H^{-1} and H^1. Under this notation, the second-order differential operator $A(t,y,u)$ given by (7.15) may be understood to be an operator from H^1 to H^{-1} by formally using the *Green formula* as follows:

$$\begin{aligned}\langle A(t,y,u)\varphi,\psi\rangle \stackrel{\Delta}{=} -\sum_{i,j=1}^{n}\Big(a^{ij}(t,x,y,u)\frac{\partial\varphi(x)}{\partial x_j},\frac{\partial\psi(x)}{\partial x_i}\Big)\\ +\sum_{i=1}^{n}\Big(a^{i}(t,x,y,u)\phi,\frac{\partial\psi(x)}{\partial x_i}\Big),\quad \forall\varphi,\psi\in H^1.\end{aligned}$$

A $\{\mathcal{Y}_t\}_{t\ge0}$-adapted, H^1-valued process $p\in L^2_{\mathcal{Y}}(0,T;H^1)$ defined on the filtered probability space $(\Omega,\mathcal{F},\{\mathcal{Y}_t\}_{t\ge0},\widetilde{\mathbf{P}})$ is called a solution to (7.14) if for each $\psi\in C_0^\infty(\mathbb{R}^n)$ and almost all $(t,\omega)\in[0,T]\times\Omega$,

$$\begin{aligned}(p(t),\psi)=(p_0,\psi)+\int_0^t\langle A(s,y(s),u(s))p(s),\psi\rangle\,ds\\ +\int_0^t\sum_{k=1}^{\tilde m}(M^k(s,y(s),u(s))p(s),\psi)dy_k(s).\end{aligned}$$

In terms of the new system (7.14), the original cost function (7.12) is converted to

$$\begin{aligned}J(u(\cdot)) &=\tilde E\left\{\int_0^T f(t,x(t),y(t),u(t))\rho(t)dt+h(x(T),y(T))\rho(T)\right\}\\ &=\tilde E\left\{\int_0^T \big(f(t,\cdot,y(t),u(t)),p(t)\big)dt+\big(h(\cdot,y(T)),p(T)\big)\right\}.\end{aligned}\tag{7.16}$$

Therefore, the original partially observable control problem can be reformulated as a problem of finding a U-valued, $\{\mathcal{Y}_t\}_{t\ge0}$-adapted process $u(\cdot)$ such that (7.16) is minimized subject to (7.14). The advantage of this formulation is that the underlying system is completely observable and both the system dynamics and the cost functional become *linear*, and the disadvantage is, of course, that the problem is turned into an *infinite-dimensional* one.

8. Historical Remarks

Although the name *optimal control theory* appeared in the late 1950s for the first time, the problem with essentially the same nature has a very long history; it can even be traced back to ancient times, when people discovered that the shortest path between two points is a straight line segment.

The well-accepted precursor of optimal control theory is the *calculus of variations*, which was born in the 1600s. Let us give a brief history of it here, which also will provide certain background for some material presented in later chapters. In 1662, Pierre de Fermat (1601–1665) wrote a paper using the method of calculus to minimize the passage time of a light ray through two optical media. The result is now known as *Fermat's principle of least time*. According to Goldstine [1], this probably can be regarded as the birth of the calculus of variations (although such a name was not given yet at Fermat's time). In 1685, Isaac Newton (1643–1727) studied the problem of characterizing the solid of revolution of least resistance. More interestingly, in June 1696, Johann Bernoulli (1667–1748) challenged the mathematical world at his time to solve the famous *brachistochrone problem* (which was first considered by Galilei Galileo (1564–1642) in 1638, but with an incorrect solution). The solution to this problem was given in the next year, 1697, by Johann Bernoulli himself, his brother Jacob (1654–1705), as well as Gottfried W. Leibniz (1646–1716), Newton, l'Hôpital (1661–1704), and Tschirnhaus (1651–1708). In 1744, Leonhard Euler (1707–1783) obtained the first-order necessary condition of the extremals—called the *Euler equation* (or the *Euler–Lagrange equation*). In 1755, Joseph L. Lagrange (1736–1813) introduced the so-called *δ-calculus*, which ushered in a new epoch in the area. After learning of this, Euler coined the name *calculus of variations* for the subject in 1756. In 1786, Adrien-Marie Legendre (1752–1833) introduced the second variation to study the sufficiency of an extremal to be either a maximum or a minimum. However, his paper contained a serious gap, which was later filled by Karl G. J. Jacobi (1804–1851) in 1838, leading to the so-called *Legendre–Jacobi theory*. In 1833, William R. Hamilton (1805–1865) introduced *Hamilton's principle of least action* (which was vaguely stated by Pierre L. M. Maupertuis (1698–1759) in 1744). In 1834–35, Hamilton found a system of ordinary differential equations (which is now called the *Hamiltonian canonical system*) equivalent to the Euler–Lagrange equation. He also derived the *Hamilton–Jacobi equation*, which was improved/modified by Jacobi in 1838. Afterwards, many scientists made extensive contributions to the field. Among others, we mention Karl K. Weierstrass (1815–1897), Ludwig Scheeffer, Rudolff F. A. Clebsch (1833–1872), Adolph Mayer (1839–1907), Oskar O. Bolza (1857–1942), Adolph Kneser (1862–1930), David Hilbert (1862–1943), Gilbert A. Bliss (1876–1951), Hans Hahn (1879–1934), and Constantin Carathéodory (1873–1950). It is well accepted that by the middle of the 20th century, the so-called *classical theory of calculus of variations* had been completed. For more detailed surveys and/or history of calculus of variations, see Goldstine [1], Giaquinta–Hildebrandt [1], Stampacchia [1], McShane [2], and Pesch–Bulirsch [1]. It is remarkable that among the famous 23 problems Hilbert posed at the International Congress of Mathematicians in 1900, the 23rd was on the calculus of variations, and the 19th and 20th were closely related to this subject.

According to Bellman–Kalaba [1], the history of (mathematical) control theory should probably be traced back to the work of James C. Maxwell (1831–1879) in 1868 and J. Vyshnegradskii in 1876. Among many who made influential contributions in the field, we mention Adolf Hurwitz (1859–1919), Henri H. Poincaré (1854–1912), Aleksandr M. Lyapunov (1857–1918), H. Nyquist, Norbert Wiener (1894–1964), and Andrey N. Kolmogorov (1903–1987).

Modern optimal control theory began its life at the end of World War II. The main starting point seems to be the (*differential*) *games* (for an obvious reason) studied both in the United States and the former Soviet Union. A group in the Rand Corporation, including R. Bellman, J. P. LaSalle, D. Blackwell, R. Isaacs, W. H. Fleming, and L. D. Berkovitz, were extensively working on differential games and related problems in the late 1940s and the 1950s. Such a study created a perfect environment for the birth of *Bellman's dynamic programming method*, which appeared publicly for the first time in 1952 (see Bellman [1]; Bellman claimed that it was obtained a couple of years earlier; see Bellman [5, p. 115]). On the other hand, according to L. S. Pontryagin (1908–1988) [3], his group, at the persistent request of the Board of Directors of the Steklov Institute of Mathematics, started the investigation of *optimal processes of regulation* in the fall of 1952, symbolized by the opening of a regular seminar at the Steklov Institute of Mathematics with the participants V. G. Boltyanski, R. V. Gamkrelidze, and E. F. Mishchenko. Among a large number of engineering problems, they singled out three types of problems: *singular perturbation of ordinary differential equations*, *differential pursuit and evasion games*, and *optimal control theory*. With their extensive study, *Pontryagin's maximum principle* was announced in 1956. In the late 1950s, R. E. Kalman [1] established optimal control theory for problems of linear systems with quadratic cost functional (*LQ problems*, for short). At the first IFAC (International Federation of Automatic Control) congress held in Moscow in 1960, Bellman's dynamic programming method, Pontryagin's maximum principle, and Kalman's LQ theory* were reported on, which formally announced the debut of modern optimal control theory.

Randomness was considered in the early stages of the development of modern optimal control theory. The earliest paper mentioning "stochastic control" was probably that of Bellman [7] in 1958. However, the Itô-type stochastic differential equation was not involved. The first paper that dealt with systems with close resemblance to those of diffusion type seems to be by Florentin [1] in 1961, where Bellman's dynamic programming approach was used to derive a partial differential equation associated with a continuous-time controlled Markov process. In 1962, Kushner [1] studied the stochastic optimal control with the Itô-type SDE as the state equation. For detailed remarks concerning various aspects of this field, see historical remarks in the later chapters.

In this chapter, Section 3 gives five practical examples that can be modeled as or approximated by stochastic optimal control problems. In fact, studies on these models themselves have been and remain very active. For the sake of the interested reader, here we sketch the literature of each of the 5 models.

The earliest formulation of a deterministic production planning prob-

---

* These works now have been regarded as the three milestones of optimal control theory; see Fleming [7].

lem was due to Modigliani–Hohn [1] in 1955 for the discrete-time case. Since then, various models including both deterministic and stochastic (diffusion) and both discrete-time and continuous-time have been studied. A systematic study and a comprehensive review of the literature in this area can be found in Bensoussan–Crouhy–Proth [1]. The stochastic production planning problem in Section 3.1 was formulated and studied by Bensoussan–Sethi–Vickson–Derzko [1]. Other than diffusion models, production planning problems incorporating various discrete events (such as breakdowns of machines and sudden changes in demand) can be found in Fleming–Sethi–Soner [1] and Akella–Kumar [1]. The state processes involved in these works are the so-called *piecewise deterministic process*, the theory of which has been extensively accounted for in Davis [4]. The mathematical framework for constructing near-optimal production control policies using the singular perturbation and hierarchical control approach was set up in Lehoczky–Sethi–Soner–Taksar [1] and further developed by Soner [2], Sethi–Zhang–Zhou [1,2], Sethi–Zhou [1], and Samaratunga–Sethi–Zhou [1].

Merton [1,2] was the first to formulate the continuous-time portfolio–consumption model (presented in Section 3.2) and apply stochastic control theory (basically the dynamic programming approach) to solve the problem in the late 1960s and early 1970s. Merton's results and technique have been further extended and developed by many authors since then. Among them we mention Richard [1], Breeden [1], Cox–Ingersoll–Ross [1], Merton [5], Fitzpatrick–Fleming [1], and Fleming–Zariphopoulou [1]. Another approach to solve the problem is the martingale method studied by Cox [1], Pliska [1], Karatzas–Lehoczky–Shreve [1], Cox–Huang [1], and Shreve–Soner–Xu [1].

Reinsurance and dividend management for insurance companies is a rather new and largely unexplored area. The issue was introduced by Leland–Toft [1] in 1996 in the context of constant debt repayment, although no optimization problem was treated there. An optimal dividend payout scheme in terms of optimization of the corporate policies was studied in Asmussen–Taksar [1] and Jeanblanc-Piqué–Shiryaev [1]. More complicated models involving both control of profit/risk-related activities as well as dividend payouts can be found in Hojgaard–Taksar [1], Radner–Shepp [1], and Taksar–Zhou [1]. The model introduced in Section 3.3 is based on that of Taksar–Zhou [1], the only difference being that one of the control variables, $c(\cdot)$ in (3.17), is the dividend payout *rate*, while a singular control problem is studied in Taksar–Zhou [1], where dividend distributions take place only at discrete time instants.

Discussions on the three phases in the life cycle of a new technology can be found in Cook–Mayes [1] and Rogers [1]. Studies on the technology diffusion root in those of marketing diffusion (Bass [1], Horsky–Simon [1], Dockner–Jorgensen [1] and Klepper [1]). The stochastic diffusion models of technology transfer were formulated and studied in Reinganum [1], Roberts–Weitzman [1], Kulatilaka [1], Dixit–Pindyck [1], Chi–Liu–Chen [1], and Liu–Zhou [1]. In particular, the diffusion coefficient of the form (3.22) was considered by Roberts–Weitzman [1], that of the form (3.23) was studied by Chi–Liu–Chen [1] (with u absent in (3.23)) and Liu–Zhou [1] (with u present in (3.23)).

Diffusion approximation, typically applied to queueing systems, is a technique to approximate some analytically difficult processes by diffusion processes, which have nice analytical structure and well-developed tools to handle. The simplest and easiest version of diffusion approximation dates back to 1951, when Donsker [1] put forward what is now known as *Donsker's theorem*, namely, sums of independent random variables can be approximated by Brownian motion. Prohorov [1] extended Donsker's theorem to doubly indexed families of random variables. Further, it was shown by Iglehart–Whitt [1,2] that a $G/G/1/\infty$ queue (namely, a queueing system with a single server, both interarrival time and service time being of general distributions, and infinite storage capacity) in heavy traffic can be approximated by a one-dimensional reflecting Brownian motion. This forms a foundation for diffusion approximations of more complex queueing systems. A good tutorial on the relevant literature can be found in Glynn [1]. Along another line, the problem of scheduling a queueing network can be approximated by a stochastic control problem involving a diffusion process. Once the latter is solved, an optimal policy can be obtained in terms of the original queueing system. Justifications of such a procedure mainly based on intuition and simulation were provided by Harrison–Wein [1,2] and Wein [1]. A more rigorous treatment using value functions was done by Kushner–Martins [1] and Kushner–Ramachandran [1,2]. Krichagina–Lou–Sethi–Taksar [1] applied diffusion approximation to study optimal production control of a single-machine system. The discussion in Section 3.5 is mainly based on Krichagina–Lou–Sethi–Taksar [1]. It should be noted, however, that a controlled queueing system is usually approximated by a *singular* control problem rather than the regular one proposed in (3.30). The one in (3.30) is a further approximation of the singular control problem. The main motivation for putting the diffusion approximation as one of the examples was to illustrate how diffusion processes are useful in various situations.

At the very early stage of development of the stochastic control theory (e.g., Florentin [1] and Kushner [1]), the formulation of the problem was completely parallel to the deterministic case, and the strong formulation was taken for granted. However, people soon realized that one needs to consider the probability space and the Brownian motion as part of the control (i.e., the weak formulation) when they were studying the existence of optimality (e.g., Fleming–Nisio [1]). Since then, most people have employed the weak formulation in discussing stochastic control problems (see Fleming–Rishel [1] and Fleming–Soner [1] and the references therein), although the importance of such a formulation may not be explained or spelled out in every work. In this book we emphasize the relation and difference of the strong and weak formulations. We point out that while the strong formulation is natural from the practical point of view, the weak formulation, as an *auxiliary* mathematical setup, is important and *necessary* not only in treating the existence of optimal controls but also in correctly applying the dynamic programming principle. Fortunately, for the problems that we study, the weak formulation does not change the *probability distribution* of the state process, hence nor the expected objective due to the weak uniqueness of the state equation (under certain reasonable conditions). This is the essence behind the weak formulation.

The existence of deterministic optimal control (with merely measurable controls) was first studied by Roxin [1] based on the so-called Roxin condition and a measurable selection theorem. See Berkovitz [1] and Cesari [1] for extensive study of the deterministic existence problems. The existence of optimal controls for processes of diffusion type had been a quite active research area during the early stage of the development of the stochastic optimal control theory. The results obtained really depended on the different formulations of the problem, but the key was to employ and/or explore certain compactness properties. Kushner [4] considered the class of controls that are functions of the time and current state and Lipschitz continuous in state. Fleming–Nisio [1], using Prohorov's topology, studied the case where the control was separated from the state in the dynamics. Beneš [1,2] and Duncan–Varaiya [1] derived the existence for controls that depended on the whole past history of the state based on the Girsanov transformation. Almost at the same time, Davis [1] employed dynamic programming and the verification theorem to derive the existence of optimality. Kushner [8] made use of the martingale representation theorem to handle the problem. Along a different line, the concept of *relaxed control* as a compactification device, first introduced by Young [1] for the deterministic case (see also Warga [1] and Balakrishnan [1, pp. 31–36]) for a detailed exposition), was applied by Fleming [5] to handle the stochastic existence. It was followed by El Karoui–Huu Nguyen–Jeanblanc-Picqué [1], who extensively investigated the existence of optimal controls by this approach. As with the deterministic case, a relaxed control reduced to a normal control under the Roxin condition. The proof of Theorem 5.3 seems to be new (as the diffusion is allowed to depend on the control), albeit one with combined ideas from Fleming–Nisio [1], Kushner [8], and Fleming [5].

Materials presented in Section 6 are based on Yong [10]. Example 6.4 is due to S. Tang. The results in this section are used to demonstrate the difference between the stochastic and deterministic systems in terms of the reachable sets, which play an essential role in studying the problems of controllability and existence of optimality.

Again, the models introduced in Section 7 have been and are still being widely studied. Let us give brief historical remarks for them for the convenience of the reader.

The first to introduce the optimal stopping problem was Chernoff [1], in 1968. It was soon discovered to be related to the free boundary problem of ice melted in water, the so-called *Stefan problem* (see van Moorbecke [1]). Bensoussan–Lions [1] employed the variational inequality technique to solve the problem. For a detailed account of such a treatment, see Bensoussan [2] and Friedman [1]. Another approach mainly based on the probabilistic argument, can be found in Dynkin [1], Shyriaev [1], and Krylov [2].

Bather–Chernoff [1] were the first to formulate a stochastic singular control problem in their study of spacecraft control in 1967. Beneš–Shepp–Witsenhaussen [1] proposed what is now known as the *principle of smooth fit* to explicitly solve a one-dimensional singular control problem based on the dynamic programming equation and the smoothness conjecture of the value function. Using the same idea, the one-dimensional case was solved completely by taking the form of the so-called *monotone follower problem*; see Karatzas [2], Karatzas–Shreve [1,2], Harrison–Taksar [1], Menaldi–Robin [1], and Chow–Menaldi–Robin [1]. For higher dimensions, the problem is much more difficult. Smoothness of the value functions was studied by Evans [1], Ishii–Koike [1], and Soner–Shreve [1,2]. When the value function is smooth, it was shown in Menaldi–Taksar [1] and Soner–Shreve [1,2] that construction of an optimal control (verification theorem) has to rely on the solution to the *Skorohod problem* (Skorohod [1], El Karoui–Chaleyat-Maurel [1]). There has been substantial research effort devoted to the Skorohod problem; see Strook–Varadhan [1], Tanaka [1], Lions–Sznitman [1], Varadhan–Williams [1], Saisho–Tanaka [1], Kurtz [1], Dupuis–Ishii [1], and Taksar [1]. Another interesting problem is the relationship between stochastic singular control and optimal stopping in the one-dimensional case, which states that the value function of the latter is nothing but the derivative in the state variable of the former. See Bather–Chernoff [1] and Karatzas–Shreve [1,2]. On the other hand, the existence of stochastic singular controls was investigated by Krylov [2], Menaldi–Taksar [1], and Haussmann–Suo [1]. The maximum principle was studied in Cadenillas–Haussmann [1], and the viscosity solution to the HJB equation was discussed in Lenhart [1], Tang–Yong [1], and Ma–Yong [2]. The (stochastic) impulse control problem was initiated by Bensoussan–Lions [2] in 1973, and a standard reference is Bensoussan–Lions [3]. For the deterministic counterpart, see Barles [1] and Yong [1,2,6,7]. A closely related problem is the so-called *optimal switching problem*. For this, we mention the works by Capuzzo-Dolcetta–Evans [1], Stojanovic–Yong [1,2], Lenhart–Yamada [1], and Yong [3–5]. The presentation of Section 7.3 is based on Ma–Yong [2].

Risk-sensitive control has been a very active research area recently. It was first put forth (with the exponential criterion) by Jacobson [1] in 1973 for linear systems, and then followed by Whittle [1], Speyer–Deyst–Jacobson [1], Speyer [1], Kumar–Van Schuppen [1], and Bensoussan–Van Schuppen [1] in the early stage of this theory. Research on the related dynamic programming equation can be found in Nagai [1] and Bensoussan–Frehse–Nagai [1], who mainly employed some PDE methods. One of the most significant features of risk-sensitive control is that it provides a link between optimality and robustness. More specifically, when the system noise approaches zero, the risk-sensitive control problem converges to a deterministic differential game problem related to H^∞ or robust control. This idea was initiated by Whittle [1,2] in an intuitive way, and was made rigorous using large deviation and viscosity solution theory by Fleming–McEneaney [1,2] and independently by James [1]. More work on the small noise limit can be found in Bensoussan–Nagai [2], Fleming–James [1], Runolfsson [1], Fleming–Hernández-Hernández [1,2], and James–Baras–Elliott [1]. Note that the H^∞ control theory has been extensively developed in parallel; see Başar–Bernhard [1].

Study on ergodic control dates back to 1957 when Bellman [6] dealt with the discrete-time case. The primary approach to continuous-time ergodic control is dynamic programming; see for example Lasry [1], Tarres [1,2], and Cox–Karatzas [1]. See also a survey paper by Robin [1]. It is interesting to note that the corresponding dynamic programming (HJB) equation, involving an unknown constant and an unknown function, has a close relation with the eigenvalue problem of a *Schrödinger operator* (an elliptic operator). See Karatzas [1], Bensoussan–Nagai [1], and Bensoussan–Frehse [1].

Optimal control of partially observable systems consists of two components: estimation and control. The estimation part is related to the filtering problem. The linear filtering problem was elegantly solved by the ubiquitous Kalman–Bucy filter [1] in 1961. For nonlinear filtering, which is much more difficult, there have been two essentially different approaches. One is based on the so-called innovation process, which was initially developed by Kailath [1] and Kailath–Frost [1] in 1968. This theory achieved its peak with the famous paper by Fujisaki–Kallianpur–Kunita [1]. See also the books by Liptser–Shiryayev [1] and Kallianpur [1] for a systematic exposition of this approach. Another approach was introduced during the middle to late 1960s by Duncan [1], Mortensen [1], and Zakai [1] independently, who derived a stochastic partial differential equation (widely known as the Duncan–Montensen–Zakai or DMZ equation) satisfied by the unnormalized conditional density function of the state. This approach was inspired by the function space integration introduced by Kac [2] and Ray [1], and has some deep relation with the path integral formulation of Feynman–Hibbs [1] in quantum physics (see a detailed discussion on this point in Mitter [1]). In fact, it was this relation that motivated Mitter [1], along with Brockett [1] and Brockett–Clark [1], to introduce the concept of *estimation algebra* and use Lie-algebraic and geometric techniques to solve the nonlinear filtering problem. For details on the geometric method see an earlier survey by Marcus [1] and a recent survey by Wong–Yau [1]. Unlike the finite-dimensional Kalman–Bucy filter, nonlinear filtering in general gives rise to infinite-dimensional filters (Hazewinkel–Marcus–Sussmann [1], Chaleyat-Maurel–Michel [1]), which are computationally insurmountable. People have been working on finding examples with finite-dimensional filters, including the most notable Beneš filter [3], as well as those proposed by Ocone [1], Wong [1], Haussmann–Pardoux [1], Charalambous–Elliott [1], and Yau–Yau [1]. As for control of partially observed processes, in the case of linear systems there is the so-called *separation principle* or *cer-*

tainty equivalence principle, originally discovered by Wonham [1] in 1968. This principle allows one to solve the estimation (filtering) problem first, and then to solve a completely observable control problem driven by the observer dynamics (*sufficient statistics* of the filtering problem). See the book of Davis [2] for a systematic account of linear filtering and control theory. For nonlinear partially observable control problems, however, the separation principle may not hold. One may then identify an *information state* given by the Duncan–Mortensen–Zakai equation and control a completely observable process in terms of the information state. This approach was sketched in Section 7.6, which nevertheless leads to an infinite-dimensional control problem. There is a vast amount of research on the nonlinear partially observed problem. We mention in particular Fleming [2], Rishel [1], Davis–Varaiya [1], Beneš–Karatzas [1], Bensoussan [4,6], Haussmann [5], Lions [3], Baras–Elliott–Kohlmann [1], Bensoussan–Nisio [1], Karatzas–Ocone [2,3], Zhang [1], Zhou [6,7], and Tang [1]. For study on stochastic partial differential equations including DMZ equations, see Pardoux [1], Kunita [1,3], Walsh [1], Krylov–Rozovskii [1–4], Rozovskii [1], and Zhou [8].

There are some other important stochastic control models and problems that are not mentioned in this book, including the problem of maximizing the probability of a certain event (Kulldorff [1] and Karatzas [4]) and the related Monge–Ampère equation (Pogorelov [1] and Krylov [1]), piecewise deterministic processes (Davis [4]), hybrid control problems (Branicky–Borkar–Mitter [1]), anticipative controls (Davis [3]), stochastic controls in Riemannian manifolds (Duncan [2] and Duncan–Upmeier [1]), controls of backward stochastic differential equations (Peng [5], Chen–Zhou [1], and Dokuchaev–Zhou [1]), stochastic near-optimal controls (Elliott–Kohlmann [1] and Zhou [12]), and numerical methods (Kushner–Dupuis [1] and Kushner–Yin [1]).

Chapter 3

Maximum Principle and Stochastic Hamiltonian Systems

1. Introduction

One of the principal approaches in solving optimization problems is to derive a set of *necessary conditions* that must be satisfied by any optimal solution. For example, in obtaining an optimum of a finite-dimensional function, one relies on the *zero-derivative condition* (for the unconstrained case) or the *Kuhn–Tucker condition* (for the constrained case), which are necessary conditions for optimality. These necessary conditions become sufficient under certain convexity conditions on the objective/constraint functions. Optimal control problems may be regarded as optimization problems in *infinite*-dimensional spaces; thus they are substantially difficult to solve. The maximum principle, formulated and derived by Pontryagin and his group in the 1950s, is truly a milestone of optimal control theory. It states that any optimal control along with the optimal state trajectory must solve the so-called (extended) Hamiltonian system, which is a two-point boundary value problem (and can also be called a *forward–backward differential equation*, to be able to compare with the stochastic case), plus a maximum condition of a function called the Hamiltonian. The mathematical significance of the maximum principle lies in that maximizing the Hamiltonian is much easier than the original control problem that is infinite-dimensional. This leads to closed-form solutions for certain classes of optimal control problems, including the linear quadratic case.

The original version of Pontryagin's maximum principle was for deterministic problems, with its key idea coming from the classical calculus of variations. In deriving the maximum principle, one first slightly perturbs an optimal control by means of the so-called *spike variation*, then considers the first-order term in a sort of Taylor expansion with respect to this perturbation. By sending the perturbation to zero, one obtains a kind of variational inequality. The final desired result (the maximum principle) then follows from the duality. However, one encounters an essential difficulty when trying to mimic this idea for optimal control problems of stochastic (diffusion) systems if the diffusion terms also depend on the controls. The main difficulty, roughly speaking, is that the Itô stochastic integral $\int_t^{t+\varepsilon} \sigma dW$ is only of order $\sqrt{\varepsilon}$ (rather than ε as with the normal Lebesgue integral); thus the usual first-order variation method fails. To overcome this difficulty, one needs to study both the first-order and second-order terms in the Taylor expansion of the spike variation and come up with a stochastic maximum principle involving a stochastic Hamiltonian system that consists of two *forward–backward stochastic differential equations* and a maximum condition with an additional term *quadratic* in the diffusion coefficient.

The rest of this chapter is organized as follows. Section 2 reviews the deterministic case with both necessary and sufficient conditions for optimality presented. In Section 3, a statement of the stochastic maximum principle is given in which the stochastic Hamiltonian system is introduced. Examples are presented to illustrate the result. Section 4 is devoted to a detailed proof of the stochastic maximum principle, while Section 5 discusses the sufficiency of the principle. Problems with terminal state constraints are dealt with in Section 6. Finally, some historical remarks are given in Section 7. We recommend that the reader skip Sections 4 and 6 at first reading.

2. The Deterministic Case Revisited

Let us first recall the formulation of the deterministic optimal control problem (as presented in Chapter 2, Section 2). We will also introduce some standard assumptions. Consider the following control system:

$$\begin{cases} \dot{x}(t) = b(t, x(t), u(t)), & \text{a.e. } t \in [0, T], \\ x(0) = x_0, \end{cases} \tag{2.1}$$

with the cost functional

$$J(u(\cdot)) = \int_0^T f(t, x(t), u(t))dt + h(x(T)). \tag{2.2}$$

Let us assume the following:

(D1) (U, d) is a separable metric space and $T > 0$.

(D2) The maps $b : [0, T] \times \mathbb{R}^n \times U \to \mathbb{R}^n$, $f : [0, T] \times \mathbb{R}^n \times U \to \mathbb{R}$, and $h : \mathbb{R}^n \to \mathbb{R}$ are measurable, and there exist a constant $L > 0$ and a modulus of continuity $\bar{\omega} : [0, \infty) \to [0, \infty)$ such that for $\varphi(t, x, u) = b(t, x, u), f(t, x, u), h(x)$, we have

$$\begin{cases} |\varphi(t, x, u) - \varphi(t, \widehat{x}, \widehat{u})| \leq L|x - \widehat{x}| + \bar{\omega}(d(u, \widehat{u})), \\ \qquad\qquad \forall t \in [0, T], x, \widehat{x} \in \mathbb{R}^n, \ u, \widehat{u} \in U, \\ |\varphi(t, 0, u)| \leq L, \qquad \forall (t, u) \in [0, T] \times U. \end{cases} \tag{2.3}$$

(D3) The maps b, f, and h are C^1 in x, and there exists a modulus of continuity $\bar{\omega} : [0, \infty) \to [0, \infty)$ such that for $\varphi(t, x, u) = b(t, x, u), f(t, x, u), h(x)$, we have

$$\begin{aligned} &|\varphi_x(t, x, u) - \varphi_x(t, \widehat{x}, \widehat{u})| \leq \bar{\omega}(|x - \widehat{x}| + d(u, \widehat{u})), \\ &\qquad\qquad \forall t \in [0, T], x, \widehat{x} \in \mathbb{R}^n, u, \widehat{u} \in U. \end{aligned} \tag{2.4}$$

Let $\mathcal{V}[0, T] = \{u(\cdot) : [0, T] \to U \mid u(\cdot) \text{ measurable } \}$. It is clear that under (D1)–(D2), for any $u(\cdot) \in \mathcal{V}[0, T]$, equation (2.1) admits a unique solution $x(\cdot) \stackrel{\Delta}{=} x(\cdot\,; u(\cdot))$ and (2.2) is well-defined. Our deterministic optimal control problem can be stated as follows.

Problem (D). Minimize (2.2) over $\mathcal{V}[0,T]$.

Any $\bar{u}(\cdot) \in \mathcal{V}[0,T]$ satisfying

$$J(\bar{u}(\cdot)) = \inf_{u(\cdot)\in\mathcal{V}[0,T]} J(u(\cdot)) \tag{2.5}$$

is called an *optimal control.* The corresponding state trajectory $\bar{x}(\cdot) \stackrel{\Delta}{=} x(\cdot\,; \bar{u}(\cdot))$ and $(\bar{x}(\cdot), \bar{u}(\cdot))$ are called an *optimal state trajectory* and *optimal pair*, respectively. Note that unlike the case in Chapter 2, Section 2, we consider here only the case where there is no state constraint, for which the feasibility and the admissibility of a control coincide.

The following is the well-known *Pontryagin's maximum principle*, which gives a set of *first-order necessary conditions* for optimal pairs.

Theorem 2.1. (Deterministic Maximum Principle) *Let (D1)–(D3) hold. Let $(\bar{x}(\cdot), \bar{u}(\cdot))$ be an optimal pair of Problem (D). Then there exists a $p(\cdot) : [0,T] \to \mathbb{R}^n$ satisfying the following:*

$$\begin{cases} \dot{p}(t) = -b_x(t, \bar{x}(t), \bar{u}(t))^\top p(t) + f_x(t, \bar{x}(t), \bar{u}(t)), & \text{a.e. } t \in [0,T], \\ p(T) = -h_x(\bar{x}(T)), \end{cases} \tag{2.6}$$

and

$$H(t, \bar{x}(t), \bar{u}(t), p(t)) = \max_{u\in U} H(t, \bar{x}(t), u, p(t)), \quad \text{a.e. } t \in [0,T], \tag{2.7}$$

where

$$\begin{aligned} H(t,x,u,p) &\stackrel{\Delta}{=} \langle\, p, b(t,x,u)\,\rangle - f(t,x,u), \\ &\qquad (t,x,u,p) \in [0,T] \times \mathbb{R}^n \times U \times \mathbb{R}^n. \end{aligned} \tag{2.8}$$

We call $p(\cdot)$ the *adjoint variable/function* and (2.6) the *adjoint equation*, respectively (corresponding to the pair $(\bar{x}(\cdot), \bar{u}(\cdot))$), and (2.7) the *maximum condition.* The function H defined by (2.8) is called the *Hamiltonian.* The state equation (2.1), the corresponding adjoint equation (2.6), along with the maximum condition (2.7), can be written as

$$\begin{cases} \dot{x}(t) = H_p(t, x(t), u(t), p(t)), \quad \text{a.e. } t \in [0,T], \\ \dot{p}(t) = -H_x(t, x(t), u(t), p(t)), \quad \text{a.e. } t \in [0,T], \\ x(0) = x_0, \qquad p(T) = -h_x(x(T)), \\ H(t, x(t), u(t), p(t)) = \max_{u\in U} H(t, x(t), u, p(t)), \quad \text{a.e. } t \in [0,T]. \end{cases} \tag{2.9}$$

The above system is called an (*extended*) *Hamiltonian system.* It partially characterizes the optimality of the problem. In cases where certain convexity conditions are presented, the Hamiltonian system of the above form fully characterizes an optimal control. In what follows, if $(x(\cdot), u(\cdot))$ is an optimal pair and $p(\cdot)$ is the corresponding adjoint function, then $(x(\cdot), u(\cdot), p(\cdot))$ will be called an *optimal triple.*

In the rest of this section we give a proof of Theorem 2.1. The main ingredients are the Taylor expansion of the state trajectory and the cost functional with respect to the perturbation of the control variable, and the duality between the variational equation (or the linearized state equation) and the adjoint equation. In doing this, we need the so-called *spike variation technique.* Let us make it more precise below.

Let $(\bar{x}(\cdot), \bar{u}(\cdot))$ be the given optimal pair. Let $\varepsilon > 0$ and $E_\varepsilon \subseteq [0,T]$ be a measurable set whose Lebesgue measure is $|E_\varepsilon| = \varepsilon$. Let $u(\cdot) \in \mathcal{V}[0,T]$ be any given control. We define the following:

$$u^\varepsilon(t) = \begin{cases} \bar{u}(t), & \text{if } t \in [0,T] \setminus E_\varepsilon, \\ u(t), & \text{if } t \in E_\varepsilon. \end{cases} \tag{2.10}$$

It is clear that $u^\varepsilon(\cdot) \in \mathcal{V}[0,T]$. Here, we should point out that since U is just a metric space, it does not necessarily have a linear structure. Thus, in general, a perturbation like $\bar{u}(t) + \delta u(t)$ is meaningless. We refer to $u^\varepsilon(\cdot)$ as a *spike* (or *needle*) *variation* of the control $\bar{u}(\cdot)$. The following result will be very useful in proving our maximum principle.

Lemma 2.2. *Let (D1)–(D3) hold. Let $x^\varepsilon(\cdot) \equiv x(\cdot\,; u^\varepsilon(\cdot))$ be the solution to (2.1) under the control $u^\varepsilon(\cdot)$, and let $y^\varepsilon(\cdot)$ be the solution of the following:*

$$\begin{cases} \dot{y}^\varepsilon(t) = b_x(t, \bar{x}(t), \bar{u}(t)) y^\varepsilon(t) \\ \qquad\qquad + \big\{ b(t, \bar{x}(t), u(t)) - b(t, \bar{x}(t), \bar{u}(t)) \big\} \chi_{E_\varepsilon}(t), \quad \text{a.e. } t \in [0,T], \\ y^\varepsilon(0) = 0. \end{cases} \tag{2.11}$$

Then

$$\begin{cases} \max\limits_{t \in [0,T]} |x^\varepsilon(t) - \bar{x}(t)| = O(\varepsilon), \\ \max\limits_{t \in [0,T]} |y^\varepsilon(t)| = O(\varepsilon), \end{cases} \tag{2.12}$$

$$\max_{t \in [0,T]} |x^\varepsilon(t) - \bar{x}(t) - y^\varepsilon(t)| = o(\varepsilon), \tag{2.13}$$

and

$$\begin{aligned} J(u^\varepsilon(\cdot)) - J(\bar{u}(\cdot)) &= \langle\, h_x(\bar{x}(T)), y^\varepsilon(T) \,\rangle \\ &\quad + \int_0^T \Big\{ \langle\, f_x(t, \bar{x}(t), \bar{u}(t)), y^\varepsilon(t) \,\rangle \\ &\quad + \big\{ f(t, \bar{x}(t), u(t)) - f(t, \bar{x}(t), \bar{u}(t)) \big\} \chi_{E_\varepsilon}(t) \Big\} dt + o(\varepsilon). \end{aligned} \tag{2.14}$$

We refer to (2.11) as the *variational equation* along the optimal pair $(\bar{x}(\cdot), \bar{u}(\cdot))$, and to (2.13) and (2.14) as the *Taylor expansions* of the state and the cost functional with respect to the spike variation of the control variable, respectively.

Proof. Let $\xi^\varepsilon(t) \stackrel{\Delta}{=} x^\varepsilon(t) - \bar{x}(t)$. Then, by (2.3), we have

$$|\xi^\varepsilon(t)| \leq \int_0^t L|\xi^\varepsilon(s)|ds + K\varepsilon, \qquad \forall t \in [0,T]. \tag{2.15}$$

Hereafter, K represents a generic constant, which may differ at different places. By Gronwall's inequality, we obtain the first relation in (2.12). The second relation in (2.12) can be proved similarly. Next, we set

$$\eta^\varepsilon(t) \stackrel{\Delta}{=} x^\varepsilon(t) - \bar{x}(t) - y^\varepsilon(t) \equiv \xi^\varepsilon(t) - y^\varepsilon(t). \tag{2.16}$$

Then (noting (2.10))

$$\begin{aligned}
\dot{\eta}^\varepsilon(t) &= b(t, x^\varepsilon(t), u^\varepsilon(t)) - b(t, \bar{x}(t), \bar{u}(t)) - b_x(t, \bar{x}(t), \bar{u}(t))y^\varepsilon(t) \\
&\quad - \big\{b(t, \bar{x}(t), u(t)) - b(t, \bar{x}(t), \bar{u}(t))\big\}\chi_{E_\varepsilon}(t) \\
&= \int_0^1 b_x(t, \bar{x}(t) + \theta\xi^\varepsilon(t), u^\varepsilon(t))d\theta \cdot \xi^\varepsilon(t) - b_x(t, \bar{x}(t), \bar{u}(t))y^\varepsilon(t) \\
&= \Big[\int_0^1 \big\{b_x(t, \bar{x}(t) + \theta\xi^\varepsilon(t), u^\varepsilon(t)) - b_x(t, \bar{x}(t), u^\varepsilon(t))\big\}d\theta\Big]\xi^\varepsilon(t) \\
&\quad + \big\{b_x(t, \bar{x}(t), u(t)) - b_x(t, \bar{x}(t), \bar{u}(t))\big\}\xi^\varepsilon(t)\chi_{E_\varepsilon}(t) \\
&\quad + b_x(t, \bar{x}(t), \bar{u}(t))\eta^\varepsilon(t).
\end{aligned}$$

Thus, it follows that (noting (2.12) and (2.4))

$$|\eta^\varepsilon(t)| \leq \int_0^t L|\eta^\varepsilon(s)|ds + K\varepsilon \int_0^T \bar{\omega}(|\xi^\varepsilon(s)|)ds + K\varepsilon^2. \tag{2.17}$$

Hence, (2.13) follows from Gronwall's inequality. In the same manner, we can prove (2.14). $\square$

Based on the above lemma, we now carry out a proof for Theorem 2.1.

Proof of Theorem 2.1. Let $p(\cdot)$ be the solution of (2.6). Then applying the differential chain rule to $\langle p(t), y^\varepsilon(t) \rangle$, we have the following duality relation:

$$\begin{aligned}
-\langle h_x(\bar{x}(T)), y^\varepsilon(T) \rangle &= \langle p(T), y^\varepsilon(T) \rangle - \langle p(0), y^\varepsilon(0) \rangle \\
&= \int_0^T \Big\{\langle f_x(t, \bar{x}(t), \bar{u}(t)), y^\varepsilon(t) \rangle \\
&\qquad + \langle p(t), b(t, \bar{x}(t), u(t)) - b(t, \bar{x}(t), \bar{u}(t)) \rangle \chi_{E_\varepsilon}(t)\Big\}dt.
\end{aligned} \tag{2.18}$$

The above is called the *duality relation* between $y^\varepsilon(\cdot)$ and $p(\cdot)$ (or that between the variational equation (2.11) and the adjoint equation (2.6)). Fix a time $\bar{t} \in [0,T)$ and a $u \in U$. Let $u(t) \equiv u$ and $E_\varepsilon = [\bar{t}, \bar{t}+\varepsilon]$ with $\varepsilon > 0$ being sufficiently small such that $E_\varepsilon \subseteq [0,T]$. Combining (2.18) with

(2.14) and noting the optimality of $\overline{u}(\cdot)$, we obtain

$$\begin{aligned}0 \le J(u^\varepsilon(\cdot)) - J(\overline{u}(\cdot)) &= \langle h_x(\overline{x}(T)), y^\varepsilon(T) \rangle \\ &+ \int_0^T \Big\{ \langle f_x(t,\overline{x}(t),\overline{u}(t)), y^\varepsilon(t) \rangle \\ &\qquad + \{ f(t,\overline{x}(t),u(t)) - f(t,\overline{x}(t),\overline{u}(t)) \} \chi_{E_\varepsilon}(t) \Big\} dt + o(\varepsilon) \\ &= - \int_{\bar{t}}^{\bar{t}+\varepsilon} \Big\{ H(t,\overline{x}(t),u(t),p(t)) - H(t,\overline{x}(t),\overline{u}(t),p(t)) \Big\} dt + o(\varepsilon).\end{aligned}$$

This, together with the separability of the metric space U, leads to the maximum condition (2.7). □

We have seen that in the above proof, the starting point is the optimality of the optimal control $\overline{u}(\cdot)$:

$$J(u^\varepsilon(\cdot)) - J(\overline{u}(\cdot)) \ge 0. \tag{2.19}$$

To obtain the first-order necessary conditions from this, we need some sort of Taylor expansion. Since $\mathcal{V}[0,T]$ does not necessarily have a linear structure, such an expansion has to be developed carefully. Once this is achieved, i.e., once (2.14) is obtained, we then have a necessary condition of the optimal pair $(\overline{x}(\cdot),\overline{u}(\cdot))$ in terms of $y^\varepsilon(\cdot)$ (the solution of the variational equation) that depends on the choice of $u(\cdot) \in \mathcal{V}[0,T]$. However, (2.14) is a sort of *implicit* necessary condition that is not easily applicable, as it involves $y^\varepsilon(\cdot)$. Thanks to the duality relation (2.18), we are able to get rid of $y^\varepsilon(\cdot)$. Then our desired result follows. In summary, the maximum principle is proved via the Taylor expansion and the duality relation.

We point out that Theorem 2.1 remains true if f and h are continuously differentiable in x with a polynomial growth in x; in particular, f and h are allowed to have quadratic growth in x. The reason is that the state trajectory $x(\cdot)$ obtained from the state equation (2.1) is bounded as long as b has linear growth.

The above maximum principle gives some necessary conditions for the optimal controls. In other words, it gives a minimum qualification for the candidates of optimal controls. One will naturally ask whether a given control satisfying the necessary conditions is indeed optimal. We now investigate *sufficient conditions* for optimality. To this end, let us introduce the following assumption.

(D4) The control domain U is a *convex body* (i.e., U is convex and has a nonempty interior) in $\mathbb{R}^k$, and the maps b and f are locally Lipschitz in u.

The following lemmas will be useful in the sequel and in later context.

Lemma 2.3. *Let $G \subseteq \mathbb{R}^n$ be a region and $\varphi : G \to \mathbb{R}$ be a locally Lipschitz continuous function. For any $x \in G$, define*

$$\partial\varphi(x) \triangleq \Big\{ \xi \in \mathbb{R}^n \mid \langle \xi, y \rangle \le \varlimsup_{\substack{z \to x, z \in G \\ t \downarrow 0}} \frac{\varphi(z+ty) - \varphi(z)}{t} \Big\}, \tag{2.20}$$

which is called Clarke's generalized gradient of φ *at* x. *Then:*

(i) $\partial\varphi(x)$ *is a nonempty, convex, and compact set in* $\mathbb{R}^n$.

(ii) $\partial(-\varphi)(x) = -\partial\varphi(x)$.

(iii) $0 \in \partial\varphi(x)$ *if* φ *attains a local minimum or maximum at* x.

(iv) *Let* $\mathcal{D}(\varphi_x) = \{z \in G \mid \varphi_x(z) \text{ exists }\}$. *Then*

$$\partial\varphi(x) = \overline{\text{co}}\{\lim_{k\to\infty} \varphi_x(x_k) \mid x_k \in \mathcal{D}(\varphi_x),\ x_k \to x,\ \lim_{k\to\infty} \varphi_x(x_k) \text{ exists}\}. \tag{2.21}$$

(v) *If in addition we assume that* $G \subseteq \mathbb{R}^n$ *is a convex body and* $\varphi : G \to \mathbb{R}$ *is a convex function (which implies the local Lipschitz continuity), then*

$$\partial\varphi(x) = \partial_c\varphi(x) \stackrel{\Delta}{=} \{\xi \in \mathbb{R}^n \mid \langle \xi, y \rangle \le \varphi(x+y) - \varphi(x),\ \forall y \in \mathbb{R}^n,\ \textit{with } x + y \in G\}. \tag{2.22}$$

Proof. (i) By assumption, for the given $x \in G$ and a small $\delta > 0$ with $B_\delta(x) \stackrel{\Delta}{=} \{z \in \mathbb{R}^n \mid |z - x| \le \delta\} \subseteq G$, there exists a constant $L > 0$ such that

$$|\varphi(z) - \varphi(\widehat{z})| \le L|z - \widehat{z}|, \qquad \forall z, \widehat{z} \in B_\delta(x).$$

Thus, for any fixed $y \in \mathbb{R}^n$,

$$\frac{|\varphi(z + ty) - \varphi(z)|}{t} \le L|y|, \qquad \forall z \in B_{\delta/2}(x),$$

provided that $t > 0$ is small enough. Hence, the following exists:

$$\varphi^0(x; y) \stackrel{\Delta}{=} \varlimsup_{\substack{z\to x,\, z\in G \\ t \downarrow 0}} \frac{\varphi(z + ty) - \varphi(z)}{t}.$$

It is easy to check that

$$\begin{cases} \varphi^0(x; \lambda y) = \lambda\varphi^0(x; y), & \forall y \in \mathbb{R}^n,\ \lambda \ge 0; \\ \varphi^0(x; y + z) \le \varphi^0(x; y) + \varphi^0(x; z), & \forall y, z \in \mathbb{R}^n. \end{cases} \tag{2.23}$$

Thus, $y \mapsto \varphi^0(x; y)$ is a convex function. Also, the second relation in (2.23) implies that

$$-\varphi^0(x; y) \le \varphi^0(x; -y), \quad \forall y \in \mathbb{R}^n. \tag{2.24}$$

Next, we fix a $z \in \mathbb{R}^n$ and define $F : \{\lambda z \mid \lambda \in \mathbb{R}\} \to \mathbb{R}$ by

$$F(\lambda z) = \lambda\varphi^0(x; z), \quad \forall \lambda \in \mathbb{R}.$$

Then for $\lambda \ge 0$ (noting (2.23) and (2.24)),

$$\begin{cases} F(\lambda z) \equiv \lambda\varphi^0(x; z) = \varphi^0(x; \lambda z), \\ F(-\lambda z) \equiv -\lambda\varphi^0(x; z) = -\varphi^0(x; \lambda z) \le \varphi^0(x, -\lambda z), \end{cases}$$

which yields

$$F(\lambda z) \le \varphi^0(x;\lambda z), \quad \forall \lambda \in \mathbb{R}. \tag{2.25}$$

Therefore, F is a linear functional defined on the linear space spanned by z, and it is dominated by the convex function $\varphi^0(x;\cdot)$. By the Hahn–Banach theorem (see Yosida [1, p. 102]), there exists a $\xi \in \mathbb{R}^n$ such that

$$\begin{cases} \langle \xi, \lambda z \rangle = F(\lambda z) \equiv \lambda\varphi^0(x;z), & \forall \lambda \in \mathbb{R}, \\ \langle \xi, y \rangle \le \varphi^0(x;y), & \forall y \in \mathbb{R}^n. \end{cases} \tag{2.26}$$

This implies $\xi \in \partial\varphi(x)$.

The convexity and compactness of $\partial\varphi(x)$ follow easily from (2.20).

(ii) We note that

$$\begin{aligned} (-\varphi)^0(x;y) &= \varlimsup_{\substack{z\to x\\ t\downarrow 0}} \frac{-\varphi(z+ty)+\varphi(z)}{t} \\ &= \varlimsup_{\substack{z'\to x\\ t\downarrow 0}} \frac{-\varphi(z')+\varphi(z'-ty)}{t} = \varphi^0(x;-y). \end{aligned}$$

Thus, $\xi \in \partial(-\varphi)(x)$ if and only if

$$\langle -\xi, y \rangle = \langle \xi, -y \rangle \le (-\varphi)^0(x;-y) = \varphi^0(x;y), \quad \forall y \in \mathbb{R}^n,$$

which is equivalent to $-\xi \in \partial\varphi(x)$.

(iii) Suppose φ attains a local minimum at x. Then

$$\begin{aligned} \varphi^0(x;y) &= \varlimsup_{\substack{z\to x\\ t\downarrow 0}} \frac{\varphi(z+ty)-\varphi(z)}{t} \\ &\ge \varlimsup_{t\downarrow 0} \frac{\varphi(x+ty)-\varphi(x)}{t} \ge 0 = \langle 0, y \rangle, \quad \forall y \in \mathbb{R}^n. \end{aligned}$$

This implies $0 \in \partial\varphi(x)$.

If φ attains a local maximum at x, then the conclusion follows from the fact that $-\varphi$ attains a local minimum at x along with (ii).

(iv) Define

$$D^*\varphi(x) = \{\lim_{k\to\infty} \varphi_k(x_k) \mid x_k \in \mathcal{D}(\varphi_x),\ x_k \to x,\ \lim_{k\to\infty} \varphi_x(x_k) \text{ exists}\}.$$

For any $\xi \in D^*\varphi(x)$ and $y \in \mathbb{R}^n$, there exists a sequence $x_k \in \mathcal{D}(\varphi_x)$ such that $x_k \to x$ and $\varphi_x(x_k) \to \xi$. Then

$$\begin{aligned} \langle \xi, y \rangle &= \lim_{k\to\infty} \langle \varphi_x(x_k), y \rangle \\ &= \lim_{k\to\infty} \lim_{t\to 0} \frac{\varphi(x_k+ty)-\varphi(x_k)}{t} \le \varphi^0(x;y). \end{aligned}$$

This implies $\xi \in \partial\varphi(x)$. Hence, by the convexity and closeness of $\partial\varphi(x)$,

$$\overline{\text{co}}\, D^*\varphi(x) \subseteq \partial\varphi(x). \tag{2.27}$$

Next, we define

$$(2.28)\qquad \widehat{\varphi}^0(x;y) \stackrel{\Delta}{=} \max_{\xi\in \overline{\mathrm{co}}\, D^*\varphi(x)} \langle \xi, y\rangle = \varlimsup_{\mathcal{D}(\varphi_x)\ni z\to x} \langle \varphi_x(z), y\rangle .$$

Note that by (2.26) (with $\lambda = 1$) and (2.27), one has

$$(2.29)\qquad \varphi^0(x;y) = \max_{\xi\in\partial\varphi(x)} \langle \xi, y\rangle \geq \widehat{\varphi}^0(x;y).$$

We claim that equality holds in (2.29). In fact, since φ is (locally) Lipschitz, by Rademacher's theorem (Evans–Gariepy [1, p. 81]), $\mathcal{D}(\varphi_x)$ is of full measure, or $|\mathcal{D}(\varphi_x)^c| = 0$. Now, let $y \neq 0$ be fixed. By the definition of $\widehat{\varphi}^0(x;y)$, for any $\varepsilon > 0$ there exists a $\delta > 0$ such that

$$(2.30)\qquad \langle \varphi_x(z), y\rangle \leq \widehat{\varphi}^0(x;y) + \varepsilon, \qquad \forall z \in B_{2\delta}(x)\cap \mathcal{D}(\varphi_x).$$

Next, for any $\zeta \in B_\delta(x)$, let

$$(2.31)\qquad R_\eta \stackrel{\Delta}{=} \{\zeta + \theta + ty \mid \theta \in \{y\}^\perp \cap B_\eta(0),\ 0 < t < \eta\},$$

with $\eta > 0$ small enough so that $R_\eta \subseteq B_{2\delta}(x)$. Here, $\{y\}^\perp \stackrel{\Delta}{=} \{z \in \mathbb{R}^n \mid \langle z, y\rangle = 0\}$. Now, by Fubini's theorem,

$$0 = |\mathcal{D}(\varphi_x)^c \cap R_\eta| = \int_{\{y\}^\perp\cap B_\eta(0)} \int_0^\eta I_{\mathcal{D}(\varphi_x)^c}(\zeta + \theta + ty)\,dt\,d\theta.$$

Thus, we obtain

$$\zeta + \theta + ty \in \mathcal{D}(\varphi_x), \qquad \text{a.e. } t\in(0,\eta), \quad \text{a.e. } \theta \in \{y\}^\perp \cap B_\eta(0).$$

Consequently, for almost all $\theta \in \{y\}^\perp \cap B_\eta(0)$ (in an $(n-1)$-dimensional Euclidean space), along the line segment $\{\zeta + \theta + ty \mid t \in (0,\eta)\}$, we have (noting (2.30) and $R_\eta \subseteq B_{2\delta}(x)$)

$$\begin{aligned}\varphi(\zeta+\theta+ty) - \varphi(\zeta+\theta) &= \int_0^t \langle \varphi_x(\zeta+\theta+sy), y\rangle\,ds\\ &\leq (\widehat{\varphi}^0(x;y) + \varepsilon)t.\end{aligned}$$

Since φ is continuous, the above implies that

$$(2.32)\qquad \varphi(z+ty) - \varphi(z) \leq (\widehat{\varphi}^0(x;y)+\varepsilon)t, \qquad \forall z \in B_\delta(x),\ t\in(0,\eta).$$

This results in $\varphi^0(x;y) \leq \widehat{\varphi}^0(x;y)$, which further leads to the equality in (2.29). Then, the convexity and closeness of the sets $\overline{\mathrm{co}}\, D^*\varphi(x)$ and $\partial\varphi(x)$ yields that they have to be the same, proving (iv).

(v) Let us first prove that φ is locally Lipschitz. We claim that for any $x_0 \in G$ and $r > 0$ with

$$B_{2r}(x_0) \stackrel{\Delta}{=} \{x \in \mathbb{R}^n \mid |x - x_0| \leq 2r\} \subseteq G,$$

one has

$$(2.33)\qquad |\varphi(x)-\varphi(y)| \le \frac{2}{r}\Big(\sup_{z\in B_{2r}(x_0)}|\varphi(z)|\Big)|x-y|, \quad \forall x,y\in B_r(x_0).$$

To show (2.33), we note that if $\theta:[\alpha,\beta]\to\mathbb{R}$ is a convex function and $\alpha\le s_1<s_2<s_3\le\beta$, then

$$(2.34)\qquad \frac{\theta(s_2)-\theta(s_1)}{s_2-s_1}\le\frac{\theta(s_3)-\theta(s_1)}{s_3-s_1}.$$

Indeed, noting $s_2=\frac{s_3-s_2}{s_3-s_1}s_1+\frac{s_2-s_1}{s_3-s_1}s_3$, we obtain, by the convexity of θ, that

$$\theta(s_2)\le\frac{s_3-s_2}{s_3-s_1}\theta(s_1)+\frac{s_2-s_1}{s_3-s_1}\theta(s_3).$$

Rearranging the above gives (2.34).

Now we prove (2.33). The case $x=y$ is trivial. Thus, let $x\ne y$. It is clear that there exists a unique $\bar t>|y-x|$ such that

$$\bar z\stackrel{\Delta}{=}x+\bar t\frac{y-x}{|y-x|}\in\partial B_{2r}(x_0)\stackrel{\Delta}{=}\{x\in\mathbb{R}^n\mid |x-x_0|=2r\}.$$

Then one must have $\bar t=|\bar z-x|\ge r$. Define

$$\theta(t)=\varphi\Big(x+t\frac{y-x}{|y-x|}\Big),\qquad 0\le t\le\bar t.$$

Clearly, $\theta:[0,\bar t]\to\mathbb{R}$ is convex, and

$$\theta(0)=\varphi(x),\quad \theta(|y-x|)=\varphi(y),\quad \theta(\bar t)=\varphi(\bar z).$$

Now, by (2.34), we obtain

$$\begin{aligned}\frac{\varphi(y)-\varphi(x)}{|y-x|}&=\frac{\theta(|y-x|)-\theta(0)}{|y-x|}\\&\le\frac{\theta(\bar t)-\theta(0)}{\bar t}=\frac{\varphi(\bar z)-\varphi(x)}{|\bar z-x|}\le\frac{2}{r}\sup_{z\in B_{2r}(x_0)}|\varphi(z)|.\end{aligned}$$

Exchanging x and y, we obtain (2.33). This proves the local Lipschitz continuity of $\varphi(\cdot)$.

Next, we prove (2.22). To this end, we first note that for any $y\in\mathbb{R}^n$, there exists a $\bar t>0$ such that $x+ty\in G$ for all $t\in[-\bar t,\bar t]$. The function $\theta(t)\stackrel{\Delta}{=}\varphi(x+ty)$ is convex on $[-\bar t,\bar t]$. Thus, by (2.34), the mapping $t\mapsto\frac{\varphi(x+ty)-\varphi(x)}{t}$ is nondecreasing for $t\in(0,\bar t]$ and bounded from below. This implies that the limit

$$D\varphi(x)(y)\stackrel{\Delta}{=}\lim_{t\downarrow0}\frac{\varphi(x+ty)-\varphi(x)}{t}$$

exists. Now let

$$\widehat\partial\varphi(x)\stackrel{\Delta}{=}\{\xi\in\mathbb{R}^n\mid\langle\xi,y\rangle\le D\varphi(x)(y),\quad\forall y\in\mathbb{R}^n\}.$$

We claim that $\partial_c\varphi(x) = \widehat{\partial}\varphi(x)$. In fact, $\xi \in \partial_c\varphi(x)$ if and only if for any $y \in \mathbb{R}^n$ and all small enough $t > 0$,

$$\langle \xi, y \rangle \leq \frac{\varphi(x+ty) - \varphi(x)}{t} \to D\varphi(x)(y),$$

which is equivalent to $\xi \in \widehat{\partial}\varphi(x)$. Thus, to prove (2.22) it suffices to prove

$$\varphi^0(x;y) = D\varphi(x)(y), \qquad \forall y \in \mathbb{R}^n. \tag{2.35}$$

First of all, by definition, we have

$$D\varphi(x)(y) \leq \varphi^0(x;y), \qquad \forall y \in \mathbb{R}^n.$$

To prove the reverse inequality, we note that the function $(t,z) \mapsto \frac{\varphi(z+ty)-\varphi(z)}{t}$ is continuous at (s,x) (with $s > 0$ and small). Hence for any $\varepsilon > 0$, there exists a $\delta > 0$ such that

$$\begin{gathered}\frac{\varphi(z+ty) - \varphi(z)}{t} \leq \frac{\varphi(x+sy) - \varphi(x)}{s} + \varepsilon,\\ \forall (t,z), \text{ with } t > 0,\ |t-s| + |z-x| < \delta.\end{gathered}$$

By the monotonicity of $t \mapsto \frac{\varphi(z+ty)-\varphi(z)}{t}$, the above inequality holds for all $t > 0$ small enough. Now, we first let $(t,z) \to (0,x)$, and then $s \to 0$ in the above inequality to get

$$\varphi^0(x;y) \leq \frac{\varphi(x+sy) - \varphi(x)}{s} + \varepsilon \to D\varphi(x)(y) + \varepsilon, \text{ as } s \to 0.$$

Hence, (2.35) holds, and (2.22) follows. □

In convex analysis, for a convex function $\varphi : G \to \mathbb{R}^n$, the set defined on the right-hand side of (2.22) is called the *subgradient* of φ at x (cf. Rockafellar [1]). Thus, the above (v) implies that for convex functions, Clarke's generalized gradient coincides with the subgradient.

By fixing some arguments of a function, one may naturally define its *partial generalized gradient.* For example, if $\varphi(x,u)$ is locally Lipschitz, by $\partial_x\varphi(x,u)$ we mean the (partial) generalized gradient of φ in x at (x,u).

Let us present one more technical lemma.

Lemma 2.4. *Let φ be a convex or concave function on $\mathbb{R}^n \times U$ with $U \subseteq \mathbb{R}^k$ being a convex body. Assume that $\varphi(x,u)$ is differentiable in x and $\varphi_x(x,u)$ is continuous in (x,u). Then*

$$\begin{aligned}\{(\varphi_x(x^*,u^*), r) \mid r \in \partial_u\varphi(x^*,u^*)\} \subseteq \partial_{x,u}\varphi(x^*,u^*),&\\ \forall (x^*,u^*) \in \mathbb{R}^n \times U.&\end{aligned} \tag{2.36}$$

Proof. First we assume that φ is convex. For any $\xi \in \mathbb{R}^n$ and $u \in \mathbb{R}^k$, we choose a sequence $\{(x_i, h_i)\} \subseteq \mathbb{R}^n \times \mathbb{R}$ in the following way:

$$\begin{aligned}&(x_i, u^*) \in \mathbb{R}^n \times U, \quad (x_i + h_i\xi, u^* + h_i u) \in \mathbb{R}^n \times U,\\ &h_i \downarrow 0, \text{ as } i \to \infty, \text{ and } |x_i - x^*| \leq h_i^2.\end{aligned}$$

Making use of the convexity of φ, we have

$$\begin{aligned}\lim_{i\to\infty} &\frac{\varphi(x_i+h_i\xi,u^*+h_iu)-\varphi(x^*,u^*+h_iu)}{h_i}\\ &\geq \lim_{i\to\infty}\frac{\langle\,\varphi_x(x^*,u^*+h_iu),x_i-x^*+h_i\xi\,\rangle}{h_i}=\langle\,\varphi_x(x^*,u^*),\xi\,\rangle.\end{aligned}$$

Similarly,

$$\lim_{i\to\infty}\frac{\varphi(x^*,u^*+h_iu)-\varphi(x^*,u^*)}{h_i}\geq\langle\,r,u\,\rangle.$$

Also,

$$\lim_{i\to\infty}\frac{\varphi(x^*,u^*)-\varphi(x_i,u^*)}{h_i}\geq\lim_{i\to\infty}\frac{\langle\,\varphi_x(x_i,u^*),x^*-x_i\,\rangle}{h_i}=0.$$

Adding up the above three inequalities, we get

$$\lim_{i\to\infty}\frac{\varphi(x_i+h_i\xi,u^*+h_iu)-\varphi(x_i,u^*)}{h_i}\geq\langle\,\varphi_x(x^*,u^*),\xi\,\rangle+\langle\,r,u\,\rangle.$$

Then $(\varphi_x(x^*,u^*),r)\in\partial_{x,u}\varphi(x^*,u^*)$ by the definition of the generalized gradient.

In the case that φ is concave, the desired result follows immediately by noting that $-\varphi$ is convex and Lemma 2.3-(ii). $\square$

Theorem 2.5. (Sufficient Conditions of Optimality) *Let (D1)–(D4) hold. Let $(\bar{x}(\cdot),\bar{u}(\cdot))$ be an admissible pair and let $p(\cdot)$ be the corresponding adjoint variable. Assume that $h(\cdot)$ is convex and $H(t,\cdot,\cdot,p(t))$ is concave. Then $(\bar{x}(\cdot),\bar{u}(\cdot))$ is optimal if*

$$H(t,\bar{x}(t),\bar{u}(t),p(t))=\max_{u\in U}H(t,\bar{x}(t),u,p(t)),\quad \text{a.e. } t\in[0,T]. \tag{2.37}$$

Proof. By (2.37) and Lemma 2.3-(iii), we have

$$0\in\partial_uH(t,\bar{x}(t),\bar{u}(t),p(t)). \tag{2.38}$$

From the assumption and Lemma 2.4, we conclude that

$$\Big(H_x(t,\bar{x}(t),\bar{u}(t),p(t)),0\Big)\in\partial_{x,u}H(t,\bar{x}(t),\bar{u}(t),p(t)). \tag{2.39}$$

Thus, by the concavity of $H(t,\cdot,\cdot,p(t))$, we have

$$\begin{aligned}&\int_0^T\{H(t,x(t),u(t),p(t))-H(t,\bar{x}(t),\bar{u}(t),p(t))\}dt\\ &\leq\int_0^T\langle\,H_x(t,\bar{x}(t),\bar{u}(t),p(t)),x(t)-\bar{x}(t)\,\rangle\,dt,\end{aligned} \tag{2.40}$$

for any admissible pair $(x(\cdot),u(\cdot))$. Define $\xi(t)=x(t)-\bar{x}(t)$, which satisfies

$$\begin{cases}\dot{\xi}(t)=b_x(t,\bar{x}(t),\bar{u}(t))\xi(t)+\alpha(t), & \text{a.e. } t\in[0,T],\\ \xi(0)=0,\end{cases} \tag{2.41}$$

where

$$\alpha(t) = -b_x(t, \bar{x}(t), \bar{u}(t))\xi(t) + b(t, x(t), u(t)) - b(t, \bar{x}(t), \bar{u}(t)).$$

By the duality relation between (2.41) and (2.6), we have

$$\begin{aligned}
\langle h_x(\bar{x}(T)), \xi(T) \rangle &= -\langle p(T), \xi(T) \rangle + \langle p(0), \xi(0) \rangle \\
&= -\int_0^T \{ \langle f_x(t, \bar{x}(t), \bar{u}(t)), \xi(t) \rangle + \langle p(t), \alpha(t) \rangle \} dt \\
&= \int_0^T \langle H_x(t, \bar{x}(t), \bar{u}(t), p(t)), \xi(t) \rangle \, dt \\
&\quad - \int_0^T \langle p(t), b(t, x(t), u(t)) - b(t, \bar{x}(t), \bar{u}(t)) \rangle \, dt \\
&\geq \int_0^T \{ H(t, x(t), u(t), p(t)) - H(t, \bar{x}(t), \bar{u}(t), p(t)) \} dt \\
&\quad - \int_0^T \langle p(t), b(t, x(t), u(t)) - b(t, \bar{x}(t), \bar{u}(t)) \rangle \, dt \\
&= -\int_0^T \{ f(t, x(t), u(t)) - f(t, \bar{x}(t), \bar{u}(t)) \} dt.
\end{aligned}$$

On the other hand, the convexity of h yields

$$\langle h_x(\bar{x}(T)), \xi(T) \rangle \leq h(x(T)) - h(\bar{x}(T)).$$

Combining the above two, we arrive at

$$J(\bar{u}(\cdot)) \leq J(u(\cdot)).$$

Since $u(\cdot)$ is arbitrary, the desired result follows. □

3. Statement of the Stochastic Maximum Principle

We first recall the *strong formulation* of the stochastic optimal control problem (see Chapter 2, Section 4.1), then introduce some assumptions.

Let $(\Omega, \mathcal{F}, \{\mathcal{F}_t\}_{t\geq 0}, \mathbf{P})$ be a *given* filtered probability space satisfying the usual conditions (see Chapter 1, Definition 2.6), on which an m-dimensional standard Brownian motion $W(t)$ (with $W(0) = 0$) is given. We consider the following stochastic controlled system:

$$\text{(3.1)} \qquad \begin{cases} dx(t) = b(t, x(t), u(t))dt + \sigma(t, x(t), u(t))dW(t), & t \in [0, T], \\ x(0) = x_0, \end{cases}$$

with the cost functional

$$\text{(3.2)} \qquad J(u(\cdot)) = E\Big\{ \int_0^T f(t, x(t), u(t))dt + h(x(T)) \Big\}.$$

In the above, $b : [0,T] \times \mathbb{R}^n \times U \to \mathbb{R}^n$, $\sigma : [0,T] \times \mathbb{R}^n \times U \to \mathbb{R}^{n\times m}$, $f : [0,T] \times \mathbb{R}^n \times U \to \mathbb{R}$, and $h : \mathbb{R}^n \to \mathbb{R}$. We define

$$
(3.3)\qquad \begin{cases} b(t,x,u) = \begin{pmatrix} b^1(t,x,u) \\ \vdots \\ b^n(t,x,u) \end{pmatrix}, \\ \sigma(t,x,u) = (\sigma^1(t,x,u),\ldots,\sigma^m(t,x,u)), \\ \sigma^j(t,x,u) = \begin{pmatrix} \sigma^{1j}(t,x,u) \\ \vdots \\ \sigma^{nj}(t,x,u) \end{pmatrix}, \quad 1 \le j \le m. \end{cases}
$$

Let us make the following assumptions (compare with (D1)–(D3)).

(S0) $\{\mathcal{F}_t\}_{t\ge0}$ is the natural filtration generated by $W(t)$, augmented by all the $\mathbf{P}$-null sets in $\mathcal{F}$.

(S1) (U,d) is a separable metric space and $T > 0$.

(S2) The maps b, σ, f, and h are measurable, and there exist a constant $L > 0$ and a modulus of continuity $\bar\omega : [0,\infty) \to [0,\infty)$ such that for $\varphi(t,x,u) = b(t,x,u), \sigma(t,x,u), f(t,x,u), h(x)$, we have

$$
(3.4)\qquad \begin{cases} |\varphi(t,x,u) - \varphi(t,\hat x,\hat u)| \le L|x-\hat x| + \bar\omega(d(u,\hat u)), \\ \qquad\qquad \forall t \in [0,T], \quad x,\hat x \in \mathbb{R}^n, \ u,\hat u \in U, \\ |\varphi(t,0,u)| \le L, \qquad \forall (t,u) \in [0,T]\times U. \end{cases}
$$

(S3) The maps b, σ, f, and h are C^2 in x. Moreover, there exist a constant $L > 0$ and a modulus of continuity $\bar\omega : [0,\infty) \to [0,\infty)$ such that for $\varphi = b, \sigma, f, h$, we have

$$
(3.5)\qquad \begin{cases} |\varphi_x(t,x,u) - \varphi_x(t,\hat x,\hat u)| \le L|x-\hat x| + \bar\omega(d(u,\hat u)), \\ |\varphi_{xx}(t,x,u) - \varphi_{xx}(t,\hat x,\hat u)| \le \bar\omega(|x-\hat x| + d(u,\hat u)), \\ \qquad\qquad \forall t \in [0,T], \ x,\hat x \in \mathbb{R}^n, \ u,\hat u \in U. \end{cases}
$$

Assumption (S0) signifies that the system noise is the only source of uncertainty in the problem, and the past information about the noise is available to the controller. This assumption will be crucial below.

Now we define (recall Convention 2.9 of Chapter 1)

$$
(3.6)\qquad \mathcal{U}[0,T] \stackrel{\Delta}{=} \{u : [0,T]\times\Omega \to U \mid u \text{ is } \{\mathcal{F}_t\}_{t\ge0}\text{-adapted}\}.
$$

Given $u(\cdot) \in \mathcal{U}[0,T]$, equation (3.1) is an SDE with random coefficients. From Chapter 1, Section 6.4, we see that under (S1)–(S2), for any $u(\cdot) \in \mathcal{U}[0,T]$, the state equation (3.1) admits a unique solution $x(\cdot) \equiv x(\cdot\,;u(\cdot))$ (in the sense of Definition 6.15 of Chapter 1) and the cost functional (3.2) is well-defined. In the case that $x(\cdot)$ is the solution of (3.1) corresponding to $u(\cdot) \in \mathcal{U}[0,T]$, we call $(x(\cdot),u(\cdot))$ an *admissible pair*, and $x(\cdot)$ an *admissible state process* (*trajectory*). Our optimal control problem can be stated as follows.

Problem (S). Minimize (3.2) over $\mathcal{U}[0,T]$.

Any $\overline{u}(\cdot) \in \mathcal{U}[0,T]$ satisfying

$$J(\overline{u}(\cdot)) = \inf_{u(\cdot)\in\mathcal{U}[0,T]} J(u(\cdot)) \tag{3.7}$$

is called an *optimal control.* The corresponding $\overline{x}(\cdot) \equiv x(\cdot\,;\overline{u}(\cdot))$ and $(\overline{x}(\cdot),\overline{u}(\cdot))$ are called an *optimal state process/trajectory* and *optimal pair*, respectively.

Notice that the *strong* formulation is adopted here for the optimal control problem (see Chapter 2, Section 4.1 for details). However, all the discussions and results in this chapter are readily adaptable to the *weak* formulation. The reason is that only the necessary and sufficient conditions of optimality are concerned here; hence an optimal pair (no matter whether in the strong or weak formulation) is given as a starting point, and all the results are valid for this given optimal pair on the probability space it attaches to.

Also, unlike the situation in Chapter 2, Section 4, Problem (S) here has no state constraint. Hence we do not need to distinguish the notions of feasibility and admissibility of a control. The problem with state constraint(s) will be discussed in Section 6.

Our next goal is to derive a set of necessary conditions for stochastic optimal controls, similar to the maximum principle for the deterministic case. To this end, we need some preparations.

3.1. Adjoint equations

In this subsection we will introduce adjoint equations involved in a stochastic maximum principle and the associated stochastic Hamiltonian system. Recall that $\mathcal{S}^n = \{A \in \mathbb{R}^{n\times n} \mid A^\top = A\}$ and let $(x(\cdot),u(\cdot))$ be a given optimal pair.

We have seen that in the deterministic case the adjoint variable $p(\cdot)$ plays a central role in the maximum principle. The adjoint equation that $p(\cdot)$ satisfies is a *backward* ordinary differential equation (meaning that the terminal value is specified). It is nevertheless equivalent to a forward equation if we reverse the time. In the stochastic case, however, one cannot simply reverse the time, as it may destroy the nonanticipativeness of the solutions. Instead, we introduce the following terminal value problem for a stochastic differential equation:

$$\begin{cases} dp(t) = -\Big\{b_x(t,\overline{x}(t),\overline{u}(t))^\top p(t) + \displaystyle\sum_{j=1}^m \sigma_x^j(t,\overline{x}(t),\overline{u}(t))^\top q_j(t) \\ \qquad\qquad - f_x(t,\overline{x}(t),\overline{u}(t))\Big\}dt + q(t)dW(t), \qquad t\in[0,T], \\ p(T) = -h_x(\overline{x}(T)). \end{cases} \tag{3.8}$$

Here the unknown is a *pair* of $\{\mathcal{F}_t\}_{t\geq 0}$-adapted processes $(p(\cdot), q(\cdot))$. We call the above a *backward stochastic differential equation* (BSDE, for short). The key issue here is that the equation is to be solved *backwards* (since the terminal value is given); however, the solution $(p(\cdot), q(\cdot))$ is required to be $\{\mathcal{F}_t\}_{t\geq 0}$-adapted. Any pair of processes $(p(\cdot), q(\cdot)) \in L^2_{\mathcal{F}}(0,T;\mathbb{R}^n) \times \left(L^2_{\mathcal{F}}(0,T;\mathbb{R}^n)\right)^m$ satisfying (3.8) is called an *adapted solution* of (3.8). A systematic study of such equations will be carried out in Chapter 7, from which, among other results, we will have the following: Under (S0)–(S3), for any $(\bar{x}(\cdot), \bar{u}(\cdot)) \in L^2_{\mathcal{F}}(0,T;\mathbb{R}^n) \times \mathcal{U}[0,T]$, (3.8) admits a *unique* adapted solution $(p(\cdot), q(\cdot))$. Note that (S0) cannot be omitted.

The adjoint variable $p(\cdot)$ in the deterministic case corresponds to the so-called *shadow price* or the *marginal value* of the resource represented by the state variable in economic theory. The maximum principle is nothing but the so-called *duality principle*: *Minimizing the total cost amounts to maximizing the total contribution of the marginal value*. See Chapter 5, Section 3.2, for a detailed discussion. Nevertheless, in the stochastic situation, the controller has to balance carefully the scale of control and the degree of uncertainty if a control made is going to affect the volatility of the system (i.e., if the diffusion coefficient depends on the control variable). Therefore, the marginal value alone may not be able to fully characterize the trade-off between the cost and control gain in an uncertain environment. One has to introduce another variable to reflect the uncertainty or the risk factor in the system. This is done by introducing an additional adjoint equation as follows:

$$
(3.9)\qquad \begin{cases} dP(t) = -\Big\{ b_x(t,\bar{x}(t),\bar{u}(t))^\top P(t) + P(t) b_x(t,\bar{x}(t),\bar{u}(t)) \\ \qquad\qquad + \displaystyle\sum_{j=1}^m \sigma_x^j(t,\bar{x}(t),\bar{u}(t))^\top P(t)\sigma_x^j(t,\bar{x}(t),\bar{u}(t)) \\ \qquad\qquad + \displaystyle\sum_{j=1}^m \{\sigma_x^j(t,\bar{x}(t),\bar{u}(t))^\top Q_j(t) + Q_j(t)\sigma_x^j(t,\bar{x}(t),\bar{u}(t)) \\ \qquad\qquad + H_{xx}(t,\bar{x}(t),\bar{u}(t),p(t),q(t))\Big\} dt + \displaystyle\sum_{j=1}^m Q_j(t) dW^j(t), \\ P(T) = -h_{xx}(\bar{x}(T)), \end{cases}
$$

where the *Hamiltonian* H is defined by

$$
(3.10)\qquad \begin{aligned} H(t,x,u,p,q) &= \langle\, p, b(t,x,u)\,\rangle + \operatorname{tr}\left[q^\top \sigma(t,x,u)\right] - f(t,x,u), \\ &\qquad (t,x,u,p,q) \in [0,T]\times\mathbb{R}^n\times U\times\mathbb{R}^n\times\mathbb{R}^{n\times m}, \end{aligned}
$$

and $(p(\cdot), q(\cdot))$ is the solution to (3.8). In the above (3.9), the unknown is again a pair of processes $(P(\cdot), Q(\cdot)) \in L^2_{\mathcal{F}}(0,T;\mathcal{S}^n) \times \left(L^2_{\mathcal{F}}(0,T;\mathcal{S}^n)\right)^m$.

Incidentally, the Hamiltonian $H(t,x,u,p,q)$ defined by (3.10) coincides with $H(t,x,u,p)$ defined by (2.8) when $\sigma = 0$.

Note that equation (3.9) is also a BSDE with matrix-valued unknowns. As with (3.8), under assumptions (S0)–(S3), there exists a unique adapted solution $(P(\cdot),Q(\cdot))$ to (3.9). We refer to (3.8) (resp. (3.9)) as the *first-order* (resp. *second-order*) *adjoint equations*, and to $p(\cdot)$ (resp. $P(\cdot)$) as the *first-order* (resp. *second-order*) *adjoint process.* In what follows, if $(\bar{x}(\cdot),\bar{u}(\cdot))$ is an optimal (resp. admissible) pair, and $(p(\cdot),q(\cdot))$ and $(P(\cdot),Q(\cdot))$ are adapted solutions of (3.8) and (3.9), respectively, then $(\bar{x}(\cdot),\bar{u}(\cdot),p(\cdot),q(\cdot),P(\cdot),Q(\cdot))$ is called an *optimal 6-tuple* (resp. *admissible 6-tuple*).

3.2. The maximum principle and stochastic Hamiltonian systems

Now we are going to state the Pontryagin-type maximum principle for optimal stochastic controls. At first glance, it might be quite natural for one to expect that a stochastic maximum principle should maximize the Hamiltonian H defined by (3.10). Unfortunately, this is *not* true if the diffusion coefficient σ depends on the control. Here is an example.

Example 3.1. Consider the following control system ($n=m=1$):

$$\begin{cases} dx(t)=u(t)dW(t), & t\in[0,1],\\ x(0)=0, \end{cases} \tag{3.11}$$

with the control domain being $U=[-1,1]$ and the cost functional being

$$J(u(\cdot))=E\Big\{\int_0^1[x(t)^2-\frac{1}{2}u(t)^2]dt+x(1)^2\Big\}. \tag{3.12}$$

Substituting $x(t)=\int_0^t u(s)dW(s)$ into the cost functional, we obtain via a simple calculation

$$J=E\int_0^1\Big(\frac{3}{2}-t\Big)u(t)^2dt. \tag{3.13}$$

Hence, the optimal control is $\bar{u}(t)\equiv 0$ with the optimal state trajectory $\bar{x}(t)\equiv 0$. However, the corresponding Hamiltonian is

$$H(t,\bar{x}(t),u,p(t),q(t))=\frac{1}{2}u^2+q(t)u. \tag{3.14}$$

This is a *convex* function in u, which does not attain a *maximum* at $\bar{u}(t)=0$ for any time t.

To obtain the correct form of the stochastic maximum principle, one should add the risk adjustment, which is related to the diffusion coefficient, to the Hamiltonian to reflect the controller's risk-seeking or risk-averse attitude. To do this, let us introduce the *generalized Hamiltonian*

$$\begin{aligned} &G(t,x,u,p,P)\\ &\quad\triangleq\frac{1}{2}\mathrm{tr}\Big(\sigma(t,x,u)^\top P\sigma(t,x,u)\Big)+\langle p,b(t,x,u)\rangle-f(t,x,u),\\ &\qquad\qquad (t,x,u,p,P)\in[0,T]\times\mathbb{R}^n\times U\times\mathbb{R}^n\times\mathcal{S}^n. \end{aligned} \tag{3.15}$$

Note that $G(t,x,u,p,P)$ depends on P, but not on q. Comparing (3.10), (3.15), and (2.8), we see that

$$\begin{cases} G(t,x,u,p,P) = H(t,x,u) + \frac{1}{2}\text{tr}\left[\sigma(t,x,u)^\top P\sigma(t,x,u)\right], \\ H(t,x,u,p,q) = H(t,x,u) + \text{tr}\left[q^\top \sigma(t,x,u)\right]. \end{cases}$$

We need to pay attention to the difference between $G(t,x,u,p,P)$ and $H(t,x,u,p,q)$. The function $G(t,x,u,p,P)$ will appear in the HJB equation of stochastic optimal control problems studied in the next chapter.

Next, associated with an optimal 6-tuple $(\bar{x}(\cdot),\bar{u}(\cdot),p(\cdot),q(\cdot),P(\cdot),Q(\cdot))$, we define an $\mathcal{H}$*-function*

$$\begin{aligned} &\mathcal{H}(t,x,u) \\ &\quad \triangleq H(t,x,u,p(t),q(t)) - \frac{1}{2}\text{tr}\left[\sigma(t,\bar{x}(t),\bar{u}(t))^\top P(t)\sigma(t,\bar{x}(t),\bar{u}(t))\right] \\ &\qquad + \frac{1}{2}\text{tr}\Big\{\left[\sigma(t,x,u) - \sigma(t,\bar{x}(t),\bar{u}(t))\right]^\top P(t) \\ &\qquad\qquad \cdot \left[\sigma(t,x,u) - \sigma(t,\bar{x}(t),\bar{u}(t))\right]\Big\} \\ &\quad \equiv \frac{1}{2}\text{tr}\left[\sigma(t,x,u)^\top P(t)\sigma(t,x,u)\right] + \langle p(t), b(t,x,u)\rangle - f(t,x,u) \\ &\qquad + \text{tr}\left[q(t)^\top\sigma(t,x,u)\right] - \text{tr}\left[\sigma(t,x,u)^\top P(t)\sigma(t,\bar{x}(t),\bar{u}(t))\right] \\ &\quad \equiv G(t,x,u,p(t),P(t)) + \text{tr}\Big\{\sigma(t,x,u)^\top\left[q(t) - P(t)\sigma(t,\bar{x}(t),\bar{u}(t))\right]\Big\}. \end{aligned} \tag{3.16}$$

Notice that an $\mathcal{H}$-function may be defined similarly associated with any *admissible* 6-tuple $(x(\cdot),u(\cdot),p(\cdot),q(\cdot),P(\cdot),Q(\cdot))$.

Theorem 3.2. (Stochastic Maximum Principle) *Let (S0)–(S3) hold. Let $(\bar{x}(\cdot),\bar{u}(\cdot))$ be an optimal pair of Problem (S). Then there are pairs of processes*

$$\begin{cases} (p(\cdot),q(\cdot)) \in L^2_{\mathcal{F}}(0,T;\mathbb{R}^n) \times \left(L^2_{\mathcal{F}}(0,T;\mathbb{R}^n)\right)^m, \\ (P(\cdot),Q(\cdot)) \in L^2_{\mathcal{F}}(0,T;\mathcal{S}^n) \times \left(L^2_{\mathcal{F}}(0,T;\mathcal{S}^n)\right)^m, \end{cases} \tag{3.17}$$

where

$$\begin{cases} q(\cdot) = (q_1(\cdot),\ldots,q_m(\cdot)), \quad Q(\cdot) = (Q_1(\cdot),\ldots,Q_m(\cdot)), \\ q_j(\cdot) \in L^2_{\mathcal{F}}(0,T;\mathbb{R}^n), \quad Q_j(\cdot) \in L^2_{\mathcal{F}}(0,T;\mathcal{S}^n), \quad 1 \le j \le m, \end{cases} \tag{3.18}$$

satisfying the first-order and second-order adjoint equations (3.8) and (3.9), respectively, such that

$$\begin{aligned} &H(t,\bar{x}(t),\bar{u}(t),p(t),q(t)) - H(t,\bar{x}(t),u,p(t),q(t)) \\ &\quad - \frac{1}{2}\text{tr}\Big(\left\{\sigma(t,\bar{x}(t),\bar{u}(t)) - \sigma(t,\bar{x}(t),u)\right\}^\top P(t) \\ &\qquad \cdot \left\{\sigma(t,\bar{x}(t),\bar{u}(t)) - \sigma(t,\bar{x}(t),u)\right\}\Big) \ge 0, \\ &\qquad\qquad \forall u \in U, \quad \text{a.e. } t \in [0,T], \quad \mathbf{P}\text{-a.s.}, \end{aligned} \tag{3.19}$$

or, equivalently,

$$(3.20)\qquad \mathcal{H}(t,\bar{x}(t),\bar{u}(t)) = \max_{u\in U}\mathcal{H}(t,\bar{x}(t),u), \quad \text{a.e. } t\in[0,T], \quad \mathbf{P}\text{-a.s.}$$

The inequality (3.19) is called the *variational inequality*, and (3.20) is called the *maximum condition*. Note that the third term on the left-hand side of (3.19) reflects the risk adjustment, which must be present when σ depends on u.

Now let us look at Example 3.1 again. The first-order and second-order adjoint equations associated with the optimal pair $(\bar{x}(t),\bar{u}(t)) \equiv (0,0)$ are

$$(3.21)\qquad \begin{cases} dp(t) = q(t)dW(t), \\ p(1) = 0, \end{cases}$$

and

$$(3.22)\qquad \begin{cases} dP(t) = 2dt + Q(t)dW(t), \\ P(1) = -2. \end{cases}$$

Thus, $(p(t),q(t)) = (0,0)$ and $(P(t),Q(t)) = (2t-4,0)$ are the (unique) adapted solutions. The left-hand side of (3.19) now reads

$$\begin{aligned} &\frac{1}{2}\bar{u}(t)^2 + q(t)\bar{u}(t) - \frac{1}{2}u^2 - q(t)u - \frac{1}{2}\big[\bar{u}(t)-u\big]^2 P(t) \\ &\quad = \Big(\frac{3}{2}-t\Big)u^2 \geq 0, \qquad \forall t\in[0,1]. \end{aligned}$$

We may also calculate

$$(3.23)\qquad \mathcal{H}(t,\bar{x}(t),u) = \frac{1}{2}(P(t)+1)u^2 + q(t)u = \frac{1}{2}(2t-3)u^2.$$

The above function $\mathcal{H}$ is *concave* in u for any $t\in[0,1]$, and $\bar{u}(t)\equiv 0$ does *maximize* $\mathcal{H}$. Note that for the present case (see (3.14)),

$$H(t,\bar{x}(t),u,p(t),q(t)) = \frac{1}{2}u^2.$$

Thus, one sees clearly how the second-order adjoint process (which represents the risk factor) plays a role in turning the convex function $u\mapsto H(t,\bar{x}(t),u,p(t),q(t))$ into the concave one $u\mapsto \mathcal{H}(t,\bar{x}(t),u)$ shown above.

Let us single out two important special cases.

Case 1. The diffusion does not contain the control variable, i.e.,

$$(3.24)\qquad \sigma(t,x,u)\equiv\sigma(t,x), \qquad \forall (t,x,u)\in[0,T]\times\mathbb{R}^n\times U.$$

In this case, the maximum condition (3.20) reduces to

$$(3.25)\qquad \begin{aligned} &H(t,\bar{x}(t),\bar{u}(t),p(t),q(t)) = \max_{u\in U} H(t,\bar{x}(t),u,p(t),q(t)), \\ &\qquad \text{a.e. } t\in[0,T], \quad \mathbf{P}\text{-a.s.}, \end{aligned}$$

which is parallel to the deterministic case (no risk adjustment is required). We note that in this case, equation (3.9) for $(P(\cdot), Q(\cdot))$ is not needed. Thus, the twice differentiability of the functions b, σ, f, and h in x is not necessary here.

Case 2. The control domain $U \subseteq \mathbb{R}^k$ is convex and all the coefficients are C^1 in u. Then (3.19) implies

$$\begin{aligned} &\langle H_u(t, \bar{x}(t), \bar{u}(t), p(t), q(t)), u - \bar{u}(t) \rangle \leq 0, \\ &\qquad\qquad \forall u \in U, \quad \text{a.e. } t \in [0, T], \quad \mathbf{P}\text{-a.s.} \end{aligned} \tag{3.26}$$

This is called a *local form* (in contrast to the *global form* (3.19) or (3.20)) of the maximum principle. Note that the local form does not involve the second-order adjoint process $P(\cdot)$ either.

Analogous to the deterministic case, the system (3.1) along with its first-order adjoint system can be written as follows:

$$\begin{cases} dx(t) = H_p(t, x(t), u(t), p(t), q(t))dt \\ \qquad\qquad + H_q(t, x(t), u(t), p(t), q(t))dW(t), \\ dp(t) = -H_x(t, x(t), u(t), p(t), q(t))dt + q(t)dW(t), \quad t \in [0, T], \\ x(0) = x_0, \qquad p(T) = -h_x(x(T)). \end{cases} \tag{3.27}$$

The combination of (3.27), (3.9), and (3.19) (or (3.20)) is called an *(extended) stochastic Hamiltonian system*, with its solution being a 6-tuple $(x(\cdot), u(\cdot), p(\cdot), q(\cdot), P(\cdot), Q(\cdot))$. Therefore, we can rephrase Theorem 3.2 as the following.

Theorem 3.3. *Let (S0)–(S3) hold. Let Problem (S) admit an optimal pair $(\bar{x}(\cdot), \bar{u}(\cdot))$. Then the optimal 6-tuple $(\bar{x}(\cdot), \bar{u}(\cdot), p(\cdot), q(\cdot), P(\cdot), Q(\cdot))$ of Problem (S) solves the stochastic Hamiltonian system (3.27), (3.9), and (3.19) (or (3.20)).*

It is seen from the above result that optimal control theory can be used to solve stochastic Hamiltonian systems. System (3.27) (with $u(\cdot)$ given) is also called a *forward-backward stochastic differential equation* (FBSDE, for short). In Chapter 7 we will study more about this type of equation.

3.3. A worked-out example

Let us now give a more substantial example to demonstrate how Theorem 3.2 can help in finding an optimal control.

Example 3.4. Consider the following control system ($n = m = 1$):

$$\begin{cases} dx(t) = u(t)dt + u(t)dW(t), \qquad t \in [0, 1], \\ x(0) = 1, \end{cases} \tag{3.28}$$

with the control domain being $U = \mathbb{R}$ and the cost functional being

$$J(u(\cdot)) = E\Big\{ - \int_0^1 \frac{1}{2} r u(t)^2 dt + \frac{1}{2} x(1)^2 \Big\}, \tag{3.29}$$

where $r \in (0,1)$ satisfies

$$\ln r + 2 - r < 0 \qquad (\text{or } 0 < r < 0.1586 \text{ approximately}). \tag{3.30}$$

Suppose $(\bar{x}(\cdot), \bar{u}(\cdot))$ is an optimal pair (which we are going to identify). Then the corresponding adjoint equations are

$$\begin{cases} dp(t) = q(t)dW(t), & t \in [0,1], \\ p(1) = -\bar{x}(1), \end{cases} \tag{3.31}$$

and

$$\begin{cases} dP(t) = Q(t)dW(t), & t \in [0,1], \\ P(1) = -1. \end{cases} \tag{3.32}$$

Clearly, $(P(t), Q(t)) = (-1, 0)$ is the only adapted solution to (3.32). The corresponding $\mathcal{H}$-function is

$$\mathcal{H}(t,x,u) = -\frac{1}{2}(1-r)u^2 + (p(t) + q(t) + \bar{u}(t))u. \tag{3.33}$$

Note that the above is a concave function of u due to the fact that $r < 1$. So by Theorem 3.2, a necessary condition for $\bar{u}(\cdot)$ to be optimal is

$$-(1-r)\bar{u}(t) + p(t) + q(t) + \bar{u}(t) = 0,$$

or

$$\bar{u}(t) = -\frac{p(t) + q(t)}{r}. \tag{3.34}$$

By plugging (3.34) into (3.28) and putting it together with (3.31), we obtain the corresponding stochastic Hamiltonian system:

$$\begin{cases} d\bar{x}(t) = -\dfrac{p(t)+q(t)}{r}dt - \dfrac{p(t)+q(t)}{r}dW(t), \\ dp(t) = q(t)dW(t), \\ \bar{x}(0) = 1, \qquad p(1) = -\bar{x}(1). \end{cases} \tag{3.35}$$

Once we find an adapted solution $(\bar{x}(\cdot), p(\cdot), q(\cdot))$ of the above, a candidate for the optimal control can then be given by (3.34). We now try to find an adapted solution of (3.35). Suppose $(\bar{x}(\cdot), p(\cdot), q(\cdot))$ is an adapted solution such that the following relation holds:

$$p(t) = \Theta(t)\bar{x}(t), \qquad \forall t \in [0,1], \ \mathbf{P}\text{-a.s.}, \tag{3.36}$$

for some deterministic function $\Theta(\cdot)$. We would like to determine the equation that $\Theta(\cdot)$ should satisfy. To this end, we differentiate (3.36) using Itô's formula:

$$dp(t) = \Big[\dot{\Theta}(t)\bar{x}(t) - \Theta(t)\frac{p(t)+q(t)}{r}\Big]dt - \Theta(t)\frac{p(t)+q(t)}{r}dW(t).$$

Comparing this with (3.31), we obtain

$$
\begin{cases} \dot{\Theta}(t)\bar{x}(t) = \dfrac{\Theta(t)(p(t)+q(t))}{r}, \\ q(t) = -\dfrac{\Theta(t)(p(t)+q(t))}{r}. \end{cases} \tag{3.37}
$$

Consequently,

$$
\begin{cases} q(t) = -\dfrac{p(t)\Theta(t)}{r+\Theta(t)} = -\dfrac{\Theta^2(t)\bar{x}(t)}{r+\Theta(t)}, \\ \dfrac{p(t)+q(t)}{r} = \dfrac{\Theta(t)\bar{x}(t)}{r+\Theta(t)}. \end{cases} \tag{3.38}
$$

Then, combining the first relation in (3.37) and the second in (3.38), we see that if $\Theta(\cdot)$ is the right choice, it should satisfy the following:

$$
\begin{cases} \dot{\Theta}(t) = \dfrac{\Theta(t)^2}{r+\Theta(t)}, \qquad t \in [0,1], \\ \Theta(1) = -1. \end{cases} \tag{3.39}
$$

Now, suppose for the time being that (3.39) admits a solution $\Theta(\cdot)$ on $[0,1]$. Then it is necessary that (noting that $r+\Theta(1) < 0$)

$$
r+\Theta(t) < 0, \qquad \forall t \in [0,1]. \tag{3.40}
$$

In this case, we have a candidate for the optimal control:

$$
\bar{u}(t) = -\frac{\Theta(t)}{r+\Theta(t)}\bar{x}(t), \quad t \in [0,1], \ \mathbf{P}\text{-a.s.}, \tag{3.41}
$$

which is in state feedback form. The corresponding state process is exactly $\bar{x}(t)$, which is the solution to the first equation in (3.35), or to the following:

$$
\begin{cases} d\bar{x}(t) = -\dfrac{\Theta(t)}{r+\Theta(t)}\bar{x}(t)dt - \dfrac{\Theta(t)}{r+\Theta(t)}\bar{x}(t)dW(t), \\ \bar{x}(0) = 1. \end{cases} \tag{3.42}
$$

Let us now show that $(\bar{x}(\cdot), \bar{u}(\cdot))$ obtained *is* indeed an optimal pair. In fact, a simple calculation via Itô's formula yields that for any admissible pair $(x(\cdot), u(\cdot))$,

$$
\begin{aligned} d(\Theta(t)x(t)^2) = &\Big\{\Theta(t)u(t)^2 + 2\Theta(t)u(t)x(t) + \frac{\Theta(t)^2x(t)^2}{r+\Theta(t)}\Big\}dt \\ &+ \{\cdots\}dW(t). \end{aligned}
$$

Integrating from 0 to 1, taking expectations, and noting (3.39) and (3.35), it follows that

$$
Ex(1)^2 = -\Theta(0) - E\int_0^1 \Big\{\Theta(t)u(t)^2 + 2\Theta(t)u(t)x(t) + \frac{\Theta(t)^2x(t)^2}{r+\Theta(t)}\Big\}dt.
$$

Substituting this into the right-hand side of (3.29) yields

$$J(u(\cdot)) = -\frac{1}{2}E\int_0^1 (r+\Theta(t))\Big(u(t)+\frac{\Theta(t)x(t)}{r+\Theta(t)}\Big)^2 dt - \frac{1}{2}\Theta(0). \tag{3.43}$$

Therefore, by (3.40), the minimum value of the cost functional will be achieved at $\overline{u}(t) = -\frac{\Theta(t)\overline{x}(t)}{r+\Theta(t)}$, which is exactly given by (3.41).

We now show that equation (3.39) does admit a solution $\Theta(\cdot)$ with the property (3.40). To this end, first note that (3.39) is equivalent to

$$\ln(-\Theta(t)) - \frac{r}{\Theta(t)} = t - 1 + r. \tag{3.44}$$

Define $f^t(\lambda) = \ln\lambda + \frac{r}{\lambda} - t + 1 - r$, $\lambda \in (0, +\infty)$. Noting (3.30), for all $t \in [0,1)$, it follows that

$$\begin{aligned} &f^t(r) = \ln r + 2 - t - r < 0,\\ &f^t(1) = 1 - t > 0, \end{aligned} \tag{3.45}$$

we conclude that there is $-\Theta(t) \in (r,1)$ (i.e., $r + \Theta(t) < 0$) such that $f^t(-\Theta(t)) = 0$ for $t \in [0,1)$. Moreover,

$$\frac{d}{d\lambda}f^t(\lambda) = \frac{\lambda - r}{\lambda^2}\begin{cases} < 0, & \text{if } \lambda < r,\\ > 0, & \text{if } \lambda > r,\\ = 0, & \text{if } \lambda = r. \end{cases} \tag{3.46}$$

Thus the $\Theta(t)$ satisfying $f^t(-\Theta(t)) = 0$ and $r + \Theta(t) < 0$ is unique. Finally, $f^1(1) = 0$ implies $\Theta(1) = -1$. This establishes the existence of solutions to (3.39) subject to (3.40).

To conclude, an optimal feedback control is given by (3.41), and the optimal value is $-\frac{1}{2}\Theta(0)$ (see (3.43) with $\Theta(\cdot)$ solved via (3.39)).

4. A Proof of the Maximum Principle

In this section we are going to give a proof of the stochastic maximum principle, Theorem 3.2. Our main idea is still the variational technique. However, due to the presence of the diffusion coefficient, which may contain the control variable, the method that works for deterministic case has to be substantially modified to fit the stochastic case.

Let us look at the following simple example, from which we can see some significant differences between the stochastic and deterministic cases.

Example 4.1. Consider the following control system ($n = m = 1$):

$$\begin{cases} dx(t) = u(t)dW(t), & t \in [0,T],\\ x(0) = 0, \end{cases} \tag{4.1}$$

with the control domain being $U = \{0, 1\}$ and the cost functional being

$$J(u(\cdot)) = Ex(T)^2. \tag{4.2}$$

Then the optimal pair is clearly given by $(\bar{x}(t), \bar{u}(t)) \equiv (0,0)$. Now let us try to make a perturbation similar to what we have done for the deterministic case. Let $E_\varepsilon = [s, s+\varepsilon] \subseteq [0,T]$ and $u(t) \equiv 1$. Set

$$u^\varepsilon(t) = \begin{cases} \bar{u}(t) \equiv 0, & \text{if } t \in [0,T] \setminus E_\varepsilon, \\ u(t) \equiv 1, & \text{if } t \in E_\varepsilon. \end{cases} \tag{4.3}$$

Let $x^\varepsilon(\cdot)$ be the solution of

$$\begin{cases} dx^\varepsilon(t) = u^\varepsilon(t)dW(t), & t \in [0,T], \\ x^\varepsilon(0) = 0. \end{cases} \tag{4.4}$$

Then, by Itô's formula, for $t \geq s+\varepsilon$,

$$E|x^\varepsilon(t) - \bar{x}(t)|^2 = E|x^\varepsilon(t)|^2 = \frac{1}{2}\int_0^t |u^\varepsilon(r)|^2 dr = \frac{1}{2}\varepsilon. \tag{4.5}$$

This means that the analogue of (2.12) no longer holds for the stochastic case. The basic reason is, roughly speaking, that the stochastic integral $\int_t^{t+\varepsilon} \sigma dW$ is only of order $\sqrt{\varepsilon}$, so the usual first-order Taylor expansion is inadequate. The resolution to this is to consider the *second-order* Taylor expansion.

4.1. A moment estimate

In this subsection we prove an elementary lemma, which will be useful in the sequel.

Lemma 4.2. *Let $Y(t) \in L^2_{\mathcal{F}}(0,T;\mathbb{R}^n)$ be the solution of the following:*

$$\begin{cases} dY(t) = \{A(t)Y(t) + \alpha(t)\}dt + \displaystyle\sum_{j=1}^m \{B^j(t)Y(t) + \beta^j(t)\}dW^j(t), \\ Y(0) = Y_0, \end{cases} \tag{4.6}$$

where $A, B^j : [0,T] \times \Omega \to \mathbb{R}^{n\times n}$ and $\alpha, \beta^j : [0,T]\times\Omega \to \mathbb{R}^n$ are $\{\mathcal{F}_t\}_{t\geq 0}$-adapted, and

$$\begin{cases} |A(t)|, |B^j(t)| \leq L, \quad \text{a.e. } t \in [0,T], \ \mathbf{P}\text{-a.s.}, \ 1 \leq j \leq m, \\ \displaystyle\int_0^T \{E|\alpha(s)|^{2k}\}^{\frac{1}{2k}} ds + \int_0^T \{E|\beta^j(s)|^{2k}\}^{\frac{1}{k}} ds < \infty, \quad 1 \leq j \leq m, \end{cases} \tag{4.7}$$

for some $k \geq 1$. Then

$$\begin{aligned} \sup_{t\in[0,T]} E|Y(t)|^{2k} \leq K\Big\{E|Y_0|^{2k} &+ \Big(\int_0^T \{E|\alpha(s)|^{2k}\}^{\frac{1}{2k}} ds\Big)^{2k} \\ &+ \sum_{j=1}^m \Big(\int_0^T \{E|\beta^j(s)|^{2k}\}^{\frac{1}{k}} ds\Big)^k\Big\}. \end{aligned} \tag{4.8}$$

Proof. For notational simplicity, we prove only the case $m = 1$ (i.e., the Brownian motion $W(t)$ is one-dimensional), leaving the case $m > 1$ to the interested reader. Thus, the index j in $B^j(\cdot)$ and $\beta^j(\cdot)$ will be dropped. We first assume that $\alpha(\cdot)$ and $\beta(\cdot)$ are bounded. Let $\varepsilon > 0$ and define

$$\langle Y \rangle_\varepsilon \stackrel{\Delta}{=} \sqrt{|Y|^2 + \varepsilon^2}, \qquad \forall Y \in \mathbb{R}^n. \tag{4.9}$$

Note that for any $\varepsilon > 0$, the map $Y \mapsto \langle Y \rangle_\varepsilon$ is smooth and $\langle Y \rangle_\varepsilon \to |Y|$ as $\varepsilon \to 0$. The purpose of using such a function is to avoid some difficulties that might be encountered in differentiating functions like $|Y|^{2k}$ for noninteger k. Applying Itô's formula to $\langle Y(t) \rangle_\varepsilon$, we have

$$\begin{aligned} E\langle Y(t)\rangle_\varepsilon^{2k} &\le E\langle Y(0)\rangle_\varepsilon^{2k} \\ &\quad + 2kE\int_0^t \langle Y(s)\rangle_\varepsilon^{2k-1}\{|A(s)|\langle Y(s)\rangle_\varepsilon + |\alpha(s)|\}ds \\ &\quad + k(2k-1)E\int_0^t \langle Y(s)\rangle_\varepsilon^{2k-2}\{|B(s)|\langle Y(s)\rangle_\varepsilon + |\beta(s)|\}^2 ds \\ &\le E\langle Y(0)\rangle_\varepsilon^{2k} + K_0 E\int_0^t \Big\{\langle Y(s)\rangle_\varepsilon^{2k} + |\alpha(s)|\langle Y(s)\rangle_\varepsilon^{2k-1} \\ &\qquad + |\beta(s)|^2\langle Y(s)\rangle_\varepsilon^{2k-2}\Big\}ds. \end{aligned} \tag{4.10}$$

Here $K_0 = K_0(k, L)$ is independent of t. Applying Young's inequality, we obtain

$$\begin{aligned} E\langle Y(t)\rangle_\varepsilon^{2k} \le& E\langle Y(0)\rangle_\varepsilon^{2k} \\ &+ KE\int_0^t \Big\{\langle Y(s)\rangle_\varepsilon^{2k} + |\alpha(s)|^{2k} + |\beta(s)|^{2k}\Big\}ds. \end{aligned} \tag{4.11}$$

Hence, it follows from Gronwall's inequality that

$$\begin{aligned} E\langle Y(t)\rangle_\varepsilon^{2k} \le K\Big\{E\langle Y(0)\rangle_\varepsilon^{2k} + E\int_0^T \big[|\alpha(s)|^{2k} + |\beta(s)|^{2k}\big]ds\Big\}, \\ t \in [0,T]. \end{aligned} \tag{4.12}$$

Here $K = K(L, k, T)$. Note that since we assume for the time being that $\alpha(\cdot)$ and $\beta(\cdot)$ are bounded, the above procedure goes through (otherwise the integration on the right-hand side of (4.12) may not exist; see (4.7)). Next, we want to refine the above estimate so that (4.8) will follow. To this end, note that (4.12) implies that its left-hand side is bounded uniformly in $t \in [0,T]$. Thus, we may set

$$\varphi(t) = \Big\{\sup_{0\le s\le t} E\langle Y(s)\rangle_\varepsilon^{2k}\Big\}^{1/2k}, \qquad t \in [0,T]. \tag{4.13}$$

We now return to (4.10), using (4.13). Define $\delta = (4K_0)^{-1}$. Then, for any

$t \in [0, \delta]$, applying Hölder's inequality and Young's inequality, we obtain

$$
\begin{aligned}
\varphi(t)^{2k} &\leq \varphi(0)^{2k} + K_0 \Big\{ \varphi(t)^{2k} t + \varphi(t)^{2k-1} \int_0^t \left(E|\alpha(s)|^{2k}\right)^{\frac{1}{2k}} ds \\
&\qquad + \varphi(t)^{2k-2} \int_0^t \left(E|\beta(s)|^{2k}\right)^{\frac{1}{k}} ds \Big\} \\
&\leq \varphi(0)^{2k} + \frac{1}{2}\varphi(t)^{2k} + K_1 \Big\{ \Big[\int_0^t \left(E|\alpha(s)|^{2k}\right)^{\frac{1}{2k}} ds \Big]^{2k} \\
&\qquad + \Big[\int_0^t \left(E|\beta(s)|^{2k}\right)^{\frac{1}{k}} ds \Big]^{k} \Big\}.
\end{aligned} \tag{4.14}
$$

The constant $K_1 = K_1(k, L, \delta)$ in (4.14) is independent of t. From (4.14), we obtain

$$
\begin{aligned}
\varphi(t)^{2k} &\leq 2\varphi(0)^{2k} + 2K_1 \Big\{ \Big[\int_0^t \left(E|\alpha(s)|^{2k}\right)^{\frac{1}{2k}} ds \Big]^{2k} \\
&\qquad + \Big[\int_0^t \left(E|\beta(s)|^{2k}\right)^{\frac{1}{k}} ds \Big]^{k} \Big\}, \quad \forall t \in [0, \delta].
\end{aligned} \tag{4.15}
$$

Now we can do the same thing on $[\delta, 2\delta]$ and on $[2\delta, 3\delta]$, and so on. Finally, we end up with

$$
\begin{aligned}
\varphi(T)^{2k} &\leq K \Big\{ \varphi(0)^{2k} + \Big[\int_0^T \left(E|\alpha(s)|^{2k}\right)^{\frac{1}{2k}} ds \Big]^{2k} \\
&\qquad + \Big[\int_0^T \left(E|\beta(s)|^{2k}\right)^{\frac{1}{k}} ds \Big]^{k} \Big\},
\end{aligned} \tag{4.16}
$$

with the constant $K = K(L, k, T, \delta)$. By (4.13), the definition of $\varphi(t)$, we conclude that

$$
\begin{aligned}
\sup_{t \in [0,T]} E \langle Y(t) \rangle_\varepsilon^{2k} &\leq K \Big\{ \langle Y(0) \rangle_\varepsilon^{2k} + \Big[\int_0^T \left(E|\alpha(s)|^{2k}\right)^{\frac{1}{2k}} ds \Big]^{2k} \\
&\qquad + \Big[\int_0^T \left(E|\beta(s)|^{2k}\right)^{\frac{1}{k}} ds \Big]^{k} \Big\}.
\end{aligned} \tag{4.17}
$$

Letting $\varepsilon \to 0$, we obtain (4.8). Finally, in the case that we only have (4.7) (instead of α and β being bounded), we can use the usual approximation. □

4.2. Taylor expansions

The following elementary lemma will be used below, whose proof is straightforward and is left to the reader.

Lemma 4.3. *Let $g \in C^2(\mathbb{R}^n)$. Then, for any $x, \bar{x} \in \mathbb{R}^n$,*

$$
\begin{aligned}
g(x) &= g(\bar{x}) + \langle g_x(\bar{x}), x - \bar{x} \rangle \\
&\qquad + \int_0^1 \langle \theta g_{xx}(\theta \bar{x} + (1-\theta)x)(x - \bar{x}), x - \bar{x} \rangle \, d\theta.
\end{aligned} \tag{4.18}
$$

Now, let $(\bar{x}(\cdot),\bar{u}(\cdot))$ be the given optimal pair. Then the following is satisfied:

$$
(4.19)\qquad \begin{cases} d\bar{x}(t)=b(t,\bar{x}(t),\bar{u}(t))dt+\sigma(t,\bar{x}(t),\bar{u}(t))dW(t), & t\in[0,T],\\ \bar{x}(0)=x_0. \end{cases}
$$

Fix any $u(\cdot)\in\mathcal{U}[0,T]$ and $\varepsilon>0$. Define

$$
(4.20)\qquad u^\varepsilon(t)=\begin{cases} \bar{u}(t), & t\in[0,T]\setminus E_\varepsilon,\\ u(t), & t\in E_\varepsilon, \end{cases}
$$

where $E_\varepsilon\subseteq[0,T]$ is a measurable set with $|E_\varepsilon|=\varepsilon$. Let $(x^\varepsilon(\cdot),u^\varepsilon(\cdot))$ satisfy the following:

$$
(4.21)\qquad \begin{cases} dx^\varepsilon(t)=b(t,x^\varepsilon(t),u^\varepsilon(t))dt+\sigma(t,x^\varepsilon(t),u^\varepsilon(t))dW(t), \ t\in[0,T],\\ x^\varepsilon(0)=x_0. \end{cases}
$$

Next, for $\varphi=b^i,\sigma^{ij},f$ $(1\le i\le n,\ 1\le j\le m)$, we define

$$
(4.22)\qquad \begin{cases} \varphi_x(t)\stackrel{\Delta}{=}\varphi_x(t,\bar{x}(t),\bar{u}(t)), \quad \varphi_{xx}(t)\stackrel{\Delta}{=}\varphi_{xx}(t,\bar{x}(t),\bar{u}(t)),\\ \delta\varphi(t)\stackrel{\Delta}{=}\varphi(t,\bar{x}(t),u(t))-\varphi(t,\bar{x}(t),\bar{u}(t)),\\ \delta\varphi_x(t)\stackrel{\Delta}{=}\varphi_x(t,\bar{x}(t),u(t))-\varphi_x(t,\bar{x}(t),\bar{u}(t)),\\ \delta\varphi_{xx}(t)\stackrel{\Delta}{=}\varphi_{xx}(t,\bar{x}(t),u(t))-\varphi_{xx}(t,\bar{x}(t),\bar{u}(t)). \end{cases}
$$

Let $y^\varepsilon(\cdot)$ and $z^\varepsilon(\cdot)$ be respectively the solution of the following stochastic differential equations:

$$
(4.23)\qquad \begin{cases} dy^\varepsilon(t)=b_x(t)y^\varepsilon(t)dt+\displaystyle\sum_{j=1}^m\big\{\sigma_x^j(t)y^\varepsilon(t)+\delta\sigma^j(t)\chi_{E_\varepsilon}(t)\big\}dW^j(t),\\ \qquad\qquad\qquad\qquad\qquad\qquad t\in[0,T],\\ y^\varepsilon(0)=0, \end{cases}
$$

and

$$
(4.24)\qquad \begin{cases} dz^\varepsilon(t)=\big\{b_x(t)z^\varepsilon(t)+\delta b(t)\chi_{E_\varepsilon}(t)+\frac{1}{2}b_{xx}(t)y^\varepsilon(t)^2\big\}dt\\ \qquad\qquad +\displaystyle\sum_{j=1}^m\big\{\sigma_x^j(t)z^\varepsilon(t)+\delta\sigma_x^j(t)y^\varepsilon(t)\chi_{E_\varepsilon}(t)\\ \qquad\qquad +\frac{1}{2}\sigma_{xx}^j(t)y^\varepsilon(t)^2\big\}dW^j(t), \quad t\in[0,T],\\ z^\varepsilon(0)=0, \end{cases}
$$

where

$$
(4.25)\qquad
\begin{aligned}
&b_{xx}(t)y^\varepsilon(t)^2 \triangleq \begin{pmatrix} \operatorname{tr}\{b^1_{xx}(t)y^\varepsilon(t)y^\varepsilon(t)^\top\} \\ \vdots \\ \operatorname{tr}\{b^n_{xx}(t)y^\varepsilon(t)y^\varepsilon(t)^\top\}\end{pmatrix},\\
&\sigma^j_{xx}(t)y^\varepsilon(t)^2 \triangleq \begin{pmatrix} \operatorname{tr}\{\sigma^{1j}_{xx}(t)y^\varepsilon(t)y^\varepsilon(t)^\top\} \\ \vdots \\ \operatorname{tr}\{\sigma^{nj}_{xx}(t)y^\varepsilon(t)y^\varepsilon(t)^\top\}\end{pmatrix}, \qquad 1 \le j \le m.
\end{aligned}
$$

The following result gives the *Taylor expansion* of the state with respect to the control perturbation.

Theorem 4.4. *Let (S1)–(S3) hold. Then, for any $k \ge 1$,*

$$
(4.26)\qquad \sup_{t\in[0,T]} E|x^\varepsilon(t) - \bar{x}(t)|^{2k} = O(\varepsilon^k),
$$

$$
(4.27)\qquad \sup_{t\in[0,T]} E|y^\varepsilon(t)|^{2k} = O(\varepsilon^k),
$$

$$
(4.28)\qquad \sup_{t\in[0,T]} E|z^\varepsilon(t)|^{2k} = O(\varepsilon^{2k}),
$$

$$
(4.29)\qquad \sup_{t\in[0,T]} E|x^\varepsilon(t) - \bar{x}(t) - y^\varepsilon(t)|^{2k} = O(\varepsilon^{2k}),
$$

$$
(4.30)\qquad \sup_{t\in[0,T]} E|x^\varepsilon(t) - \bar{x}(t) - y^\varepsilon(t) - z^\varepsilon(t)|^{2k} = o(\varepsilon^{2k}).
$$

Moreover, the following expansion holds for the cost functional:

$$
(4.31)\qquad
\begin{aligned}
J(u^\varepsilon(\cdot)) &= J(\bar{u}(\cdot)) + E\,\langle\, h_x(\bar{x}(T)), y^\varepsilon(T) + z^\varepsilon(T)\,\rangle \\
&\quad + \frac{1}{2} E\,\langle\, h_{xx}(\bar{x}(T)) y^\varepsilon(T), y^\varepsilon(T)\,\rangle \\
&\quad + E\int_0^T \Big\{ \langle\, f_x(t), y^\varepsilon(t) + z^\varepsilon(t)\,\rangle + \frac{1}{2}\langle\, f_{xx}(t) y^\varepsilon(t), y^\varepsilon(t)\,\rangle \\
&\qquad + \delta f(t)\chi_{E_\varepsilon}(t)\Big\} dt + o(\varepsilon).
\end{aligned}
$$

The proof of the above theorem is rather lengthy and technical. The reader may skip it at first reading and proceed to the next subsection directly.

Proof. For simplicity of presentation, we carry out the proof only for the case $n = m = 1$ (thus, the indices i and j will be omitted below). The interested reader is encouraged to give a proof for the general case.

1. Proof of (4.26) and (4.27).

Let $\xi^\varepsilon(t) \triangleq x^\varepsilon(t) - \bar{x}(t)$. Then we have

$$
(4.32) \qquad \begin{cases} d\xi^\varepsilon(t) = \Big\{\widetilde{b}_x(t)\xi^\varepsilon(t) + \delta b(t)\chi_{E_\varepsilon}(t)\Big\}dt \\ \qquad\qquad + \Big\{\widetilde{\sigma}_x^\varepsilon(t)\xi^\varepsilon(t) + \delta\sigma(t)\chi_{E_\varepsilon}(t)\Big\}dW(t), \quad t \in [0,T], \\ \xi^\varepsilon(0) = 0, \end{cases}
$$

where

$$
(4.33) \qquad \begin{cases} \widetilde{b}_x^\varepsilon(t) \triangleq \displaystyle\int_0^1 b_x(t, \bar{x}(t) + \theta(x^\varepsilon(t) - \bar{x}(t)), u^\varepsilon(t))d\theta, \\ \widetilde{\sigma}_x^\varepsilon(t) \triangleq \displaystyle\int_0^1 \sigma_x(t, \bar{x}(t) + \theta(x^\varepsilon(t) - \bar{x}(t)), u^\varepsilon(t))d\theta. \end{cases}
$$

By Lemma 4.2, we obtain

$$
(4.34) \qquad \begin{aligned} \sup_{t\in[0,T]} E|\xi^\varepsilon(t)|^{2k} &\leq K\Big\{\Big(\int_0^T \{E\big|\delta b(s)\chi_{E_\varepsilon}(s)\big|^{2k}\}^{\frac{1}{2k}} ds\Big)^{2k} \\ &\qquad + \Big(\int_0^T \{E\big|\delta\sigma(s)\chi_{E_\varepsilon}(s)\big|^{2k}\}^{\frac{1}{k}} ds\Big)^{k}\Big\} \\ &\leq K\{\varepsilon^{2k} + \varepsilon^k\} \leq K\varepsilon^k. \end{aligned}
$$

This proves (4.26). Similarly, we can prove (4.27).

2. Proof of (4.28).

Form (4.24), Lemma 4.2 and (4.27), we have

$$
(4.35) \qquad \begin{aligned} &\sup_{t\in[0,T]} E|z^\varepsilon(t)|^{2k} \\ &\leq K\Big\{\Big(\int_0^T \{E\big|\delta b(s)\chi_{E_\varepsilon}(s) + \frac{1}{2}b_{xx}(s)y^\varepsilon(s)^2\big|^{2k}\}^{\frac{1}{2k}} ds\Big)^{2k} \\ &\quad + \Big(\int_0^T \{E\big|\delta\sigma_x(s)\chi_{E_\varepsilon}(s)y^\varepsilon(s) + \frac{1}{2}\sigma_{xx}(s)y^\varepsilon(s)^2\big|^{2k}\}^{\frac{1}{k}} ds\Big)^{k}\Big\} \\ &\leq K\Big\{\Big(\int_0^T \{\chi_{E_\varepsilon}(s) + (E\big|y^\varepsilon(s)\big|^{4k})^{\frac{1}{2k}}\}ds\Big)^{2k} \\ &\quad + \Big(\int_0^T \{\chi_{E_\varepsilon}(s)(E\big|y^\varepsilon(s)\big|^{2k})^{\frac{1}{k}} + (E\big|y^\varepsilon(s)\big|^{4k})^{\frac{1}{k}}\}ds\Big)^{k}\Big\} \\ &\leq K\varepsilon^{2k}. \end{aligned}
$$

This gives (4.28).

3. Proof of (4.29).

Set

$$
(4.36) \qquad \eta^\varepsilon(t) \triangleq x^\varepsilon(t) - \bar{x}(t) - y^\varepsilon(t) \equiv \xi^\varepsilon(t) - y^\varepsilon(t).
$$

By (4.32) and (4.23), we have

$$
\begin{aligned}
d\eta^\varepsilon(t) &= \Big\{\widetilde{b}_x^\varepsilon(t)\xi^\varepsilon(t) + \delta b(t)\chi_{E_\varepsilon}(t) - b_x(t)y^\varepsilon(t)\Big\}dt \\
&\quad + \Big\{\widetilde{\sigma}_x^\varepsilon(t)\xi^\varepsilon(t) - \sigma_x(t)y^\varepsilon(t)\Big\}dW(t) \\
&= \Big\{\widetilde{b}_x^\varepsilon(t)\eta^\varepsilon(t) + \delta b(t)\chi_{E_\varepsilon}(t) + \big[\widetilde{b}_x^\varepsilon(t) - b_x(t)\big]y^\varepsilon(t)\Big\}dt \\
&\quad + \Big\{\widetilde{\sigma}_x^\varepsilon(t)\eta^\varepsilon(t) + \big[\widetilde{\sigma}_x^\varepsilon(t) - \sigma_x(t)\big]y^\varepsilon(t)\Big\}dW(t).
\end{aligned}
\tag{4.37}
$$

Thus, it follows from Lemma 4.2 that

$$
\begin{aligned}
&E\big|\eta^\varepsilon(t)\big|^{2k} \\
&\le K\Big\{\Big(\int_0^T \{E\big|\delta b(s)\chi_{E_\varepsilon}(s) + \big[\widetilde{b}_x^\varepsilon(s) - b_x(s)\big]y^\varepsilon(s)\big|^{2k}\}^{\frac{1}{2k}}ds\Big)^{2k} \\
&\qquad + \Big(\int_0^T \{E\big|\big[\widetilde{\sigma}_x^\varepsilon(s) - \sigma_x(s)\big]y^\varepsilon(s)\big|^{2k}\}^{\frac{1}{k}}ds\Big)^k\Big\} \\
&\le K\Big\{\Big(\varepsilon + \int_0^T \{E\big|y^\varepsilon(s)\big|^{4k}\}^{\frac{1}{4k}}\{E|\widetilde{b}_x^\varepsilon(s) - b_x(s)|^{4k}\}^{\frac{1}{4k}}ds\Big)^{2k} \\
&\qquad + \Big(\int_0^T \{E\big|y^\varepsilon(s)\big|^{4k}\}^{\frac{1}{2k}}\{E|\widetilde{\sigma}_x^\varepsilon(s) - \sigma_x(s)|^{4k}\}^{\frac{1}{2k}}ds\Big)^k\Big\} \\
&\le K\Big\{\varepsilon^{2k} + \varepsilon^k\Big(\int_0^T \{E|\widetilde{b}_x^\varepsilon(s) - b_x(s)|^{4k}\}^{\frac{1}{4k}}ds\Big)^{2k} \\
&\qquad + \varepsilon^k\Big(\int_0^T \{E|\widetilde{\sigma}_x^\varepsilon(s) - \sigma_x(s)|^{4k}\}^{\frac{1}{2k}}ds\Big)^k\Big\}.
\end{aligned}
\tag{4.38}
$$

Note that (by (S3) and (4.26))

$$
\begin{aligned}
&\int_0^T \Big\{E|\widetilde{b}_x^\varepsilon(s) - b_x(s)|^{4k}\Big\}^{\frac{1}{4k}}ds \\
&\quad = \int_0^T \Big\{E\Big|\int_0^1 \big[b_x(s,\overline{x}(s) + \theta(x^\varepsilon(s) - \overline{x}(s)), u^\varepsilon(s)) \\
&\qquad\qquad - b_x(s,\overline{x}(s),\overline{u}(s))\big]d\theta\Big|^{4k}\Big\}^{\frac{1}{4k}}ds \\
&\quad \le K\int_0^T \Big\{E\Big|L|x^\varepsilon(s) - \overline{x}(s)| + \delta b_x(s)\chi_{E_\varepsilon}(s)\Big|^{4k}\Big\}^{\frac{1}{4k}}ds \\
&\quad \le K\Big\{\varepsilon + \int_0^T \{E|x^\varepsilon(s) - \overline{x}(s)|^{4k}\}^{\frac{1}{4k}}ds\Big\} \le K\sqrt{\varepsilon}.
\end{aligned}
\tag{4.39}
$$

Similarly, we have

$$
\int_0^T \{E|\widetilde{\sigma}_x^\varepsilon(s) - \sigma_x(s)|^{4k}\}^{\frac{1}{2k}}ds \le K\varepsilon.
\tag{4.40}
$$

Then (4.29) follows from (4.38).

4. Proof of (4.30).

Set

$$\zeta^\varepsilon(t) \triangleq x^\varepsilon(t) - \overline{x}(t) - y^\varepsilon(t) - z^\varepsilon(t) \equiv \xi^\varepsilon(t) - y^\varepsilon(t) - z^\varepsilon(t) \equiv \eta^\varepsilon(t) - z^\varepsilon(t). \tag{4.41}$$

It is clear that

$$\begin{cases} d\zeta^\varepsilon(t) = B(t)dt + \Sigma(t)dW(t), \\ \zeta^\varepsilon(0) = 0, \end{cases} \tag{4.42}$$

where (noting (4.19)–(4.24) and Lemma 4.3)

$$\begin{aligned} B(t) &= b(t, x^\varepsilon(t), u^\varepsilon(t)) - b(t, \overline{x}(t), u^\varepsilon(t)) - b_x(t)\big[y^\varepsilon(t) + z^\varepsilon(t)\big] \\ &\quad - \frac{1}{2} b_{xx}(t) y^\varepsilon(t)^2 \\ &= b_x(t, \overline{x}(t), u^\varepsilon(t))\xi^\varepsilon(t) + \frac{1}{2}\widetilde{b}^\varepsilon_{xx}(t)\xi^\varepsilon(t)^2 \\ &\quad - b_x(t)\big[y^\varepsilon(t) + z^\varepsilon(t)\big] - \frac{1}{2} b_{xx}(t) y^\varepsilon(t)^2 \\ &= b_x(t)\zeta^\varepsilon(t) + \delta b_x(t)\chi_{E_\varepsilon}(t)\xi^\varepsilon(t) \\ &\quad + \frac{1}{2}\big[\widetilde{b}^\varepsilon_{xx}(t) - b_{xx}(t, \overline{x}(t), u^\varepsilon(t))\big]\xi^\varepsilon(t)^2 \\ &\quad + \frac{1}{2}\delta b_{xx}(t)\chi_{E_\varepsilon}(t)\xi^\varepsilon(t)^2 + \frac{1}{2} b_{xx}(t)\big[\xi^\varepsilon(t)^2 - y^\varepsilon(t)^2\big] \\ &\equiv b_x(t)\zeta^\varepsilon(t) + \alpha^\varepsilon(t), \end{aligned} \tag{4.43}$$

and

$$\begin{aligned} \Sigma(t) &= \sigma(t, x^\varepsilon(t), u^\varepsilon(t)) - \sigma(t, \overline{x}(t), u^\varepsilon(t)) - \sigma_x(t)\big[y^\varepsilon(t) + z^\varepsilon(t)\big] \\ &\quad - \frac{1}{2}\sigma_{xx}(t) y^\varepsilon(t)^2 - \delta\sigma_x(t)\chi_{E_\varepsilon}(t) y^\varepsilon(t) \\ &= \sigma_x(t, \overline{x}(t), u^\varepsilon(t))\xi^\varepsilon(t) + \frac{1}{2}\widetilde{\sigma}^\varepsilon_{xx}(t)\xi^\varepsilon(t)^2 - \sigma_x(t)\big[y^\varepsilon(t) + z^\varepsilon(t)\big] \\ &\quad - \frac{1}{2}\sigma_{xx}(t) y^\varepsilon(t)^2 - \delta\sigma_x(t)\chi_{E_\varepsilon}(t) y^\varepsilon(t) \\ &= \sigma_x(t)\zeta^\varepsilon(t) + \delta\sigma_x(t)\chi_{E_\varepsilon}(t)\eta^\varepsilon(t) \\ &\quad + \frac{1}{2}\big[\widetilde{\sigma}^\varepsilon_{xx}(t) - \sigma_{xx}(t, \overline{x}(t), u^\varepsilon(t))\big]\xi^\varepsilon(t)^2 \\ &\quad + \frac{1}{2}\delta\sigma_{xx}(t)\chi_{E_\varepsilon}(t)\xi^\varepsilon(t)^2 + \frac{1}{2}\sigma_{xx}(t)\big[\xi^\varepsilon(t)^2 - y^\varepsilon(t)^2\big] \\ &\equiv \sigma_x(t)\zeta^\varepsilon(t) + \beta^\varepsilon(t), \end{aligned} \tag{4.44}$$

with

$$\begin{cases} \widetilde{b}^\varepsilon_{xx}(t) \triangleq 2\displaystyle\int_0^1 \theta b_{xx}(t, \theta\overline{x}(t) + (1-\theta)x^\varepsilon(t), u^\varepsilon(t))d\theta, \\ \widetilde{\sigma}^\varepsilon_{xx}(t) \triangleq 2\displaystyle\int_0^1 \theta \sigma_{xx}(t, \theta\overline{x}(t) + (1-\theta)x^\varepsilon(t), u^\varepsilon(t))d\theta, \end{cases} \tag{4.45}$$

and

$$
(4.46)\qquad \begin{cases} \alpha^\varepsilon(t) \stackrel{\Delta}{=} \delta b_x(t)\chi_{E_\varepsilon}(t)\xi^\varepsilon(t) + \frac{1}{2}\big[\widetilde{b}^\varepsilon_{xx}(t) - b_{xx}(t,\bar{x}(t),u^\varepsilon(t))\big]\xi^\varepsilon(t)^2 \\ \qquad\qquad + \frac{1}{2}\delta b_{xx}(t)\chi_{E_\varepsilon}(t)\xi^\varepsilon(t)^2 + \frac{1}{2}b_{xx}(t)\big[\xi^\varepsilon(t)^2 - y^\varepsilon(t)^2\big], \\ \beta^\varepsilon(t) \stackrel{\Delta}{=} \delta \sigma_x(t)\chi_{E_\varepsilon}(t)\eta^\varepsilon(t) + \frac{1}{2}\big[\widetilde{\sigma}^\varepsilon_{xx}(t) - \sigma_{xx}(t,\bar{x}(t),u^\varepsilon(t))\big]\xi^\varepsilon(t)^2 \\ \qquad\qquad + \frac{1}{2}\delta \sigma_{xx}(t)\chi_{E_\varepsilon}(t)\xi^\varepsilon(t)^2 + \frac{1}{2}\sigma_{xx}(t)\big[\xi^\varepsilon(t)^2 - y^\varepsilon(t)^2\big]. \end{cases}
$$

In order to use Lemma 4.2, we need to estimate $\alpha^\varepsilon(\cdot)$ and $\beta^\varepsilon(\cdot)$. To this end, recall that $\bar{\omega}$ appearing in (S3) is a modulus of continuity for $b_{xx}(t,\cdot,u)$ (uniform in $t \in [0,T]$ and $u \in U$). Thus for any $\rho > 0$, there exists a constant $K_\rho > 0$ such that

$$
(4.47)\qquad \bar{\omega}(r) \le \rho + rK_\rho, \qquad \forall r \ge 0.
$$

Consequently,

$$
(4.48)\qquad \big|\widetilde{b}^\varepsilon_{xx}(t) - b_{xx}(t,\bar{x}(t),u^\varepsilon(t))\big| \le \bar{\omega}(|\xi^\varepsilon(t)|) \le \rho + K_\rho|\xi^\varepsilon(t)|.
$$

Recalling (4.36) and (4.46), as well as (4.26)–(4.29), we can estimate $\alpha^\varepsilon(\cdot)$ as follows:

$$
(4.49)\qquad \begin{aligned}
&\int_0^T \Big(E\big|\alpha^\varepsilon(t)\big|^{2k}\Big)^{\frac{1}{2k}} dt \\
&\le \int_0^T \Big\{\Big(E\big|\delta b_x(t)\chi_{E_\varepsilon}(t)\xi^\varepsilon(t)\big|^{2k}\Big)^{\frac{1}{2k}} \\
&\qquad + \Big(E\Big|\frac{1}{2}\big[\widetilde{b}^\varepsilon_{xx}(t) - b_{xx}(t,\bar{x}(t),u^\varepsilon(t))\big]\xi^\varepsilon(t)^2\Big|^{2k}\Big)^{\frac{1}{2k}} \\
&\qquad + \Big(E\Big|\frac{1}{2}\delta b_{xx}(t)\chi_{E_\varepsilon}(t)\xi^\varepsilon(t)^2\Big|^{2k}\Big)^{\frac{1}{2k}} \\
&\qquad + \Big(E\Big|\frac{1}{2}b_{xx}(t)\big[\xi^\varepsilon(t)^2 - y^\varepsilon(t)^2\big]\Big|^{2k}\Big)^{\frac{1}{2k}}\Big\}dt \\
&\le K\int_0^T \Big\{\chi_{E_\varepsilon}(t)\Big(E|\xi^\varepsilon(t)|^{2k}\Big)^{\frac{1}{2k}} + \chi_{E_\varepsilon}(t)\Big(E|\xi^\varepsilon(t)|^{4k}\Big)^{\frac{1}{2k}} \\
&\qquad + \Big(E\big[\rho + K_\rho|\xi^\varepsilon(t)|\big]^{4k}\Big)^{\frac{1}{4k}}\Big(E|\xi^\varepsilon(t)|^{8k}\Big)^{\frac{1}{4k}} \\
&\qquad + \Big(E|\eta^\varepsilon(t)|^{4k}\Big)^{\frac{1}{4k}}\Big(E|\xi^\varepsilon(t) + y^\varepsilon(t)|^{4k}\Big)^{\frac{1}{4k}}\Big\}dt \\
&\le K\big\{\varepsilon^{3/2} + \varepsilon(\rho + \sqrt{\varepsilon}K_\rho) + \varepsilon^2 + \varepsilon^{3/2}\big\}.
\end{aligned}
$$

This implies

$$
(4.50)\qquad \int_0^T \Big(E\big|\alpha^\varepsilon(t)\big|^{2k}\Big)^{\frac{1}{2k}} dt = o(\varepsilon).
$$

Similar to (4.48), we have

$$|\widetilde{\sigma}^{\varepsilon}_{xx}(t)-\sigma_{xx}(t,\overline{x}(t),u^{\varepsilon}(t))|\le\overline{\omega}(|\xi^{\varepsilon}(t)|)\le\rho+K_{\rho}|\xi^{\varepsilon}(t)|. \tag{4.51}$$

As with (4.49), we can estimate $\beta^{\varepsilon}(\cdot)$ as follows.

$$\begin{aligned}
&\int_0^T\Big(E|\beta^{\varepsilon}(t)|^{2k}\Big)^{\frac{1}{k}}dt\\
&\le\int_0^T\Big\{\Big(E\big|\delta\sigma_x(t)\chi_{E_\varepsilon}(t)\eta^{\varepsilon}(t)\big|^{2k}\Big)^{\frac{1}{k}}\\
&\qquad+\Big(E\Big|\frac{1}{2}[\widetilde{\sigma}^{\varepsilon}_{xx}(t)-\sigma_{xx}(t,\overline{x}(t),u^{\varepsilon}(t))]\xi^{\varepsilon}(t)^2\Big|^{2k}\Big)^{\frac{1}{k}}\\
&\qquad+\Big(E\Big|\frac{1}{2}\delta\sigma_{xx}(t)\chi_{E_\varepsilon}(t)\xi^{\varepsilon}(t)^2\Big|^{2k}\Big)^{\frac{1}{k}}\\
&\qquad+\Big(E\Big|\frac{1}{2}\sigma_{xx}(t)\big[\xi^{\varepsilon}(t)^2-y^{\varepsilon}(t)^2\big]\Big|^{2k}\Big)^{\frac{1}{k}}\Big\}dt\\
&\le K\int_0^T\Big\{\chi_{E_\varepsilon}(t)\Big(E|\eta^{\varepsilon}(t)|^{2k}\Big)^{\frac{1}{k}}+\chi_{E_\varepsilon}(t)\Big(E|\xi^{\varepsilon}(t)|^{4k}\Big)^{\frac{1}{k}}\\
&\qquad+\Big(E\big[\rho+K_{\rho}|\xi^{\varepsilon}(t)|\big]^{4k}\Big)^{\frac{1}{2k}}\Big(E|\xi^{\varepsilon}(t)|^{8k}\Big)^{\frac{1}{2k}}\\
&\qquad+\Big(E|\eta^{\varepsilon}(t)|^{4k}\Big)^{\frac{1}{2k}}\Big(E|\xi^{\varepsilon}(t)+y^{\varepsilon}(t)|^{4k}\Big)^{\frac{1}{2k}}\Big\}dt\\
&\le K\big\{\varepsilon^3+\varepsilon^2(\rho^2+\varepsilon K_{\rho}^2)+\varepsilon^3+\varepsilon^3\big\}.
\end{aligned}\tag{4.52}$$

Thus,

$$\int_0^T\Big(E|\beta^{\varepsilon}(t)|^{2k}\Big)^{\frac{1}{k}}dt=o(\varepsilon^2). \tag{4.53}$$

Then, by Lemma 4.2, we obtain (4.30).

5. Proof of (4.31).

By Lemma 4.3, we have

$$\begin{aligned}
&J(u^{\varepsilon}(\cdot))-J(\overline{u}(\cdot))\\
&=E\Big\{h(x^{\varepsilon}(T))-h(\overline{x}(T))\Big\}\\
&\quad+E\int_0^T\Big\{f(t,x^{\varepsilon}(t),u^{\varepsilon}(t))-f(t,\overline{x}(t),\overline{u}(t))\Big\}dt\\
&=E\,\langle\, h_x(\overline{x}(T)),\xi^{\varepsilon}(T)\,\rangle\\
&\quad+E\int_0^1\langle\,\theta h_{xx}(\theta\overline{x}(T)+(1-\theta)x^{\varepsilon}(T))\xi^{\varepsilon}(T),\xi^{\varepsilon}(T)\,\rangle\,d\theta\\
&\quad+E\int_0^T\Big\{\delta f(t)\chi_{E_\varepsilon}(t)+\langle\, f_x(t,\overline{x}(t),u^{\varepsilon}(t)),\xi^{\varepsilon}(t)\,\rangle\\
&\quad+\int_0^1\langle\,\theta f_{xx}(t,\theta\overline{x}(t)+(1-\theta)x^{\varepsilon}(t),u^{\varepsilon}(t))\xi^{\varepsilon}(t),\xi^{\varepsilon}(t)\,\rangle\,d\theta\Big\}dt.
\end{aligned}$$

Now, recalling the definitions of ξ^ε, η^ε, and ζ^ε, we have

$$
\begin{aligned}
&J(u^\varepsilon(\cdot)) - J(\overline{u}(\cdot)) \\
&\quad = E\,\langle h_x(\overline{x}(T)), y^\varepsilon(T) + z^\varepsilon(T)\,\rangle \\
&\qquad + E\,\langle h_x(\overline{x}(T)), \zeta^\varepsilon(T)\,\rangle + \frac{1}{2}E\,\langle h_{xx}(\overline{x}(T))y^\varepsilon(T), y^\varepsilon(T)\,\rangle \\
&\qquad + \frac{1}{2}E\,\langle h_{xx}(\overline{x}(T))\eta^\varepsilon(T), \xi^\varepsilon(T) + y^\varepsilon(T)\,\rangle \\
&\qquad + E\int_0^1 \langle\,\theta\big\{h_{xx}(\theta\overline{x}(T) + (1-\theta)x^\varepsilon(T)) \\
&\qquad\qquad\qquad - h_{xx}(\overline{x}(T))\big\}\xi^\varepsilon(T), \xi^\varepsilon(T)\,\rangle\,d\theta \\
&\qquad + E\int_0^T \Big\{\delta f(t)\chi_{E_\varepsilon}(t) + \langle\,\delta f_x(t), \xi^\varepsilon(t)\,\rangle\,\chi_{E_\varepsilon}(t) \\
&\qquad\qquad + \langle\, f_x(t), y^\varepsilon(t) + z^\varepsilon(t)\,\rangle + \langle\, f_x(t), \zeta^\varepsilon(t)\,\rangle \\
&\qquad + \int_0^1 \langle\,\theta\big[f_{xx}(t, \theta\overline{x}(t) + (1-\theta)x^\varepsilon(t), u^\varepsilon(t)) \\
&\qquad\qquad\qquad - f_{xx}(t, \overline{x}(t), u^\varepsilon(t))\big]\xi^\varepsilon(t), \xi^\varepsilon(t)\,\rangle\,d\theta \\
&\qquad + \frac{1}{2}\langle\,\delta f_{xx}(t)\xi^\varepsilon(t), \xi^\varepsilon(t)\,\rangle\,\chi_{E_\varepsilon}(t) \\
&\qquad + \frac{1}{2}\langle\, f_{xx}(t)y^\varepsilon(t), y^\varepsilon(t)\,\rangle + \frac{1}{2}\langle\, f_{xx}(t)\eta^\varepsilon(t), \xi^\varepsilon(t) + y^\varepsilon(t)\,\rangle\Big\}dt.
\end{aligned}
$$

Then, by (4.26)–(4.30), we can show that

$$
\begin{aligned}
&J(u^\varepsilon(\cdot)) - J(\overline{u}(\cdot)) \\
&\quad = E\,\langle h_x(\overline{x}(T)), y^\varepsilon(T) + z^\varepsilon(T)\rangle + \frac{1}{2}E\,\langle h_{xx}(\overline{x}(T))y^\varepsilon(T), y^\varepsilon(T)\rangle \\
&\qquad + E\int_0^T \Big\{\langle f_x(t), y^\varepsilon(t) + z^\varepsilon(t)\rangle + \frac{1}{2}\langle f_{xx}(t)y^\varepsilon(t), y^\varepsilon(t)\rangle \\
&\qquad\qquad + \delta f(t)\chi_{E_\varepsilon}(t)\Big\}dt + R(\varepsilon),
\end{aligned}
\tag{4.54}
$$

where $R(\varepsilon)$ is of order $o(\varepsilon)$. Hence, our conclusion follows. □

4.3. Duality analysis and completion of the proof

From Theorem 4.4, we conclude that a necessary condition for a given optimal pair $(\overline{x}(\cdot), \overline{u}(\cdot))$ is the following:

$$
\begin{aligned}
0 &\le E\Big\{\langle h_x(\overline{x}(T)), y^\varepsilon(T) + z^\varepsilon(T)\rangle + \frac{1}{2}\langle h_{xx}(\overline{x}(T))y^\varepsilon(T), y^\varepsilon(T)\rangle\Big\} \\
&\quad + E\int_0^T \Big\{\langle\, f_x(t), y^\varepsilon(t) + z^\varepsilon(t)\,\rangle + \frac{1}{2}\langle\, f_{xx}(t)y^\varepsilon(t), y^\varepsilon(t)\,\rangle \\
&\qquad + \delta f(t)\chi_{E_\varepsilon}(t)\Big\}dt + o(\varepsilon), \quad \forall u(\cdot) \in \mathcal{U}[0,T], \forall \varepsilon > 0,
\end{aligned}
\tag{4.55}
$$

where $y^\varepsilon(\cdot)$ and $z^\varepsilon(\cdot)$ are solutions to the (approximate) variational systems (4.23) and (4.24), respectively. As in the deterministic case, we are now in a position to get rid of $y^\varepsilon(\cdot)$ and $z^\varepsilon(\cdot)$, and then pass to the limit. To this end, we need some duality relations between the variational systems (4.23)–(4.24) and the adjoint equations (3.8) and (3.9).

Lemma 4.5. *Let (S0)–(S3) hold. Let $y^\varepsilon(\cdot)$ and $z^\varepsilon(\cdot)$ be the solutions of (4.23) and (4.24), respectively. Let $(p(\cdot), q(\cdot))$ be the adapted solution of (3.8). Then*

$$\tag{4.56}\begin{aligned}&E\,\langle\, p(T), y^\varepsilon(T)\,\rangle\\&\quad= E\int_0^T \big\{\,\langle\, f_x(t), y^\varepsilon(t)\,\rangle + \mathrm{tr}\,\big[q(t)^\top \delta\sigma(t)\big]\chi_{E_\varepsilon}(t)\big\}dt,\end{aligned}$$

and

$$\tag{4.57}\begin{aligned}E\,\langle\, p(T), z^\varepsilon(T)\,\rangle = E\int_0^T &\big\{\,\langle\, f_x(t), z^\varepsilon(t)\,\rangle\\&+\frac{1}{2}\Big[\,\langle\, p(t), b_{xx}(t) y^\varepsilon(t)^2\,\rangle + \sum_{j=1}^m \langle\, q_j(t), \sigma^j_{xx}(t) y^\varepsilon(t)^2\,\rangle\,\Big]\\&+\Big[\,\langle\, p(t), \delta b(t)\,\rangle + \sum_{j=1}^m \langle\, q_j(t), \delta\sigma^j_x(t) y^\varepsilon(t)\,\rangle\,\Big]\chi_{E_\varepsilon}(t)\big\}dt.\end{aligned}$$

The proof follows immediately from Itô's formula, and we leave it to the reader (see also Chapter 1, Corollary 5.6).

Adding (4.56) and (4.57), and appealing to the Taylor expansions given in Theorem 4.4, we get

$$\tag{4.58}\begin{aligned}&-E\,\langle\, h_x(\overline{x}(T)), y^\varepsilon(T) + z^\varepsilon(T)\,\rangle\\&= E\int_0^T \big\{\,\langle\, f_x(t), y^\varepsilon(t) + z^\varepsilon(t)\,\rangle\\&\qquad+\frac{1}{2}\,\langle\, p(t), b_{xx}(t) y^\varepsilon(t)^2\,\rangle + \frac{1}{2}\sum_{j=1}^m \langle\, q_j(t), \sigma^j_{xx}(t) y^\varepsilon(t)^2\,\rangle\\&\qquad+\big[\,\langle\, p(t), \delta b(t)\,\rangle + \mathrm{tr}\,\big(q(t)^\top \delta\sigma(t)\big)\big]\chi_{E_\varepsilon}(t)\big\}dt + o(\varepsilon).\end{aligned}$$

Thus, by (4.31) and the optimality of $\overline{u}(\cdot)$, we have

$$
\begin{aligned}
0 &\geq J(\bar{u}(\cdot)) - J(u^\varepsilon(\cdot)) \\
&= -\frac{1}{2}E\,\langle\, h_{xx}(\bar{x}(T))y^\varepsilon(T), y^\varepsilon(T)\,\rangle \\
&\quad + \frac{1}{2}E\int_0^T \Big\{ -\langle\, f_{xx}(t)y^\varepsilon(t), y^\varepsilon(t)\,\rangle + \langle\, p(t), b_{xx}(t)y^\varepsilon(t)^2\,\rangle \\
&\qquad\qquad + \sum_{j=1}^m \langle\, q_j(t), \sigma^j_{xx}(t)y^\varepsilon(t)^2\,\rangle\Big\}dt \\
&\quad + E\int_0^T \Big\{ -\delta f(t) + \langle\, p(t), \delta b(t)\,\rangle \\
&\qquad\qquad + \sum_{j=1}^m \langle\, q_j(t), \delta\sigma^j(t)\,\rangle\Big\}\chi_{E_\varepsilon}(t)dt + o(\varepsilon) \\
&= \frac{1}{2}E\mathrm{tr}\,\{P(T)Y^\varepsilon(T)\} \\
&\quad + E\int_0^T \Big\{\frac{1}{2}\mathrm{tr}\,\big[H_{xx}(t)Y^\varepsilon(t)\big] + \delta H(t)\chi_{E_\varepsilon}(t)\Big\}dt + o(\varepsilon),
\end{aligned}
\tag{4.59}
$$

where

$$
\begin{cases}
Y^\varepsilon(t) = y^\varepsilon(t)y^\varepsilon(t)^\top, \\
H_{xx}(t) = H_{xx}(t, \bar{x}(t), \bar{u}(t), p(t), q(t)), \\
\delta H(t) = H(t, \bar{x}(t), u(t), p(t), q(t)) - H(t, \bar{x}(t), \bar{u}(t), p(t), q(t)).
\end{cases}
\tag{4.60}
$$

We see that (4.59) no longer contains the first-order terms in $y^\varepsilon(\cdot)$ and $z^\varepsilon(\cdot)$. But, unlike the deterministic case, there are left some second-order terms in $y^\varepsilon(\cdot)$, which are written in terms of the first-order in $Y^\varepsilon(\cdot)$. Hence, we want further to get rid of $Y^\varepsilon(\cdot)$. To this end, we need some duality relation between the equation satisfied by $Y^\varepsilon(\cdot)$ and the second-order adjoint equation (3.9) (which is exactly where the second-order adjoint equation comes in). Let us now derive the SDE satisfied by $Y^\varepsilon(\cdot)$. Applying Itô's formula to $y^\varepsilon(t)y^\varepsilon(t)^\top$ and noting (4.23), one has

$$
\begin{aligned}
dY^\varepsilon(t) &= \Big\{b_x(t)Y^\varepsilon(t) + Y^\varepsilon(t)b_x(t)^\top \\
&\quad + \sum_{j=1}^m \sigma^j_x(t)Y^\varepsilon(t)\sigma^j_x(t)^\top + \sum_{j=1}^m \delta\sigma^j(t)\delta\sigma^j(t)^\top\chi_{E_\varepsilon}(t) \\
&\quad + \sum_{j=1}^m \Big(\sigma^j_x(t)y^\varepsilon(t)\delta\sigma^j(t)^\top + \delta\sigma^j(t)y^\varepsilon(t)^\top\sigma^j_x(t)^\top\Big)\chi_{E_\varepsilon}(t)\Big\}dt \\
&\quad + \sum_{j=1}^m \Big(\sigma^j_x(t)Y^\varepsilon(t) + Y^\varepsilon(t)\sigma^j_x(t)^\top\Big)dW^j(t) \\
&\quad + \sum_{j=1}^m \Big(\delta\sigma^j(t)y^\varepsilon(t)^\top + y^\varepsilon(t)\delta\sigma^j(t)^\top\Big)\chi_{E_\varepsilon}(t)dW^j(t).
\end{aligned}
\tag{4.61}
$$

To establish the duality relation between (4.61) and (3.9), we need the following lemma, whose proof follows directly from Itô's formula (which is also parallel to Chapter 1, Corollary 5.6).

Lemma 4.6. *Let $Y(\cdot), P(\cdot) \in L^2_{\mathcal{F}}(0,T;\mathbb{R}^{n\times n})$ satisfy the following:*

$$\left\{\begin{aligned} dY(t) &= \Phi(t)dt + \sum_{j=1}^{m} \Psi_j(t)dW^j(t), \\ dP(t) &= \Theta(t)dt + \sum_{j=1}^{m} Q_j(t)dW^j(t), \end{aligned}\right. \tag{4.62}$$

with $\Phi(\cdot)$, $\Psi_j(\cdot)$, $\Theta(\cdot)$, and $Q_j(\cdot)$ all being elements in $L^2_{\mathcal{F}}(0,T;\mathbb{R}^{n\times n})$. Then

$$\begin{aligned} &E\{\mathrm{tr}\,[P(t)Y(t)] - \mathrm{tr}\,[P(0)Y(0)]\} \\ &\quad = E\int_0^T \{\mathrm{tr}\,[\Theta(t)Y(t) + P(t)\Phi(t) + \sum_{j=1}^{m} Q_j(t)\Psi_j(t)]\}dt. \end{aligned} \tag{4.63}$$

Now we apply the above lemma to (4.61) and (3.9) to get the following (using Theorem 4.4, and noting $\mathrm{tr}\,[AB] = \mathrm{tr}\,[BA]$ and $Y(0) = 0$):

$$\begin{aligned} &E\Big\{\mathrm{tr}\,[P(T)Y^\varepsilon(T)]\Big\} \\ &\quad = E\int_0^T \mathrm{tr}\,\big[\delta\sigma(t)^\top P(t)\delta\sigma(t)\chi_{E_\varepsilon}(t) - H_{xx}(t)Y^\varepsilon(t)\big]dt + o(\varepsilon), \end{aligned} \tag{4.64}$$

where

$$\left\{\begin{aligned} &H_{xx}(t) \triangleq H_{xx}(t,\bar{x}(t),\bar{u}(t),p(t),q(t)), \\ &\delta\sigma(t) \triangleq \sigma(t,\bar{x}(t),u(t)) - \sigma(t,\bar{x}(t),\bar{u}(t)). \end{aligned}\right.$$

Hence, (4.59) can be written as

$$o(\varepsilon) \geq E\int_0^T \Big\{\delta H(t) + \frac{1}{2}\mathrm{tr}\,\big[\delta\sigma(t)^T P(t)\delta\sigma(t)\big]\Big\}\chi_{E_\varepsilon}(t)dt. \tag{4.65}$$

Then we can easily obtain the variational inequality (3.19). Easy manipulation shows that (3.19) is equivalent to (3.20). This completes the proof of Theorem 3.2. □

We point out that Theorem 3.2 remains true if f and h are C^2 in x and allowed to have polynomial (in particular, quadratic) growth in x, provided that b and σ have linear growth. The reader is encouraged to complete a proof for this case.

5. Sufficient Conditions of Optimality

In this section we will show that for the general controlled stochastic systems formulated earlier, the maximum condition in terms of the $\mathcal{H}$-function (see (3.20)) plus some convexity conditions constitute sufficient conditions for optimality.

Let us first introduce an additional assumption.

(S4) The control domain U is a convex body in $\mathbb{R}^k$. The maps b, σ, and f are locally Lipschitz in u, and their derivatives in x are continuous in (x, u).

The following lemma will play an interesting role below.

Lemma 5.1. *Let (S0)–(S4) hold. Let* $(\bar{x}(\cdot), \bar{u}(\cdot), p(\cdot), q(\cdot), P(\cdot), Q(\cdot))$ *be a given admissible 6-tuple and let* $\mathcal{H}$ *be the corresponding* $\mathcal{H}$*-function. Then for any* $t \in [0, T]$ *and* $\omega \in \Omega$,

$$\partial_u H(t, \bar{x}(t), \bar{u}(t), p(t), q(t)) = \partial_u \mathcal{H}(t, \bar{x}(t), \bar{u}(t)). \tag{5.1}$$

Proof. Fix a $t \in [0, T]$ and $\omega \in \Omega$. Define

$$\begin{cases} H(u) \stackrel{\Delta}{=} H(t, \bar{x}(t), u, p(t), q(t)), \\ \mathcal{H}(u) \stackrel{\Delta}{=} \mathcal{H}(t, \bar{x}(t), u), \\ \sigma(u) \stackrel{\Delta}{=} \sigma(t, \bar{x}(t), u), \\ \psi(u) \stackrel{\Delta}{=} \frac{1}{2} \operatorname{tr} \left[\sigma(u)^\top P(t) \sigma(u)\right] - \operatorname{tr} \left[\sigma(u)^\top P(t) \sigma(\bar{u}(t))\right]. \end{cases}$$

Then, by the second equality of (3.16), we have

$$\mathcal{H}(u) = H(u) + \psi(u).$$

Note that for any $r \downarrow 0$, $u, v \in U$, with $u \to \bar{u}(t)$,

$$\begin{aligned} &\psi(u + rv) - \psi(u) \\ &= \frac{1}{2} \operatorname{tr} \left\{ \left[\sigma(u + rv) - \sigma(u)\right]^\top P(t) \left[\sigma(u + rv) + \sigma(u) - 2\sigma(\bar{u}(t))\right] \right\} \\ &= o(r). \end{aligned}$$

Thus,

$$\varlimsup_{\substack{u \to \bar{u}(t), u \in U \\ r \downarrow 0}} \frac{\mathcal{H}(u + rv) - \mathcal{H}(u)}{r} = \varlimsup_{\substack{u \to \bar{u}(t), u \in U \\ r \downarrow 0}} \frac{H(u + rv) - H(u)}{r}.$$

Consequently, by (2.20), the desired result follows. □

Now we present the following sufficient condition of optimality.

Theorem 5.2. (Sufficient Conditions of Optimality) *Let (S0)–(S4) hold. Let* $(\bar{x}(\cdot), \bar{u}(\cdot), p(\cdot), q(\cdot), P(\cdot), Q(\cdot))$ *be an admissible 6-tuple. Suppose that* $h(\cdot)$ *is convex,* $H(t, \cdot, \cdot, p(t), q(t))$ *is concave for all* $t \in [0, T]$ *almost surely, and*

$$\mathcal{H}(t, \bar{x}(t), \bar{u}(t)) = \max_{u \in U} \mathcal{H}(t, \bar{x}(t), u), \quad \text{a.e. } t \in [0, T], \ \mathbf{P}\text{-a.s.} \tag{5.2}$$

Then $(\bar{x}(\cdot), \bar{u}(\cdot))$ is an optimal pair of Problem (S).

Proof. By the maximum condition (5.2), Lemmas 5.1 and 2.3-(iii), we have

$$0 \in \partial_u \mathcal{H}(t, \bar{x}(t), \bar{u}(t)) = \partial_u H(t, \bar{x}(t), \bar{u}(t), p(t), q(t)). \tag{5.3}$$

By Lemma 2.4, we further conclude that

$$\Big(H_x(t, \bar{x}(t), \bar{u}(t), p(t), q(t)), 0\Big) \in \partial_{x,u} H(t, \bar{x}(t), \bar{u}(t), p(t), q(t)). \tag{5.4}$$

Thus, by the concavity of $H(t, \cdot, \cdot, p(t), q(t))$, one obtains

$$\begin{aligned} &\int_0^T \{H(t, x(t), u(t), p(t), q(t)) - H(t, \bar{x}(t), \bar{u}(t), p(t), q(t))\} dt \\ &\qquad \leq \int_0^T \langle\, H_x(t, \bar{x}(t), \bar{u}(t), p(t), q(t)), x(t) - \bar{x}(t)\,\rangle\, dt, \end{aligned} \tag{5.5}$$

for any admissible pair $(x(\cdot), u(\cdot))$. Define $\xi(t) \stackrel{\Delta}{=} x(t) - \bar{x}(t)$, which satisfies the following equation:

$$\begin{cases} d\xi(t) = \{b_x(t, \bar{x}(t), \bar{u}(t))\xi(t) + \alpha(t)\} dt \\ \qquad\qquad + \displaystyle\sum_{j=1}^m [\sigma_x^j(t, \bar{x}(t), \bar{u}(t))\xi(t) + \beta^j(t)] dW^j(t), \quad t \in [0, T], \\ \xi(0) = 0, \end{cases} \tag{5.6}$$

where

$$\begin{cases} \alpha(t) \stackrel{\Delta}{=} -b_x(t, \bar{x}(t), \bar{u}(t))\xi(t) + b(t, x(t), u(t)) - b(t, \bar{x}(t), \bar{u}(t)), \\ \beta^j(t) \stackrel{\Delta}{=} -\sigma_x^j(t, \bar{x}(t), \bar{u}(t))\xi(t) + \sigma^j(t, x(t), u(t)) - \sigma^j(t, \bar{x}(t), \bar{u}(t)), \\ \qquad\qquad 1 \leq j \leq m. \end{cases}$$

By the duality relation between (5.6) and (3.8), we have

$$
\begin{aligned}
&E\langle h_x(\bar{x}(T)),\xi(T)\rangle\\
&\quad=-E\langle p(T),\xi(T)\rangle+E\langle p(0),\xi(0)\rangle\\
&\quad=-E\int_0^T\{\langle f_x(t,\bar{x}(t),\bar{u}(t)),\xi(t)\rangle+\langle p(t),\alpha(t)\rangle+\sum_{j=1}^m\langle q_j(t),\beta^j(t)\rangle\}dt\\
&\quad=E\int_0^T\langle H_x(t,\bar{x}(t),\bar{u}(t),p(t),q(t)),\xi(t)\rangle dt\\
&\qquad-E\int_0^T\Big\{\langle p(t),b(t,x(t),u(t))-b(t,\bar{x}(t),\bar{u}(t))\rangle\\
&\qquad\qquad+\sum_{j=1}^m\langle q_j(t),\sigma^j(t,\bar{x}(t),u(t))-\sigma^j(t,\bar{x}(t),\bar{u}(t))\rangle\Big\}dt\\
&\quad\ge E\int_0^T\{H(t,x(t),u(t),p(t),q(t))-H(t,\bar{x}(t),\bar{u}(t),p(t),q(t))\}dt\\
&\qquad-E\int_0^T\Big\{\langle p(t),b(t,x(t),u(t))-b(t,\bar{x}(t),\bar{u}(t))\rangle\\
&\qquad\qquad+\sum_{j=1}^m\langle q_j(t),\sigma^j(t,\bar{x}(t),u(t))-\sigma^j(t,\bar{x}(t),\bar{u}(t))\rangle\Big\}dt\\
&\quad=-E\int_0^T\{f(t,x(t),u(t))-f(t,\bar{x}((t),\bar{u}(t))\}dt.
\end{aligned}
$$

On the other hand, the convexity of h implies

$$E\langle h_x(\bar{x}(T)),\xi(T)\rangle\le Eh(x(T))-Eh(\bar{x}(T)).$$

Combining the above two, we arrive at

$$J(\bar{u}(\cdot))\le J(u(\cdot)). \tag{5.7}$$

Since $u(\cdot)\in\mathcal{U}[0,T]$ is arbitrary, the desired result follows. □

Let us give an example.

Example 5.3. Consider the following control system ($n=m=1$):

$$\begin{cases}dx(t)=u(t)dW(t), & t\in[0,1],\\ x(0)=0,\end{cases} \tag{5.8}$$

with the control domain being $U=[0,1]$ and the cost functional being

$$J(u(\cdot))=E\Big\{-\int_0^1 u(t)dt+\frac{1}{2}x(1)^2\Big\}. \tag{5.9}$$

Suppose $(\overline{x}(\cdot), \overline{u}(\cdot))$ is an optimal pair (which we are going to identify). Then the corresponding adjoint equations are

$$
(5.10) \qquad \begin{cases} dp(t) = q(t)dW(t), & t \in [0,1], \\ p(1) = -\overline{x}(1), \end{cases}
$$

and

$$
(5.11) \qquad \begin{cases} dP(t) = Q(t)dW(t), & t \in [0,1], \\ P(1) = -1. \end{cases}
$$

Clearly, $(P(t), Q(t)) = (-1, 0)$ is the unique adapted solution to (5.11). The corresponding $\mathcal{H}$-function is

$$
(5.12) \qquad \mathcal{H}(t, \overline{x}(t), u) = -\frac{1}{2}u^2 + (1 + q(t) + \overline{u}(t))u.
$$

The function $\mathcal{H}(t, \overline{x}(t), \cdot)$ attains its maximum at $u = 1 + q(t) + \overline{u}(t)$. Hence the necessary condition of optimality as specified by the stochastic maximum principle (Theorem 3.2) would be satisfied if one could find a control $\overline{u}(\cdot)$ such that the corresponding $q(t) \equiv -1$. By the first-order adjoint equation (5.10), it is clear that if we take $\overline{u}(t) \equiv 1$ with the corresponding state $\overline{x}(t) = W(t)$, then the unique solution of (5.10) is $(p(t), q(t)) = (-W(t), -1)$.

In order to see that $\overline{u}(t) \equiv 1$ is *indeed* optimal, let us calculate

$$
H(t, x, u, p(t), q(t)) = H(t, x, u, -W(t), -1) = u - u = 0,
$$

which is concave in (x, u). Moreover, $h(x) = \frac{1}{2}x^2$ is convex. So the optimality follows from the sufficient condition (Theorem 5.2).

One may realize now that a key fact in proving the sufficient condition (Theorem 5.2) is (5.3). Based on this, the concavity of the Hamiltonian H and the maximum condition for the $\mathcal{H}$-function, roughly speaking, imply a maximum condition for H. So one may derive the optimality in the same way as in the deterministic case (Theorem 2.5). We have seen in Examples 3.1 and 3.4 how *convex* Hamiltonians (in u) are turned into *concave* $\mathcal{H}$-functions by adding the risk adjustment involving $P(t)$. However, when the Hamiltonian $H(t, x, u, p, q)$ is already concave in u, the second-order adjoint process $P(\cdot)$ plays no role at all, as was shown in the proof of Theorem 5.2. This is because the concavity of H reflects the risk adjustment already.

6. Problems with State Constraints

In many applications, the state process $x(\cdot)$ is subject to some constraints. This section is devoted to a study of some of these situations.

6.1. Formulation of the problem and the maximum principle

We consider the state equation (3.1) with the cost functional (3.2). In addition, we require that the state process $x(\cdot)$ satisfy

$$Eh(x(T)) + E\int_0^T f(t, x(t), u(t))dt \in \Gamma, \tag{6.1}$$

where $\mathbf{h} : \mathbb{R}^n \to \mathbb{R}^\ell$ and $\mathbf{f} : [0, T] \times \mathbb{R}^n \times U \to \mathbb{R}^\ell$ are given maps, and Γ is a subset in $\mathbb{R}^\ell$. We call (6.1) a *state constraint.* Let us recall the following notions (see Chapter 2, Section 4.1). Any $u(\cdot) \in \mathcal{U}[0, T]$ is called a *feasible control.* The corresponding solution $x(\cdot)$ of (3.1) and the pair $(x(\cdot), u(\cdot))$ are called a *feasible state process (trajectory)* and a *feasible pair,* respectively. If in addition, the state constraint (6.1) is satisfied and the integrals in (3.2) and (6.1) are integrable, then $u(\cdot)$, $x(\cdot)$, and $(x(\cdot), u(\cdot))$ are called an *admissible control, admissible state process (trajectory),* and *admissible pair,* respectively. The set of all admissible controls is denoted by $\mathcal{U}_{ad}[0, T]$. We note that the situation of the previous sections can be regarded as the case where $\Gamma = \mathbb{R}^\ell$ with both $\mathbf{h}$ and $\mathbf{f}$ being zero. In that case, a pair $(x(\cdot), u(\cdot))$ is admissible if and only if it is feasible. However, we do need to distinguish these two notions in this section.

In what follows, we define

$$\mathbf{h}(x) = \begin{pmatrix} h^1(x) \\ \vdots \\ h^\ell(x) \end{pmatrix}, \quad \mathbf{f}(t, x, u) = \begin{pmatrix} f^1(t, x, u) \\ \vdots \\ f^\ell(t, x, u) \end{pmatrix}, \tag{6.2}$$

with h^i and f^i being scalar functions. For notational convenience, we will also define

$$h^0(x) = h(x), \quad f^0(t, x, u) = f(t, x, u), \tag{6.3}$$

where h and f are the functions appearing in the cost functional (3.2).

Before going further, let us look at some special cases of (6.1).

Case 1. Suppose we are faced with a *multiobjective optimal control problem:*

$$\text{Minimize } (J_0(u(\cdot)), J_1(u(\cdot)), \ldots, J_\ell(u(\cdot))),$$

where

$$J_i(u(\cdot)) = E\Big\{\int_0^T f^i(t, x(t), u(t))dt + h^i(x(T))\Big\}, \qquad i = 0, 1, \ldots, \ell.$$

A control $\overline{u}(\cdot)$ is called an *efficient control (solution)* of the problem if there does not exist any control $u(\cdot)$ such that

$$J_i(u(\cdot)) \leq J_i(\overline{u}(\cdot)), \qquad i = 0, 1, \ldots, \ell,$$

with at least one strict inequality. The set of all efficient controls is called an *efficient frontier.* By the multiobjective optimization theory (see Zeleny

[1]), the efficient frontier can be obtained by solving the *constrained* single-objective optimal control problem

$$\begin{cases} \text{Minimize } J_0(u(\cdot)), \\ \text{subject to } J_i(u(\cdot)) \leq r_i, \qquad i = 1, \ldots, \ell, \end{cases}$$

and varying the values of $r_i \in (-\infty, +\infty)$ for $i = 1, \ldots, \ell$. It is clear that the above constraints specialize (6.1).

Case 2. $\ell = n$, $\mathbf{h}(x) = x$, $\mathbf{f} = 0$, and

$$\Gamma = \{x \equiv (x_1, \ldots, x_n) \mid x_i \geq a_i, \quad 1 \leq i \leq n\}.$$

In this case, (6.1) reads

$$Ex_i(T) \geq a_i, \qquad i = 1, 2, \ldots, n.$$

A typical example of this kind of constraint is the following: In a manufacturing system, say, the expected inventory of the ith product should be at least a_i at the end of the time period $t = T$.

Case 3. $\ell = n$, $\mathbf{h} = 0$, $\mathbf{f}(t, x, u) = (c_1(t)x_1, \ldots, c_n(t)x_n)^\top$, and

$$\Gamma = \{x \equiv (x_1, \ldots, x_n) \mid x_i \leq a_i, \quad 1 \leq i \leq n\}.$$

In this case, (6.1) becomes

$$E \int_0^T c_i(t)x_i(t)dt \leq a_i, \qquad i = 1, 2, \ldots, n.$$

In the manufacturing system example, this could be interpreted as meaning that the total expected running cost of inventory for the ith product cannot exceed a given value a_i.

More special cases may be considered. We leave the details to the reader.

Now we introduce the following assumption concerning the set Γ and the functions $\mathbf{h}$ and $\mathbf{f}$.

(S5) Γ is convex and closed in $\mathbb{R}^\ell$. Moreover, for each $i = 1, 2, \ldots, \ell$, h^i and f^i satisfy (3.4)–(3.5).

Under (S1)–(S2) and (S5), for any $u(\cdot) \in \mathcal{U}[0, T]$, there exists a unique solution $x(\cdot)$ to (3.1), i.e., $(x(\cdot), u(\cdot))$ is feasible, and both the cost functional (3.2) and the left-hand side of (6.1) are well-defined. Then one may tell whether this pair is admissible by checking (6.1). Our optimal control problem with state constraint can then be stated as follows.

Problem (SC). Minimize (3.2) over $\mathcal{U}_{ad}[0, T]$.

Our goal in this section is to prove a stochastic maximum principle for the optimal controls of the above problem. We point out that the difficulty in investigating the problem is that we can only compare the values of the cost functional among all *admissible* controls (rather than merely *feasible* controls).

To state our maximum principle, let us introduce the following *Hamiltonian* (compare with (3.10)):

$$
\begin{aligned}
H(t,x,u,p,q,\psi^0,\psi) &\triangleq -\psi^0 f(t,x,u) - \langle \psi, \mathbf{f}(t,x,u)\rangle \\
&\qquad + \langle p, b(t,x,u)\rangle + \mathrm{tr}\,[q^\top \sigma(t,x,u)], \\
(t,x,u,p,q,\psi^0,\psi) &\in [0,T]\times\mathbb{R}^n\times U\times\mathbb{R}^n\times\mathbb{R}^{n\times m}\times\mathbb{R}\times\mathbb{R}^\ell.
\end{aligned}
\tag{6.4}
$$

It is seen that if $\psi^0 = 1$ and $\psi = 0$, then (6.4) reduces to (3.10).

Theorem 6.1. *Let (S0)–(S3) and (S5) hold. Let $(\bar{x}(\cdot),\bar{u}(\cdot))$ be an optimal pair of Problem (SC). Then there exist a $(\psi^0,\psi)\in\mathbb{R}^{1+\ell}$ satisfying*

$$
\psi^0 \ge 0, \quad |\psi^0|^2 + |\psi|^2 = 1,
\tag{6.5}
$$

$$
\langle \psi, z - E\mathbf{h}(\bar{x}(T)) + \int_0^T E\mathbf{f}(t,\bar{x}(t),\bar{u}(t))dt\rangle \ge 0, \quad \forall z\in\Gamma,
\tag{6.6}
$$

and adapted solutions

$$
\begin{cases}
(p(\cdot),q(\cdot)) \in L^2(0,T;\mathbb{R}^n)\times\left[L^2(0,T;\mathbb{R}^n)\right]^m, \\
(P(\cdot),Q(\cdot)) \in L^2(0,T;\mathcal{S}^n)\times\left[L^2(0,T;\mathcal{S}^n)\right]^m
\end{cases}
$$

of the following first- and second-order adjoint equations:

$$
\begin{cases}
dp(t) = -H_x(t,\bar{x}(t),\bar{u}(t),p(t),q(t),\psi^0,\psi)dt + q(t)dW(t), \\
p(T) = -\displaystyle\sum_{i=0}^{\ell} \psi^i h^i_x(\bar{x}(T)),
\end{cases}
\tag{6.7}
$$

$$
\begin{cases}
dP(t) = -\Big\{ b_x(t,\bar{x}(t),\bar{u}(t))^\top P(t) + P(t) b_x(t,\bar{x}(t),\bar{u}(t)) \\
\qquad + \displaystyle\sum_{j=1}^m \sigma^j_x(t,\bar{x}(t),\bar{u}(t))^\top P(t)\sigma^j_x(t,\bar{x}(t),\bar{u}(t)) \\
\qquad + \displaystyle\sum_{j=1}^m \left[\sigma^j_x(t,\bar{x}(t),\bar{u}(t))^\top Q_j(t) + Q_j(t)\sigma^j_x(t,\bar{x}(t),\bar{u}(t))\right] \\
\qquad + H_{xx}(t,\bar{x}(t),\bar{u}(t),p(t),q(t),\psi^0,\psi)\Big\}dt \\
\qquad + \displaystyle\sum_{j=1}^m Q_j(t)dW^j(t), \\
P(T) = -\displaystyle\sum_{i=0}^{\ell} \psi^i h^i_{xx}(\bar{x}(T)),
\end{cases}
\tag{6.8}
$$

such that

$$
\begin{aligned}
&H(t,\overline{x}(t),\overline{u}(t),p(t),q(t),\psi^0,\psi) - H(t,\overline{x}(t),u,p(t),q(t),\psi^0,\psi)\\
&\quad -\frac{1}{2}\mathrm{tr}\Big(\big\{\sigma(t,\overline{x}(t),\overline{u}(t)) - \sigma(t,\overline{x}(t),u)\big\}^\top P(t)\\
&\quad \cdot\big\{\sigma(t,\overline{x}(t),\overline{u}(t)) - \sigma(t,\overline{x}(t),u)\big\}\Big) \geq 0,\\
&\qquad\qquad\qquad\qquad \forall u \in U, \quad \text{a.e. } t \in [0,T], \quad \mathbf{P}\text{-a.s.}
\end{aligned}
\tag{6.9}
$$

We call (6.6) the *transversality condition.* In the case where both $\mathbf{h}$ and $\mathbf{f}$ are zero and $\Gamma = \mathbb{R}^\ell$, namely, when there is no state constraint, condition (6.6) implies $\psi = 0$. Thus, (6.5) leads to $\psi^0 = 1$, and the Hamiltonian H defined by (6.4) coincides with (3.10). Consequently, Theorem 6.1 reduces to Theorem 3.2. In general, however, ψ^0 could be zero. When $\psi^0 \neq 0$, we say that the *qualification condition* is satisfied.

One may also give a maximum condition, equivalent to the variational inequality (6.9), in terms of an $\mathcal{H}$-function similar to (3.16). We leave the details to the interested reader.

The following subsections are devoted to a proof of Theorem 6.1. The reader may skip them at first reading.

6.2. Some preliminary lemmas

In this section we present some preliminary results that are necessary for the proof of Theorem 6.1. The first one is called *Ekeland's variational principle.*

Lemma 6.2. *Let (V,d) be a complete metric space and let $F : V \to (-\infty,+\infty]$ be a proper (i.e., $F \not\equiv +\infty$), lower semicontinuous function bounded from below. Let $v_0 \in \mathcal{D}(F) \stackrel{\Delta}{=} \{v \in V \mid F(v) < \infty\}$ and let $\lambda > 0$ be fixed. Then there exists a $\bar{v} \in V$ such that*

$$
F(\bar{v}) + \lambda d(\bar{v}, v_0) \leq F(v_0),
\tag{6.10}
$$

$$
F(\bar{v}) < F(v) + \lambda d(v,\bar{v}), \qquad \forall v \neq \bar{v}.
\tag{6.11}
$$

Proof. Without loss of generality, we may assume $\lambda = 1$, since otherwise we may change the metric to $\lambda d(\cdot\,,\cdot)$. Also, by considering the function $F(\cdot) - \inf_{v\in V} F(v)$ instead of $F(\cdot)$, we may assume that $F(\cdot)$ is nonnegative valued. Next, we define a set-valued function

$$
G(v) = \{w \in V \mid F(w) + d(w,v) \leq F(v)\}.
\tag{6.12}
$$

Since F is lower semicontinuous, for any $v \in V$, $G(v)$ is a closed set in V and

$$
v \in G(v), \qquad \forall v \in V.
\tag{6.13}
$$

Next, we claim that

$$w \in G(v) \quad \Rightarrow \quad G(w) \subseteq G(v). \tag{6.14}$$

In fact, (6.14) is true if $F(v) = +\infty$, as $G(v) = V$ in this case. Now let $F(v) < \infty$. Then $w \in G(v)$ implies

$$F(w) + d(w, v) \leq F(v), \tag{6.15}$$

and for any $u \in G(w)$, we have

$$F(u) + d(u, w) \leq F(w). \tag{6.16}$$

Thus, combining (6.15) and (6.16), using the triangle inequality, one has

$$F(u) + d(u, v) \leq F(u) + d(u, w) + d(w, v) \leq F(v). \tag{6.17}$$

Thus (6.14) holds. Define

$$f(v) = \inf_{w \in G(v)} F(w), \qquad \forall v \in \mathcal{D}(F). \tag{6.18}$$

Then, for any $w \in G(v)$, we have $f(v) \leq F(w) \leq F(v) - d(w, v)$. Hence,

$$d(w, v) \leq F(v) - f(v). \tag{6.19}$$

So the diameter $\operatorname{diam} G(v)$ of the set $G(v)$ satisfies

$$\operatorname{diam} G(v) \stackrel{\Delta}{=} \sup_{w, u \in G(v)} d(w, u) \leq 2(F(v) - f(v)). \tag{6.20}$$

Now we define a sequence in the following way: $v_{n+1} \in G(v_n)$, $n \geq 0$, such that

$$F(v_{n+1}) \leq f(v_n) + \frac{1}{2^n}. \tag{6.21}$$

By (6.14), we know that $G(v_{n+1}) \subseteq G(v_n)$. Thus,

$$f(v_n) \leq f(v_{n+1}), \qquad n \geq 0. \tag{6.22}$$

On the other hand, $f(w) \leq F(w)$, since $w \in G(w)$ (see (6.13)). Therefore, in view of (6.21) and (6.22), we obtain

$$0 \leq F(v_{n+1}) - f(v_{n+1}) \leq f(v_n) + \frac{1}{2^n} - f(v_{n+1}) \leq \frac{1}{2^n}. \tag{6.23}$$

By (6.20), the diameter of $G(v_n)$ goes to 0 as $n \to \infty$. Since $G(v_n)$ is a sequence of nested closed sets in V (i.e., $G(v_{n+1}) \subseteq G(v_n)$, $\forall n \geq 0$) and V is complete, we must have some point $\bar{v} \in V$ such that

$$\bigcap_{n \geq 0} G(v_n) = \{\bar{v}\}. \tag{6.24}$$

In particular, $\bar{v} \in G(v_0)$, which gives (6.10) with $\lambda = 1$. Also, since $\bar{v} \in G(v_n)$ for any $n \geq 0$, we conclude from (6.14) that

$$G(\bar{v}) \subseteq \bigcap_{n\geq 0} G(v_n) = \{\bar{v}\}. \tag{6.25}$$

This implies that $G(\bar{v}) = \{\bar{v}\}$. Hence, for any $v \neq \bar{v}$, we have $v \notin G(\bar{v})$, which gives (6.11) (with $\lambda = 1$). □

Corollary 6.3. *Let the assumptions of Lemma 6.2 hold. Let $\rho > 0$ and $v_0 \in V$ be such that*

$$F(v_0) \leq \inf_{v\in V} F(v) + \rho. \tag{6.26}$$

Then there exists a $v_\rho \in V$ such that

$$F(v_\rho) \leq F(v_0), \qquad d(v_\rho, v_0) \leq \sqrt{\rho}, \tag{6.27}$$

and for all $v \in V$,

$$-\sqrt{\rho}\, d(v, v_\rho) \leq F(v) - F(v_\rho). \tag{6.28}$$

Proof. We take $\lambda = \sqrt{\rho}$. By Lemma 6.2, there exists a $v_\rho \in V$ such that (noting (6.26))

$$F(v_\rho) + \sqrt{\rho}\, d(v_\rho, v_0) \leq F(v_0) \leq \inf_{v\in V} F(v) + \rho \leq F(v_\rho) + \rho, \tag{6.29}$$

and

$$F(v_\rho) < F(v) + \sqrt{\rho}\, d(v, v_\rho), \qquad \forall v \neq v_\rho. \tag{6.30}$$

Then, (6.27) follows from (6.29), and (6.28) follows from (6.30). □

Lemma 6.4. *Let $\bar{d}$ be defined by the following:*

$$\begin{aligned} \bar{d}(u(\cdot), \widehat{u}(\cdot)) &\stackrel{\Delta}{=} \big|\{(t,\omega) \in [0,T] \times \Omega \mid u(t,\omega) \neq \widehat{u}(t,\omega)\}\big|, \\ &\forall u(\cdot), \widehat{u}(\cdot) \in \mathcal{U}[0,T], \end{aligned} \tag{6.31}$$

where $|A|$ denotes the product measure of the Lebesgue measure and the probability $\mathbf{P}$ of a set $A \subseteq [0,T] \times \Omega$. Then $\bar{d}$ is a metric under which $\mathcal{U}[0,T]$ is a complete metric space.

Proof. Let $\{u_n(\cdot)\}$ be a Cauchy sequence in $\mathcal{U}[0,T]$ under the metric $\bar{d}$, i.e.,

$$\bar{d}(u_n(\cdot), u_m(\cdot)) \to 0, \quad n, m \to \infty. \tag{6.32}$$

Then there exists a subsequence $\{u_{n_k}(\cdot)\}$, such that

$$\bar{d}(u_{n_k}(\cdot), u_{n_{k+1}}(\cdot)) \leq 2^{-k}, \qquad k \geq 2. \tag{6.33}$$

Let

$$\text{(6.34)}\quad \begin{cases} E_{nm} = \{(t,\omega) \in [0,T] \times \Omega \mid u_n(t,\omega) \neq u_m(t,\omega)\}, & n, m \geq 1, \\ A_k = \bigcup_{p \geq k} E_{n_p, n_{p+1}}, \quad k \geq 2. \end{cases}$$

Clearly, $A_k \supseteq A_{k+1}$, $\forall k \geq 1$, and

$$\text{(6.35)}\qquad |A_k| \leq \sum_{p=k}^{\infty} 2^{-p} = 2^{1-k}, \qquad k \geq 2.$$

Consequently, $\left| \bigcup_{k \geq 1} A_k^c \right| = T$. Now we define

$$\text{(6.36)}\qquad \bar{u}(t,\omega) = u_{n_k}(t,\omega), \qquad t \in A_k^c, \quad k \geq 2.$$

From the definition of A_k, we see that $\bar{u}(\cdot)$ is well-defined and $\bar{u}(\cdot) \in \mathcal{U}[0,T]$. Moreover,

$$\text{(6.37)}\qquad \bar{d}(u_{n_k}(\cdot), \bar{u}(\cdot)) \leq |A_k| \leq 2^{1-k} \to 0.$$

Therefore, $\bar{d}(u_n(\cdot), \bar{u}(\cdot)) \to 0$, proving the completeness of $\mathcal{U}[0,T]$. □

Next, we consider the convex and closed set Γ in $\mathbb{R}^\ell$. Define

$$\text{(6.38)}\qquad d_\Gamma(z) = \inf_{z' \in \Gamma} |z - z'|, \quad \forall z \in \mathbb{R}^\ell.$$

We call d_Γ the *distance function* (to the set Γ).

Lemma 6.5. *Let $\Gamma \subseteq \mathbb{R}^\ell$ be convex and closed. Then:*

(i) *$d_\Gamma : \mathbb{R}^\ell \to \mathbb{R}$ is convex and Lipschitz continuous with Lipschitz constant 1.*

(ii) *For any $z \notin \Gamma$, $\partial d_\Gamma(z)$ has exactly one element with its length being 1.*

(iii) *$d_\Gamma(\cdot)^2$ is C^1.*

Proof. The proof of (i) is immediate.

(ii) Let $z \notin \Gamma$. Since $d_\Gamma(\cdot)$ is Lipschitz continuous with Lipschitz constant 1, from (2.20), it follows that if $\xi \in \partial d_\Gamma(z)$, then

$$\langle \xi, y \rangle \leq \overline{\lim_{\substack{z' \to z \\ t \downarrow 0}}} \frac{d_\Gamma(z' + ty) - d_\Gamma(z')}{t} \leq |y|, \quad \forall y \in \mathbb{R}^\ell.$$

This implies

$$\text{(6.39)}\qquad |\xi| \leq 1, \qquad \forall \xi \in \partial d_\Gamma(z).$$

On the other hand, since $z \notin \Gamma$, for any $0 < \delta < 1$, there exists a $z_\delta \in \Gamma$ such that

$$\text{(6.40)}\qquad d_\Gamma(z) \geq (1 - \delta)|z - z_\delta| > 0.$$

As $d_\Gamma(\cdot)$ is convex, by Lemma 2.3-(v) we have

$$\partial d_\Gamma(z) = \{\xi \in \mathbb{R}^\ell \mid d_\Gamma(z') - d_\Gamma(z) \geq \langle \xi, z' - z \rangle, \quad \forall z' \in \mathbb{R}^\ell\}. \tag{6.41}$$

Thus, for any $\xi \in \partial d_\Gamma(z)$, we have (noting $d_\Gamma(z_\delta) = 0$)

$$-d_\Gamma(z) \geq \langle \zeta, z_\delta - z \rangle. \tag{6.42}$$

Combining (6.40) and (6.41), we obtain

$$0 < (1-\delta)|z - z_\delta| \leq d_\Gamma(z) \leq -\langle \xi, z_\delta - z \rangle \leq |\xi||z - z_\delta|. \tag{6.43}$$

Hence, $|\xi| \geq 1 - \delta$. By sending $\delta \to 0$, we obtain that

$$|\xi| = 1, \qquad \forall \xi \in \partial d_\Gamma(z), \tag{6.44}$$

provided that $z \notin \Gamma$. However, $\partial d_\Gamma(z)$ is convex by Lemma 2.3-(i). Hence, $\partial d_G(z)$ is a (nonempty) convex and closed set on the unit sphere in $\mathbb{R}^\ell$ (with the usual Euclidean metric). Thus, $\partial d_\Gamma(z)$ must be a singleton.

(iii) A direct computation using the definition of derivative yields

$$\nabla\big[d_\Gamma(z)^2\big] = \begin{cases} 2d_\Gamma(z)\partial d_\Gamma(z), & z \notin \Gamma, \\ 0, & z \in \Gamma. \end{cases} \tag{6.45}$$

Here, due to (ii), we may denote the unique element in $\partial d_\Gamma(z)$ by $\partial d_\Gamma(z)$ when $z \notin \Gamma$. Clearly, the right-hand side of (6.45) is continuous. □

Since $d_\Gamma(z) = 0$ when $z \in \Gamma$, it will be convenient and without any ambiguity to write (instead of (6.45))

$$\nabla\big[d_\Gamma(z)^2\big] = 2d_\Gamma(z)\partial d_\Gamma(z), \qquad \forall z \in \mathbb{R}^\ell. \tag{6.46}$$

We will use this convention in a later context.

6.3. A proof of Theorem 6.1.

We now present a proof of Theorem 6.1. Besides the Taylor expansion and duality analysis, we need some more technique to handle the state constraints.

In what follows, for notational convenience, we denote a generic point in $\mathbb{R}^\ell$ by $\mathbf{x} = (x^1, \ldots, x^\ell)^\top$. For any feasible pair, let

$$\begin{cases} x^i(t) = h^i(x(t)) + \displaystyle\int_0^t f^i(s, x(s), u(s))ds, & 1 \leq i \leq \ell, \\ \mathbf{x}(t) = (x^1(t), \ldots, x^\ell(t))^\top, & t \in [0,T]. \end{cases} \tag{6.47}$$

By adding or subtracting a constant in the cost functional, we may assume, without loss of generality, that $J(\overline{u}(\cdot)) = 0$. For any $\rho > 0$, define

$$J_\rho(u(\cdot)) = \big\{\big[(J(u(\cdot)) + \rho)^+\big]^2 + d_\Gamma\big(E\mathbf{x}(T)\big)^2\big\}^{1/2}. \tag{6.48}$$

We call $J_\rho(u(\cdot))$ a *penalty functional* associated with our Problem (SC). It is important to point out that we will study this functional over the whole *feasible control* set $\mathcal{U}[0,T]$, rather than just the set of *admissible* controls. In other words, we are able to get rid of the state constraints by introducing such a penalty function.

It is clear that $J_\rho : (\mathcal{U}[0,T], \bar{d}) \to \mathbb{R}$ is continuous and

$$\begin{cases} J_\rho(u(\cdot)) > 0, & \forall u(\cdot) \in \mathcal{U}[0,T], \\ J_\rho(\bar{u}(\cdot)) = \rho \leq \inf\limits_{u(\cdot)\in\mathcal{U}[0,T]} J_\rho(u(\cdot)) + \rho. \end{cases} \tag{6.49}$$

Recall that by Lemma 6.4, $(\mathcal{U}[0,T], \bar{d})$ is a complete metric space. Thus, by Corollary 6.3, there exists a $u_\rho(\cdot) \in \mathcal{U}[0,T]$ such that

$$J_\rho(u_\rho(\cdot)) \leq J_\rho(\bar{u}(\cdot)) = \rho, \quad \bar{d}(u_\rho(\cdot), \bar{u}(\cdot)) \leq \sqrt{\rho}, \tag{6.50}$$

and

$$-\sqrt{\rho}\bar{d}(u_\rho(\cdot), u(\cdot)) \leq J_\rho(u(\cdot)) - J_\rho(u_\rho(\cdot)), \quad \forall u(\cdot) \in \mathcal{U}[0,T]. \tag{6.51}$$

If we call Problem $(\mathrm{SC})^\rho$ the optimal control problem with the state equation (3.1) and the cost functional $J_\rho(u(\cdot)) + \sqrt{\rho}\bar{d}(u_\rho(\cdot), u(\cdot))$ (*without* the state constraint), then (6.50) and (6.51) imply that $u_\rho(\cdot)$ is an optimal control of Problem $(\mathrm{SC})^\rho$, which is very close to $\bar{u}(\cdot)$, the optimal control of Problem (SC) for which we want to derive the necessary conditions. Our idea is to derive the necessary conditions for $u_\rho(\cdot)$ and then let $\rho \to 0$ to get proper conditions for $\bar{u}(\cdot)$.

Since Problem $(\mathrm{SC})^\rho$ does not have state constraints, and it differs from Problem (S) only in the cost functional, the derivation of necessary conditions for $u_\rho(\cdot)$ will be very similar to what we presented in Section 4. Thus, we will omit some of the similar proofs here.

Fix a $\rho > 0$ and a $u(\cdot) \in \mathcal{U}[0,T]$. For any $\varepsilon > 0$, take any Borel measurable set $E_\varepsilon \subseteq [0,T]$ with $|E_\varepsilon| = \varepsilon$. Define

$$u_\rho^\varepsilon(t) = \begin{cases} u_\rho(t), & t \in [0,T] \setminus E_\varepsilon, \\ u(t), & t \in E_\varepsilon. \end{cases} \tag{6.52}$$

Clearly,

$$\bar{d}(u_\rho^\varepsilon(\cdot), u_\rho(\cdot)) \leq |E_\varepsilon \times \Omega| = \varepsilon.$$

Thus, by taking $u(\cdot) = u^\varepsilon_\rho(\cdot)$ in (6.51), we obtain

$$
\begin{aligned}
-\sqrt{\rho}\,\varepsilon &\le J_\rho(u^\varepsilon_\rho(\cdot)) - J_\rho(u_\rho(\cdot)) \\
&= \frac{\big[\big(J(u^\varepsilon_\rho(\cdot)) + \rho\big)^+\big]^2 - \big[\big(J(u_\rho(\cdot)) + \rho\big)^+\big]^2}{J_\rho(u^\varepsilon_\rho(\cdot)) + J_\rho(u_\rho(\cdot))} \\
&\quad + \frac{d_\Gamma\big(E\mathbf{x}^\varepsilon_\rho(T)\big)^2 - d_\Gamma\big(E\mathbf{x}_\rho(T)\big)^2}{J_\rho(u^\varepsilon_\rho(\cdot)) + J_\rho(u_\rho(\cdot))} \\
&= \psi^{0,\varepsilon}_\rho \big[J(u^\varepsilon_\rho)) - J(u_\rho(\cdot))\big] + \langle\, \psi^\varepsilon_\rho, E\big[\mathbf{x}^\varepsilon_\rho(T) - \mathbf{x}_\rho(T)\big]\,\rangle \\
&= E\Big\{ \sum_{i=1}^{\ell} \psi^{0,\varepsilon}\rho\Big[h^i(x^\varepsilon_\rho(T)) - h^i(x_\rho(T)) \\
&\quad + \int_0^T \big[f^i(t, x^\varepsilon_\rho(t), u^\varepsilon_\rho(t)) - f^i(t, x_\rho(t), u_\rho(t))\big]dt\Big]\Big\},
\end{aligned}
\tag{6.53}
$$

where $x^\varepsilon_\rho(\cdot)$ and $x_\rho(\cdot)$ are the states under controls $u^\varepsilon_\rho(\cdot)$ and $u_\rho(\cdot)$, respectively, $\mathbf{x}^\varepsilon_\rho(\cdot)$ and $\mathbf{x}_\rho(\cdot)$ are the corresponding functions defined by (6.47), and

$$
\begin{cases}
\psi^{0,\varepsilon}_\rho \triangleq \dfrac{\big[J(u_\rho(\cdot)) + \rho\big]^+}{J_\rho(u_\rho(\cdot))} + o(1), \\[2ex]
\psi^\varepsilon_\rho \triangleq \dfrac{d_\Gamma\big(E\mathbf{x}_\rho(T)\big)\partial d_\Gamma\big(E\mathbf{x}_\rho(T)\big)}{J_\rho(u_\rho(\cdot))} + o(1),
\end{cases}
\qquad \text{as } \varepsilon \to 0,
\tag{6.54}
$$

with $\psi^\varepsilon_\rho = (\psi^{1,\varepsilon}_\rho, \dots, \psi^{\ell,\varepsilon}_\rho)$. Now, in the present case, we also have Taylor expansions (in terms of ε) similar to those of Theorem 4.4 along the pair $(x_\rho(\cdot), u_\rho(\cdot))$. Thus, we have

$$
\begin{aligned}
-\sqrt{\rho}\,\varepsilon \le &E\sum_{i=0}^{\ell} \Big\{ \psi^{i,\varepsilon}_\rho \Big[\langle\, h^i_x(x_\rho(T)), y^\varepsilon_\rho(T) + z^\varepsilon_\rho(T)\,\rangle \\
&+ \frac{1}{2}\langle\, h_{xx}(x_\rho(T)) y^\varepsilon_\rho(T), y^\varepsilon_\rho(T)\,\rangle \Big] \\
&+ \int_0^T \Big[\langle\, f^{i,\rho}_x(t), y^\varepsilon_\rho(t) + z^\varepsilon_\rho(t)\,\rangle \\
&+ \frac{1}{2}\langle\, f^{i,\rho}_{xx}(t) y^\varepsilon_\rho(t), y^\varepsilon_\rho(t)\,\rangle + \delta f^{i,\rho}(t)\chi_{E_\varepsilon}(t)\Big]dt\Big\} + o(\varepsilon),
\end{aligned}
\tag{6.55}
$$

where $y^\varepsilon_\rho(\cdot)$ and $z^\varepsilon_\rho(\cdot)$ satisfy variational equations similar to (4.23) and (4.24) with $(\overline{x}(\cdot), \overline{u}(\cdot))$ replaced by $(x_\rho(\cdot), u_\rho(\cdot))$, and $f^{i,\rho}_x(t)$, $f^{i,\rho}_{xx}(t)$, and $\delta f^{i,\rho}(t)$ are defined in a manner similar to (4.22) with $(\overline{x}(\cdot), \overline{u}(\cdot))$ replaced by $(x_\rho(\cdot), u_\rho(\cdot))$. Now, as before, we let $(p^\varepsilon_\rho(\cdot), q^\varepsilon_\rho(\cdot))$ and $(P^\varepsilon_\rho(\cdot), Q^\varepsilon_\rho(\cdot))$ be the adapted solutions of the following *approximate first- and second-order adjoint equations:*

$$(6.56)\qquad \begin{cases} dp_\rho^\varepsilon(t) = -H_x(t, x_\rho(t), u_\rho(t), p_\rho^\varepsilon(t), q_\rho^\varepsilon(t), \psi_\rho^{0,\varepsilon}, \psi_\rho^\varepsilon)dt \\ \qquad\qquad + q_\rho^\varepsilon(t)dW(t), \\ p_\rho^\varepsilon(T) = -\sum_{i=0}^{\ell} \psi_\rho^{i,\varepsilon} h_x^i(x_\rho(T)), \end{cases}$$

$$(6.57)\qquad \begin{cases} dP_\rho^\varepsilon(t) = -\Big\{ b_x(t, x_\rho(t), u_\rho(t))^\top P_\rho^\varepsilon(t) + P_\rho^\varepsilon(t) b_x(t, x_\rho(t), u_\rho(t)) \\ \qquad + \sum_{j=1}^{m} \sigma_x^j(t, x_\rho(t), u_\rho(t))^\top P_\rho^\varepsilon(t) \sigma_x^j(t, x_\rho(t), u_\rho(t)) \\ \qquad + \sum_{j=1}^{m} \big[\sigma_x^j(t, x_\rho(t), u_\rho(t))^\top Q_{\rho,j}^\varepsilon(t) \\ \qquad\qquad + Q_{\rho,j}^\varepsilon(t) \sigma_x^j(t, x_\rho(t), u_\rho(t))\big] \\ \qquad + H_{xx}(t, x_\rho(t), u_\rho(t), p_\rho^\varepsilon(t), q_\rho^\varepsilon(t), \psi_\rho^{0,\varepsilon}, \psi_\rho^\varepsilon) \Big\} dt \\ \qquad + \sum_{j=1}^{m} Q_{\rho,j}^\varepsilon(t) dW^j(t), \\ P(T) = -\sum_{i=0}^{\ell} \psi^i h_{xx}^i(x_\rho(T)). \end{cases}$$

Then, using duality relations similar to those of Section 4, we have

$$(6.58)\quad \sqrt{\rho}\,\varepsilon + o(\varepsilon) \geq E\int_0^T \Big\{ \delta H_\rho^\varepsilon(t) + \frac{1}{2}\mathrm{tr}\,\big[\delta\sigma_\rho(t)^\top P_\rho^\varepsilon(t)\delta\sigma_\rho(t)\big] \Big\} \chi_{E_\varepsilon}(t)dt,$$

where

$$(6.59)\qquad \begin{cases} \delta H_\rho^\varepsilon(t) \stackrel{\Delta}{=} H(t, x_\rho(t), u(t), p_\rho^\varepsilon(t), q_\rho^\varepsilon(t), \psi_\rho^{0,\varepsilon}, \psi_\rho^\varepsilon) \\ \qquad\qquad - H(t, x_\rho(t), u_\rho(t), p_\rho^\varepsilon(t), q_\rho^\varepsilon(t), \psi_\rho^{0,\varepsilon}, \psi_\rho^\varepsilon), \\ \delta\sigma_\rho(t) \stackrel{\Delta}{=} \sigma(t, x_\rho(t), u(t)) - \sigma(t, x_\rho(t), u_\rho(t)). \end{cases}$$

From (6.54) and Lemma 6.5, we have

$$(6.60)\qquad \psi_\rho^{0,\varepsilon} \geq o(1), \quad |\psi_\rho^{0,\varepsilon}|^2 + |\psi_\rho^\varepsilon|^2 = 1 + o(1), \qquad \text{as } \varepsilon \to 0.$$

Thus, there is a subsequence, still denoted by $(\psi_\rho^{0,\varepsilon}, \psi_\rho^\varepsilon)$, such that

$$\lim_{\varepsilon\to 0}(\psi_\rho^{0,\varepsilon}, \psi_\rho^\varepsilon) = (\psi_\rho^0, \psi_\rho),$$

for some $(\psi_\rho^0, \psi_\rho) \in \mathbb{R}^{1+\ell}$ with

$$(6.61)\qquad \psi_\rho^0 \geq 0, \quad |\psi_\rho^0|^2 + |\psi_\rho|^2 = 1.$$

Consequently, from (6.56)–(6.57) and the continuous dependence of the (adapted) solutions to the BSDEs on the parameters (see Chapter 7, Theorems 2.2 and 3.3), we have the following limits as $\varepsilon \to 0$:

$$\begin{cases} (p_\rho^\varepsilon(\cdot), q_\rho^\varepsilon(\cdot)) \to (p_\rho(\cdot), q_\rho(\cdot)), \\ (P_\rho^\varepsilon(\cdot), Q_\rho^\varepsilon(\cdot)) \to (P_\rho(\cdot), Q_\rho(\cdot)), \end{cases}$$

strongly in the spaces $L^2_{\mathcal{F}}(0,T;\mathbb{R}^n) \times [L^2_{\mathcal{F}}(0,T;\mathbb{R}^n)]^m$ and $L^2_{\mathcal{F}}(0,T;\mathcal{S}^n) \times [L^2_{\mathcal{F}}(0,T;\mathcal{S}^n)]^m$, respectively, where $(p_\rho(\cdot), q_\rho(\cdot))$ and $(P_\rho(\cdot), Q_\rho(\cdot))$ are adapted solutions to BSDEs like (6.56) and (6.57) with $(\psi_\rho^{0,\varepsilon}, \psi_\rho^\varepsilon)$ replaced by (ψ_ρ^0, ψ_ρ). Then, from (6.58), by properly choosing E_ε and sending $\varepsilon \to 0$, we have

$$\begin{aligned} \sqrt{\rho} &\geq E\{\delta H_\rho(t) + \frac{1}{2}\mathrm{tr}\left[\delta\sigma_\rho(t)^\top P_\rho(t)\delta\sigma_\rho(t)\right]\} \\ &\equiv E\Big\{H(t, x_\rho(t), u(t), p_\rho(t), q_\rho(t), \psi_\rho^0, \psi_\rho) \\ &\qquad - H(t, x_\rho(t), u_\rho(t), p_\rho(t), q_\rho(t), \psi_\rho^0, \psi_\rho) \\ &\qquad + \frac{1}{2}\mathrm{tr}\Big(\left[\sigma(t, x_\rho(t), u(t)) - \sigma(t, x_\rho(t), u_\rho(t))\right]^\top P_\rho(t) \\ &\qquad\qquad \cdot \left[\sigma(t, x_\rho(t), u(t)) - \sigma(t, x_\rho(t), u_\rho(t))\right]\Big)\Big\}. \end{aligned} \tag{6.62}$$

Now we let $\rho \to 0$. By choosing a subsequence (if necessary) and using (6.61), we may assume that

$$(\psi_\rho^0, \psi_\rho) \to (\psi^0, \psi) \in \mathbb{R}^{1+\ell},$$

and (6.5) holds. Then, by the second relation in (6.50) and the continuous dependence of the solutions to BSDEs, we obtain pairs $(p(\cdot), q(\cdot))$ and $(P(\cdot), Q(\cdot))$ as the limits of $(p_\rho(\cdot), q_\rho(\cdot))$ and $(P_\rho(\cdot), Q_\rho(\cdot))$, respectively, which are adapted solutions of (6.7) and (6.8), respectively. Passing to the limits in (6.62), we obtain (6.9).

Finally, by the definition of ψ_ρ^ε, we have

$$\langle \psi_\rho^\varepsilon, z - E\mathbf{x}_\rho^\varepsilon(T) \rangle \geq -o(1), \qquad \text{as } \varepsilon \to 0. \tag{6.63}$$

Sending $\varepsilon \to 0$ and then $\rho \to 0$ in the above, we obtain the transversality condition (6.6). This completes the proof. □

7. Historical Remarks

The essential idea of the *maximum principle* can be traced back to the 1930s, when McShane [1] studied the Lagrange problem by introducing a very influential method that inspired the Pontryagin school later (See Berkovitz–Fleming [1] for detailed comments). The form of the *maximum principle* that is now familiar to us was first obtained by Hestenes [1] in 1950 as a necessary condition for the solutions to calculus of variations problems. He used the result to investigate some time optimal control problems. The maximum principle obtained by Hestenes [1] turns out to be equivalent to the so-called Weierstrass condition and is a consequence of Carathéodory's necessary condition in the classical calculus of variations (see Carathéodory [1,2], Pesch–Bulirsch [1], and Boltyanski [2] for further discussions).

Boltyanski–Gamkrelidze–Pontryagin [1] announced the *maximum principle* for the first time in 1956. The first rigorous proof for the linear system case was given by Gamkrelidze [1] in 1957. The general case was proved by Boltyanski [1] in 1958. A detailed presentation appeared in Boltyanski–Gamkrelidze–Pontryagin [2] in 1960. These results were summarized in the classical book (Pontryagin–Boltyanski–Gamkrelidze–Mischenko [1]*). Since then the theory has been developed extensively by many authors, and different versions of maximum principle exist, which can be found literally in almost every book on optimal control theory. See for example, Lee–Markus [1], Young [1], Warga [1], and Berkovitz [1]. McShane [2] commented that Pontryagin's school "introduced one important new feature. They allow the set $\Omega(t, y)$ (the control set) to be closed sets, not demanding that they be open as previous researchers had." Pesch–Bulirsch [1] pointed out that the work of Pontryagin's school "led to the cutting of the umbilical cord between the calculus of variations and optimal control theory."

People have been extending the maximum principle to infinite dimensions in various aspects. For a detailed description of these, see the book by Li–Yong [2] and the references cited therein. See also Sussmann [1] for a survey of some results on manifolds and the relation to differential geometry.

Theorem 2.1 is one of the easiest versions of the maximum principle (without any state constraint), whose proof nevertheless catches the essence in the original proof of the Pontryagin school, namely, the spike variation. Sufficiency of the maximum principle in the presence of some convexity conditions was shown first by Mangasarian [1] in 1966, on which Theorem 2.5 is based, and by Arrow–Kurz [1]. Lemma 2.3 was essentially extracted from Clarke [2], where a systematic treatment of the generalized gradient can be found. It should be noted that Clarke's generalized gradient can be used to derive a maximum principle by assuming only the Lipschitz continuity of the functions b, f, and h with respect to the state variable; see Clarke [1].

Attempts were made to extend the maximum principle to the stochastic (diffusion) case shortly after the work of the Pontryagin school. The earliest paper concerning the stochastic maximum principle was published in 1964 by Kushner–Schweppe [1]. See also Kushner [2,3].

In 1972, Kushner [7] employed the spike variation and Neustadt's variational principle (Neustadt [1]) to derive a stochastic maximum principle. On the other hand, Haussmann [1] extensively investigated the necessary conditions of stochastic optimal *state feedback* controls based on the Girsanov transformation. These works were summarized in the book by Hauss-

* Recently, in a widely circulated preprint, Boltyanski [2] elaborated from his point of view the process of the birth of the maximum principle.

mann [4]. However, due to the limitation of the Girsanov transformation, this approach works only for systems with *nondegenerate* diffusion coefficients. Along another line, Bismut [1,2,3] derived the adjoint equation as what is called today the backward stochastic differential equation via the martingale representation theorem. Bensoussan [1,3] further developed this approach.

However, prior to 1988 all the results on the stochastic maximum principle were obtained basically under the assumptions that the diffusion coefficients were *independent* of the controls and/or the systems were *nondegenerate* and/or the control regions were *convex*. The results obtained were pretty much parallel to those of the deterministic systems, and one hardly saw an essential difference between the stochastic and deterministic systems from these results. During 1987–1989, in a weekly seminar led by X. Li at the Institute of Mathematics, Fudan University, a group of people, including Y. Hu, J. Ma, S. Peng, J. Yong, and X. Y. Zhou, were combating the problems with control-dependent diffusions. Their insistence on solving the control-dependent diffusion cases was based on the following belief: Only when the controls/decisions could or would influence the scale of uncertainty (as indeed is the case in many practical systems, especially in the area of finance) do the stochastic problems differ from the deterministic ones. In terms of the necessary conditions of stochastic optimality, Peng [1] first considered the second-order term in the "Taylor expansion" of the variation and obtained a stochastic maximum principle for systems that are possibly degenerate, with control-dependent diffusions and not necessarily convex control regions. The form of his maximum principle is quite different from the deterministic ones and reflects the stochastic nature of the problem. Peng's proof was simplified by Zhou [4]. Later, Cadenillas–Karatzas [1] extended the result to systems with random coefficients, and Elliott–Kohlmann [2] employed stochastic flows to obtain similar results. The proof presented in Section 4 is based on Peng [1] and Zhou [4]. At the same time as Peng obtained the stochastic maximum principle, Zhou [4] established the relationship between the stochastic maximum principle and dynamic programming via (once again!) the second-order variations. See Chapter 5 for more details. It is also interesting to note that it was the study of the stochastic maximum principle that motivated Peng to formulate nonlinear backward stochastic differential equations (BSDEs, for short), because the adjoint equations are by nature BSDEs. See more comments in Chapter 7.

The sufficiency of the stochastic maximum principle was investigated by Bismut [3], in which the conditions derived were rather complicated. Zhou [10] proved that Peng's maximum condition is also sufficient in the presence of certain convexity. Section 5 is mainly based on Zhou [10].

Problems with state constraints are very difficult in general. The result in Section 6 is an extension of Peng [1], with some ideas adopted from the proof of the maximum principle for deterministic infinite-dimensional systems as in Li–Yong [1,2]. In particular, the results in Section 6.2 are extracted from Ekeland [1,2] and Li–Yong [1]. Here, we should mention that the constraints that we can treat are mainly of *finite codimension*. Otherwise, there is no guarantee of the nontriviality of the *multiplier* (ψ^0, ψ). Note that in general, ψ has to live in some infinite-dimensional spaces. See Li–Yong [1,2] for some extensive discussion on this issue, although it is not particularly for stochastic problems.

There is a very extensive literature on the necessary conditions of optimality for other types of stochastic optimal control problems. Let us just mention a few: For partially observed systems, see Fleming [2], Bensoussan [4], Baras–Elliott–Kohlmann [1], Haussmann [5], Zhou [7], and Tang [1]; for systems with jumps, see Tang–Li [1]; and for stochastic impulse control, see Hu–Yong [1].

Chapter 4

Dynamic Programming and HJB Equations

1. Introduction

In this chapter we turn to study another powerful approach to solving optimal control problems, namely, the method of *dynamic programming.* Dynamic programming, originated by R. Bellman in the early 1950s, is a mathematical technique for making a sequence of interrelated decisions, which can be applied to many optimization problems (including optimal control problems). The basic idea of this method applied to optimal controls is to consider a *family* of optimal control problems with *different* initial times and states, to establish relationships among these problems via the so-called *Hamilton–Jacobi–Bellman equation* (HJB, for short), which is a *nonlinear* first-order (in the deterministic case) or second-order (in the stochastic case) partial differential equation. If the HJB equation is solvable (either analytically or numerically), then one can obtain an optimal feedback control by taking the maximizer/minimizer of the Hamiltonian or generalized Hamiltonian involved in the HJB equation. This is the so-called *verification technique.* Note that this approach actually gives solutions to the *whole family* of problems (with different initial times and states), and in particular, the original problem.

However, there was a major drawback in the classical dynamic programming approach: It required that the HJB equation admit *classical solutions*, meaning that the solutions be smooth enough (to the order of derivatives involved in the equation). Unfortunately, this is not necessarily the case even for some very simple situations. In the stochastic case where the diffusion is possibly degenerate, the HJB equation may in general have no classical solutions either. To overcome this difficulty, Crandall and Lions introduced the so-called *viscosity solutions* in the early 1980s. This new notion is a kind of nonsmooth solutions to partial differential equations, whose key feature is to replace the conventional derivatives by the (set-valued) super-/subdifferentials while maintaining the uniqueness of solutions under very mild conditions. These make the theory a powerful tool in tackling optimal control problems. The viscosity solutions that we are going to discuss in this book can be merely continuous (not necessarily differentiable). In a more general framework, viscosity solutions can even be discontinuous. Such a situation will not be discussed in this book (see Section 7 for more biographical remarks on this topic).

The rest of this chapter is organized as follows. Section 2 reviews the deterministic case. The notion of viscosity solutions to the first-order HJB equations is introduced. Existence and uniqueness of the viscosity solutions are discussed. Section 3 proves a stochastic version of Bellman's principle of optimality, based on which the second-order HJB equation is derived.

In Section 4 some properties of the value function are discussed. Viscosity solutions to the second-order HJB equations are defined in Section 5, and some basic properties of viscosity solutions are presented. Section 6 deals with the uniqueness of the viscosity solutions. This section is very technical, and the reader is advised to skip it at first reading. Finally, Section 7 gives some historical remarks.

2. The Deterministic Case Revisited

For convenience, let us recall the formulation of deterministic optimal control problems. Given $x_0 \in \mathbb{R}^n$, consider the following control system:

$$\begin{cases} \dot{x}(t) = b(t, x(t), u(t)), & \text{a.e. } t \in [0, T], \\ x(0) = x_0, \end{cases} \tag{2.1}$$

where the control $u(\cdot)$ belongs to

$$\mathcal{V}[0, T] = \{u(\cdot) : [0, T] \to U \mid u(\cdot) \text{ is measurable}\},$$

with U being a metric space, $T > 0$, and $b : [0, T] \times \mathbb{R}^n \times U \to \mathbb{R}^n$ a given map. The cost functional associated with (2.1) is the following:

$$J(u(\cdot)) = \int_0^T f(t, x(t), u(t))dt + h(x(T)), \tag{2.2}$$

for some given maps f and h. In what follows, proper conditions will be imposed such that for any $u(\cdot) \in \mathcal{V}[0, T]$, the state equation (2.1) admits a unique solution $x(\cdot) \in C([0, T]; \mathbb{R}^n)$ and (2.2) is well-defined. The optimal control problem is stated as follows:

Problem (D). Minimize (2.2) over $\mathcal{V}[0, T]$.

Such a formulation of the (deterministic) optimal control problem has been already given in Chapters 2 and 3. Note that the initial time ($t = 0$) and the initial state ($x(0) = x_0$) are fixed in the formulation. The basic idea of the *dynamic programming* method is, however, to consider a family of optimal control problems with *different* initial times and states, to establish *relationships* among these problems, and finally to solve *all of them.* Let us now make this precise.

Let $(s, y) \in [0, T) \times \mathbb{R}^n$, and consider the following control system over $[s, T]$ (compare with (2.1)):

$$\begin{cases} \dot{x}(t) = b(t, x(t), u(t)), & \text{a.e. } t \in [s, T], \\ x(s) = y. \end{cases} \tag{2.3}$$

Here, the control $u(\cdot)$ is in

$$\mathcal{V}[s, T] \stackrel{\Delta}{=} \{u(\cdot) : [s, T] \to U \mid u(\cdot) \text{ is measurable}\}.$$

The cost functional is the following (compare with (2.2)):

$$J(s,y;u(\cdot)) = \int_s^T f(t,x(t),u(t))dt + h(x(T)). \tag{2.4}$$

We state the corresponding optimal control problem as follows:

Problem (D_{sy}). For given $(s,y) \in [0,T) \times \mathbb{R}^n$, minimize (2.4) subject to (2.3) over $\mathcal{V}[s,T]$.

The above is a *family* of optimal control problems parameterized by $(s,y) \in [0,T) \times \mathbb{R}^n$, in which the original Problem (D) is "embedded". Actually, Problem (D_{sy}) with $s = 0$ and $y = x_0$ coincides with Problem (D). Once again, it is essential that we allow the initial (s,y) to vary in $[0,T) \times \mathbb{R}^n$. This will bring us useful "dynamic" information of the family of problems that constitute Problem (D_{sy}).

Before going further, let us recall first Assumption (D1) stated in Chapter 3, Section 2:

(D1) (U,d) is a separable metric space and $T > 0$.

In addition, we introduce

(D2)$'$ The maps $b : [0,T] \times \mathbb{R}^n \times U \to \mathbb{R}^n$, $f : [0,T] \times \mathbb{R}^n \times U \to \mathbb{R}$, and $h : \mathbb{R}^n \to \mathbb{R}$ are uniformly continuous, and there exists a constant $L > 0$ such that for $\varphi(t,x,u) = b(t,x,u), f(t,x,u), h(x)$,

$$\begin{cases} |\varphi(t,x,u) - \varphi(t,\widehat{x},u)| \le L|x - \widehat{x}|, \\ \qquad\qquad\qquad\qquad \forall t \in [0,T], x, \widehat{x} \in \mathbb{R}^n, \ u \in U, \\ |\varphi(t,0,u)| \le L, \qquad \forall (t,u) \in [0,T] \times U. \end{cases} \tag{2.5}$$

We note that unlike (D2) introduced in Chapter 3, Section 2, (D2)$'$ requires the continuity of the given functions in (t,x,u), including t.

Clearly, under (D1) and (D2)$'$, for any $(s,y) \in [0,T) \times \mathbb{R}^n$ and $u(\cdot) \in \mathcal{V}[s,T]$, (2.3) admits a unique solution $x(\cdot) \equiv x(\cdot\,; s,y,u(\cdot))$, and (2.4) is well-defined. Further, we can define the following function:

$$\begin{cases} V(s,y) = \inf\limits_{u(\cdot)\in\mathcal{V}[s,T]} J(s,y;u(\cdot)), \qquad \forall (s,y) \in [0,T) \times \mathbb{R}^n, \\ V(T,y) = h(y), \qquad \forall y \in \mathbb{R}^n, \end{cases} \tag{2.6}$$

which is called the *value function* of Problem (D). The function $V(\cdot\,,\cdot)$ will play an important role in obtaining optimal controls. Thus, we would like to study $V(\cdot\,,\cdot)$ in great detail.

First of all, we present the following result, which is called *Bellman's principle of optimality.*

Theorem 2.1. *Let (D1) and (D2)$'$ hold. Then for any* $(s,y) \in [0,T) \times \mathbb{R}^n$,

$$\begin{aligned} V(s,y) = \inf_{u(\cdot)\in\mathcal{V}[s,T]} \Big\{ \int_s^{\hat{s}} f(t,x(t;s,y,u(\cdot)),u(t))dt \\ + V(\hat{s},x(\hat{s};s,y,u(\cdot)))\Big\}, \quad \forall 0 \le s \le \hat{s} \le T. \end{aligned} \tag{2.7}$$

Proof. Let us denote the right-hand side of (2.7) by $\overline{V}(s,y)$. By (2.6), we have

$$V(s,y) \leq J(s,y;u(\cdot)) = \int_s^{\hat{s}} f(t,x(t),u(t))dt + J(\hat{s},x(\hat{s});u(\cdot)), \qquad \forall u(\cdot) \in \mathcal{V}[s,T].$$

Thus, taking the infimum over $u(\cdot) \in \mathcal{V}[s,T]$ we get

$$V(s,y) \leq \overline{V}(s,y). \tag{2.8}$$

Conversely, for any $\varepsilon > 0$, there exists a $u_\varepsilon(\cdot) \in \mathcal{V}[s,T]$ such that

$$\begin{aligned} V(s,y) + \varepsilon &\geq J(s,y;u_\varepsilon(\cdot)) \\ &\geq \int_s^{\hat{s}} f(t,x_\varepsilon(t),u_\varepsilon(t))dt + V(\hat{s},x_\varepsilon(\hat{s})) \geq \overline{V}(s,y), \end{aligned} \tag{2.9}$$

where $x_\varepsilon(\cdot) = x(\cdot\,;s,y,u_\varepsilon(\cdot))$. Combining (2.8)–(2.9), we obtain (2.7). □

Let us make an observation on (2.7). Suppose $(\overline{x}(\cdot),\overline{u}(\cdot))$ is an optimal pair of Problem (D_{sy}) and $\hat{s} \in (s,T)$. Then

$$\begin{aligned} V(s,y) = J(s,y;\overline{u}(\cdot)) &= \int_s^{\hat{s}} f(t,\overline{x}(t),\overline{u}(t))dt + J(\hat{s},\overline{x}(\hat{s});\overline{u}(\cdot)) \\ &\geq \int_s^{\hat{s}} f(t,\overline{x}(t),\overline{u}(t))dt + V(\hat{s},\overline{x}(\hat{s})) \geq V(s,y). \end{aligned} \tag{2.10}$$

The last inequality in (2.10) is due to (2.7). Thus, all the equalities in (2.10) hold, in particular,

$$V(\hat{s},\overline{x}(\hat{s})) = J(\hat{s},\overline{x}(\hat{s});\overline{u}(\cdot)) \equiv \int_{\hat{s}}^T f(t,\overline{x}(t),\overline{u}(t))dt + h(\overline{x}(T)). \tag{2.11}$$

This means that

$$\begin{aligned} &\overline{u}(\cdot) \text{ is optimal on } [s,T] \text{ (with the initial } (s,y)) \\ &\qquad \Rightarrow \quad \overline{u}\big|_{[\hat{s},T]}(\cdot) \text{ is optimal on } [\hat{s},T] \ \big(\text{with the initial } (\hat{s},\overline{x}(\hat{s}))\big). \end{aligned} \tag{2.12}$$

The above may be roughly interpreted as the following "principle":

$$\text{globally optimal} \quad \Rightarrow \quad \text{locally optimal}. \tag{2.13}$$

This is the essence of Bellman's principle of optimality. Also, this gives a necessary condition for $\overline{u}(\cdot)$ to be optimal.

We refer to (2.7) as the *dynamic programming equation.* This equation gives a relationship among the family Problem (D_{sy}) via the value function. Unfortunately, this equation is very difficult to handle, since the operation involved on the right-hand side of (2.7) is too complicated. Thus, we would like to explore (2.7) further, trying to get an equation for $V(s,y)$ with a simpler form. It turns out that this can be done to some extent. The following result gives a partial differential equation that a continuously differentiable value function should satisfy. Hereafter, we let $C^1([0,T]\times\mathbb{R}^n)$ be the set of all continuously differentiable functions $v:[0,T]\times\mathbb{R}^n\to\mathbb{R}$.

Proposition 2.2. *Let (D1) and (D2)′ hold. Suppose $V\in C^1([0,T]\times\mathbb{R}^n)$. Then V is a solution to the following terminal value problem of a first-order partial differential equation:*

$$\begin{cases} -v_t+\sup_{u\in U}H(t,x,u,-v_x)=0, & (t,x)\in[0,T)\times\mathbb{R}^n,\\ v\big|_{t=T}=h(x), & x\in\mathbb{R}^n, \end{cases}\tag{2.14}$$

where

$$\begin{aligned} H(t,x,u,p)&\stackrel{\Delta}{=}\langle\, p,b(t,x,u)\,\rangle-f(t,x,u),\\ &\qquad\forall(t,x,u,p)\in[0,T]\times\mathbb{R}^n\times U\times\mathbb{R}^n. \end{aligned}\tag{2.15}$$

We call (2.14) the *Hamilton–Jacobi–Bellman* (*HJB*, for short) *equation* associated with Problem (D), and the function $H(t,x,u,p)$ defined by (2.15) the *Hamiltonian*, which is the same as that defined in Chapter 3, Section 2.

Proof. Fix a $u\in U$. Let $x(\cdot)$ be the state trajectory corresponding to the control $u(t)\equiv u$. By (2.7) with $\hat s\downarrow s$,

$$\begin{aligned} 0&\geq-\frac{V(\hat s,x(\hat s))-V(s,y)}{\hat s-s}-\frac{1}{\hat s-s}\int_s^{\hat s}f(t,x(t),u)dt\\ &\to-V_t(s,y)-\langle\,V_x(s,y),b(s,y,u)\,\rangle-f(s,y,u),\qquad\forall u\in U, \end{aligned}$$

which results in

$$0\geq-V_t(s,y)+\sup_{u\in U}H(s,y,u,-V_x(s,y)).\tag{2.16}$$

On the other hand, for any $\varepsilon>0$, $0\leq s<\hat s\leq T$ with $\hat s-s>0$ small enough, there exists a $u(\cdot)\equiv u_{\varepsilon,\hat s}(\cdot)\in\mathcal{V}[s,T]$ such that

$$V(s,y)+\varepsilon(\hat s-s)\geq\int_s^{\hat s}f(t,x(t),u(t))dt+V(\hat s,x(\hat s)).$$

Thus, it follows that (noting $V \in C^1([0,T] \times \mathbb{R}^n)$)

$$
\begin{aligned}
-\varepsilon &\leq -\frac{V(\hat{s}, x(\hat{s})) - V(s,y)}{\hat{s}-s} - \frac{1}{\hat{s}-s}\int_s^{\hat{s}} f(t,x(t),u(t))dt \\
&= \frac{1}{\hat{s}-s}\int_s^{\hat{s}} \Big\{ - V_t(t,x(t)) - \langle V_x(t,x(t)), b(t,x(t),u(t)) \rangle \\
&\qquad\qquad - f(t,x(t),u(t))\Big\} dt \\
&= \frac{1}{\hat{s}-s}\int_s^{\hat{s}} \Big\{ - V_t(t,x(t)) + H\big(t,x(t),u(t),-V_x(t,x(t))\big)\Big\} dt \\
&\leq \frac{1}{\hat{s}-s}\int_s^{\hat{s}} \Big\{ - V_t(t,x(t)) + \sup_{u\in U} H\big(t,x(t),u,-V_x(t,x(t))\big)\Big\} dt \\
&\to -V_t(s,y) + \sup_{u\in U} H\big(s,y,u,-V_x(s,y)\big), \qquad \text{as } \hat{s}\downarrow s.
\end{aligned}
\tag{2.17}
$$

In proving the last limit above, we have used the fact that

$$
\lim_{t\downarrow s} \sup_{y\in\mathbb{R}^n, u\in U} |\varphi(t,y,u) - \varphi(s,y,u)| = 0, \tag{2.18}
$$

for $\varphi = b, f$. This is implied by the uniform continuity of functions b and f as assumed in (D2)$'$. Combining (2.16) and (2.17), we obtain our conclusion. □

Now, let us briefly discuss how the solutions to the HJB equation might help us in finding an optimal control. Suppose we have found the value function $V \in C^1([0,T] \times \mathbb{R}^n)$ via solving the HJB equation (2.14). Moreover, assume that for any $(t,x) \in [0,T] \times \mathbb{R}^n$, the supremum in (2.14) is achieved at $u = \mathbf{\bar{u}}(t,x)$, i.e.,

$$
\begin{aligned}
H(t,x,\mathbf{\bar{u}}(t,x), -V_x(t,x)) &= \sup_{u\in U} H(t,x,u,-V_x(t,x)), \\
&\qquad \forall (t,x) \in [0,T] \times \mathbb{R}^n.
\end{aligned}
\tag{2.19}
$$

In addition, suppose that for any $(s,y) \in [0,T) \times \mathbb{R}^n$, there exists a solution $\bar{x}(\cdot)$ to the following:

$$
\begin{cases}
\dot{\bar{x}}(t) = b(t,\bar{x}(t),\mathbf{\bar{u}}(t,\bar{x}(t))), & \text{a.e. } t \in [s,T], \\
\bar{x}(s) = y.
\end{cases}
\tag{2.20}
$$

Then, by setting $\bar{u}(t) = \mathbf{\bar{u}}(t,\bar{x}(t))$, we have

$$
\begin{aligned}
\frac{d}{dt}V(t,\bar{x}(t)) &= V_t(t,\bar{x}(t)) + \langle V_x(t,\bar{x}(t)), b(t,\bar{x}(t),\bar{u}(t)) \rangle \\
&= -f(t,\bar{x}(t),\bar{u}(t)), \qquad t \in [s,T].
\end{aligned}
\tag{2.21}
$$

Integrating (2.21) from s to T, we obtain

$$
V(s,y) = h(\bar{x}(T)) + \int_s^T f(t,\bar{x}(t),\bar{u}(t))dt = J(s,y;\bar{u}(\cdot)). \tag{2.22}
$$

This means that $(\bar{x}(\cdot), \bar{u}(\cdot))$ is an optimal pair of Problem (D_{sy}).

The above implies that if we could obtain the value function V by solving the HJB equation, then we could, at least formally, construct an optimal pair for each of Problem (D_{sy}) and, in particular, for the original Problem (D), via (2.19). This is called the *verification theorem* (*technique*).

The verification technique in principle involves the following steps to solve the original Problem (D):

Step 1. Solve the HJB equation (2.14) to find the value function $V(t,x)$.

Step 2. Find $\bar{\mathbf{u}}(t,x)$ through (2.19).

Step 3. Solve (2.20) (with $(s,y) = (0,x_0)$) to get the optimal pair $(\bar{x}(\cdot), \bar{u}(\cdot))$.

In Chapter 5 we will discuss the verification theorem in a more rigorous way.

To summarize, the key points in the dynamic programming approach are as follows: (1) Introduce Problem (D_{sy}); (2) define the value function $V(s,y)$; (3) derive the principle of optimality; (4) derive the HJB equation and then follow the above Steps 1–3 to obtain an optimal pair.

It is clear that one of the central issues in the above approach is to determine the value function $V(\cdot\,,\cdot)$. If (i) $V \in C^1([0,T] \times \mathbb{R}^n)$ and (ii) the HJB equation (2.14) admits a unique (classical) solution, then V is characterized by the solution to the HJB equation (2.14). Unfortunately, neither (i) nor (ii) is true in general. Here is an example.

Example 2.3. Consider a one-dimensional control system,

$$\left\{\begin{aligned} &\dot{x}(t) = u(t)x(t), \qquad \text{a.e. } t \in [s,T], \\ &x(s) = y, \end{aligned}\right. \tag{2.23}$$

with the control $u(t)$ taking values in $U \equiv [-1,1]$ and the cost functional being

$$J(u(\cdot)) = x(T). \tag{2.24}$$

One sees easily that

$$x(t) = ye^{\int_s^t u(r)dr}, \qquad \forall t \in [s,T].$$

Thus, the value function is given by the following:

$$V(s,y) = \begin{cases} ye^{T-s}, & y \le 0, \\ ye^{s-T}, & y \ge 0, \end{cases} \tag{2.25}$$

which is only Lipschitz and *not* $C^1([0,T] \times \mathbb{R})$, since $V_x(s,y)$ has a jump at $(s,0)$ for all $0 \le s < T$.

On the other hand, in the present case, we have

$$\sup_{u \in U} H(s,y,u,p) = \sup_{|u| \le 1} \{puy\} = |py|.$$

Thus, the associated HJB equation is

$$(2.26)\qquad \begin{cases} -v_t+|xv_x|=0, & (t,x)\in[0,T)\times\mathbb{R},\\ v\big|_{t=T}=x, & x\in\mathbb{R}.\end{cases}$$

We claim that this equation does not admit any $C^1([0,T]\times\mathbb{R})$ solution. To see this, suppose $v\in C^1([0,T]\times\mathbb{R})$ is a solution of (2.26). Then it is necessary that

$$v_x(T,x)=1,\qquad \forall x\in\mathbb{R}.$$

By the continuity of v_x, we have

$$(2.27)\qquad v_x(t,x)>0,\qquad \forall(t,x)\in\mathcal{N}\stackrel{\Delta}{=}\{(t,x)\mid\varphi(x)\le t\le T\},$$

for some continuous function $\varphi:\mathbb{R}\to[0,T)$, and φ can be chosen symmetric in x (with respect to $x=0$) and nondecreasing in $|x|$. Then, by (2.26), v satisfies

$$(2.28)\qquad \begin{cases} v_t=xv_x, & (t,x)\in\mathcal{N}^+\stackrel{\Delta}{=}\mathcal{N}\bigcap\{(t,x)\mid x>0,t\in[0,T]\},\\ v\big|_{t=T}=x, & x\in\mathbb{R}.\end{cases}$$

Let

$$(2.29)\qquad \begin{cases}\tau=t,\\ z=xe^t,\end{cases}\qquad \Phi(\tau,z)\stackrel{\Delta}{=}v(\tau,ze^{-\tau}).$$

We have

$$\Phi_\tau=v_t+v_x[-ze^{-\tau}]=v_t-xv_x=0.$$

This implies that Φ does not depend on the argument τ. So we may write $\Phi(\tau,z)=\Phi(z)$. It then follows from (2.29) that

$$(2.30)\qquad v(t,x)=\Phi(xe^t),\qquad \forall(t,x)\in\mathcal{N}^+.$$

By the terminal condition in (2.28), we have

$$(2.31)\qquad xe^{-T}=v(T,xe^{-T})=\Phi(x),\qquad \forall x\in\mathbb{R}.$$

Hence, combining (2.30)–(2.31), we obtain

$$(2.32)\qquad v(t,x)=xe^{t-T},\qquad \forall(t,x)\in\mathcal{N}^+.$$

Similarly, we can show that

$$(2.33)\qquad v(t,x)=xe^{T-t},\qquad \forall(t,x)\in\mathcal{N}\setminus\mathcal{N}^+.$$

Thus, v coincides with V in $\mathcal{N}$. Since V is not in $C^1(\mathcal{N})$, neither is v. This leads to a contradiction. □

From this example, we have to admit that the assumptions of Proposition 2.2 are too restrictive. To find a rigorous assertion similar in nature to

Proposition 2.2 but without restrictive assumptions, we need the following notion.

Definition 2.4. A function $v \in C([0,T] \times \mathbb{R}^n)$ is called a *viscosity subsolution* of (2.14) if

$$v(T,x) \leq h(x), \qquad \forall x \in \mathbb{R}^n, \tag{2.34}$$

and for any $\varphi \in C^1([0,T] \times \mathbb{R})$, whenever $v - \varphi$ attains a local maximum at $(t,x) \in [0,T) \times \mathbb{R}^n$, we have

$$-\varphi_t(t,x) + \sup_{u \in U} H(t,x,u,-\varphi_x(t,x)) \leq 0. \tag{2.35}$$

A function $v \in C([0,T] \times \mathbb{R}^n)$ is called a *viscosity supersolution* of (2.14) if in (2.34)–(2.35) the inequalities "$\leq$" are changed to "$\geq$" and "local maximum" is changed to "local minimum." In the case that v is both a viscosity subsolution and supersolution of (2.14), it is called a *viscosity solution* of (2.14).

We point out that in the above definition, one may replace "local maximum (minimum)" by "local strict maximum (minimum)" or even "global strict maximum (minimum)." We encourage the reader to give a proof of such equivalences.

The following gives a characterization of the value function in terms of viscosity solutions to the HJB equation.

Theorem 2.5. *Let (D1) and (D2)′ hold. Then the value function $V(\cdot\,,\cdot)$ satisfies*

$$\begin{aligned} |V(s,y) - V(\bar{s},\bar{y})| \leq & K\{|y-\bar{y}| + (1+|y| \vee |\bar{y}|)|s-\bar{s}|\}, \\ & \forall (s,y),(\bar{s},\bar{y}) \in [0,T] \times \mathbb{R}^n, \end{aligned} \tag{2.36}$$

for some $K > 0$. Moreover, V is the only viscosity solution of (2.14) in the class $C([0,T] \times \mathbb{R}^n)$.

Proof. We split the proof into several steps.

Step 1. $V(\cdot\,,\cdot)$ satisfies (2.36).

By Gronwall's inequality, taking into account (D2)′, we easily get

$$\begin{cases} |x(t;s,y,u(\cdot))| \leq K(1+|y|), \\ \qquad\qquad \forall t \in [s,T],\ u(\cdot) \in \mathcal{V}[s,T],\ (s,y) \in [0,T) \times \mathbb{R}^n, \\ |x(t;s,y,u(\cdot)) - x(t;\bar{s},\bar{y},u(\cdot))| \\ \qquad \leq K\{|y-\bar{y}| + (1+|y| \vee |\bar{y}|)|s-\bar{s}|\}, \\ \qquad \forall s,\bar{s} \in [0,T],\ y,\bar{y} \in \mathbb{R}^n,\ t \in [s \vee \bar{s},T],\ u(\cdot) \in \mathcal{V}[s \wedge \bar{s},T]. \end{cases} \tag{2.37}$$

Thus, it follows that

$$
(2.38)\qquad \begin{cases} |J(s,y;u(\cdot))-J(s,\bar y;u(\cdot))|\le K|y-\bar y|, \\ \qquad\qquad \forall s\in[0,T),\ y,\bar y\in\mathbb{R}^n,\ u(\cdot)\in\mathcal{V}[s,T], \\ |J(s,y;u(\cdot))-J(\bar s,y;u(\cdot))|\le K(1+|y|\vee|\bar y|)|s-\bar s|, \\ \qquad\qquad \forall s,\bar s\in[0,T),\ y\in\mathbb{R}^n,\ u(\cdot)\in\mathcal{V}[s\wedge\bar s,T]. \end{cases}
$$

These yield (2.36). Note that in proving (2.36) at $t=T$, we have used the Lipschitz continuity of h.

Step 2. $V(\cdot\,,\cdot)$ is a viscosity solution of the HJB equation (2.14).

Suppose $\varphi\in C^1([0,T]\times\mathbb{R})$ such that $V-\varphi$ attains a local maximum at $(s,y)\in[0,T)\times\mathbb{R}^n$. Fix a $u\in U$. Let $x(\cdot)=x(\cdot\,;s,y,u)$ be the state trajectory under the control $u(t)\equiv u$. Then, by Theorem 2.1, we have (for $\hat s>s$ with $\hat s-s>0$ small enough)

$$
(2.39)\qquad \begin{aligned} 0&\le\frac{V(s,y)-\varphi(s,y)-V(\hat s,x(\hat s))+\varphi(\hat s,x(\hat s))}{\hat s-s}\\ &\le\frac{1}{\hat s-s}\Big\{\int_s^{\hat s}f(t,x(t),u)dt-\varphi(s,y)+\varphi(\hat s,x(\hat s))\Big\}\\ &\to f(s,y,u)+\varphi_t(s,y)+\langle\,\varphi_x(s,y),b(s,y,u)\,\rangle,\qquad \text{as } \hat s\to s. \end{aligned}
$$

This leads to

$$
-\varphi_t(s,y)+H(s,y,u,-\varphi_x(s,y))\le 0,\qquad \forall u\in U.
$$

Hence,

$$
(2.40)\qquad -\varphi_t(s,y)+\sup_{u\in U}H(s,y,u,-\varphi_x(s,y))\le 0.
$$

On the other hand, if $V-\varphi$ attains a local minimum at $(s,y)\in[0,T)\times\mathbb{R}^n$, then, for any $\varepsilon>0$ and $\hat s>s$ (with $\hat s-s>0$ small enough), we can find a $u(\cdot)=u_{\varepsilon,\hat s}(\cdot)\in\mathcal{V}[s,T]$ such that (noting (2.7))

$$
(2.41)\qquad \begin{aligned} 0&\ge V(s,y)-\varphi(s,y)-V(\hat s,x(\hat s))+\varphi(\hat s,x(\hat s))\\ &\ge-\varepsilon(\hat s-s)+\int_s^{\hat s}f(t,x(t),u(t))dt+\varphi(\hat s,x(\hat s))-\varphi(s,y). \end{aligned}
$$

Dividing by $(\hat s-s)$, we get

$$
(2.42)\qquad \begin{aligned} -\varepsilon&\le\frac{1}{\hat s-s}\int_s^{\hat s}\Big\{-f(t,x(t),u(t))-\varphi_t(t,x(t))\\ &\qquad\qquad -\langle\,\varphi_x(t,x(t)),b(t,x(t),u(t))\,\rangle\Big\}dt\\ &\le\frac{1}{\hat s-s}\int_s^{\hat s}\{-\varphi_t(t,x(t))+\sup_{u\in U}H(t,x(t),u,-\varphi_x(t,x(t))\}dt\\ &\to-\varphi_t(s,y)+\sup_{u\in U}H(s,y,u,-\varphi_x(s,y)),\qquad \text{as } \hat s\to s. \end{aligned}
$$

As with (2.17), in obtaining the limit above, we have used the uniform continuity of the functions f and b, which leads to (2.18).

Combining (2.40) and (2.42), we conclude that V is a viscosity solution of the HJB equation (2.14).

Step 3. The HJB equation (2.14) admits at most one viscosity solution.*

Let us define

$$H(t,x,p) = \sup_{u\in U} H(t,x,u,p), \quad \forall (t,x,p) \in [0,T] \times \mathbb{R}^n \times \mathbb{R}^n. \tag{2.43}$$

By (D1) and (D2)′, the following inequalities hold:

$$\begin{cases} |H(t,x,p) - H(t,x,q)| \le L(1+|x|)|p-q|, \\ |H(t,x,p) - H(t,y,p)| \le \bar\omega(|x| \vee |y|, |x-y|(1+|p|)), \\ \qquad\qquad\qquad \forall t \in [0,T],\ x,y,p,q \in \mathbb{R}^n, \end{cases} \tag{2.44}$$

where $\bar\omega : [0,\infty) \times [0,\infty) \to [0,\infty)$ is continuous, nondecreasing in each of its arguments, and $\bar\omega(r,0) = 0$ for all $r \ge 0$.

Let $v(s,y)$ and $\widehat v(s,y)$ be two viscosity solutions of (2.14). To prove the uniqueness it suffices to prove

$$v(s,y) \le \widehat v(s,y), \qquad \forall (s,y) \in [0,T] \times \mathbb{R}^n. \tag{2.45}$$

Suppose (2.45) is not true. Then we may assume that

$$\sup_{(s,y)\in\mathcal{N}} [v(s,y) - \widehat v(s,y)] \ge \bar\gamma > 0, \tag{2.46}$$

where

$$\begin{cases} \mathcal{N} = \{(t,x) \in (T-T_0,T) \times \mathbb{R}^n \mid |x| < L_0(t-T+T_0)\}, \\ 0 < T_0 < T,\ T_0 < \dfrac{1}{L},\ L_0 = \dfrac{L}{1-LT_0}. \end{cases} \tag{2.47}$$

By (2.44) and (2.47), for all $(t,x) \in \mathcal{N}$ and $p,q \in \mathbb{R}^n$,

$$\begin{aligned} |H(t,x,p) - H(t,x,q)| &\le L(1+|x|)|p-q| \\ &\le L(1+L_0T_0)|p-q| = L_0|p-q|. \end{aligned} \tag{2.48}$$

Now, take $\varepsilon, \delta > 0$ such that

$$\varepsilon + \delta < L_0 T_0. \tag{2.49}$$

Take $K > 0$ and $\zeta \in C^\infty(\mathbb{R})$ satisfying

$$K > \sup_{(t,x,s,y)\in\mathcal{N}\times\mathcal{N}} \{v(t,x) - \widehat v(s,y)\}, \tag{2.50}$$

---

* The reader may skip this part at first reading, since it is rather technical.

and

$$\zeta(r)=\begin{cases}0, & r\le -\delta,\\ -K, & r\ge 0,\end{cases}\qquad \zeta'(r)\le 0,\quad \forall r\in\mathbb{R}. \tag{2.51}$$

For any $\alpha,\beta,\gamma>0$, we define

$$\begin{aligned}\Phi(t,x,s,y) &= v(t,x)-\widehat{v}(s,y)-\frac{1}{\alpha}|x-y|^2\\ &\quad -\frac{1}{\beta}|t-s|^2+\zeta(\langle x\rangle_\varepsilon-L_0(t-T+T_0))\\ &\quad +\zeta(\langle y\rangle_\varepsilon-L_0(t-T+T_0))+\gamma(t+s)-2\gamma T,\\ &\qquad\qquad \forall (t,x,s,y)\in\mathcal{N}\times\mathcal{N},\end{aligned} \tag{2.52}$$

where $\langle\cdot\rangle_\varepsilon$ is defined as $\langle z\rangle_\varepsilon \overset{\Delta}{=} (|z|^2+\varepsilon^2)^{1/2}$, $\forall z\in\mathbb{R}^n$. Since Φ is continuous and $\overline{\mathcal{N}\times\mathcal{N}}$ is compact, we may assume that Φ attains its maximum over $\overline{\mathcal{N}\times\mathcal{N}}$ at $(t_0,x_0,s_0,y_0)\in\overline{\mathcal{N}\times\mathcal{N}}$. Then, from

$$\Phi(T,0,T,0)\le \Phi(t_0,x_0,s_0,y_0),$$

one obtains (noting (2.49)–(2.51))

$$\begin{aligned}0 &= v(T,0)-\widehat{v}(T,0)+2\zeta(\varepsilon-L_0T_0)\\ &\le v(t_0,x_0)-\widehat{v}(s_0,y_0)-\frac{1}{\alpha}|x_0-y_0|^2\\ &\quad -\frac{1}{\beta}|t_0-s_0|^2+\zeta(\langle x_0\rangle_\varepsilon-L_0(t_0-T+T_0))\\ &\quad +\zeta(\langle y_0\rangle_\varepsilon-L_0(s_0-T+T_0))+\gamma(t_0+s_0)-2\gamma T\\ &< K+\zeta(\langle x_0\rangle_\varepsilon-L_0(t_0-T+T_0))\\ &\quad +\zeta(\langle y_0\rangle_\varepsilon-L_0(s_0-T+T_0))+\gamma(t_0+s_0)-2\gamma T.\end{aligned} \tag{2.53}$$

Now we claim that

$$\begin{cases}|x_0|<L_0(t_0-T+T_0),\\ |y_0|<L_0(s_0-T+T_0).\end{cases} \tag{2.54}$$

Indeed, if (2.54) is not true, then either (note $|z|<\langle z\rangle_\varepsilon$)

$$\langle x_0\rangle_\varepsilon-L_0(t_0-T+T_0)>0$$

or

$$\langle y_0\rangle_\varepsilon-L_0(s_0-T+T_0)>0.$$

Thus, by (2.51), it follows from (2.53) that

$$0<K-K+\gamma(t_0+s_0)-2\gamma T\le 0,$$

which is a contradiction. Hence, (2.54) holds.

On the other hand, by the fact that

$$\Phi(t_0,x_0,t_0,x_0)+\Phi(s_0,y_0,s_0,y_0)\le 2\Phi(t_0,x_0,s_0,y_0),$$

we obtain

$$\begin{aligned}
&v(t_0,x_0)-\widehat v(t_0,x_0)+2\zeta(\langle x_0\rangle_\varepsilon-L_0(t_0-T+T_0))\\
&\quad+v(s_0,y_0)-\widehat v(s_0,y_0)+2\zeta(\langle y_0\rangle_\varepsilon-L_0(s_0-T+T_0))\\
&\quad+2\gamma(t_0+s_0)-4\gamma T\\
&< 2v(t_0,x_0)-2\widehat v(s_0,y_0)-\frac{2}{\alpha}|x_0-y_0|^2-\frac{2}{\beta}|t_0-s_0|^2\\
&\quad+2\zeta(\langle x_0\rangle_\varepsilon-L_0(t_0-T+T_0))\\
&\quad+2\zeta(\langle y_0\rangle_\varepsilon-L_0(s_0-T+T_0))+2\gamma(t_0+s_0)-4\gamma T,
\end{aligned}$$

which yields

$$\begin{aligned}
\frac{2}{\alpha}|x_0-y_0|^2+\frac{2}{\beta}|t_0-s_0|^2&\le v(t_0,x_0)-v(s_0,y_0)\\
&\quad+\widehat v(t_0,x_0)-\widehat v(s_0,y_0)\le 2\eta(|t_0-s_0|+|x_0-y_0|),
\end{aligned}\tag{2.55}$$

where

$$\eta(r)=\frac12\sup_{\substack{|t-s|+|x-y|\le r\\(t,x,s,y)\in\mathcal N\times\mathcal N}}\big\{|v(t,x)-v(s,y)|+|\widehat v(t,x)-\widehat v(s,y)|\big\}.$$

Clearly, $\lim_{r\to0}\eta(r)=0$. By the boundedness of $\mathcal N$, we have

$$\eta_0\overset{\Delta}{=}\sup_{r>0}\eta(r)<\infty.\tag{2.56}$$

Then it follows from (2.55) that

$$|x_0-y_0|\le\sqrt{\alpha\eta_0},\quad |t_0-s_0|\le\sqrt{\beta\eta_0}.\tag{2.57}$$

Combining (2.55) with (2.57), we further have

$$\frac1\alpha|x_0-y_0|^2+\frac1\beta|t_0-s_0|^2\le\eta(\sqrt{\alpha\eta_0}+\sqrt{\beta\eta_0}).\tag{2.58}$$

Set

$$\Delta_{\varepsilon,\delta}\overset{\Delta}{=}\{(t,x)\in\mathcal N\,|\,\langle x\rangle_\varepsilon\le L_0(t-T+T_0)-\delta\}.$$

By (2.46) and (2.51), for $\varepsilon,\delta,\gamma>0$ small enough, we may assume

$$\begin{aligned}
&\sup_{(t,x)\in\Delta_{\varepsilon,\delta}}\Phi(t,x,t,x)\\
&\quad=\sup_{(t,x)\in\Delta_{\varepsilon,\delta}}[v(t,x)-\widehat v(t,x)+2\gamma(t-T)]\ge\frac{\overline\gamma}{2}>0.
\end{aligned}\tag{2.59}$$

Next,

$$
\begin{aligned}
(2.60)\qquad \sup_{(t,x)\in\Delta_{\varepsilon,\delta}} \Phi(t,x,t,x) &\le \sup_{\mathcal{N}\times\mathcal{N}} \Phi(t,x,s,y)\\
&= \Phi(t_0,x_0,s_0,y_0) \le v(t_0,x_0)-\widehat{v}(s_0,y_0).
\end{aligned}
$$

We claim that there exists an $r_0>0$ such that for any $0<\alpha,\beta<r_0$,

$$
(2.61)\qquad (t_0,x_0,s_0,y_0)\in\mathcal{N}\times\mathcal{N}.
$$

Note that the point (t_0,x_0,s_0,y_0) depends on the choice of $(\alpha,\beta,\varepsilon,\delta,\gamma)$. We now prove (2.61). If it is false, then by virtue of (2.54), there is only one possibility, i.e., for some sequence $(\alpha_m,\beta_m)\to(0,0)$, the corresponding maximum point (t_m,x_m,s_m,y_m) of Φ over $\overline{\mathcal{N}\times\mathcal{N}}$ satisfies

$$
(2.62)\qquad t_m=T,\qquad \text{or}\qquad s_m=T,\qquad \forall m\ge 1.
$$

By (2.57), we obtain

$$
|x_m-y_m|\to 0,\quad t_m,s_m\to T,\qquad \text{as } m\to\infty.
$$

Hence, it follows from (2.59)–(2.60) that

$$
0<\frac{\overline{\gamma}}{2}\le \lim_{m\to\infty}[v(t_m,x_m)-\widehat{v}(s_m,y_m)]=0,
$$

which is a contradiction. Thus, (2.61) holds.

Now we are going to use the definition of viscosity solution. For $0<\alpha,\beta<r_0$, by the definition of (t_0,x_0,s_0,y_0), the function

$$
\begin{aligned}
(t,x)\mapsto v(t,x)-\Big\{\widehat{v}(s_0,y_0)&+\frac{1}{\alpha}|x-y_0|^2\\
&+\frac{1}{\beta}|t-s_0|^2-\zeta(\langle x\rangle_\varepsilon-L_0(t-T+T_0))\\
&-\zeta(\langle y_0\rangle_\varepsilon-L_0(s_0-T+T_0))-\gamma(t+s_0)+2\gamma T\Big\}
\end{aligned}
$$

attains a local maximum at $(t_0,x_0)\in\mathcal{N}$. Hence, we have

$$
\begin{aligned}
(2.63)\qquad &\frac{2}{\beta}(s_0-t_0)-\zeta'(X)L_0+\gamma\\
&\quad+H\big(t_0,x_0,\frac{2}{\alpha}(y_0-x_0)+\zeta'(X)\frac{x_0}{\langle x_0\rangle_\varepsilon}\big)\le 0,
\end{aligned}
$$

where $X\triangleq\langle x_0\rangle_\varepsilon-L_0(t_0-T+T_0)$. Similarly, the function

$$
\begin{aligned}
(s,y)\mapsto \widehat{v}(s,y)-\Big\{v(t_0,x_0)&-\frac{1}{\alpha}|x_0-y|^2\\
&-\frac{1}{\beta}|t_0-s|^2+\zeta(\langle x_0\rangle_\varepsilon-L_0(t_0-T+T_0))\\
&+\zeta(\langle y\rangle_\varepsilon-L_0(s-T+T_0))+\gamma(t_0+s)-2\gamma T\Big\}
\end{aligned}
$$

attains a local minimum at $(s_0, y_0) \in \mathcal{N}$. Hence, we have

$$
\begin{aligned}
&\frac{2}{\beta}(s_0 - t_0) + \zeta'(Y)L_0 - \gamma \\
&\qquad + H\big(s_0, y_0, \frac{2}{\alpha}(y_0 - x_0) - \zeta'(Y)\frac{y_0}{\langle y_0 \rangle_\varepsilon})\big) \geq 0,
\end{aligned} \tag{2.64}
$$

where $Y \triangleq \langle y_0 \rangle_\varepsilon - L_0(s_0 - T + T_0)$. Combining (2.63)–(2.64), we obtain

$$
\begin{aligned}
2\gamma \leq\ & L_0[\zeta'(X) + \zeta'(Y)] - H\Big(t_0, x_0, \frac{2}{\alpha}(y_0 - x_0) + \zeta'(X)\frac{x_0}{\langle x_0 \rangle_\varepsilon}\Big) \\
&+ H\Big(s_0, y_0, \frac{2}{\alpha}(y_0 - x_0) - \zeta'(Y)\frac{y_0}{\langle y_0 \rangle_\varepsilon}\Big).
\end{aligned} \tag{2.65}
$$

Choose some sequence $\beta \downarrow 0$ such that (t_0, x_0, s_0, y_0) converges. For notational simplicity, we still denote the limit by (t_0, x_0, s_0, y_0) itself. Note that for this limit, by (2.57), we must have $s_0 = t_0$, and (2.58) becomes

$$
\frac{1}{\alpha}|x_0 - y_0|^2 \leq \eta(\sqrt{\alpha \eta_0}). \tag{2.66}
$$

Then (2.65) implies (using (2.44) and (2.48))

$$
\begin{aligned}
2\gamma \leq\ & L_0[\zeta'(X) + \zeta'(Y)] \\
&- H\Big(t_0, x_0, \frac{2}{\alpha}(y_0 - x_0) + \zeta'(X)\frac{x_0}{\langle x_0 \rangle_\varepsilon}\Big) \\
&+ H\Big(t_0, y_0, \frac{2}{\alpha}(y_0 - x_0) - \zeta'(Y)\frac{y_0}{\langle y_0 \rangle_\varepsilon}\Big) \\
\leq\ & L_0[\zeta'(X) + \zeta'(Y)] + L_0\Big|\zeta'(X)\frac{x_0}{\langle x_0 \rangle_\varepsilon} + \zeta'(Y)\frac{y_0}{\langle y_0 \rangle_\varepsilon}\Big| \\
&+ \bar{\omega}\Big(|x_0| \vee |y_0|, |x_0 - y_0|\Big[1 + \Big|\frac{2}{\alpha}(y_0 - x_0) + \zeta'(X)\frac{x_0}{\langle x_0 \rangle_\varepsilon}\Big|\Big]\Big) \\
\leq\ & L_0[\zeta'(X) + \zeta'(Y)] + L_0[|\zeta'(X)| + |\zeta'(Y)|] \\
&+ \bar{\omega}\big(|x_0| \vee |y_0|, |x_0 - y_0|(1 + |\zeta'(X)|) + \frac{2}{\alpha}|x_0 - y_0|^2\big).
\end{aligned}
$$

Since $\zeta'(r) \leq 0$, we obtain

$$
0 < 2\gamma \leq \bar{\omega}\big(|x_0| \vee |y_0|, |x_0 - y_0|(1 + |\zeta'(X)|) + \frac{2}{\alpha}|x_0 - y_0|^2\big).
$$

Sending $\alpha \to 0$ and using (2.66), we obtain a contradiction. Therefore, (2.45) holds, and our conclusion follows. □

From the above proof we see that the uniqueness part alone does not require the linear growth of f and h on x (whereas (2.36) does). Therefore, as far as the uniqueness of viscosity solutions is concerned, the functions f and h are allowed to have more general growth in x. A typical example is the linear quadratic optimal control problem.

Next, we introduce *super-* and *subdifferentials*. For any function $v : [0,T] \times \mathbb{R}^n \to \mathbb{R}$ and $(t,x) \in [0,T) \times \mathbb{R}^n$, define

$$
(2.67) \qquad
\begin{aligned}
D_{t,x}^{1,+}v(t,x) &= \Big\{(q,p) \in \mathbb{R} \times \mathbb{R}^n \ \Big| \\
&\qquad \varlimsup_{\substack{s\to t,\, s\in[0,T)\\ y\to x}} \frac{v(s,y)-v(t,x)-q(s-t)-\langle p, y-x\rangle}{|s-t|+|y-x|} \le 0\Big\},\\
D_{t,x}^{1,-}v(t,x) &= \Big\{(q,p) \in \mathbb{R} \times \mathbb{R}^n \ \Big| \\
&\qquad \varliminf_{\substack{s\to t,\, s\in[0,T)\\ y\to x}} \frac{v(s,y)-v(t,x)-q(s-t)-\langle p, y-x\rangle}{|s-t|+|y-x|} \ge 0\Big\}.
\end{aligned}
$$

Furthermore, we define $D_{t+,x}^{1,+}v(t,x)$ and $D_{t+,x}^{1,-}v(t,x)$ by restricting $s \downarrow t$ in (2.67). We call $D_{t,x}^{1,+}v(t,x)$ (resp. $D_{t,x}^{1,-}v(t,x)$) the *first-order super-* (resp. *sub-*) *differential* of v at (t,x), and $D_{t+,x}^{1,+}v(t,x)$ (resp. $D_{t+,x}^{1,-}v(t,x)$) the *first-order right super-* (resp. *sub-*) *differential* of v at (t,x).

Some basic properties of the super- and subdifferentials are collected in the following proposition. Their proofs are straightforward and hence left to the reader.

Proposition 2.6. *The super- and subdifferentials have the following properties:*

(i) *The sets $D_{t,x}^{1,\pm}v(t,x)$ and $D_{t+,x}^{1,\pm}v(t,x)$ are convex and closed, satisfying*

$$
(2.68) \qquad
\begin{cases}
D_{t,x}^{1,+}v(t,x) \subseteq D_{t+,x}^{1,+}v(t,x),\\
D_{t,x}^{1,-}v(t,x) \subseteq D_{t+,x}^{1,-}v(t,x),\\
D_{t,x}^{1,+}(-v)(t,x) = -D_{t,x}^{1,-}v(t,x),\\
D_{t+,x}^{1,+}(-v)(t,x) = -D_{t+,x}^{1,-}v(t,x).
\end{cases}
\qquad \forall (t,x) \in [0,T) \times \mathbb{R}^n.
$$

(ii) *$v(t,x)$ is differentiable at $(t_0,x_0) \in (0,T) \times \mathbb{R}^n$ if and only if*

$$
(2.69) \qquad D_{t,x}^{1,+}v(t_0,x_0) \bigcap D_{t,x}^{1,-}v(t_0,x_0) \ne \phi;
$$

in this case, it is necessary that both $D_{t,x}^{1,+}v(t_0,x_0)$ and $D_{t,x}^{1,-}v(t_0,x_0)$ be singletons and

$$
(2.70) \qquad D_{t,x}^{1,+}v(t_0,x_0) = D_{t,x}^{1,-}v(t_0,x_0) = \big\{\big(v_t(t_0,x_0), v_x(t_0,x_0)\big)\big\}.
$$

(iii) *If $v(t,x)$ is locally Lipschitz in (t,x) at (t_0,x_0), then*

$$
(2.71) \qquad D_{t,x}^{1,+}v(t_0,x_0) \bigcup D_{t,x}^{1,-}v(t_0,x_0) \subseteq \partial v(t_0,x_0),
$$

where $\partial v(t_0,x_0)$ is Clarke's generalized gradient (see Chapter 3, (2.20)).

(iv) *If $v(t,x)$ is convex in (t,x), then*

$$
(2.72) \qquad D_{t,x}^{1,-}v(t,x) = \partial v(t,x) = \partial_c v(t,x), \quad \forall (t,x) \in [0,T) \times \mathbb{R}^n,
$$

where $\partial_c v(t,x)$ is the subgradient in convex analysis (see Chapter 3, (2.22)).

The following results link the super- and subdifferentials to the notion of viscosity solutions.

Lemma 2.7. *Let $v \in C([0,T] \times \mathbb{R}^n)$ and $(t_0, x_0) \in [0,T) \times \mathbb{R}^n$ be given. Then*

(i) $(q,p) \in D^{1,+}_{t,x} v(t_0,x_0)$ *if and only if there exists a function $\varphi \in C^1(\mathbb{R} \times \mathbb{R}^n)$ such that $v - \varphi$ attains a strict maximum at (t_0,x_0) and*

$$(\varphi(t_0,x_0), \varphi_t(t_0,x_0), \varphi_x(t_0,x_0)) = (v(t_0,x_0), q, p). \tag{2.73}$$

(ii) $(q,p) \in D^{1,+}_{t+,x} v(t_0,x_0)$ *if and only if there exists a function $\varphi \in C^1(\mathbb{R} \times \mathbb{R}^n)$ such that*

$$\begin{cases} (\varphi(t_0,x_0), \varphi_t(t_0,x_0), \varphi_x(t_0,x_0)) = (v(t_0,x_0), q, p), \\ \varphi(t,x) > v(t,x), \quad \forall (t_0,x_0) \neq (t,x) \in [t_0,T] \times \mathbb{R}^n. \end{cases} \tag{2.74}$$

Proof. We prove (ii). The proof of (i) is similar. Suppose $(q,p) \in D^{1,+}_{t+,x} v(t_0,x_0)$. Let

$$\Phi(t,x) = \begin{cases} \dfrac{\big(v(t,x) - v(t_0,x_0) - q(t-t_0) - \langle\, p, x-x_0 \,\rangle\big) \vee 0}{t - t_0 + |x-x_0|}, \\ \qquad\qquad \text{if } (t_0,x_0) \neq (t,x) \in [t_0,T] \times \mathbb{R}^n, \\ 0, \qquad\qquad \text{otherwise}, \end{cases}$$

and

$$\varepsilon(r) = \begin{cases} \sup\{\Phi(t,x) : (t,x) \in [t_0,T] \times \mathbb{R}^n, s - t + |y-x| \leq r\}, \\ \qquad\qquad \text{if } r > 0, \\ 0, \qquad\qquad \text{if } r \leq 0. \end{cases}$$

Then it follows from (2.67) that $\varepsilon : \mathbb{R} \to [0,\infty)$ is a continuous nondecreasing function with $\varepsilon(0) = 0$. Further,

$$\begin{aligned} &v(t,x) - v(t_0,x_0) - q(t-t_0) - \langle\, p, x - x_0 \,\rangle \\ &\quad \leq \big(t - t_0 + |x-x_0|\big)\varepsilon(t - t_0 + |x-x_0|), \\ &\qquad\qquad \forall (t,x) \in [t_0,T] \times \mathbb{R}^n. \end{aligned} \tag{2.75}$$

Set

$$\psi(t,x) = \begin{cases} \displaystyle\int_0^{2(t-t_0+|x-x_0|)} \varepsilon(\rho)d\rho + \big(t - t_0 + |x-x_0|\big)^2, \\ \qquad\qquad \text{if } (t,x) \in [t_0,T] \times \mathbb{R}^n, \\ 0, \qquad\qquad \text{if } (t,x) \in [0,t_0) \times \mathbb{R}^n. \end{cases}$$

Clearly, $\psi \in C^1(\mathbb{R} \times \mathbb{R}^n)$ with

$$\psi(t_0,x_0) = 0, \quad \psi_t(t_0,x_0) = 0, \quad \psi_x(t_0,x_0) = 0,$$

and (noting (2.75))

$$\begin{aligned}\psi(t,x) &\geq \int_{t-t_0+|x-x_0|}^{2(t-t_0+|x-x_0|)} \varepsilon(\rho)d\rho + \big(t-t_0+|x-x_0|\big)^2\\ &\geq (t-t_0+|x-x_0|)\varepsilon(t-t_0+|x-x_0|) + \big(t-t_0+|x-x_0|\big)^2\\ &> (t-t_0+|x-x_0|)\varepsilon(t-t_0+|x-x_0|)\\ &\geq v(t,x) - v(t_0,x_0) - q(t-t_0) - \langle\, p, x-x_0\,\rangle,\\ &\qquad\qquad \forall (t_0,x_0) \neq (t,x) \in [t_0,T]\times \mathbb{R}^n.\end{aligned}$$

By defining

$$\varphi(t,x) = v(t_0,x_0) + q(t-t_0) + \langle\, p, x-x_0\,\rangle + \psi(t,x), \qquad (t,x) \in \mathbb{R}\times\mathbb{R}^n,$$

we obtain (2.74). This proves the "only if" part. The "if" part is clear.

□

The following are the parallel results for $D^{1,-}_{t,x}v(t,x)$ and $D^{1,-}_{t+,x}v(t,x)$.

Lemma 2.8. *Let $v \in C([0,T]\times\mathbb{R}^n)$ and $(t_0,x_0) \in [0,T)\times\mathbb{R}^n$ be given. Then*

(i) *$(q,p) \in D^{1,-}_{t,x}v(t_0,x_0)$ if and only if there exists a function $\varphi \in C^1(\mathbb{R}\times\mathbb{R}^n)$ such that $v-\varphi$ attains a strict minimum at (t_0,x_0) and*

$$(2.76)\qquad \big(\varphi(t_0,x_0), \varphi_t(t_0,x_0), \varphi_x(t_0,x_0)\big) = (v(t_0,x_0), q, p).$$

(ii) *$(q,p) \in D^{1,-}_{t+,x}v(t_0,x_0)$ if and only if there exists a function $\varphi \in C^1(\mathbb{R}\times\mathbb{R}^n)$ such that*

$$(2.77)\qquad \begin{cases} \big(\varphi(t_0,x_0), \varphi_t(t_0,x_0), \varphi_x(t_0,x_0)\big) = (v(t_0,x_0), q, p),\\ \varphi(t,x) < v(t,x), \quad \forall (t_0,x_0) \neq (t,x) \in [t_0,T]\times\mathbb{R}^n.\end{cases}$$

Note that Lemmas 2.7 and 2.8 remain true if in (i) the word "strict" is omitted and in (ii) the inequality ">" or "<" is replaced by "≥" or "≤", respectively.

Let us say a few words about the geometric meaning of Lemmas 2.7 and 2.8. Suppose (q,p) is taken from $D^{1,+}_{t,x}(t_0,x_0)$ (resp. $D^{1,-}_{t,x}(t_0,x_0)$). Then one can find a C^1 function φ whose graph touches that of v from above (resp. below) only at the point (t_0,x_0) in a neighborhood of this point, and the gradient (φ_t,φ_x) of φ at (t_0,x_0) coincides with (q,p).

In terms of the super- and subdifferentials, we can give an equivalent definition of viscosity solution.

Theorem 2.9. *A function $v \in C([0,T]\times\mathbb{R}^n)$ is a viscosity solution of (2.14) if and only if $v(T,x) = h(x)$ for all $x\in\mathbb{R}^n$, and for all $(t,x) \in [0,T)\times\mathbb{R}^n$,*

$$(2.78)\qquad \begin{cases} -q + \sup\limits_{u\in U} H(t,x,u,-p) \leq 0, \quad \forall (q,p) \in D^{1,+}_{t,x}v(t,x),\\ -q + \sup\limits_{u\in U} H(t,x,u,-p) \geq 0, \quad \forall (q,p) \in D^{1,-}_{t,x}v(t,x).\end{cases}$$

Proof. Suppose v is a viscosity solution of (2.14). Then, for any $(q,p) \in D^{1,+}_{t,x}v(t,x)$, by Lemma 2.7 we can find a $\varphi \in C^1([0,T] \times \mathbb{R}^n)$ such that $v - \varphi$ attains a strict local maximum at (t,x), at which (2.73) holds. By Definition 2.4, (2.35) holds. This gives the first relation in (2.78) by noting (2.73). One can prove the second relation in (2.78) similarly.

Now, suppose (2.78) holds. If $\varphi \in C^1([0,T] \times \mathbb{R}^n)$ is such that $v - \varphi$ attains a local maximum at $(t,x) \in [0,T) \times \mathbb{R}^n$, then one can show that $(\varphi_t(t,x), \varphi_x(t,x)) \in D^{1,+}_{t,x}v(t,x)$. Hence, using the first relation in (2.78), we obtain (2.35), which implies that v is a viscosity subsolution of (2.14). In the same fashion, we can show that v is also a viscosity supersolution of (2.14). □

We have introduced $D^{1,\pm}_{t+,x}v(t,x)$, the right super-/subdifferential. Note that Theorem 2.9 is not necessarily true if $D^{1,\pm}_{t,x}v(t,x)$ is replaced by $D^{1,\pm}_{t+,x}v(t,x)$, as the former could be strictly smaller than the latter. However, under (D1) and (D2)′, we have the following stronger result for the value function V.

Theorem 2.10. *Let (D1) and (D2)′ hold. Then the value function $V \in C([0,T] \times \mathbb{R}^n)$ is the only function that satisfies the following: For all $(t,x) \in [0,T) \times \mathbb{R}^n$,*

$$
\begin{cases}
-q + \sup_{u \in U} H(t,x,u,-p) \le 0, & \forall (q,p) \in D^{1,+}_{t+,x}V(t,x), \\
-q + \sup_{u \in U} H(t,x,u,-p) \ge 0, & \forall (q,p) \in D^{1,-}_{t+,x}V(t,x), \\
V(T,x) = h(x).
\end{cases}
\tag{2.79}
$$

Proof. For any $(q,p) \in D^{1,\pm}_{t+,x}V(t,x)$, take the function φ as specified by Lemma 2.7-(ii) or Lemma 2.8-(ii) (with v replaced by V). Then we can use exactly the same argument as in Step 2 of the proof of Theorem 2.5 (note that only the *right* limit in time was used there) to obtain (2.79). On the other hand, the uniqueness of such functions satisfying (2.79) comes from the uniqueness part of Theorem 2.5 along with Theorem 2.9, in view of the facts that $D^{1,+}_{t,x}V(t,x) \subseteq D^{1,+}_{t+,x}V(t,x)$ and $D^{1,-}_{t,x}V(t,x) \subseteq D^{1,-}_{t+,x}V(t,x)$. □

From the above, we see that under (D1) and (D2)′, the conditions (2.78) and (2.79) are equivalent in defining the (unique) viscosity solution (which is nothing but the value function V) of the HJB equation (2.14).

3. The Stochastic Principle of Optimality and the HJB Equation

We now proceed to consider the stochastic case. Though the idea of dynamic programming is similar to the deterministic case, the analysis in the stochastic case will be much more delicate.

3.1. A stochastic framework for dynamic programming

Let $(\Omega, \mathcal{F}, \{\mathcal{F}_t\}_{t\ge 0}, \mathbf{P})$ be a *given* filtered probability space satisfying the usual condition, on which is defined an m-dimensional standard Brownian motion $W(t)$. We consider the following stochastic controlled system:

$$
(3.1) \qquad \begin{cases} dx(t) = b(t, x(t), u(t))dt + \sigma(t, x(t), u(t))dW(t), & t \in [0, T], \\ x(0) = x_0, \end{cases}
$$

with the cost functional

$$
(3.2) \qquad J(u(\cdot)) = E\Big\{ \int_0^T f(t, x(t), u(t))dt + h(x(T)) \Big\}.
$$

Define (recall Convention 2.9 in Chapter 1)

$$
(3.3) \qquad \begin{aligned} \mathcal{U}^s[0,T] \stackrel{\Delta}{=} \{u : [0,T] \times \Omega \to U \;|\, & u \text{ is measurable and} \\ & \{\mathcal{F}_t\}_{t\geq 0}\text{-adapted}\}, \end{aligned}
$$

where the superscript "s" indicates that the strong formulation is being considered. Under certain assumptions (which will be specified below), for any $u(\cdot) \in \mathcal{U}^s[0,T]$ equation (3.1) admits a unique solution, and the cost (3.2) is well-defined. The optimal control problem can be stated as follows.

Problem (S). Minimize (3.2) subject to the state equation (3.1) over $\mathcal{U}^s[0,T]$.

As introduced in Chapter 2, Problem (S) is in a strong formulation, which is the one we would like to solve eventually. It is observed in the previous section that one needs to consider a *family* of optimal control problems with different initial times and states along a given state trajectory in order to apply the dynamic programming technique. However, in the stochastic case the states along a state trajectory become random variables in the *original* probability space. More specifically, if $x(\cdot)$ is a state trajectory starting from x_0 at time 0 in a probability space $(\Omega, \mathcal{F}, \mathbf{P})$ along with a filtration $\{\mathcal{F}_t\}_{t\geq 0}$, then for any $t > 0$, $x(t)$ is a random variable in $(\Omega, \mathcal{F}, \mathbf{P})$ rather than a deterministic point in $\mathbb{R}^n$. Nevertheless, an admissible control $u(\cdot)$ is $\{\mathcal{F}_t\}_{t\geq 0}$-adapted, meaning that at any time instant t the controller knows about all the relevant past information of the system up to time t (as specified by $\mathcal{F}_t$) and in particular about $x(t)$. This implies that $x(t)$ is actually *not* uncertain for the controller at time t. In mathematical terms, $x(t)$ is almost surely deterministic under a *different* probability measure $\mathbf{P}(\cdot\,|\mathcal{F}_t)$ (see Chapter 1, Proposition 2.13). Thus the above idea requires us to vary the probability spaces as well in order to employ dynamic programming. Very naturally, we need to consider the *weak* formulation of the stochastic control problem as an auxiliary tool.

Now we set up the framework. Let $T > 0$ be given and let U be a metric space. For any $(s, y) \in [0, T) \times \mathbb{R}^n$, consider the state equation:

$$
(3.4) \qquad \begin{cases} dx(t) = b(t, x(t), u(t))dt + \sigma(t, x(t), u(t))dW(t), & t \in [s, T], \\ x(s) = y, \end{cases}
$$

along with the cost functional

$$J(s,y;u(\cdot)) = E\Big\{\int_s^T f(t,x(t),u(t))dt + h(x(T))\Big\}. \tag{3.5}$$

Fixing $s \in [0,T)$, we denote by $\mathcal{U}^w[s,T]$ the set of all 5-tuples $(\Omega,\mathcal{F},\mathbf{P},W(\cdot),u(\cdot))$ satisfying the following:

(i) $(\Omega,\mathcal{F},\mathbf{P})$ is a complete probability space (see Chapter, Section 1.1).

(ii) $\{W(t)\}_{t\geq s}$ is an m-dimensional standard Brownian motion defined on $(\Omega,\mathcal{F},\mathbf{P})$ over $[s,T]$ (with $W(s)=0$ almost surely), and $\mathcal{F}_t^s = \sigma\{W(r) : s\leq r\leq t\}$ augmented by all the $\mathbf{P}$-null sets in $\mathcal{F}$.

(iii) $u : [s,T]\times\Omega\to U$ is an $\{\mathcal{F}_t^s\}_{t\geq s}$-adapted process on $(\Omega,\mathcal{F},\mathbf{P})$.

(iv) Under $u(\cdot)$, for any $y\in\mathbb{R}^n$ equation (3.4) admits a unique solution (in the sense of Chapter 1, Definition 6.15) $x(\cdot)$ on $(\Omega,\mathcal{F},\{\mathcal{F}_t^s\}_{t\geq s},\mathbf{P})$.

(v) $f(\cdot\,,x(\cdot),u(\cdot)) \in L^1_{\mathcal{F}}(0,T;\mathbb{R})$ and $h(x(T)) \in L^1_{\mathcal{F}_T}(\Omega;\mathbb{R})$. Here, the spaces $L^1_{\mathcal{F}}(0,T;\mathbb{R})$ and $L^1_{\mathcal{F}_T}(\Omega;\mathbb{R})$ are defined on the given filtered probability space $(\Omega,\mathcal{F},\{\mathcal{F}_t^s\}_{t\geq s},\mathbf{P})$ (associated with the given 5-tuple).

Sometimes, we simply write $u(\cdot)\in\mathcal{U}^w[s,T]$ instead of $(\Omega,\mathcal{F},\mathbf{P},W(\cdot),u(\cdot)) \in \mathcal{U}^w[s,T]$ if the probability space $(\Omega,\mathcal{F},\mathbf{P})$ and the Brownian motion $W(t)$ are clear from the context. Notice that in the weak formulation defined in Chapter 2, Section 4.2, the filtration $\{\mathcal{F}_t^s\}_{t\geq s}$ is also part of the control. However, here we *restrict* $\{\mathcal{F}_t^s\}_{t\geq s}$ to be generated by the Brownian motion. Hence the formulation in this chapter is more special than the one in Chapter 2.

We emphasize that in (3.4) the initial state y is a *deterministic* (almost surely) variable under $(\Omega,\mathcal{F},\mathbf{P})$. Also note that in (3.5) the mathematical expectation E is with respect to the probability $\mathbf{P}$. Our optimal control problem can be stated as follows:

Problem (S$_{sy}$). For given $(s,y)\in[0,T)\times\mathbb{R}^n$, find a 5-tuple $\bar u(\cdot)\equiv(\bar\Omega,\bar{\mathcal{F}},\bar{\mathbf{P}},\bar W(\cdot),\bar u(\cdot))\in\mathcal{U}^w[s,T]$ such that

$$J(s,y;\bar u(\cdot)) = \inf_{u(\cdot)\in\mathcal{U}^w[s,T]} J(s,y;u(\cdot)). \tag{3.6}$$

We now introduce some assumptions (compare with those imposed in Chapter 3).

(S1)$'$ (U,d) is a Polish space and $T>0$.

(S2)$'$ The maps $b:[0,T]\times\mathbb{R}^n\times U\to\mathbb{R}^n$, $\sigma:[0,T]\times\mathbb{R}^n\times U\to\mathbb{R}^{n\times m}$, $f:[0,T]\times\mathbb{R}^n\times U\to\mathbb{R}$, and $h:\mathbb{R}^n\to\mathbb{R}$ are uniformly continuous, and there exists a constant $L>0$ such that for $\varphi(t,x,u)=b(t,x,u)$, $\sigma(t,x,u)$, $f(t,x,u)$, $h(x)$,

$$\begin{cases} |\varphi(t,x,u)-\varphi(t,\hat x,u)|\leq L|x-\hat x|, & \forall t\in[0,T], x,\hat x\in\mathbb{R}^n, u\in U,\\ |\varphi(t,0,u)|\leq L, & \forall (t,u)\in[0,T]\times U.\end{cases} \tag{3.7}$$

Note that unlike (S1)–(S2) introduced in Chapter 3, Section 3, (S1)′–(S2)′ require U to be complete and the functions involved to be continuous in (t, x, u), including t.

Under (S1)′–(S2)′, for any $(s, y) \in [0, T) \times \mathbb{R}^n$ and $u(\cdot) \in \mathcal{U}^w[s, T]$, (3.4) admits a unique solution $x(\cdot) \equiv x(\cdot\,; s, y, u(\cdot))$ (see Chapter 1, Theorem 6.16), and the cost functional (3.5) is well-defined. Thus, we can define the following function:

$$\left\{\begin{aligned} &V(s,y) = \inf_{u(\cdot)\in\mathcal{U}^w[s,T]} J(s,y;u(\cdot)), \qquad \forall (s,y) \in [0,T) \times \mathbb{R}^n, \\ &V(T,y) = h(y), \qquad \forall y \in \mathbb{R}^n, \end{aligned}\right. \tag{3.8}$$

which is called the *value function* of the original Problem (S).

The following result gives some basic properties of the value function.

Proposition 3.1. *Let (S1)′–(S2)′ hold. Then the value function $V(s, y)$ satisfies the following:*

$$|V(s,y)| \le K(1+|y|), \quad \forall (s,y) \in [0,T] \times \mathbb{R}^n, \tag{3.9}$$

$$\begin{aligned} |V(s,y) - V(\hat{s},\hat{y})| \le K\{|y-\hat{y}| + (1+|y| \vee |\hat{y}|)|s-\hat{s}|^{1/2}\}, \\ \forall s,\hat{s} \in [0,T], \ y,\hat{y} \in \mathbb{R}^n. \end{aligned} \tag{3.10}$$

Comparing (2.36) with (3.10), we see that unlike the value function $V(s, y)$ of Problem (D), which is (locally) Lipschitz continuous in the time variable s, the value function $V(s, y)$ of Problem (S) is not necessarily (local) Lipschitz continuous in s. This is an essential difference between the deterministic and stochastic cases.

Proof. Let $(s, y) \in [0, T) \times \mathbb{R}^n$ be fixed. For any $(\Omega, \mathcal{F}, \mathbf{P}, W(\cdot), u(\cdot)) \in \mathcal{U}^w[s, T]$, by Theorem 6.16 of Chapter 1, we have

$$E \sup_{t\in[s,T]} |x(t)| \le K(1+|y|). \tag{3.11}$$

Thus, by (S2)′,

$$|J(s,y;u(\cdot))| \le K(1+|y|), \qquad \forall u(\cdot) \in \mathcal{U}^w[s,T]. \tag{3.12}$$

This implies (3.9).

Next, we let $0 \le s \le \hat{s} \le T$ and $y, \hat{y} \in \mathbb{R}^n$. For any w-admissible control $(\Omega, \mathcal{F}, \mathbf{P}, W(\cdot), u(\cdot)) \in \mathcal{U}^w[s, T]$, let $x(\cdot)$ and $\hat{x}(\cdot)$ be the states corresponding to $(s, y, u(\cdot))$ and $(\hat{s}, \hat{y}, u(\cdot))$, respectively. Then, again by Theorem 6.16 of Chapter 1, we have

$$E \sup_{t\in[\hat{s},T]} |x(t) - \hat{x}(t)| \le K\{|y-\hat{y}| + (1+|y| \vee |\hat{y}|)|s-\hat{s}|^{1/2}\}. \tag{3.13}$$

Hence by using (S2)′, we obtain

$$|J(s,y;u(\cdot)) - J(s,\hat{y};u(\cdot))| \le K\{|y-\hat{y}| + (1+|y| \vee |\hat{y}|)|s-\hat{s}|^{1/2}\}. \tag{3.14}$$

Taking the infimum in $u(\cdot) \in \mathcal{U}^w[s,T]$, we get (3.10). $\square$

Now let us present a technical result. For any $s \in [0,T)$ and any $(\Omega, \mathcal{F}, \mathbf{P}, W(\cdot), u(\cdot)) \in \mathcal{U}^w[s,T]$, take $\hat{s} \in [s,T)$ and an $\mathcal{F}^s_{\hat{s}}$-measurable random variable ξ. Under (S1)$'$–(S2)$'$, we can solve the following SDE on $(\Omega, \mathcal{F}, \{\mathcal{F}^s_t\}_{t\geq s}, \mathbf{P})$:

$$
(3.15) \qquad \begin{cases} dx(t) = b(t,x(t),u(t))dt + \sigma(t,x(t),u(t))dW(t), & t \in [\hat{s},T], \\ x(\hat{s}) = \xi. \end{cases}
$$

Denote the solution by $x(\cdot\,; \hat{s}, \xi, u(\cdot))$. We have the following lemma.

Lemma 3.2. *Let $(s,y) \in [0,T) \times \mathbb{R}^n$ and $(\Omega, \mathcal{F}, \mathbf{P}, W(\cdot), u(\cdot)) \in \mathcal{U}^w[s,T]$. Then, for any $\hat{s} \in [s,T)$ and $\mathcal{F}^s_{\hat{s}}$-measurable random variable ξ,*

$$
(3.16) \qquad \begin{aligned} J(\hat{s}, \xi(\omega); u(\cdot)) = E\Big\{ \int_{\hat{s}}^T & f(t, x(t; \hat{s}, \xi, u(\cdot)), u(t))dt \\ & + h(x(T; \hat{s}, \xi, u(\cdot))) \big| \mathcal{F}^s_{\hat{s}} \Big\}(\omega), \quad \mathbf{P}\text{-a.s.}\, \omega. \end{aligned}
$$

Proof. Since $u(\cdot)$ is $\{\mathcal{F}^s_t\}_{t\geq s}$-adapted with $\mathcal{F}^s_t = \sigma\{W(r) : s \leq r \leq t\}$, by Theorem 2.10 in Chapter 1 there is a function $\psi \in \mathcal{A}^m_T(U)$ such that

$$
u(t,\omega) = \psi(t, W(\cdot \wedge t, \omega)), \qquad \mathbf{P}\text{-a.s.}\, \omega \in \Omega, \quad \forall t \in [s,T].
$$

Therefore, (3.15) can be rewritten as

$$
(3.17) \qquad \begin{cases} dx(t) = b(t, x(t), \psi(t, W(\cdot \wedge t)))dt \\ \qquad\qquad + \sigma(t, x(t), \psi(t, W(\cdot \wedge t)))dW(t), & t \in [\hat{s},T], \\ x(\hat{s}) = \xi. \end{cases}
$$

Due to (S1)$'$–(S2)$'$ and Chapter 1, Theorem 6.17, this equation has a unique strong solution, and weak uniqueness holds (i.e., unique in the distribution sense; see Chapter 1, Definition 6.6). On the other hand, by Proposition 2.13 of Chapter 1,

$$
\mathbf{P}\Big\{\bar{\omega} : \xi(\bar{\omega}) = \xi(\omega) \Big| \mathcal{F}^s_{\hat{s}} \Big\}(\omega) = 1, \quad \mathbf{P}\text{-a.s.}
$$

This means that there is an $\Omega_0 \in \mathcal{F}$ with $\mathbf{P}(\Omega_0) = 1$, so that for any fixed $\omega_0 \in \Omega_0$, ξ becomes a *deterministic* constant $\xi(\omega_0)$ under the new probability space $(\Omega, \mathcal{F}, \mathbf{P}(\cdot | \mathcal{F}^s_{\hat{s}})(\omega_0))$. In addition, for any $t \geq \hat{s}$,

$$
u(t,\omega) = \psi(t, W(\cdot \wedge t, \omega)) = \psi(t, \widetilde{W}(\cdot \wedge t, \omega) + W(\hat{s}, \omega)),
$$

where $\widetilde{W}(t) = W(t) - W(\hat{s})$ is a standard Brownian motion. Note that $W(\hat{s})$ almost surely equals a constant $W(\hat{s}, \omega_0)$ under the probability measure $\mathbf{P}(\cdot | \mathcal{F}^s_{\hat{s}})(\omega_0)$. It follows then that $u(t)$ is adapted to (the standard Brownian motion) $\widetilde{W}(t)$ for $t \geq \hat{s}$. Hence by the definition of admissible controls,

$$
(\Omega, \mathcal{F}, \mathbf{P}(\cdot | \mathcal{F}^s_{\hat{s}})(\omega_0), \widetilde{W}(t), u(t)) \in \mathcal{U}^w[\hat{s}, T].
$$

Thus if we work under the probability space $(\Omega, \mathcal{F}, \mathbf{P}(\cdot|\mathcal{F}_{\hat{s}}^s)(\omega_0))$ and notice the weak uniqueness of (3.15) or (3.17), we obtain our result. □

3.2. Principle of optimality

The following is the stochastic version of *Bellman's principle of optimality* (compare Theorem 2.1 in the previous section).

Theorem 3.3. *Let (S1)′–(S2)′ hold. Then for any* $(s,y) \in [0,T) \times \mathbb{R}^n$,

$$\begin{aligned} V(s,y) = \inf_{u(\cdot)\in\mathcal{U}^w[s,T]} E\Big\{ \int_s^{\hat{s}} f(t, x(t;s,y,u(\cdot)), u(t))dt \\ + V\big(\hat{s}, x(\hat{s};s,y,u(\cdot))\big)\Big\}, \quad \forall 0 \le s \le \hat{s} \le T. \end{aligned} \tag{3.18}$$

Proof. Denote the right-hand side of (3.18) by $\overline{V}(s,y)$. For any $\varepsilon > 0$, there exists an $(\Omega, \mathcal{F}, \mathbf{P}, W(\cdot), u(\cdot)) \in \mathcal{U}^w[s,T]$ such that (noting Lemma 3.2)

$$\begin{aligned} &V(s,y) + \varepsilon > J(s,y;u(\cdot)) \\ &= E\Big\{ \int_s^T f(t,x(t;s,y,u(\cdot)),u(t))dt + h(x(T;s,y,u(\cdot)))\Big\} \\ &= E\Big\{ \int_s^{\hat{s}} f(t,x(t;s,y,u(\cdot)),u(t))dt \\ &\qquad + E\Big[\int_{\hat{s}}^T f(t,x(t;s,y,u(\cdot)),u(t))dt + h(x(T;s,y,u(\cdot)))\Big|\mathcal{F}_{\hat{s}}^s\Big]\Big\} \\ &= E\Big\{ \int_s^{\hat{s}} f(t,x(t;s,y,u(\cdot)),u(t))dt \\ &\qquad + E\Big[\int_{\hat{s}}^T f(t,x(t;\hat{s},x(\hat{s}),u(\cdot)),u(t))dt + h(x(T;\hat{s},x(\hat{s}),u(\cdot)))\Big|\mathcal{F}_{\hat{s}}^s\Big]\Big\} \\ &= E\Big\{ \int_s^{\hat{s}} f(t,x(t;s,y,u(\cdot)),u(t))dt + J\big(\hat{s},x(\hat{s};s,y,u(\cdot));u(\cdot)\big)\Big\} \\ &\ge E\Big\{ \int_s^{\hat{s}} f(t,x(t;s,y,u(\cdot)),u(t))dt + V(\hat{s},x(\hat{s};s,y,u(\cdot)))\Big\} \ge \overline{V}(s,y). \end{aligned}$$

Conversely, for any $\varepsilon > 0$, by Proposition 3.1 and its proof, there is a $\delta = \delta(\varepsilon)$ such that whenever $|y - \hat{y}| < \delta$,

$$\begin{aligned} |J(\hat{s},y;u(\cdot)) - J(\hat{s},\hat{y};u(\cdot))| + |V(\hat{s},y) - V(\hat{s},\hat{y})| \le \varepsilon, \\ \forall u(\cdot) \in \mathcal{U}^w[\hat{s},T]. \end{aligned} \tag{3.19}$$

Let $\{D_j\}_{j\ge1}$ be a *Borel partition* of $\mathbb{R}^n$ (meaning that $D_j \in \mathcal{B}(\mathbb{R}^n)$, $\bigcup_{j\ge1} D_j = \mathbb{R}^n$, and $D_i \cap D_j = \phi$ if $i \ne j$) with diameter $\operatorname{diam}(D_j) < \delta$. Choose $x_j \in D_j$. For each j, there is $(\Omega_j, \mathcal{F}_j, \mathbf{P}_j, W_j(\cdot), u_j(\cdot)) \in \mathcal{U}^w[\hat{s},T]$ such that

$$J(\hat{s},x_j;u_j(\cdot)) \le V(\hat{s},x_j) + \varepsilon. \tag{3.20}$$

Hence for any $x \in D_j$, combining (3.19)–(3.20), we have

$$J(\hat{s}, x; u_j(\cdot)) \leq J(\hat{s}, x_j; u_j(\cdot)) + \varepsilon \leq V(\hat{s}, x_j) + 2\varepsilon \leq V(\hat{s}, x) + 3\varepsilon. \tag{3.21}$$

By the definition of the w-admissible 5-tuple $(\Omega_j, \mathcal{F}_j, \mathbf{P}_j, W_j(\cdot), u_j(\cdot))$, there is a function $\psi_j \in \mathcal{A}_T^m(U)$ such that

$$u_j(t, \omega) = \psi_j(t, W_j(\cdot \wedge t, \omega)), \quad \mathbf{P}_j\text{-a.s. } \omega \in \Omega_j, \ \forall t \in [\hat{s}, T].$$

Now, for any $(\Omega, \mathcal{F}, \mathbf{P}, W(\cdot), u(\cdot)) \in \mathcal{U}^w[s, T]$, let $x(\cdot) \equiv x(\cdot\,; s, y, u(\cdot))$ denote the corresponding state trajectory for Problem (S_{sy}). Define a new control

$$\widetilde{u}(t, \omega) = \begin{cases} u(t, \omega), & \text{if } t \in [s, \hat{s}); \\ \psi_j(t, W(\cdot \wedge t, \omega) - W(\hat{s}, \omega)), & \text{if } t \in [\hat{s}, T] \text{ and } x(t, \omega) \in D_j. \end{cases}$$

Clearly, $(\Omega, \mathcal{F}, \mathbf{P}, W(\cdot), \widetilde{u}(\cdot)) \in \mathcal{U}^w[s, T]$. Thus,

$$\begin{aligned} V(s, y) &\leq J(s, y; \widetilde{u}(\cdot)) \\ &= E\Big\{ \int_s^T f(t, x(t; s, y, \widetilde{u}(\cdot)), \widetilde{u}(t))dt + h(x(T; s, y, \widetilde{u}(\cdot))) \Big\} \\ &= E\Big\{ \int_s^{\hat{s}} f(t, x(t; s, y, u(\cdot)), u(t))dt \\ &\qquad + E\Big[\int_{\hat{s}}^T f(t, x(t; \hat{s}, x(\hat{s}), u(\cdot)), \widetilde{u}(t))dt + h(x(T; \hat{s}, x(\hat{s}), \widetilde{u}(\cdot))) \big| \mathcal{F}_{\hat{s}}^s \Big] \Big\} \\ &= E\Big\{ \int_s^{\hat{s}} f(t, x(t; s, y, u(\cdot)), u(t))dt + J(\hat{s}, x(\hat{s}; s, y, u(\cdot)); \widetilde{u}(\cdot)) \Big\} \\ &\leq E\Big\{ \int_s^{\hat{s}} f(t, x(t; s, y, u(\cdot)), u(t))dt + V(\hat{s}, x(\hat{s}; s, y, u(\cdot))) + 3\varepsilon \Big\}, \end{aligned}$$

where the last inequality is due to (3.21). Note that in deriving the above we have employed Lemma 3.2 and the weak uniqueness of solutions. Consequently, we obtain our conclusion by taking the infimum over $u(\cdot) \in \mathcal{U}^w[s, T]$. $\square$

Theorem 3.4. *Let (S1)′–(S2)′ hold. If $(\bar{x}(\cdot), \bar{u}(\cdot))$ is optimal for Problem (S_{sy}), then*

$$V(t, \bar{x}(t)) = E\Big\{ \int_t^T f(r, \bar{x}(r), \bar{u}(r))dr + h(\bar{x}(T)) \big| \mathcal{F}_t^s \Big\}, \quad \mathbf{P}\text{-a.s.}, \quad \forall t \in [s, T]. \tag{3.22}$$

Proof. By the same argument as that in the proof of Theorem 3.3, we have

$$
\begin{aligned}
V(s,y) &= J(s,y;\overline{u}(\cdot)) \\
&= E\Big\{\int_s^t f(r,\overline{x}(r),\overline{u}(r))dr \\
&\qquad + E\Big[\int_t^T f(r,\overline{x}(r),\overline{u}(r))dr + h(\overline{x}(T))\Big|\mathcal{F}_t^s\Big]\Big\} \\
&= E\int_t^T f(r,\overline{x}(r),\overline{u}(r))dr + EJ\big(t,\overline{x}(t);\overline{u}(\cdot)\big) \\
&\geq E\int_t^T f(r,\overline{x}(r),\overline{u}(r))dr + EV(t,\overline{x}(t)) \\
&\geq V(s,y),
\end{aligned}
\tag{3.23}
$$

where the last inequality is due to (3.18). Hence all the inequalities in (3.23) become equalities. In particular,

$$
EJ\big(t,\overline{x}(t);\overline{u}(\cdot)\big) = EV(t,\overline{x}(t)).
$$

However, by definition one has $V(t,\overline{x}(t)) \leq J\big(t,\overline{x}(t);\overline{u}(\cdot)\big)$, $\mathbf{P}$-a.s. Thus

$$
V(t,\overline{x}(t)) = J\big(t,\overline{x}(t);\overline{u}(\cdot)\big), \qquad \mathbf{P}\text{-a.s.},
$$

which gives (3.22). □

3.3. The HJB equation

As in the deterministic case, we call (3.18) the *dynamic programming equation.* This equation is very complicated, and it seems impossible to solve such an equation directly. This subsection is devoted to a *formal* derivation of a partial differential equation that the value function $V(\cdot\,,\cdot)$ should satisfy, based on equation (3.18). We let $C^{1,2}([0,T]\times\mathbb{R}^n)$ be the set of all continuous functions $v:[0,T]\times\mathbb{R}^n\to\mathbb{R}$ such that v_t, v_x, and v_{xx} are all continuous in (t,x).

Proposition 3.5. *Suppose (S1)′–(S2)′ hold and the value function $V\in C^{1,2}([0,T]\times\mathbb{R}^n)$. Then V is a solution of the following terminal value problem of a (possibly degenerate) second-order partial differential equation:*

$$
\begin{cases}
-v_t + \sup\limits_{u\in U} G(t,x,u,-v_x,-v_{xx}) = 0, & (t,x)\in[0,T)\times\mathbb{R}^n, \\
v\big|_{t=T} = h(x), & x\in\mathbb{R}^n,
\end{cases}
\tag{3.24}
$$

where

$$
\begin{aligned}
G(t,x,u,p,P) &\stackrel{\Delta}{=} \frac{1}{2}\operatorname{tr}\big(P\sigma(t,x,u)\sigma(t,x,u)^\top\big) \\
&\qquad + \langle\, p, b(t,x,u)\,\rangle - f(t,x,u), \\
&\qquad \forall (t,x,u,p,P)\in[0,T]\times\mathbb{R}^n\times U\times\mathbb{R}^n\times\mathcal{S}^n.
\end{aligned}
\tag{3.25}
$$

Proof. Fix $(s, y) \in [0, T) \times \mathbb{R}^n$ and $u \in U$. Let $x(\cdot)$ be the state trajectory corresponding to the control $u(\cdot) \in \mathcal{U}^w[s, T]$ for Problem (S_{sy}) with $u(t) \equiv u$. By (3.18) with $\hat{s} \downarrow s$ and Itô's formula, we obtain

$$\begin{aligned} 0 &\geq -\frac{E\{V(\hat{s}, x(\hat{s})) - V(s, y)\}}{\hat{s} - s} - \frac{1}{\hat{s} - s} E \int_s^{\hat{s}} f(t, x(t), u) dt \\ &= -\frac{1}{\hat{s} - s} E \int_s^{\hat{s}} \{ - V_t(t, x(t)) \\ &\qquad + G\big(t, x(t), u(t), -V_x(t, x(t)), -V_{xx}(t, x(t))\big) \} dt \\ &\to -V_t(s, y) + G(s, y, u, -V_x(s, y), -V_{xx}(s, y)), \qquad \forall u \in U. \end{aligned}$$

This results in

$$0 \geq -V_t(s, y) + \sup_{u \in U} G(s, y, u, -V_x(s, y), -V_{xx}(s, y)). \tag{3.26}$$

On the other hand, for any $\varepsilon > 0$, $0 \leq s < \hat{s} \leq T$ with $\hat{s} - s > 0$ small enough, there exists a $u(\cdot) \equiv u_{\varepsilon, \hat{s}}(\cdot) \in \mathcal{U}^w[s, T]$ such that

$$V(s, y) + \varepsilon(\hat{s} - s) \geq E\Big\{ \int_s^{\hat{s}} f(t, x(t), u(t)) dt + V(\hat{s}, x(\hat{s})) \Big\}.$$

Thus, it follows from Itô's formula that as $\hat{s} \downarrow s$,

$$\begin{aligned} -\varepsilon &\leq -\frac{E\{V(\hat{s}, x(\hat{s})) - V(s, y)\}}{\hat{s} - s} - \frac{1}{\hat{s} - s} E \int_s^{\hat{s}} f(t, x(t), u(t)) dt \\ &= \frac{1}{\hat{s} - s} \int_s^{\hat{s}} \{ - V_t(t, x(t)) \\ &\qquad + G\big(t, x(t), u(t), -V_x(t, x(t)), -V_{xx}(t, x(t))\big) \} dt \\ &\leq \frac{1}{\hat{s} - s} \int_s^{\hat{s}} \{ - V_t(t, x(t)) \\ &\qquad + \sup_{u \in U} G\big(t, x(t), u, -V_x(t, x(t)), -V_{xx}(t, x(t))\big) \} dt \\ &\to -V_t(s, y) + \sup_{u \in U} G\big(s, y, u, -V_x(s, y), -V_{xx}(s, y)\big). \end{aligned} \tag{3.27}$$

In proving the last limit above, we have used the fact that

$$\lim_{t \downarrow s} \sup_{y \in \mathbb{R}^n, u \in U} |\varphi(t, y, u) - \varphi(s, y, u)| = 0, \tag{3.28}$$

for $\varphi = b, \sigma, f$. This is implied by the uniform continuity of the functions b, σ, and f as assumed in (S2)′. Combining (3.26) and (3.27), we obtain our conclusion. □

We call (3.24) the *Hamilton–Jacobi–Bellman equation* (HJB equation, for short) of Problem (S). The function $G(t, x, u, p, P)$ defined by (3.25) is called the *generalized Hamiltonian*, which was defined earlier in (3.15) of Chapter 3. By comparing (3.24) with (2.14), we see that (3.24) is of second order, allowing degeneracy. In the case that $\sigma(t,x,u) \equiv 0$, (3.24) reduces to (2.14). On the other hand, when $(\sigma\sigma^\top)(t,x,u)$ is uniformly positive definite, along with other mild conditions, (3.24) admits a classical solution (see Fleming–Soner [1, p. 169]).

4. Other Properties of the Value Function

In this section we are going to explore more properties of the value function V defined in the previous section.

4.1. Continuous dependence on parameters

Suppose we have a family of parametrized optimal control problems, and we investigate whether the value function depends continuously on the parameters under proper conditions. Such dependence will be useful in approximation and/or regularization.

Consider a family of problems parameterized by $\varepsilon \in [0,1]$, for which the state equations, each called $(3.4)_\varepsilon$, are of the form (3.4) with $b(t,x,u)$ and $\sigma(t,x,u)$ replaced by $b^\varepsilon(t,x,u)$ and $\sigma^\varepsilon(t,x,u)$, respectively, and the cost functionals, each denoted by $J^\varepsilon(s,y;u(\cdot))$, are of the form (3.5) with $f(t,x,u)$ and $h(x)$ replaced by $f^\varepsilon(t,x,u)$ and $h^\varepsilon(x)$, respectively. The corresponding optimal control problems are called Problem $(\mathrm{S}^\varepsilon_{sy})$ (and Problem (S^ε) if $(s,y)=(0,x_0)$). Also, we assume that Problem (S^0_{sy}) coincides with Problem (S_{sy}) stated in Section 3.1 (i.e., $b^0=b$, $\sigma^0=\sigma$, $f^0=f$, and $h^0=h$). We can define the value function, denoted by $V^\varepsilon(s,y)$, of Problem (S^ε) in the same way as that for Problem (S), provided that (S2)′ holds for b^ε, σ^ε, f^ε, and h^ε. We then have the following convergence result.

Proposition 4.1. *Suppose (S1)′–(S2)′ hold for $(b^\varepsilon,\sigma^\varepsilon,f^\varepsilon,h^\varepsilon)$ with the constants in (3.7) uniformly in all $\varepsilon \in [0,1]$. Suppose further that*

$$\lim_{\varepsilon\to 0} |\varphi^\varepsilon(t,x,u) - \varphi^0(t,x,u)| = 0, \tag{4.1}$$

uniformly in $(t,u)\in[0,T]\times U$ and x in compact sets of $\mathbb{R}^n$, where $\varphi^\varepsilon = b^\varepsilon,\sigma^\varepsilon,f^\varepsilon,h^\varepsilon$. Then

$$\lim_{\varepsilon\to 0} V^\varepsilon(s,y) = V^0(s,y) \equiv V(s,y), \tag{4.2}$$

uniformly in (s,y) in any compact set of $[0,T]\times\mathbb{R}^n$.

Proof. By the assumption (4.1), we can find a continuous function $\eta:[0,\infty)\times[0,\infty)\to[0,\infty)$ that is nondecreasing in each of its arguments, with $\eta(0,R)=0$ for all $R\geq 0$, such that

$$|\varphi^\varepsilon(t,x,u)-\varphi^0(t,x,u)| \leq \eta(\varepsilon,|x|), \quad \forall (t,x,u)\in[0,T]\times\mathbb{R}^n\times U. \tag{4.3}$$

Let $(s,y)\in[0,T]\times\mathbb{R}^n$ be fixed and take any $u(\cdot)\in\mathcal{U}^w[s,T]$. Define $x^\varepsilon(\cdot) = x^\varepsilon(\cdot\,;s,y,u(\cdot))$ to be the solution of $(3.4)_\varepsilon$ for $\varepsilon\in[0,1]$, and $x(\cdot)=x^0(\cdot)$ the solution of (3.4). Then, using Itô's formula, Burkholder–Davis–Gundy's inequality, and Gronwall's inequality, we have

$$
\begin{aligned}
E\Big[\sup_{s\le t\le T}|x^{\varepsilon}(t)-x(t)|^2\Big]
&\le KE\int_s^T|b^{\varepsilon}(t,x(t),u(t))-b(t,x(t),u(t))|^2dt\\
&\quad+KE\int_s^T|\sigma^{\varepsilon}(t,x(t),u(t))-\sigma(t,x(t),u(t))|^2dt\\
&\le K\int_s^T E\eta(\varepsilon;|x(t)|)^2dt.
\end{aligned}\tag{4.4}
$$

By (S2)$'$,

$$
\eta(\varepsilon;|x|)\le 2L(1+|x|),\qquad \forall x\in\mathbb{R}^n,\ \varepsilon>0.
$$

Thus, for any $R>0$,

$$
\begin{aligned}
E\eta(\varepsilon;|x(t)|)^2&\le\eta(\varepsilon;R)^2+E\big[I_{\{|x(t)|>R\}}4L^2(1+|x(t)|^2)\big]\\
&\le\eta(\varepsilon;R)^2+K\mathbf{P}(|x(t)|>R)^{1/2}\big[E(1+|x(t)|^4)\big]^{1/2}\\
&\le\eta(\varepsilon;R)^2+K\frac{\big(E|x(t)|^4\big)^{1/2}}{R^2}(1+|y|^2)\\
&\le\eta(\varepsilon;R)^2+\frac{K(1+|y|^4)}{R^2}.
\end{aligned}\tag{4.5}
$$

Combining (4.4)–(4.5), we obtain

$$
\lim_{\varepsilon\to0}E\Big[\sup_{s\le t\le T}|x^{\varepsilon}(t)-x(t)|\Big]=0.\tag{4.6}
$$

Consequently,

$$
\begin{aligned}
&\big|J^{\varepsilon}(s,y;u(\cdot))-J(s,y;u(\cdot))\big|\\
&\le KE\Big\{\sup_{s\le t\le T}|x^{\varepsilon}(t)-x(t)|+|h^{\varepsilon}(x(T))-h(x(T))|\\
&\qquad+\int_s^T|f^{\varepsilon}(t,x(t),u(t))-f(t,x(t),u(t))|dt\Big\}\\
&\le KE\Big\{\sup_{s\le t\le T}|x^{\varepsilon}(t)-x(t)|+\eta(\varepsilon;|x(T)|)+\int_s^T\eta(\varepsilon;|x(t)|)dt\Big\}\\
&\le K\Big[\eta(\varepsilon;R)+\frac{1+|y|^2}{R}\Big],\qquad\forall R>0.
\end{aligned}\tag{4.7}
$$

It follows that

$$
\lim_{\varepsilon\to0}|J^{\varepsilon}(s,y;u(\cdot))-J(s,y;u(\cdot))|=0,\tag{4.8}
$$

uniformly in (s,y) on compact sets of $[0,T]\times\mathbb{R}^n$ and $u(\cdot)\in\mathcal{U}^w[s,T]$. This leads to (4.2). □

Let us look at a special case of the above result. Let

$$\text{(4.9)}\qquad \begin{cases} b^\varepsilon(t,x,u)=b(t,x,u), \ f^\varepsilon(t,x,u)=f(t,x,u), \ h^\varepsilon(x)=h(x), \\ \sigma^\varepsilon(t,x,u)=(\sigma(t,x,u),\sqrt{2\varepsilon}\, I_n)_{n\times(m+n)}. \end{cases}$$

Denote the value function of corresponding Problem $(\text{S}^\varepsilon_{sy})$ by $V^\varepsilon(s,y)$. Problem (S^0_{sy}) is basically the same as Problem (S_{sy}) in view of the uniqueness of the weak solutions to the systems. Thus $V^0(s,y)$ coincides with $V(s,y)$. By Proposition 4.1, we have the following corollary.

Corollary 4.2. *Let (S1)′–(S2)′ hold. Then there exists a constant $K>0$ such that*

$$\text{(4.10)}\qquad |V^\varepsilon(s,y)-V(s,y)|\le K\sqrt{\varepsilon},\quad \forall (s,y)\in[0,T]\times\mathbb{R}^n.$$

Proof. In the present case, (4.4) can be improved to

$$\text{(4.11)}\qquad E\Big[\sup_{s\le t\le T}|x^\varepsilon(t)-x(t)|^2\Big]\le K\varepsilon,$$

and (4.7) is replaced by

$$\text{(4.12)}\qquad |J^\varepsilon(s,y;u(\cdot))-J(s,y;u(\cdot))|\le K\sqrt{\varepsilon},\quad \forall u(\cdot)\in\mathcal{U}^w[s,T].$$

Then (4.10) follows. □

We know that $V(s,y)$ is a solution of (3.24) if it is smooth. Also, if $V^\varepsilon(s,y)$ is smooth, then it is a solution of the following HJB equation:

$$\text{(4.13)}\qquad \begin{cases} -v^\varepsilon_t-\varepsilon\Delta v^\varepsilon+\sup\limits_{u\in U}G(t,x,u,-v^\varepsilon_x,-v^\varepsilon_{xx})=0, \\ \qquad\qquad (t,x)\in[0,T)\times\mathbb{R}^n, \\ v^\varepsilon\Big|_{t=T}=h(x),\qquad x\in\mathbb{R}^n. \end{cases}$$

This is a terminal value problem of a *nondegenerate* second-order parabolic partial differential equation. According to some standard theory of PDEs (see Krylov [2,3]), when (S1)′–(S2)′ holds, (4.13) admits a unique classical solution. Thus, the above result roughly says that the classical solution V^ε of (4.13) approaches V, which is hopefully a "solution" (in a certain sense) of the *degenerate* parabolic equation (3.24). This will be made rigorous a little later.

4.2. Semiconcavity

In this subsection we present another important property of the value function. Let us start with the following definition.

Definition 4.3. A function $\varphi:\mathbb{R}^n\to\mathbb{R}$ is said to be *semiconcave* if there exists a constant $K\ge 0$ such that $\psi(x)\stackrel{\Delta}{=}\varphi(x)-K|x|^2$ is *concave*, i.e.,

$$\text{(4.14)}\quad \psi(\lambda x+(1-\lambda)y)\ge\lambda\psi(x)+(1-\lambda)\psi(y),\quad \forall x,y\in\mathbb{R}^n,\ \lambda\in[0,1].$$

A function $\varphi : \mathbb{R}^n \to \mathbb{R}$ is said to be *semiconvex* if $-\varphi$ is semiconcave. Finally, A family of functions $\varphi^\alpha(x) : \mathbb{R}^n \to \mathbb{R}$ is said to be semiconcave in $x \in \mathbb{R}^n$ uniformly in α if there exists a constant $K > 0$, independent of α, such that $\psi^\alpha(x) \triangleq \varphi^\alpha(x) - K|x|^2$ is concave.

Note that if a function $\varphi : \mathbb{R}^n \to \mathbb{R}$ is C^2 with φ_{xx} bounded from above (resp. below), then it is semiconcave (resp. semiconvex). These are the cases, in particular, when φ_{xx} is bounded.

Proposition 4.4. *A function $\varphi : \mathbb{R}^n \to \mathbb{R}$ is semiconcave if and only if for some constant $K \geq 0$,*

$$\begin{aligned} &\lambda\varphi(x) + (1-\lambda)\varphi(y) - \varphi(\lambda x + (1-\lambda)y) \\ &\qquad \leq K\lambda(1-\lambda)|x-y|^2, \quad \forall x, y \in \mathbb{R}^n, \ \lambda \in [0,1]. \end{aligned} \tag{4.15}$$

Proof. Note that φ is semiconcave in the sense of Definition 4.3 if and only if there exists a constant $K \geq 0$ such that for any $\lambda \in [0,1]$ and $x, y \in \mathbb{R}^n$,

$$\begin{aligned} &\varphi(\lambda x + (1-\lambda)y) - K|\lambda x + (1-\lambda)y|^2 \\ &\qquad \geq \lambda\varphi(x) - \lambda K|x|^2 + (1-\lambda)\varphi(y) - (1-\lambda)K|y|^2. \end{aligned} \tag{4.16}$$

Thus, the fact that

$$\lambda|x|^2 + (1-\lambda)|y|^2 - |\lambda x + (1-\lambda)y|^2 = \lambda(1-\lambda)|x-y|^2 \tag{4.17}$$

gives the desired result. □

Let us introduce an additional assumption.

(S3)′ $h(x)$ is semiconcave, $f(t,x,u)$ is semiconcave in x uniformly in $(t,u) \in [0,T] \times U$, and $b(t,x,u)$ and $\sigma(t,x,u)$ are differentiable in x with

$$\begin{cases} |b_x(t,x,u) - b_x(t,\widehat{x},u)| \leq L|x-\widehat{x}|, \\ |\sigma_x(t,x,u) - \sigma_x(t,\widehat{x},u)| \leq L|x-\widehat{x}|, \\ \qquad\qquad \forall t \in [0,T], \ x, \widehat{x} \in \mathbb{R}^n, \ u \in U. \end{cases} \tag{4.18}$$

Proposition 4.5. *Let (S1)′–(S3)′ hold. Then the value function $V(s,y)$ of Problem (S) is semiconcave in $y \in \mathbb{R}^n$ uniformly in $s \in [0,T]$.*

Proof. As our convention, K is a generic constant that may differ in different places. Let $y_0, y_1 \in \mathbb{R}^n$, and $\lambda \in [0,1]$. Define $y_\lambda \triangleq \lambda y_1 + (1-\lambda)y_0$. For any $\varepsilon > 0$, there exists a $u_\varepsilon(\cdot) \in \mathcal{U}^w[s,T]$ such that

$$J(s, y_\lambda; u_\varepsilon(\cdot)) < V(s, y_\lambda) + \varepsilon. \tag{4.19}$$

We define

$$\begin{cases} x_\lambda(\cdot) \triangleq x(\cdot\,; s, y_\lambda, u_\varepsilon(\cdot)), \quad \lambda \in [0,1], \\ x^\lambda(\cdot) \triangleq \lambda x_1(\cdot) + (1-\lambda)x_0(\cdot), \quad \lambda \in [0,1]. \end{cases} \tag{4.20}$$

Note that since our state equation (3.4) is not necessarily linear in $x(\cdot)$, the above-defined $x_\lambda(\cdot)$ and $x^\lambda(\cdot)$ are different in general. Now, by the semiconcavity of $f(t,x,u)$ and $h(x)$ in x, together with (4.19), we have the following:

$$\begin{aligned}
&\lambda V(s,y_1)+(1-\lambda)V(s,y_0)-V(s,y_\lambda)-\varepsilon\\
&\quad\le \lambda J(s,y_1;u_\varepsilon(\cdot))+(1-\lambda)J(s,y_0;u_\varepsilon(\cdot))-J(s,y_\lambda;u_\varepsilon(\cdot))\\
&\quad= E\Big\{\int_s^T[\lambda f(t,x_1(t),u_\varepsilon(t))+(1-\lambda)f(t,x_0(t),u_\varepsilon(t))\\
&\qquad\qquad -f(t,x_\lambda(t),u_\varepsilon(t))]dt\\
&\qquad\qquad +\lambda h(x_1(T))+(1-\lambda)h(x_0(T))-h(x_\lambda(T))\Big\}\\
&\quad\le K\lambda(1-\lambda)E\Big\{\int_s^T|x_1(t)-x_0(t)|^2dt+|x_1(T)-x_0(T)|^2\Big\}\qquad(4.21)\\
&\qquad +E\Big\{\int_s^T|f(t,x^\lambda(t),u_\varepsilon(t))-f(t,x_\lambda(t),u_\varepsilon(t))|dt\\
&\qquad\qquad +|h(x^\lambda(T))-h(x_\lambda(T))|\Big\}\\
&\quad\le K\lambda(1-\lambda)|y_1-y_0|^2\\
&\qquad +LE\Big\{\int_s^T|x^\lambda(t)-x_\lambda(t)|dt+|x^\lambda(T)-x_\lambda(T)|\Big\}.
\end{aligned}$$

To estimate the last two terms on the right-hand side of (4.21), let us use (4.18). It follows that (defining $x^\lambda \triangleq \lambda x_1+(1-\lambda)x_0$)

$$\begin{aligned}
&\big|\lambda b(t,x_1,u)+(1-\lambda)b(t,x_0,u)-b(t,x^\lambda,u)\big|\\
&\quad=\Big|\lambda\int_0^1 b_x(t,x^\lambda+\theta(1-\lambda)(x_1-x_0),u)d\theta(1-\lambda)(x_1-x_0)\\
&\qquad+(1-\lambda)\int_0^1 b_x(t,x^\lambda+\theta\lambda(x_0-x_1),u)d\theta\lambda(x_0-x_1)\Big|\qquad(4.22)\\
&\quad=\lambda(1-\lambda)|x_1-x_0|\Big|\int_0^1 b_x(t,x^\lambda+\theta(1-\lambda)(x_1-x_0),u)\\
&\qquad\qquad -b_x(t,x^\lambda+\theta\lambda(x_0-x_1),u)d\theta\Big|\\
&\quad\le K\lambda(1-\lambda)|x_1-x_0|^2.
\end{aligned}$$

Similarly, we have

$$\begin{aligned}
&\big|\lambda\sigma(t,x_1,u)+(1-\lambda)\sigma(t,x_0,u)-\sigma(t,x^\lambda,u)\big|\\
&\qquad\le K\lambda(1-\lambda)|x_1-x_0|^2.
\end{aligned}\tag{4.23}$$

Thus, for any $t\in[s,T]$, by the state equation (3.4) and the Burkholder–

Davis–Gundy inequality,

$$
\begin{aligned}
&E \sup_{r\in[s,t]} \left|x^{\lambda}(r) - x_{\lambda}(r)\right|^2 \\
&\le KE \sup_{r\in[s,t]} \Big| \int_s^r \big[\lambda b(\rho, x_1(\rho), u_\varepsilon(\rho)) + (1-\lambda) b(\rho, x_0(\rho), u_\varepsilon(\rho)) \\
&\qquad\qquad - b(\rho, x^{\lambda}(\rho), u_\varepsilon(\rho))\big] d\rho \Big|^2 \\
&\quad + KE \sup_{r\in[s,t]} \Big| \int_s^r \lambda\sigma(\rho, x_1(\rho), u_\varepsilon(\rho)) + (1-\lambda)\sigma(\rho, x_0(\rho), u_\varepsilon(\rho)) \\
&\qquad\qquad - \sigma(\rho, x^{\lambda}(\rho), u_\varepsilon(\rho))\big] dW(\rho) \Big|^2 \\
&\quad + KE \sup_{r\in[s,t]} \Big| \int_s^r \big[b(\rho, x^{\lambda}(\rho), u_\varepsilon(\rho)) - b(\rho, x_{\lambda}(\rho), u_\varepsilon(\rho))\big] d\rho \Big|^2 \qquad (4.24)\\
&\quad + KE \sup_{r\in[s,t]} \Big| \int_s^r \big[\sigma(\rho, x^{\lambda}(\rho), u_\varepsilon(\rho)) - \sigma(\rho, x_{\lambda}(\rho), u_\varepsilon(\rho))\big] dW(\rho) \Big|^2 \\
&\le K\lambda^2(1-\lambda)^2 E \int_s^t |x_1(\rho) - x_0(\rho)|^4 d\rho \\
&\quad + K\lambda^2(1-\lambda)^2 E\Big\{ \int_s^t \left|x_1(\rho) - x_0(\rho)\right|^4 d\rho \Big\} \\
&\quad + KE \int_s^t |x^{\lambda}(\rho) - x_{\lambda}(\rho)|^2 d\rho + KE\Big\{ \int_s^t |x^{\lambda}(\rho) - x_{\lambda}(\rho)|^2 d\rho \Big\} \\
&\le K\lambda^2(1-\lambda)^2 |y_1 - y_0|^4 + K \int_s^t E|x^{\lambda}(\rho) - x_{\lambda}(\rho)|^2 d\rho.
\end{aligned}
$$

Then, by Gronwall's inequality, we have

$$
E \sup_{r\in[s,T]} |x^{\lambda}(r) - x_{\lambda}(r)| \le K\lambda(1-\lambda)|y_1 - y_0|^2. \tag{4.25}
$$

Combining (4.21) and (4.25), we obtain the semiconcavity of the value function $V(s, y)$. □

With the same proof, we also have the following result for the deterministic problem.

Proposition 4.6. *Let (D1), (D2)′, and (S3)′ hold. Then the value function $V(s, y)$ of Problem (D) is semiconcave in $y \in \mathbb{R}^n$ uniformly in $s \in [0, T]$.*

5. Viscosity Solutions

As in the deterministic case, the value function V of Problem (S) is not necessarily smooth. Thus, one also introduces the notion of viscosity solutions and tries to characterize the value function as the unique viscosity solution of the corresponding HJB equation (3.24).

5.1. Definitions

Let us introduce the following definition.

Definition 5.1. A function $v \in C([0,T] \times \mathbb{R}^n)$ is called a *viscosity subsolution* of (3.24) if

$$v(T,x) \leq h(x), \qquad \forall x \in \mathbb{R}^n, \tag{5.1}$$

and for any $\varphi \in C^{1,2}([0,T] \times \mathbb{R}^n)$, whenever $v - \varphi$ attains a local maximum at $(t,x) \in [0,T) \times \mathbb{R}^n$, we have

$$-\varphi_t(t,x) + \sup_{u \in U} G(t,x,u,-\varphi_x(t,x),-\varphi_{xx}(t,x)) \leq 0. \tag{5.2}$$

A function $v \in C([0,T] \times \mathbb{R}^n)$ is called a *viscosity supersolution* of (3.24) if in (5.1)–(5.2) the inequalities "$\leq$" are changed to "$\geq$" and "local maximum" is changed to "local minimum." Further, if $v \in C([0,T] \times \mathbb{R}^n)$ is both a viscosity subsolution and viscosity supersolution of (3.24), then it is called a *viscosity solution* of (3.24).

We note that as in the deterministic case, one may replace "local maximum (minimum)" in the above definition by "global and/or strict maximum (minimum)." Thus, hereafter, when it is necessary, we will use global and/or strict maximum (minimum) in the definition of viscosity solutions. Our first result is the following.

Theorem 5.2. *Let (S1)′–(S2)′ hold. Then the value function V is a viscosity solution of (3.24).*

Proof. For any $\varphi \in C^{1,2}([0,T] \times \mathbb{R}^n)$, let $V - \varphi$ attain a local maximum at $(s,y) \in [0,T) \times \mathbb{R}^n$. Fix a $u \in U$. Let $x(\cdot) = x(\cdot\,; s,y,u)$ be the state trajectory with the control $u(t) \equiv u$. Then by Theorem 3.3 and Itô's formula, we have (for $\hat{s} > s$ with $\hat{s} - s > 0$ small enough)

$$\begin{aligned} 0 &\leq \frac{E\{V(s,y) - \varphi(s,y) - V(\hat{s},x(\hat{s})) + \varphi(\hat{s},x(\hat{s}))\}}{\hat{s}-s} \\ &\leq \frac{1}{\hat{s}-s} E\Big\{ \int_s^{\hat{s}} f(t,x(t),u)dt - \varphi(s,y) + \varphi(\hat{s},x(\hat{s}))\Big\} \\ &\to \varphi_t(s,y) - G(s,y,u,-\varphi_x(s,y),-\varphi_{xx}(s,y)). \end{aligned} \tag{5.3}$$

This leads to

$$-\varphi_t(s,y) + G(s,y,u,-\varphi_x(s,y),-\varphi_{xx}(s,y)) \leq 0, \qquad \forall u \in U.$$

Hence,

$$-\varphi_t(s,y) + \sup_{u \in U} G(s,y,u,-\varphi_x(s,y),-\varphi_{xx}(s,y)) \leq 0. \tag{5.4}$$

On the other hand, if $V - \varphi$ attains a local minimum at $(s,y) \in [0,T) \times \mathbb{R}^n$, then, for any $\varepsilon > 0$ and $\hat{s} > s$ (with $\hat{s} - s > 0$ small enough), we can find a $u(\cdot) = u_{\varepsilon,\hat{s}}(\cdot) \in \mathcal{U}^w[s,T]$ such that (noting (3.18))

$$\begin{aligned} 0 &\geq E\{V(s,y) - \varphi(s,y) - V(\hat{s},x(\hat{s})) + \varphi(\hat{s},x(\hat{s}))\} \\ &\geq -\varepsilon(\hat{s}-s) + E\Big\{\int_s^{\hat{s}} f(t,x(t),u(t))dt + \varphi(\hat{s},x(\hat{s})) - \varphi(s,y)\Big\}. \end{aligned} \tag{5.5}$$

Dividing by $(\hat{s}-s)$ and applying Itô's formula to the process $\varphi(t,x(t))$, we get

$$
\begin{aligned}
-\varepsilon &\leq \frac{1}{\hat{s}-s}E\int_s^{\hat{s}}\Big\{-\varphi_t(t,x(t)) \\
&\qquad\qquad + G(t,x(t),u,-\varphi_x(t,x(t)),-\varphi_{xx}(t,x(t)))\Big\}dt \\
&\leq \frac{1}{\hat{s}-s}E\int_s^{\hat{s}}\Big\{-\varphi_t(t,x(t)) \\
&\qquad\qquad + \sup_{u\in U} G(t,x(t),u,-\varphi_x(t,x(t)),-\varphi_{xx}(t,x(t)))\Big\}dt \\
&\to -\varphi_t(s,y)+\sup_{u\in U} G(s,y,u,-\varphi_x(s,y),-\varphi_{xx}(s,y)).
\end{aligned}
\tag{5.6}
$$

Combining (5.4) and (5.6), we conclude that V is a viscosity solution of the HJB equation (3.24). $\square$

In Section 2, we have introduced the notion of first-order super- and subdifferentials (see (2.67)) and obtained some equivalent definitions of the viscosity solution (in terms of these differentials) for the deterministic case. In the stochastic case, we need to introduce higher-order super- and subdifferentials. For $v \in C([0,T]\times\mathbb{R}^n)$ and $(t,x)\in[0,T)\times\mathbb{R}^n$, the *second-order parabolic superdifferential* of v at (t,x) is defined as follows:

$$
\begin{aligned}
D_{t,x}^{1,2,+}v(t,x) \triangleq \Big\{&(q,p,P)\in\mathbb{R}\times\mathbb{R}^n\times\mathcal{S}^n \;\Big|\\
&\varlimsup_{\substack{s\to t,\, s\in[0,T)\\ y\to x}} \frac{1}{|s-t|+|y-x|^2}\Big[v(s,y)-v(t,x)\\
&-q(s-t)-\langle p,y-x\rangle-\frac{1}{2}(y-x)^\top P(y-x)\Big]\leq 0\Big\}.
\end{aligned}
\tag{5.7}
$$

Similarly, the *second-order parabolic subdifferential* of v at (t,x) is defined as follows:

$$
\begin{aligned}
D_{t,x}^{1,2,-}v(t,x) \triangleq \Big\{&(q,p,P)\in\mathbb{R}\times\mathbb{R}^n\times\mathcal{S}^n \;\Big|\\
&\varliminf_{\substack{s\to t,\, s\in[0,T)\\ y\to x}} \frac{1}{|s-t|+|y-x|^2}\Big[v(s,y)-v(t,x)\\
&-q(s-t)-\langle p,y-x\rangle-\frac{1}{2}(y-x)^\top P(y-x)\Big]\geq 0\Big\}.
\end{aligned}
\tag{5.8}
$$

The *second-order right parabolic super-/subdifferential* $D_{t+,x}^{1,2,+}v(t,x)$ and $D_{t+,x}^{1,2,-}v(t,x)$ are defined by restricting $s\downarrow t$ in (5.7) and (5.8), respectively. As in Proposition 2.6, $D_{t,x}^{1,2,\pm}v(t,x)$ and $D_{t+,x}^{1,2,\pm}v(t,x)$ are convex. But we do not have the closedness of these sets in general (see Example 5.3 below).

Also, as in (2.68), we have

$$
(5.9)\qquad \begin{cases} D_{t,x}^{1,2,+}v(t,x) \subseteq D_{t+,x}^{1,2,+}v(t,x), \\ D_{t,x}^{1,2,-}v(t,x) \subseteq D_{t+,x}^{1,2,-}v(t,x), \\ D_{t,x}^{1,2,+}(-v)(t,x) = -D_{t,x}^{1,2,-}v(t,x), \\ D_{t+,x}^{1,2,+}(-v)(t,x) = -D_{t+,x}^{1,2,-}v(t,x), \end{cases} \qquad \forall (t,x) \in [0,T) \times \mathbb{R}^n.
$$

These inclusions could be strict (see Example 5.3 below). By (5.7) and (5.8), we see that for any $v \in C([0,T] \times \mathbb{R}^n)$ and $(t,x) \in [0,T] \times \mathbb{R}^n$,

$$
(5.10)\qquad \begin{cases} D_{t,x}^{1,2,+}v(t,x) + \{0\} \times \{0\} \times \mathcal{S}_+^n = D_{t,x}^{1,2,+}v(t,x), \\ D_{t,x}^{1,2,-}v(t,x) - \{0\} \times \{0\} \times \mathcal{S}_+^n = D_{t,x}^{1,2,-}v(t,x), \end{cases}
$$

and for $(t,x) \in [0,T) \times \mathbb{R}^n$,

$$
(5.11)\qquad \begin{cases} D_{t+,x}^{1,2,+}v(t,x) + [0,\infty) \times \{0\} \times \mathcal{S}_+^n = D_{t+,x}^{1,2,+}v(t,x), \\ D_{t+,x}^{1,2,-}v(t,x) - [0,\infty) \times \{0\} \times \mathcal{S}_+^n = D_{t+,x}^{1,2,-}v(t,x), \end{cases}
$$

where $\mathcal{S}_+^n \stackrel{\Delta}{=} \{S \in \mathcal{S}^n | S \geq 0\}$, and $A \pm B \stackrel{\Delta}{=} \{a \pm b \mid a \in A, b \in B\}$, for any subsets A and B in the same Euclidean space.

On the other hand, a function $v \in C([0,T] \times \mathbb{R}^n)$ admits v_t, v_x, and v_{xx} at $(t_0, x_0) \in (0,T) \times \mathbb{R}^n$ if and only if (compare with (2.69))

$$
(5.12)\qquad D_{t,x}^{1,2,+}v(t_0,x_0) \bigcap D_{t,x}^{1,2,-}v(t_0,x_0) \neq \phi.
$$

In this case, it is necessary that

$$
(5.13)\qquad \begin{cases} D_{t,x}^{1,2,+}v(t_0,x_0) = \big\{\big(v_t(t_0,x_0), v_x(t_0,x_0)\big)\big\} \times [v_{xx}(t_0,x_0), \infty), \\ D_{t,x}^{1,2,-}v(t_0,x_0) = \big\{\big(v_t(t_0,x_0), v_x(t_0,x_0)\big)\big\} \times (-\infty, v_{xx}(t_0,x_0)], \end{cases}
$$

where $[S, \infty) \stackrel{\Delta}{=} \{\widehat{S} \in \mathcal{S}^n \mid \widehat{S} \geq S\}$, and $(-\infty, S]$ is defined similarly. Thus,

$$
\begin{aligned} D_{t,x}^{1,2,+}v(t_0,x_0) &\cap D_{t,x}^{1,2,-}v(t_0,x_0) \\ &= \big\{\big(v_t(t_0,x_0), v_x(t_0,x_0), v_{xx}(t_0,x_0)\big)\big\}. \end{aligned}
$$

Note, however, that even if v is a smooth function, neither $D_{t+,x}^{1,2,+}v(t,x)$ nor $D_{t+,x}^{1,2,-}v(t,x)$ is a singleton. We can see the difference between (2.70) and (5.13).

It is seen that the variables t and x are treated unequally in the definitions of $D_{t,x}^{1,2,+}v(t,x)$, etc. Due to this, we have used the term "parabolic" in the above definitions (a similar situation occurs in second-order parabolic partial differential equations).

Let us now present a simple example.

Example 5.3. Let

$$
(5.14)\qquad v(t,x) = -t^+ - x^+, \qquad (t,x) \in \mathbb{R}^2.
$$

A direct computation shows that

$$\begin{cases} D_{t,x}^{1,2,+}v(0,0) = [-1,0] \times \big\{[(-1,0)\times \mathbb{R}] \cup [\{-1,0\}\times[0,\infty)]\big\}, \\ D_{t+,x}^{1,2,+}v(0,0) = [-1,\infty) \times \big\{[(-1,0)\times \mathbb{R}] \cup [\{-1,0\}\times[0,\infty)]\big\}. \end{cases}$$

Thus, $D_{t,x}^{1,2,+}v(0,0)$ is not closed, and it is strictly smaller than $D_{t+,x}^{1,2,+}v(0,0)$, even though the function $v(t,x)$ defined by (5.14) is Lipschitz in (t,x).

The following result is comparable to Lemma 2.7.

Lemma 5.4. *Let $v \in C([0,T]\times\mathbb{R}^n)$ and $(t_0,x_0)\in[0,T)\times\mathbb{R}^n$ be given. Then:*

(i) *$(q,p,P) \in D_{t,x}^{1,2,+}v(t_0,x_0)$ if and only if there exists a function $\varphi \in C^{1,2}([0,T]\times\mathbb{R}^n)$ such that $v-\varphi$ attains a strict maximum at (t_0,x_0) and*

$$\begin{aligned} &\big(\varphi(t_0,x_0),\varphi_t(t_0,x_0),\varphi_x(t_0,x_0),\varphi_{xx}(t_0,x_0)\big) \\ &\qquad = (v(t_0,x_0),q,p,P). \end{aligned} \tag{5.15}$$

(ii) *$(q,p,P) \in D_{t+,x}^{1,2,+}v(t_0,x_0)$ if and only if there exists a function $\varphi \in C^{1,2}([0,T]\times\mathbb{R}^n)$ such that*

$$\begin{cases} \big(\varphi(t_0,x_0),\varphi_t(t_0,x_0),\varphi_x(t_0,x_0),\varphi_{xx}(t_0,x_0)\big) \\ \qquad = (v(t_0,x_0),q,p,P), \\ \varphi(t,x) > v(t,x), \quad \forall (t_0,x_0)\neq(t,x)\in[t_0,T]\times\mathbb{R}^n. \end{cases} \tag{5.16}$$

Proof. We prove (ii). The proof of (i) is similar. Suppose $(q,p,P) \in D_{t+,x}^{1,2,+}v(t_0,x_0)$. Let

$$\Phi(t,x) = \begin{cases} \dfrac{1}{t-t_0+|x-x_0|^2}\Big[v(t,x)-v(t_0,x_0) \\ \qquad -q(t-t_0)-\langle p,x-x_0\rangle-\dfrac{1}{2}(x-x_0)^\top P(x-x_0)\Big]\vee 0, \\ \qquad \text{if } (t_0,x_0)\neq(t,x)\in[t_0,T]\times\mathbb{R}^n, \\ 0, \qquad \text{otherwise,} \end{cases}$$

and

$$\varepsilon(r) = \begin{cases} \sup\{\Phi(t,x) : (t,x)\in(t_0,T]\times\mathbb{R}^n, t-t_0+|x-x_0|^2\le r\}, \\ \qquad \text{if } r>0, \\ 0, \qquad \text{if } r\le 0. \end{cases}$$

Then it follows from (5.7) that $\varepsilon : \mathbb{R}\to[0,\infty)$ is a continuous nondecreasing function with $\varepsilon(0)=0$. Further, we have

$$\begin{aligned} &v(t,x)-\Big[v(t_0,x_0)+q(t-t_0)+\langle p,x-x_0\rangle+\frac{1}{2}(x-x_0)^\top P(x-x_0)\Big] \\ &\qquad \le (t-t_0+|x-x_0|^2)\varepsilon(t-t_0+|x-x_0|^2), \\ &\qquad\qquad \forall (t,x)\in[t_0,T]\times\mathbb{R}^n. \end{aligned} \tag{5.17}$$

Define

$$\Psi(\rho) = \frac{2}{\rho}\int_0^{2\rho}\int_0^r \varepsilon(\theta)d\theta\,dr, \quad \rho > 0. \tag{5.18}$$

Then

$$\Psi_\rho(\rho) = -\frac{2}{\rho^2}\int_0^{2\rho}\int_0^r \varepsilon(\theta)d\theta\,dr + \frac{4}{\rho}\int_0^{2\rho}\varepsilon(\theta)d\theta, \tag{5.19}$$

and

$$\Psi_{\rho\rho}(\rho) = \frac{4}{\rho^3}\int_0^{2\rho}\int_0^r \varepsilon(\theta)d\theta\,dr - \frac{8}{\rho^2}\int_0^{2\rho}\varepsilon(\theta)d\theta + \frac{8}{\rho}\varepsilon(2\rho). \tag{5.20}$$

Consequently,

$$|\Psi(\rho)| \le 4\rho\varepsilon(2\rho), \quad |\Psi_\rho(\rho)| \le 12\varepsilon(2\rho), \quad |\Psi_{\rho\rho}(\rho)| \le \frac{32\varepsilon(2\rho)}{\rho}. \tag{5.21}$$

Now we define

$$\psi(t,x) = \begin{cases} \Psi(\rho(t,x)) + \rho(t,x)^2, & \text{if } (t_0,x_0) \ne (t,x) \in [t_0,T]\times \mathbb{R}^n, \\ 0, & \text{otherwise,} \end{cases} \tag{5.22}$$

where $\rho(t,x) = t - t_0 + |x-x_0|^2$. Further, we set

$$\begin{aligned} \varphi(t,x) &= v(t_0,x_0) + q(t-t_0) + \langle p, x-x_0 \rangle \\ &\quad + \frac{1}{2}(x-x_0)^\top P(x-x_0) + \psi(t,x), \ \forall (t,x) \in [0,T]\times\mathbb{R}^n. \end{aligned} \tag{5.23}$$

We claim that $\varphi \in C^{1,2}([0,T]\times\mathbb{R}^n)$, satisfying (5.16). First of all, for any $(t,x)\in[t_0,T]\times\mathbb{R}^n$ with $(t,x)\ne(t_0,x_0)$, we have

$$\begin{aligned} \psi(t,x) &> \frac{2}{\rho(t,x)}\int_{\rho(t,x)}^{2\rho(t,x)}\int_{\rho(t,x)}^r \varepsilon(\theta)d\theta\,dr \\ &\ge \frac{2}{\rho(t,x)}\varepsilon(\rho(t,x))\int_{\rho(t,x)}^{2\rho(t,x)}[r-\rho(t,x)]dr \\ &= \rho(t,x)\varepsilon(\rho(t,x)). \end{aligned} \tag{5.24}$$

Moreover, for any $(t,x)\in(t_0,T]\times\mathbb{R}^n$,

$$\psi_t(t,x) = \Psi_\rho(\rho(t,x)) + 2\rho(t,x), \tag{5.25}$$

$$\psi_x(t,x) = 2\Psi_\rho(\rho(t,x))(x-x_0) + 4\rho(t,x)(x-x_0), \tag{5.26}$$

$$\begin{aligned} \psi_{xx}(t,x) &= 4\Psi_{\rho\rho}(\rho(t,x))(x-x_0)(x-x_0)^\top + 2\Psi_\rho(\rho(t,x))I \\ &\quad + 4\rho(t,x)I + 8(x-x_0)(x-x_0)^\top. \end{aligned} \tag{5.27}$$

Thus, we obtain (noting $|x-x_0|^2 \le \rho(t,x)$)

$$(5.28)\quad \begin{cases} |\psi(t,x)| \le 4\rho(t,x)\varepsilon(2\rho(t,x)) + \rho(t,x)^2, \\ |\psi_t(t,x)| \le 12\varepsilon(2\rho(t,x)) + 2\rho(t,x), \\ |\psi_x(t,x)| \le 24|x-x_0|\varepsilon(2\rho(t,x)) + 4\rho(t,x)|x-x_0|, \\ |\psi_{xx}(t,x)| \le \dfrac{128|x-x_0|^2}{\rho(t,x)}\varepsilon(2\rho(t,x)) + 24\varepsilon(2\rho(t,x)) + 12\rho(t,x) \\ \qquad\qquad \le 152\varepsilon(2\rho(t,x)) + 12\rho(t,x). \end{cases}$$

So, $\psi \in C^{1,2}([0,T]\times \mathbb{R}^n)$ with

$$(5.29)\quad \psi(t_0,x_0)=0, \quad \psi_t(t_0,x_0)=0, \quad \psi_x(t_0,x_0)=0, \quad \psi_{xx}(t_0,x_0)=0.$$

This shows that the function $\varphi(t,x)$ defined by (5.23) satisfies (5.16).

The above proves the "only if" part. The "if" part is clear. □

For the subdifferentials $D^{1,2,-}_{t,x}v(t,x)$ and $D^{1,2,-}_{t+,x}v(t,x)$, we have the following parallel results.

Lemma 5.5. *Let $v \in C([0,T]\times\mathbb{R}^n)$ and $(t_0,x_0)\in[0,T)\times\mathbb{R}^n$.*

(i) *$(q,p,P) \in D^{1,2,-}_{t,x}v(t_0,x_0)$ if and only if there exists a function $\varphi \in C^{1,2}([0,T]\times\mathbb{R}^n)$ such that $v-\varphi$ attains a strict minimum at (t_0,x_0) and*

$$(5.30)\quad \begin{aligned} &\big(\varphi(t_0,x_0),\varphi_t(t_0,x_0),\varphi_x(t_0,x_0),\varphi_{xx}(t_0,x_0)\big) \\ &\qquad = (v(t_0,x_0),q,p,P). \end{aligned}$$

(ii) *$(q,p,P) \in D^{1,2,-}_{t+,x}v(t_0,x_0)$ if and only if there exists a function $\varphi \in C^{1,2}([0,T]\times\mathbb{R}^n)$ such that*

$$(5.31)\quad \begin{cases} \big(\varphi(t_0,x_0),\varphi_t(t_0,x_0),\varphi_x(t_0,x_0),\varphi_{xx}(t_0,x_0)\big) \\ \qquad = (v(t_0,x_0),q,p,P), \\ \varphi(t,x) < v(t,x), \quad \forall (t_0,x_0)\ne(t,x)\in[t_0,T]\times\mathbb{R}^n. \end{cases}$$

As in the case of the first-order super- and subdifferentials, Lemmas 5.4 and 5.5 remain true if we omit the word "strict" and replace the inequalities "$>$" and "$<$" by "$\ge$" and "$\le$", respectively. The geometric interpretation of Lemmas 5.4 and 5.5 is similar to that of Lemmas 2.7 and 2.8 described in Section 2.

The following result is parallel to Theorem 2.9.

Proposition 5.6. *A function $v\in C([0,T]\times\mathbb{R}^n)$ is a viscosity subsolution of (3.24) if and only if $v(T,x)\le h(x)$ and*

$$(5.32)\quad -q + \sup_{u\in U} G(t,x,u,-p,-P) \le 0, \quad \forall (q,p,P)\in D^{1,2,+}_{t,x}v(t,x).$$

Respectively, a function $v \in C([0,T] \times \mathbb{R}^n)$ is a viscosity supersolution of (3.24) if and only if $v(T,x) \geq h(x)$ and

$$-q + \sup_{u \in U} G(t,x,u,-p,-P) \geq 0, \quad \forall (q,p,P) \in D^{1,2,-}_{t,x} v(t,x). \tag{5.33}$$

Proof. The result is immediate in view of Lemmas 5.4 and 5.5. □

Corollary 5.7. *Let (S1)′–(S2)′ hold. Let $v \in C([0,T] \times \mathbb{R}^n)$ be a viscosity solution of (3.24). Then there exists a constant $K > 0$ such that*

$$\begin{cases} q \geq -K\{1 + |x| + |p|(1+|x|) + |P|(1+|x|^2)\}, \\ \qquad\qquad \forall (q,p,P) \in D^{1,2,+}_{t,x} v(t,x), \\ q \leq K\{1 + |x| + |p|(1+|x|) + |P|(1+|x|^2)\}, \\ \qquad\qquad \forall (q,p,P) \in D^{1,2,-}_{t,x} v(t,x). \end{cases} \tag{5.34}$$

Proof. The result follows from Proposition 5.6 and (S1)′–(S2)′ immediately. □

The above result gives an estimate of q in terms of x, p, P for any $(q,p,P) \in D^{1,2,\pm}_{t,x} v(t,x)$ when v is a viscosity solution of (3.24).

5.2. Some properties

In this subsection we study some basic properties of viscosity solutions defined in the previous subsection.

Proposition 5.8. *Let $v \in C^{1,2}([0,T] \times \mathbb{R}^n)$. Then v is a viscosity solution of (3.24) if and only if it is a classical solution of (3.24).*

Proof. The result is clear from Proposition 5.6, (5.13), and the fact that the generalized Hamiltonian G, defined by (3.25), is nondecreasing in its last argument $P \in S^n$. □

Due to this result, a viscosity solution qualifies as a generalized solution of the HJB equation.

Proposition 5.9. *Let v^ε be a viscosity solution of the following:*

$$\begin{cases} -v^\varepsilon_t + G^\varepsilon(t,x,-v^\varepsilon_x,-v^\varepsilon_{xx}) = 0, & (t,x) \in [0,T] \times \mathbb{R}^n, \\ v^\varepsilon \big|_{t=T} = h^\varepsilon(x), \end{cases} \tag{5.35}$$

where $G^\varepsilon : [0,T] \times \mathbb{R}^n \times \mathbb{R}^n \times S^n \to \mathbb{R}$ and $h^\varepsilon : \mathbb{R}^n \to \mathbb{R}$ are continuous and

$$\begin{cases} \lim\limits_{\varepsilon \to 0} G^\varepsilon(t,x,p,P) = G^0(t,x,p,P), \\ \lim\limits_{\varepsilon \to 0} h^\varepsilon(x) = h^0(x), \\ \lim\limits_{\varepsilon \to 0} v^\varepsilon(t,x) = v^0(t,x), \end{cases} \tag{5.36}$$

uniformly in (t,x,p,P) over any compact set. Then v^0 is a viscosity solution of

$$(5.37)\qquad \begin{cases} -v_t^0 + G^0(t,x,-v_x^0,-v_{xx}^0) = 0, & (t,x)\in[0,T]\times\mathbb{R}^n,\\ v^0\big|_{t=T} = h^0(x). \end{cases}$$

Proof. First of all, $v^0 \in C([0,T]\times\mathbb{R}^n)$. Next, let $\varphi\in C^{1,2}([0,T]\times\mathbb{R}^n)$ such that $v^0-\varphi$ attains a strict local maximum at $(t_0,x_0)\in[0,T)\times\mathbb{R}^n$. Then, for $\varepsilon>0$ small enough, there exists a point $(t_\varepsilon,x_\varepsilon)\in[0,T)\times\mathbb{R}^n$ such that $v^\varepsilon-\varphi$ attains a strict local maximum at $(t_\varepsilon,x_\varepsilon)$. Moreover, we have

$$(5.38)\qquad (t_\varepsilon,x_\varepsilon)\to(t_0,x_0),\qquad \text{as } \varepsilon\to 0.$$

Since v^ε is a viscosity solution of (5.35), we have

$$-\varphi_t(t_\varepsilon,x_\varepsilon)+G^\varepsilon(t_\varepsilon,x_\varepsilon,-\varphi_x(t_\varepsilon,x_\varepsilon),-\varphi_{xx}(t_\varepsilon,x_\varepsilon))\le 0.$$

Letting $\varepsilon\to 0$, we obtain

$$(5.39)\qquad -\varphi_t(t_0,x_0)+G(t_0,x_0,-\varphi_x(t_0,x_0),-\varphi_{xx}(t_0,x_0))\le 0.$$

This implies that v^0 is a viscosity subsolution of (5.37). We can similarly prove that v^0 is a viscosity supersolution of (5.37). Hence, our conclusion follows. □

Since σ could be degenerate, the HJB equation (3.24) is a degenerate second-order nonlinear parabolic partial differential equation in general. Such an equation usually does not admit classical solutions. A natural way of *regularizing* that equation is to consider the following:

$$(5.40)\qquad \begin{cases} -v_t^\varepsilon - \varepsilon\Delta v^\varepsilon + \sup\limits_{u\in U} G(t,x,u,-v_x^\varepsilon,-v_{xx}^\varepsilon) = 0,\\ \qquad\qquad\qquad\qquad\qquad (t,x)\in[0,T]\times\mathbb{R}^n,\\ v^\varepsilon\big|_{t=T} = h(x). \end{cases}$$

Note that (5.40) is the same as (4.13), which is a nondegenerate second-order nonlinear parabolic partial differential equation. From PDE theory (see, e.g., Krylov [2,3]), (5.40) admits a unique classical solution v^ε, under some conditions, say, (S1)$'$–(S2)$'$.

Proposition 5.10. *Let v^ε be a classical solution of (5.40) and let v^0 be a viscosity solution of (3.24). Then there exists a constant $K>0$ such that*

$$(5.41)\qquad |v^\varepsilon(t,x)-v^0(t,x)|\le K\sqrt{\varepsilon},\qquad \forall (t,x)\in[0,T]\times\mathbb{R}^n,\ \varepsilon>0.$$

Proof. First of all, by virtue of Theorem 5.2, the value functions V and V^ε of Problems (S) and (S$^\varepsilon$) (see Section 4.1) are viscosity solutions of (3.24) and (5.40), respectively. Next, from Theorem 6.1, which will be proved in Section 6 below, it follows that both (5.40) and (3.24) admit unique viscosity solutions. Thus, it is necessary that $v^\varepsilon = V^\varepsilon$ and $v^0 = V$. Using Corollary 4.2, we obtain (5.41). □

The above gives a possible approximation of the viscosity solution to a degenerate parabolic PDE by solutions of a sequence of nondegenerate parabolic PDEs obtained by adding the term $\varepsilon\Delta v^\varepsilon$ to the original equation. In classical fluid mechanics, the term $\varepsilon\Delta v^\varepsilon$ usually represents the *viscosity*. Thus, the above procedure is called the *vanishing of viscosity*. Since v^0 is the limit obtained in the vanishing process of viscosity, Crandall and Lions coined the name *viscosity solution* for v^0 in the early 1980s.

6. Uniqueness of Viscosity Solutions

Generally speaking, for a notion of solutions to a (differential) equation, the weaker the notion, the easier the existence, but the harder the uniqueness. Viscosity solutions are very weak notion of solutions to HJB equations, and we have the existence of such solutions to the HJB equation (3.24) (see Theorem 5.2). The next natural question is whether the solution is unique. If this is the case, then the viscosity solution to the HJB equation (3.24) characterizes the value function V of Problem (S). This section is devoted to a proof of uniqueness of the viscosity solution to (3.24). Notice that while we had a similar result for the deterministic case or the first-order HJB equations (Theorem 2.5), the stochastic case is much more delicate.

6.1. A uniqueness theorem

We now state the main result of this section.

Theorem 6.1. *Let (S1)′–(S2)′ hold. Then the HJB equation (3.24) admits at most one viscosity solution $v(\cdot\,,\cdot)$ in the class of functions satisfying (3.9)–(3.10).*

Indeed, we have the following stronger result in terms of the *right* super-/subdifferentials (compare with Theorem 2.10).

Theorem 6.2. *Let (S1)′–(S2)′ hold. Then the value function $V(\cdot\,,\cdot) \in C([0,T]\times\mathbb{R}^n)$ of Problem (S) is the only function that satisfies (3.9)–(3.10) and the following: For all $(t,x)\in[0,T)\times\mathbb{R}^n$,*

$$
(6.1)\qquad \begin{cases} -q+\sup\limits_{u\in U} G(t,x,u,-p,-P)\le 0, & \forall (q,p,P)\in D^{1,2,+}_{t+,x}V(t,x),\\ -q+\sup\limits_{u\in U} G(t,x,u,-p,-P)\ge 0, & \forall (q,p,P)\in D^{1,2,-}_{t+,x}V(t,x),\\ V(T,x)=h(x). \end{cases}
$$

Proof. We may prove the conclusion the same way as we proved Theorem 2.10, by using Theorem 5.2, Lemma 5.4-(ii), (5.9), and Theorem 6.1. □

The basic idea of proving Theorem 6.1 is similar to that of Theorem 2.5. However, the techniques involved are much more subtle. We need some preparations. The first result is concerned with a certain kind of approximation of viscosity sub-/supersolutions.

Lemma 6.3. *Let $v \in C([0,T]\times\mathbb{R}^n)$ satisfying (3.9)–(3.10). For any $\gamma > 0$, define*

$$v^\gamma(t,x) \stackrel{\Delta}{=} \sup_{(s,y)\in[0,T]\times\mathbb{R}^n} \Big\{ v(s,y) - \frac{1}{2\gamma^2}\big[|t-s|^2 + |x-y|^2\big]\Big\}, \quad (t,x)\in[0,T]\times\mathbb{R}^n. \tag{6.2}$$

Then $v^\gamma(\cdot\,,\cdot)$ is semiconvex, satisfying

$$|v^\gamma(t,x)| \le K(1+|x|), \quad \forall (t,x)\in[0,T]\times\mathbb{R}^n,\ \gamma>0, \tag{6.3}$$

$$|v^\gamma(t,x) - v^\gamma(s,y)| \le K\big\{|x-y| + (1+|x|\vee|y|)|t-s|^{1/2}\big\}, \quad \forall t,s\in[0,T],\ x,y,\in\mathbb{R}^n,\ \gamma>0, \tag{6.4}$$

for some constant $K > 0$ (possibly different from that in (3.9)–(3.10)). Moreover, for any $(t,x)\in[0,T]\times\mathbb{R}^n$, there exists a $(\hat t,\hat x)\in[0,T]\times\mathbb{R}^n$ such that

$$v^\gamma(t,x) = v(\hat t,\hat x) - \frac{1}{2\gamma^2}\big[|t-\hat t|^2 + |x-\hat x|^2\big], \tag{6.5}$$

and for some absolute constant K (independent of (t,x) and γ)

$$\frac{1}{2\gamma^2}\big[|t-\hat t|^2 + |x-\hat x|^2\big] \le K\big\{(1+|x|)\wedge\big[(1+|x|^{\frac43})\gamma^{\frac23}\big]\big\}. \tag{6.6}$$

Consequently,

$$0 \le v^\gamma(t,x) - v(t,x) \le K\big\{(1+|x|)\wedge\big[(1+|x|^{\frac43})\gamma^{\frac23}\big]\big\}. \tag{6.7}$$

Proof. First of all, it is clear that the function $v^\gamma(t,x) + \frac{1}{2\gamma^2}(|t|^2+|x|^2)$ is convex in (t,x), which implies that $v^\gamma(t,x)$ is semiconvex. Next, by (3.9), for any $(t,x)\in[0,T]\times\mathbb{R}^n$, there exists a $(\hat t,\hat x)\in[0,T]\times\mathbb{R}^n$ such that (6.5) holds. Using (3.10), (6.2), and (6.5), we obtain (6.4). We now estimate (6.6). From (6.2), (6.5), and (3.9), one has

$$\begin{aligned}|x-\hat x|^2 &\le 2\gamma^2\big[v(\hat t,\hat x) - v(t,x)\big] \le 2\gamma^2 K(2+|\hat x|+|x|)\\ &\le 4\gamma^2K(1+|x|) + 2\gamma^2K|x-\hat x|,\end{aligned} \tag{6.8}$$

which yields

$$|\hat x| \le K(1+|x|), \quad \text{for any } (\hat t,\hat x) \text{ satisfying (6.5)}. \tag{6.9}$$

Consequently, by (6.5) and (3.10), we have

$$\begin{aligned}\frac{1}{2\gamma^2}\big[|t-\hat t|^2 + |x-\hat x|^2\big] &= v(\hat t,\hat x) - v(t,x)\\ &\le K\big[|x-\hat x| + (1+|x|)|t-\hat t|^{\frac12}\big]\\ &\le K(1+|x|)\big[|x-\hat x|^2 + |t-\hat t|^2\big]^{\frac14}.\end{aligned} \tag{6.10}$$

Then (6.6) follows. Finally, (6.7) follows from (6.2) and (6.5)–(6.6), and (6.3) follows from (3.9) and (6.7). □

Due to the above result, we call $v^\gamma(\cdot\,,\cdot)$ the *semiconvex approximation* of $v(\cdot\,,\cdot)$. Likewise, we can define the *semiconcave approximation* as follows:

$$(6.11)\qquad \begin{aligned} v_\gamma(t,x) \stackrel{\Delta}{=} \inf_{(s,y)\in[0,T]\times\mathbb{R}^n} \Big\{ v(s,y) + \frac{1}{2\gamma^2}\big[|t-s|^2 + |x-y|^2\big]\Big\},& \\ (t,x)\in[0,T]\times\mathbb{R}^n.& \end{aligned}$$

Similar to Lemma 6.3, we have

Lemma 6.4. *Let $v \in C([0,T]\times\mathbb{R}^n)$ satisfying (3.9)–(3.10). For any $\gamma > 0$, define $v_\gamma(\cdot\,,\cdot)$ by (6.11). Then $v_\gamma(\cdot\,,\cdot)$ is semiconcave, satisfying*

$$(6.12)\qquad |v_\gamma(t,x)| \le K(1+|x|), \quad \forall (t,x)\in[0,T]\times\mathbb{R}^n,\ \gamma>0,$$

$$(6.13)\qquad \begin{aligned} |v_\gamma(t,x) - v_\gamma(s,y)| \le K\big\{|x-y| + (1+|x|\vee|y|)|t-s|^{1/2}\big\},& \\ \forall t,s\in[0,T],\ x,y\in\mathbb{R}^n,\ \gamma>0,& \end{aligned}$$

for some constant $K > 0$ (possibly different from that in (3.9)–(3.10)). Moreover, for any $(t,x)\in[0,T]\times\mathbb{R}^n$, there exists a $(\hat t,\hat x)\in[0,T]\times\mathbb{R}^n$ such that

$$(6.14)\qquad v_\gamma(t,x) = v(\hat t,\hat x) + \frac{1}{2\gamma^2}\big[|t-\hat t|^2 + |x-\hat x|^2\big],$$

and (6.6) holds for some absolute constant K (independent of (t,x) and γ). Consequently,

$$(6.15)\qquad 0 \le v(t,x) - v_\gamma(t,x) \le K\big\{(1+|x|)\wedge\big[(1+|x|^{\frac43})\gamma^{\frac23}\big]\big\}.$$

Combining (6.7) and (6.15), one has

$$(6.16)\qquad 0 \le v^\gamma(t,x) - v_\gamma(t,x) \le 2K\big\{(1+|x|)\wedge\big[(1+|x|^{\frac43})\gamma^{\frac23}\big]\big\}.$$

Next, for any $(t,x,p,P)\in[0,T]\times\mathbb{R}^n\times\mathbb{R}^n\times\mathcal{S}^n$, we define

$$(6.17)\qquad \left\{\begin{aligned} G^\gamma(t,x,p,P) &\stackrel{\Delta}{=} \inf_{(s,y)\in[0,T]\times\mathbb{R}^n} \Big\{ \sup_{u\in U} G(s,y,u,p,P) \,\Big|\\ &\qquad \frac{1}{2\gamma^2}\big[|t-s|^2+|x-y|^2\big] \le K\big\{(1+|x|)\wedge\big[(1+|x|^{\frac43})\gamma^{\frac23}\big]\big\}\Big\},\\ G_\gamma(t,x,p,P) &\stackrel{\Delta}{=} \sup_{(s,y)\in[0,T]\times\mathbb{R}^n} \Big\{ \sup_{u\in U} G(s,y,u,p,P) \,\Big|\\ &\qquad \frac{1}{2\gamma^2}\big[|t-s|^2+|x-y|^2\big] \le K\big\{(1+|x|)\wedge\big[(1+|x|^{\frac43})\gamma^{\frac23}\big]\big\}\Big\}, \end{aligned}\right.$$

where the generalized Hamiltonian G is defined by (3.25) and $K > 0$ is the constant appearing in (6.6). It is clear that under (S1)′–(S2)′, one has

$$(6.18)\qquad \lim_{\gamma\to0} G^\gamma(t,x,p,P) = \lim_{\gamma\to0} G_\gamma(t,x,p,P) = \sup_{u\in U} G(t,x,u,p,P),$$

uniformly for (t, x, p, P) in compact sets. We now present a result that makes the semiconvex and semiconcave approximations very useful.

Lemma 6.5. *Let (S1)′–(S2)′ hold and $v(\cdot\,,\cdot) \in C([0,T]\times\mathbb{R}^n)$ be a viscosity subsolution of (3.24). Then, for each $\gamma > 0$, $v^\gamma(\cdot\,,\cdot)$ is a viscosity subsolution of the following:*

$$\begin{cases} -v_t + G^\gamma(t, x, -v_x, -v_{xx}) = 0, & (t,x) \in [0,T) \times \mathbb{R}^n, \\ v\big|_{t=T} = v^\gamma(T, x), & x \in \mathbb{R}^n. \end{cases} \tag{6.19}$$

Likewise, if $v(\cdot\,,\cdot) \in C([0,T] \times \mathbb{R}^n)$ is a viscosity supersolution of (3.24), then for each $\gamma > 0$, $v_\gamma(\cdot\,,\cdot)$ is a viscosity supersolution of the following:

$$\begin{cases} -v_t + G_\gamma(t, x, -v_x, -v_{xx}) = 0, & (t,x) \in [0,T) \times \mathbb{R}^n, \\ v\big|_{t=T} = v_\gamma(T, x), & x \in \mathbb{R}^n. \end{cases} \tag{6.20}$$

Proof. Let $\varphi \in C^{1,2}([0,T] \times \mathbb{R}^n)$ such that $v^\gamma - \varphi$ attains a maximum at (t, x). Suppose $(\hat{t}, \hat{x}) \in [0,T] \times \mathbb{R}^n$ satisfies (6.5). Then, for any $(s, y) \in [0,T] \times \mathbb{R}^n$, one has (noting the definition of v^γ)

$$\begin{aligned} v(\hat{t}, \hat{x}) - \varphi(t,x) &= v^\gamma(t,x) + \frac{1}{2\gamma^2}\big(|t-\hat{t}|^2 + |x-\hat{x}|^2\big) - \varphi(t,x) \\ &\geq v^\gamma(s,y) - \varphi(s,y) + \frac{1}{2\gamma^2}\big(|t-\hat{t}|^2 + |x-\hat{x}|^2\big) \\ &\geq v(s - t + \hat{t}, y - x + \hat{x}) - \varphi(s,y). \end{aligned} \tag{6.21}$$

Consequently, for any $(\tau, \zeta) \in [0,T] \times \mathbb{R}^n$, by taking $s = \tau + t - \hat{t}$ and $y = \zeta + x - \hat{x}$, we get

$$v(\hat{t}, \hat{x}) - \varphi(t,x) \geq v(\tau,\zeta) - \varphi(\tau + t - \hat{t}, \zeta + x - \hat{x}), \tag{6.22}$$

which means that the function $(\tau, \zeta) \mapsto v(\tau, \zeta) - \varphi(\tau + t - \hat{t}, \zeta + x - \hat{x})$ attains a maximum at $(\tau, \zeta) = (\hat{t}, \hat{x})$. Thus, by the definition of viscosity subsolution and (6.17), we obtain

$$\begin{aligned} &-\varphi_t(t,x) + G^\gamma(t, x, \varphi_x(t,x), -\varphi_{xx}(t,x)) \\ &\quad \leq -\varphi_t(t,x) + \sup_{u\in U} G(\hat{t}, \hat{x}, u, -\varphi_x(t,x), -\varphi_{xx}(t,x)) \leq 0. \end{aligned} \tag{6.23}$$

This proves that $v^\gamma(\cdot\,,\cdot)$ is a viscosity subsolution of (6.19). In a similar manner, we can prove that $v_\gamma(\cdot\,,\cdot)$ is a viscosity supersolution of (6.20). □

Note that we only have

$$v^\gamma(T,x) \geq h(x) \geq v_\gamma(T,x), \qquad x \in \mathbb{R}^n.$$

Thus, in (6.19) and (6.20), $v^\gamma(T,x)$ and $v_\gamma(T,x)$ are used, respectively, instead of $h(x)$ as the terminal condition. However, we will see that this will not matter due to (6.16).

In proving Theorem 6.1, the following two technical results play crucial roles. We will provide their proofs in the next subsection for the reader's convenience.

Lemma 6.6. (Alexandrov's theorem) *Let $Q \subseteq \mathbb{R}^n$ be a convex body and $\varphi : Q \to \mathbb{R}$ a semiconvex (or semiconcave) function. Then there exists a set $N \subseteq Q$ with $|N| = 0$ such that at any $x \in Q \setminus N$, φ is twice differentiable, i.e., there are $(p, P) \in \mathbb{R}^n \times \mathcal{S}^n$ such that*

$$\varphi(x+y) = \varphi(x) + \langle p, y \rangle + \frac{1}{2} \langle Py, y \rangle + o(|y|^2), \quad \text{for all } |y| \text{ small enough.} \tag{6.24}$$

Lemma 6.7. (Jensen's lemma) *Let $Q \subseteq \mathbb{R}^n$ be a convex body, $\varphi : Q \to \mathbb{R}$ a semiconvex function, and $\bar{x} \in \operatorname{Int} Q$ a local strict maximum of φ. Then, for any small $r, \delta > 0$, the set*

$$\mathcal{K} \stackrel{\Delta}{=} \{x \in B_r(\bar{x}) \mid \exists |p| \le \delta, \text{ such that } \varphi_p(\cdot) \stackrel{\Delta}{=} \varphi(\cdot) + \langle p, \cdot \rangle \text{ attains a strict local maximum at } x\} \tag{6.25}$$

has a positive Lebesgue measure, where $B_r(\bar{x}) \stackrel{\Delta}{=} \{x \in \mathbb{R}^n \mid |x - \bar{x}| \le r\}$.

Note that in Jensen's lemma the condition that φ attains a local *strict* maximum at $\bar{x}$ is crucial. It excludes the case where $\varphi(x)$ is a constant near $\bar{x}$, in which case $\mathcal{K} = \partial B_r(\bar{x})$, and the lemma does not hold.

Now, we are ready to present a proof of Theorem 6.1.

Proof of Theorem 6.1. Let v and $\widehat{v}$ be two viscosity solutions of (3.24) satisfying (3.9)–(3.10). We claim that

$$v(t,x) \le \widehat{v}(t,x), \qquad \forall (t,x) \in [T-\rho, T] \times \mathbb{R}^n, \tag{6.26}$$

with $\rho = (4L + 4L^2)^{-1}$ (L is the constant in Assumption (S2)'). If this is proved, by the symmetry of v and $\widehat{v}$, the reversed inequality must also hold, and thus v and $\widehat{v}$ are equal on $[T-\rho, T] \times \mathbb{R}^n$. Repeating this argument on $[T-2\rho, T-\rho] \times \mathbb{R}^n$, etc., we can obtain the uniqueness.

Now we prove (6.26) by contradiction. Suppose (6.26) is false. Then there exists a point $(\bar{t}, \bar{x}) \in (T-\rho, T) \times \mathbb{R}^n$ such that

$$2\eta \stackrel{\Delta}{=} v(\bar{t}, \bar{x}) - \widehat{v}(\bar{t}, \bar{x}) > 0. \tag{6.27}$$

Let $v^\gamma(\cdot\,,\cdot)$ and $\widehat{v}_\gamma(\cdot\,,\cdot)$ be the semiconvex and semiconcave approximations of $v(\cdot\,,\cdot)$ and $\widehat{v}(\cdot\,,\cdot)$, respectively (see (6.2) and (6.11)). By (6.7) and (6.15), for all small enough $\gamma > 0$,

$$v^\gamma(\bar{t}, \bar{x}) - \widehat{v}_\gamma(\bar{t}, \bar{x}) \ge \eta > 0. \tag{6.28}$$

Let $\mu = 1 + [8T(L+L^2)]^{-1} > 1$. For any $\alpha, \beta, \varepsilon, \delta, \lambda \in (0,1)$, define

$$
(6.29) \quad \begin{cases} \varphi(t,x,s,y) = \alpha\Big(\frac{2\mu T - t - s}{2\mu T}\Big)\big(|x|^2 + |y|^2\big) - \beta(t+s) \\ \qquad\qquad + \frac{1}{2\varepsilon}|t-s|^2 + \frac{1}{2\delta}|x-y|^2 + \frac{\lambda}{t-T+\rho} + \frac{\lambda}{s-T+\rho}, \\ \Phi(t,x,s,y) = v^\gamma(t,x) - \widehat{v}_\gamma(s,y) - \varphi(t,x,s,y), \\ \qquad\qquad \forall (t,x),(s,y) \in (T-\rho, T] \times \mathbb{R}^n. \end{cases}
$$

By (6.3) and (6.12), we have

$$
(6.30) \quad \begin{cases} \lim\limits_{|x|+|y|\to\infty} \Phi(t,x,s,y) = -\infty, \quad \text{uniformly in } t,s \in (T-\rho,T], \\ \lim\limits_{t\wedge s\downarrow T-\rho} \Phi(t,s,x,y) = -\infty, \quad \text{uniformly in } x,y \in \mathbb{R}^n. \end{cases}
$$

Thus, there exists a $(t_0,x_0,s_0,y_0) \in \{(T-\rho,T]\times\mathbb{R}^n\}^2$ (depending on the parameters $\alpha, \beta, \varepsilon, \delta, \lambda$, and γ) such that

$$
\begin{aligned} \Phi(t_0,x_0,s_0,y_0) &= \max_{\{(T-\rho,T]\times\mathbb{R}^2\}^2} \Phi(t,x,s,y) \ge \Phi(T,0,T,0) \\ &= v^\gamma(T,0) - \widehat{v}_\gamma(T,0) + 2\beta T - \frac{2\lambda}{\rho}. \end{aligned}
$$

This, together with (6.3) and (6.12), yields the following:

$$
\begin{aligned} &\alpha(|x_0|^2 + |y_0|^2) + \frac{1}{2\varepsilon}|t_0 - s_0|^2 + \frac{1}{2\delta}|x_0 - y_0|^2 \\ &+ \frac{\lambda}{t_0 - T + \rho} + \frac{\lambda}{s_0 - T + \rho} \le K(1 + |x_0| + |y_0|), \end{aligned}
$$

for some $K > 0$, independent of $\alpha, \beta, \varepsilon, \delta, \lambda$, and γ. Consequently, there is a constant $K_\alpha > 0$ (independent of $\beta, \varepsilon, \lambda, \delta, \gamma$) such that

$$
(6.31) \quad \begin{cases} |x_0| + |y_0| + \frac{1}{2\varepsilon}|t_0 - s_0|^2 + \frac{1}{2\delta}|x_0 - y_0|^2 \le K_\alpha, \\ t_0, s_0 \in [T - \rho + \frac{\lambda}{K_\alpha}, T]. \end{cases}
$$

Next, from the inequality

$$
2\Phi(t_0,x_0,s_0,y_0) \ge \Phi(t_0,x_0,t_0,x_0) + \Phi(s_0,y_0,s_0,y_0)
$$

along with (6.31), (6.4), and (6.13), it follows that

$$
(6.32) \quad \begin{aligned} &\frac{1}{2\varepsilon}|t_0 - s_0|^2 + \frac{1}{2\delta}|x_0 - y_0|^2 \\ &\le |v^\gamma(t_0,x_0) - v^\gamma(s_0,y_0)| + |\widehat{v}_\gamma(t_0,x_0) - \widehat{v}_\gamma(s_0,y_0)| \\ &\le K\big\{|x_0 - y_0| + (1 + |x_0| \vee |y_0|)|t_0 - s_0|^{\frac{1}{2}}\big\} \to 0, \quad \text{as } \varepsilon, \delta \to 0. \end{aligned}
$$

Note that (t_0,x_0,s_0,y_0) depends on the parameters $\alpha, \beta, \varepsilon, \delta, \lambda$, and γ. We now separate two cases.

Case 1. For some $(\beta, \varepsilon, \delta, \lambda, \gamma) \to 0$, the corresponding t_0, s_0 satisfy

$$t_0 \vee s_0 = T. \tag{6.33}$$

For this case, we observe the following:

$$\begin{aligned} & v^\gamma(\bar{t}, \bar{x}) - \widehat{v}_\gamma(\bar{t}, \bar{x}) - 2\alpha \frac{\mu T - \bar{t}}{\mu T} |\bar{x}|^2 + 2\beta \bar{t} - \frac{2\lambda}{\bar{t} - T + \rho} \\ & = \Phi(\bar{t}, \bar{x}, \bar{t}, \bar{x}) \le \Phi(t_0, x_0, s_0, y_0) \\ & \le v^\gamma(t_0, x_0) - \widehat{v}_\gamma(s_0, y_0) + \beta(t_0 + s_0). \end{aligned} \tag{6.34}$$

Now we send $\varepsilon, \delta \to 0$. By (6.31), some subsequence of (t_0, x_0, s_0, y_0), still denoted by itself, converges. By (6.32)–(6.33), the limit has to be of the form $(T, \bar{x}_0, T, \bar{x}_0)$. Then (6.34) becomes

$$\begin{aligned} & v^\gamma(\bar{t}, \bar{x}) - \widehat{v}_\gamma(\bar{t}, \bar{x}) - 2\alpha \frac{\mu T - \bar{t}}{\mu T} |\bar{x}|^2 + 2\beta \bar{t} - \frac{2\lambda}{\bar{t} - T + \rho} \\ & \le v^\gamma(T, \bar{x}_0) - \widehat{v}_\gamma(T, \bar{x}_0) + 2\beta T. \end{aligned} \tag{6.35}$$

Next, by sending $\gamma \to 0$ and using (6.16), we obtain

$$v(\bar{t}, \bar{x}) - \widehat{v}(\bar{t}, \bar{x}) - 2\alpha \frac{\mu T - \bar{t}}{\mu T} |\bar{x}|^2 + 2\beta \bar{t} - \frac{2\lambda}{\bar{t} - T + \rho} \le 2\beta T. \tag{6.36}$$

Finally, by sending $\alpha, \beta, \lambda \to 0$, we obtain a contradiction to (6.27).

Case 2. For any $\alpha, \beta, \varepsilon, \delta, \lambda, \gamma \in (0, 1)$, the corresponding t_0, s_0 satisfy

$$t_0, s_0 < T. \tag{6.37}$$

For fixed $\alpha, \lambda \in (0, 1)$, define

$$Q \stackrel{\Delta}{=} \Big\{ (t, x, s, y) \in \{[0, T] \times \mathbb{R}^n\}^2 \mid t, s \ge T - \rho + \frac{\lambda}{2K_\alpha}, \ |x|, |y| \le 2K_\alpha \Big\},$$

with K_α being the same as that appearing in (6.31). Thus, by restricting (t, x, s, y) on Q, the function $\varphi(t, x, s, y)$ is smooth with bounded derivatives, which implies its semiconcavity. Consequently, $\Phi(t, x, s, y)$ is semiconvex and attains its maximum at (t_0, x_0, s_0, y_0) in the interior of Q (noting (6.31)). Hence, for any small $r > 0$,

$$\widehat{\Phi}(t, x, s, y) \stackrel{\Delta}{=} \Phi(t, x, s, y) - r\big(|t - t_0|^2 + |s - s_0|^2 + |x - x_0|^2 + |y - y_0|^2\big)$$

is semiconvex on Q, attaining a strict maximum at (t_0, x_0, s_0, y_0). By Lemmas 6.6 and 6.7, for the above given $r > 0$, there exist $q, \hat{q} \in \mathbb{R}$ and $p, \hat{p} \in \mathbb{R}^n$ with

$$|q| + |\hat{q}| + |p| + |\hat{p}| \le r, \tag{6.38}$$

and $(\hat{t}_0, \hat{x}_0, \hat{s}_0, \hat{y}_0) \in Q$ with

$$|\hat{t}_0 - t_0| + |\hat{x}_0 - x_0| + |\hat{s}_0 - s_0| + |\hat{y}_0 - y_0| \le r, \tag{6.39}$$

such that

$$
(6.40)\qquad \begin{aligned}
&\widehat{\Phi}(t,x,s,y)+qt+\hat{q}s+\langle p,x\rangle+\langle \hat{p},y\rangle\\
&\quad\equiv v^{\gamma}(t,x)-\widehat{v}_{\gamma}(s,y)-\varphi(t,x,s,y)\\
&\qquad -r\big(|t-t_0|^2+|s-s_0|^2+|x-x_0|^2+|y-y_0|^2\big)\\
&\qquad +qt+\hat{q}s+\langle p,x\rangle+\langle \hat{p},y\rangle
\end{aligned}
$$

attains a maximum at $(\hat{t}_0,\hat{x}_0,\hat{s}_0,\hat{y}_0)$, at which $v^{\gamma}(t,x)-\widehat{v}_{\gamma}(s,y)$ is twice differentiable. For notational simplicity, we now drop γ in $v^{\gamma}(t,x)$ and $\widehat{v}_{\gamma}(s,y)$. Then, by the first- and second-order necessary conditions for a maximum point, at the point $(\hat{t}_0,\hat{x}_0,\hat{s}_0,\hat{y}_0)$, we must have

$$
(6.41)\qquad \begin{cases}
v_t=\varphi_t+2r(\hat{t}_0-t_0)-q,\\
\hat{v}_s=-\varphi_s-2r(\hat{s}_0-s_0)+\hat{q},\\
v_x=\varphi_x+2r(\hat{x}_0-x_0)-p,\\
\hat{v}_y=-\varphi_y-2r(\hat{y}_0-y_0)+\hat{p},\\
\begin{pmatrix} v_{xx} & 0\\ 0 & -\widehat{v}_{yy}\end{pmatrix}\le\begin{pmatrix}\varphi_{xx} & \varphi_{xy}\\ \varphi_{xy}^{\top} & \varphi_{yy}\end{pmatrix}+2rI_{2n},
\end{cases}
$$

where I_{2n} is the $2n\times 2n$ identity matrix. Now, at $(\hat{t}_0,\hat{x}_0,\hat{s}_0,\hat{y}_0)$, we calculate the following:

$$
(6.42)\qquad \begin{cases}
\varphi_t=-\beta-\dfrac{\lambda}{(\hat{t}_0-T+\rho)^2}-\dfrac{\alpha}{2\mu T}\big(|\hat{x}_0|^2+|\hat{y}_0|^2\big)+\dfrac{1}{\varepsilon}(\hat{t}_0-\hat{s}_0),\\
\varphi_s=-\beta-\dfrac{\lambda}{(\hat{s}_0-T+\rho)^2}-\dfrac{\alpha}{2\mu T}\big(|\hat{x}_0|^2+|\hat{y}_0|^2\big)+\dfrac{1}{\varepsilon}(\hat{s}_0-\hat{t}_0),\\
\varphi_x=\dfrac{\alpha(2\mu T-\hat{t}_0-\hat{s}_0)}{\mu T}\hat{x}_0+\dfrac{\hat{x}_0-\hat{y}_0}{\delta},\\
\varphi_y=\dfrac{\alpha(2\mu T-\hat{t}_0-\hat{s}_0)}{\mu T}\hat{y}_0+\dfrac{\hat{y}_0-\hat{x}_0}{\delta},\\
A\equiv\begin{pmatrix}\varphi_{xx} & \varphi_{xy}\\ \varphi_{xy}^{\top} & \varphi_{yy}\end{pmatrix}=\dfrac{1}{\delta}\begin{pmatrix} I_n & -I_n\\ -I_n & I_n\end{pmatrix}+\dfrac{\alpha(2\mu T-\hat{t}_0-\hat{s}_0)}{\mu T}I_{2n}.
\end{cases}
$$

On the other hand, by Lemma 6.5 and the definition of viscosity sub- and supersolutions, we have

$$
(6.43)\qquad \begin{cases}
-v_t(\hat{t}_0,\hat{x}_0)+G^{\gamma}(\hat{t}_0,\hat{x}_0,-v_x(\hat{t}_0,\hat{x}_0),-v_{xx}(\hat{t}_0,\hat{x}_0))\le 0,\\
-\widehat{v}_s(\hat{s}_0,\hat{y}_0)+G_{\gamma}(\hat{s}_0,\hat{y}_0,-\widehat{v}_y(\hat{s}_0,\hat{y}_0),-\widehat{v}_{yy}(\hat{s}_0,\hat{y}_0))\ge 0.
\end{cases}
$$

By (6.17), one can find a $(\bar{t}_0,\bar{x}_0,\bar{s}_0,\bar{y}_0)$ with

$$
(6.44)\qquad |\bar{t}_0-\hat{t}_0|+|\bar{x}_0-\hat{x}_0|+|\bar{s}_0-\hat{s}_0|+|\bar{y}_0-\hat{y}_0|\le\overline{K}_{\alpha}\gamma^2,
$$

for some $\overline{K}_\alpha > 0$ (depending only on α), such that

$$
\begin{aligned}
&\widehat{v}_s(\hat{s}_0,\hat{y}_0) - v_t(\hat{t}_0,\hat{x}_0)\\
&\quad \le G_\gamma(\hat{s}_0,\hat{y}_0,-v_y(\hat{s}_0,\hat{y}_0),-\widehat{v}_{yy}(\hat{s}_0,\hat{y}_0))\\
&\qquad - G^\gamma(\hat{t}_0,\hat{x}_0,-v_x(\hat{t}_0,\hat{x}_0),-v_{xx}(\hat{t}_0,\hat{x}_0))\\
&\quad = \sup_{u\in U} G(\overline{s}_0,\overline{y}_0,u,-\widehat{v}_y(\hat{s}_0,\hat{y}_0),-v_{yy}(\hat{s}_0,\hat{y}_0))\\
&\qquad - \sup_{u\in U} G(\overline{t}_0,\overline{x}_0,u,-v_x(\hat{t}_0,\hat{x}_0),-v_{xx}(\hat{t}_0,\hat{x}_0))\\
&\quad \le \sup_{u\in U}\Big\{G(\overline{s}_0,\overline{y}_0,u,-\widehat{v}_y(\hat{s}_0,\hat{y}_0),-\widehat{v}_{yy}(\hat{s}_0,\hat{y}_0))\\
&\qquad\qquad - G(\overline{t}_0,\overline{x}_0,u,-v_x(\hat{t}_0,\hat{x}_0),-v_{xx}(\hat{t}_0,\hat{x}_0))\Big\} \qquad (6.45)\\
&\quad = \sup_{u\in U}\Big\{\frac{1}{2}\mathrm{tr}\,\Big[v_{xx}(\hat{t}_0,\hat{x}_0)\sigma(\overline{t}_0,\overline{x}_0,u)\sigma(\overline{t}_0,\overline{x}_0,u)^\top\\
&\qquad\qquad - \widehat{v}_{yy}(\hat{s}_0,\hat{y}_0)\sigma(\overline{s}_0,\overline{y}_0,u)\sigma(\overline{s}_0,\overline{y}_0,u)^\top\Big]\\
&\qquad + \Big[\langle v_x(\hat{t}_0,\hat{x}_0),b(\overline{t}_0,\overline{x}_0)\rangle - \langle \widehat{v}_y(\hat{s}_0,\hat{y}_0),b(\overline{s}_0,\overline{y}_0,u)\rangle\Big]\\
&\qquad + \Big[f(\overline{t}_0,\overline{x}_0,u) - f(\overline{s}_0,\overline{y}_0,u)\Big]\Big\}\\
&\quad \equiv \sup_{u\in U}\big\{(I)+(II)+(III)\big\}.
\end{aligned}
$$

By (6.41)–(6.42) and (6.38)–(6.39), we have

$$
\begin{aligned}
&\widehat{v}_s(\hat{s}_0,\hat{y}_0) - v_t(\hat{t}_0,\hat{x}_0)\\
&\quad = 2\beta + \frac{\alpha}{\mu T}\big(|\hat{x}_0|^2+|\hat{y}_0|^2\big) + \frac{\lambda}{(\hat{t}_0-T+\rho)^2} + \frac{\lambda}{(\hat{s}_0-T+\rho)^2}\\
&\qquad - 2r(\hat{t}_0 - t_0 + \hat{s}_0 - s_0) + q + \hat{q}\\
&\quad \ge 2\beta + \frac{\alpha}{\mu T}\big(|\hat{x}_0|^2+|\hat{y}_0|^2\big) - Kr,
\end{aligned}
$$

for some absolute constant $K > 0$. By (6.32) and (6.39), we see that one may assume that as $\varepsilon,\delta,r \to 0$, $(\hat{t}_0,\hat{x}_0)$ and $(\hat{s}_0,\hat{y}_0)$ converge to the same limit, denoted by (t_α,x_α), to emphasize the dependence on α. Thus, letting $\varepsilon,\delta,r\to 0$ in the above leads to

$$
\hat{v}_s(t_\alpha,x_\alpha) - v_t(t_\alpha,x_\alpha) \ge 2\beta + \frac{2\alpha}{\mu T}|x_\alpha|^2. \qquad (6.46)
$$

This gives an estimate for the left-hand side of (6.45). We now estimate the terms (I), (II), and (III) on the right-hand side of (6.45) one by one. First of all, from (3.7), (6.32), (6.39), and (6.44), one obtains an estimate for (III):

$$
(III) \stackrel{\Delta}{=} f(\overline{t}_0,\overline{x}_0,u) - f(\overline{s}_0,\overline{y}_0,u) \to 0, \qquad \text{as } \varepsilon,\delta,\gamma,r\to 0, \qquad (6.47)
$$

uniformly in $u \in U$. Next,

$$
\begin{aligned}
(II) \stackrel{\Delta}{=} & \langle \frac{\alpha(2\mu T - \hat{t}_0 - \hat{s}_0)}{\mu T}\hat{x}_0 + \frac{\hat{x}_0 - \hat{y}_0}{\delta} + 2r(\hat{x}_0 - x_0) - p, b(\bar{t}_0, \bar{x}_0, u) \rangle \\
& + \langle \frac{\alpha(2\mu T - \hat{t}_0 - \hat{s}_0)}{\mu T}\hat{y}_0 + \frac{\hat{y}_0 - \hat{x}}{\delta} + 2r(\hat{y}_0 - y_0) - \hat{p}, b(\bar{s}_0, \bar{y}_0, u) \rangle \\
\le & \langle \frac{\hat{x}_0 - \hat{y}_0}{\delta}, b(\bar{t}_0, \bar{x}_0, u) - b(\bar{s}_0, \bar{y}_0, u) \rangle \\
& + \frac{\alpha(2\mu T - \hat{t}_0 - \hat{s}_0)}{\mu T} L\big[|\hat{x}_0|(1 + |\bar{x}_0|) + |\hat{y}_0|(1 + |\bar{y}_0|)\big] \\
& + rK(1 + |\bar{x}_0| + |\bar{y}_0|).
\end{aligned}
$$

Letting $\varepsilon, \gamma, r \to 0$, we may assume that $(\hat{t}_0, \hat{x}_0, \hat{s}_0, \hat{y}_0)$ and $(\bar{t}_0, \bar{x}_0, \bar{s}_0, \bar{y}_0)$ converge. Clearly, the limits of these two sequences have to be the same (see (6.6), (6.32), (6.39), and (6.44)), which is denoted by (t_0, x_0, t_0, y_0). Thus,

$$
\lim_{\varepsilon, \gamma, r \to 0} (II) \le L\frac{|x_0 - y_0|^2}{\delta} + \frac{2\alpha(\mu T - t_0)L}{\mu T}\big(|x_0| + |y_0| + |x_0|^2 + |y_0|^2\big).
$$

Then letting $\delta \to 0$, one concludes that (t_0, x_0) and (t_0, y_0) approach a common limit (see (6.32)), called (t_α, x_α). Consequently,

$$
\lim_{\delta \to 0} \lim_{\varepsilon, \gamma, r \to 0} (II) \le \frac{4\alpha(\mu T - t_\alpha)}{\mu T} L(|x_\alpha| + |x_\alpha|^2). \tag{6.48}
$$

Now we treat (I) in (6.45). By the inequality in (6.41),

$$
\begin{aligned}
(I) \stackrel{\Delta}{=} & \frac{1}{2}\text{tr}\Big[v_{xx}(\hat{t}_0, \hat{x}_0)\sigma(\bar{t}_0, \bar{x}_0, u)\sigma(\bar{t}_0, \bar{x}_0, u)^\top \\
& \qquad - \widehat{v}_{yy}(\hat{s}_0, \hat{y}_0)\sigma(\bar{s}_0, \bar{y}_0, u)\sigma(\bar{s}_0, \bar{y}_0, u)^\top\Big] \\
= & \frac{1}{2}\text{tr}\Big[\begin{pmatrix} \sigma(\bar{t}_0, \bar{x}_0, u) \\ \sigma(\bar{s}_0, \bar{y}_0, u) \end{pmatrix}^\top \begin{pmatrix} v_{xx}(\hat{t}_0, \hat{x}_0) & 0 \\ 0 & -\widehat{v}_{yy}(\hat{s}_0, \hat{y}_0) \end{pmatrix} \begin{pmatrix} \sigma(\bar{t}_0, \bar{x}_0, u) \\ \sigma(\bar{s}_0, \bar{y}_0, u) \end{pmatrix}\Big] \\
\le & \frac{1}{2}\text{tr}\Big[\begin{pmatrix} \sigma(\bar{t}_0, \bar{x}_0, u) \\ \sigma(\bar{s}_0, \bar{x}_0, u) \end{pmatrix}^\top A \begin{pmatrix} \sigma(\bar{t}_0, \bar{x}_0, u) \\ \sigma(\bar{s}_0, \bar{y}_0, u) \end{pmatrix}\Big] \\
\le & \frac{1}{2}\Big\{\frac{1}{\delta}|\sigma(\bar{t}_0, \bar{x}_0, u) - \sigma(\bar{s}_0, \bar{y}_0, u)|^2 \\
& \qquad + \frac{\alpha(2\mu T - \hat{t}_0 - \hat{s}_0)}{\mu T}\big(|\sigma(\bar{t}_0, \bar{x}_0, u)|^2 + |\sigma(\bar{s}_0, \bar{y}_0, u)|^2\big)\Big\} \\
\le & L^2\frac{|\bar{x}_0 - \bar{y}_0|^2}{2\delta} + \frac{\alpha(2\mu T - \hat{t}_0 - \hat{s}_0)}{\mu T} L^2\big[(1 + |\bar{x}_0|)^2 + (1 + |\bar{y}_0|)^2\big].
\end{aligned}
$$

As above, we first let $\varepsilon, \gamma, r \to 0$ and then let $\delta \to 0$ to get

$$
\lim_{\delta \to 0} \lim_{\varepsilon, \gamma, r \to 0} (I) \le \frac{4\alpha(\mu T - t_\alpha)}{\mu T} L^2(1 + |x_\alpha|)^2. \tag{6.49}
$$

Combining (6.45)–(6.49), we obtain

$$\beta + \frac{\alpha}{\mu T}|x_\alpha|^2 \le \frac{2\alpha(\mu T - t_\alpha)}{\mu T}\big[L(|x_\alpha| + |x_\alpha|^2) + L^2(1+|x_\alpha|)^2\big]. \tag{6.50}$$

Hence,

$$\beta \le -\frac{\alpha}{\mu T}\Big\{\big[1 - 2(\mu T - t_\alpha)(L+L^2)\big]|x_\alpha|^2 + K(1+|x_\alpha|)\big]\Big\}, \tag{6.51}$$

for some constant $K > 0$ independent of $\alpha > 0$. Recalling $\rho = (4L+4L^2)^{-1}$ and $\mu = 1 + [8T(L+L^2)]^{-1}$, for any $t \in (T-\rho, T]$, we have

$$\begin{aligned} 1 - 2(\mu T - t_\alpha)(L+L^2) &\ge 1 - 2(\mu T - T + \rho)(L+L^2) \\ &\ge \frac{1}{2} - 2(\mu - 1)T(L+L^2) \ge \frac{1}{4}. \end{aligned}$$

Thus, (6.51) becomes

$$\beta \le \alpha\Big\{-\frac{1}{4\mu T}|x_\alpha|^2 + K(1+|x_\alpha|)\Big\}. \tag{6.52}$$

It is clear that the term inside the braces on the right-hand side of (6.52) is bounded from above uniformly in α. Thus, by sending $\alpha \to 0$, we obtain $\beta \le 0$, which contradicts our assumption $\beta > 0$. This proves (6.26). $\square$

Through the HJB equation (3.24), one may obtain an optimal (feedback) control via the so-called *verification theorems*, in a way similar to that described in Section 2 for the deterministic case (see the paragraphs following Proposition 2.2). Certainly, one faces the same problem of possible nonsmoothness when applying the verification theorems. Therefore, certain *nonsmooth* verification theorems have to be developed. We defer the study on this problem to Chapter 5 for both the deterministic and stochastic situations (see Chapter 5, Sections 3.5 and 5). We remark here that the uniqueness of viscosity solutions will play an essential role there.

6.2. Proofs of Lemmas 6.6 and 6.7*

We need the following lemma.

Lemma 6.8. *Let $f = (f^1, \dots, f^n)^\top : \mathbb{R}^n \to \mathbb{R}^n$ be Lipschitz continuous. Then, for any $g \in L^1(\mathbb{R}^n; \mathbb{R}^n)$,*

$$\int_{\mathbb{R}^n} g(x)\big|\det\big[Df(x)\big]\big|dx = \int_{\mathbb{R}^n}\Big\{\sum_{x\in f^{-1}\{y\}} g(x)\Big\}dy, \tag{6.53}$$

where $Df \stackrel{\Delta}{=} (\frac{\partial f^j}{\partial x_i})_{n\times n}$ is the Jacobian matrix of f.

When $f(\cdot)$ is C^1 and bijective, the above is the well-known *change of variable formula* for definite integrals. We are not going to prove the above result. For those who are interested in a proof, see Evans–Gariepy [1].

* This subsection serves as an appendix to the previous subsection.

Proof of Lemma 6.6. Without loss of generality, we may assume that φ is convex. By Lemma 2.3-(v) of Chapter 3, φ is locally Lipschitz continuous, and thus it is differentiable almost everywhere. Let

$$F_1 = \{x \in \mathbb{R}^n \big| D\varphi(x) \text{ exists}\}. \tag{6.54}$$

Then $|F_1^c| = 0$ (or we say that F_1 is of *full measure*). On the other hand, since φ is convex, the subgradient $\partial\varphi : \mathbb{R}^n \to 2^{\mathbb{R}^n}$ of φ is *maximal monotone*, i.e.,

$$\langle\, \xi_1 - \xi_2, x_1 - x_2 \,\rangle \geq 0, \quad \forall x_i \in \partial\varphi(x_i),\ i = 1, 2,$$

and the range $\mathcal{R}(I + \partial\varphi)$ of $I + \partial\varphi$ coincides with $\mathbb{R}^n$. Thus, one can show that $J \stackrel{\Delta}{=} (I + \partial\varphi)^{-1} : \mathbb{R}^n \to \mathbb{R}^n$ is single-valued, Lipschitz continuous with Lipschitz constant 1. Also, it is surjective, since for any $z \in \mathbb{R}^n$, by $\partial\varphi(z) \neq \phi$, there exists an $x \in z + \partial\varphi(z)$, which leads to

$$z = (I + \partial\varphi)^{-1}(x) \equiv J(x).$$

Next, from the Lipschitz continuity of $J(x)$, it follows that $DJ(x)$ exists almost everywhere. Define

$$F_2 = \{x \in \mathbb{R}^n \mid DJ(x) \text{ exists and is nonsingular}\}. \tag{6.55}$$

Note that for almost every $y \in \mathbb{R}^n$, $J^{-1}(y) = y + D\varphi(y)$ is a single-valued map. By Lemma 6.8, we obtain

$$\begin{aligned} |J(F_2^c)| &= \int_{\mathbb{R}^n} \chi_{J(F_2^c)}(y)dy = \int_{\mathbb{R}^n} \chi_{F_2^c}(J^{-1}(y))dy \\ &= \int_{\mathbb{R}^n} \Big\{ \sum_{x \in J^{-1}\{y\}} \chi_{F_2^c}(x) \Big\} dy = \int_{\mathbb{R}^n} \chi_{F_2^c}(x) \big| \det\big[DJ(x)\big] \big| dx = 0. \end{aligned}$$

Thus, the surjectivity of J yields that $J(F_2)$ is of full measure. Set

$$\begin{aligned} F = F_1 \bigcap J(F_2) \equiv \{J(x) \mid x \in \mathbb{R}^n, DJ(x) &\text{ exists} \\ &\text{and is nonsingular, and } D\varphi(J(x)) \text{ exists}\}. \end{aligned} \tag{6.56}$$

This set is also of full measure. Now we take $J(x) \in F \subseteq F_1$. By the definition of J, we have

$$D\varphi(J(x)) = x - J(x). \tag{6.57}$$

Let δy be small enough, and $J(x) + \delta y \in F \subseteq F_1$. Then $D\varphi(J(x) + \delta y)$ exists. Consequently, the following equation (for δx)

$$J(x + \delta x) = J(x) + \delta y \tag{6.58}$$

admits a unique solution given by

$$\delta x = J(x) + \delta y + D\varphi(J(x) + \delta y) - x. \tag{6.59}$$

By Taylor expansion, we have from (6.58) that

$$\delta y = J(x+\delta x) - J(x) = DJ(x)\delta x + o(|\delta x|). \tag{6.60}$$

Since J is contractive and $DJ(x)$ is nonsingular, by (6.58) and (6.60), we can find a constant $K > 0$ such that

$$|\delta y| \leq |\delta x| \leq K|\delta y|. \tag{6.61}$$

Hence, $|\delta x|$ and $|\delta y|$ are of the same order. Now, noting $J(x) + \delta y \in F_1$ and (6.57)–(6.60), we obtain

$$\begin{aligned} D\varphi(J(x)+\delta y) &= D\varphi(J(x+\delta x)) = x + \delta x - J(x+\delta x) \\ &= x + \delta x - J(x) - \delta y \\ &= D\varphi(J(x)) + (I - DJ(x))\delta x + o(|\delta x|) \\ &= D\varphi(J(x)) + \big\{[DJ(x)]^{-1} - I\big\}\delta y + o(|\delta y|). \end{aligned} \tag{6.62}$$

Define

$$\begin{cases} \psi(\delta y) = \varphi(J(x)+\delta y), \\ \widetilde{\psi}(\delta y) = \varphi(J(x)) + D\varphi(J(x))\delta y + \dfrac{1}{2}\left\langle \{[DJ(x)]^{-1} - I\}\delta y, \delta y \right\rangle. \end{cases} \tag{6.63}$$

Then

$$\psi(0) = \varphi(J(x)) = \widetilde{\psi}(0),$$

and by (6.62),

$$\begin{aligned} D\psi(\delta y) &= D\varphi(J(x)+\delta y) \\ &= D\varphi(J(x)) + \{[DJ(x)]^{-1} - I\}\delta y + o(|\delta y|) \\ &= D\widetilde{\psi}(\delta y) + o(|\delta y|). \end{aligned}$$

Hence, $\Psi(\delta y) \stackrel{\Delta}{=} \psi(\delta y) - \widetilde{\psi}(\delta y)$ is locally Lipschitz and

$$\Psi(0) = 0, \quad D\Psi(\delta y) = o(|\delta y|), \qquad \text{for almost all small enough } \delta y.$$

Thus, by Taylor expansion we have

$$\Psi(\delta y) = \Psi(0) + D\Psi(0)\delta y + o(|\delta y|^2) = o(|\delta y|^2)$$

for almost all small δy. By Lipschitz continuity, the above holds for all δy small enough. Consequently,

$$\begin{aligned} \varphi(J(x)+\delta y) = \varphi(J(x)) &+ D\varphi(J(x))\delta y \\ &+ \frac{1}{2}\left\langle \{[DJ(x)]^{-1} - I\}\delta y, \delta y \right\rangle + o(|\delta y|^2), \quad \forall J(x) \in F. \end{aligned}$$

Then our conclusion follows from letting $y = J(x) \in F$. $\square$

Proof of Lemma 6.7. We first assume that φ is C^2. Take $r > 0$ small enough such that $\overline{x}$ is the strict maximum of φ in the ball $B_r(\overline{x})$. Then, for $\delta > 0$ small enough, for any $x \in \partial B_r(\overline{x})$, we have

$$\varphi_p(\overline{x}) - \varphi_p(x) \geq \varphi(\overline{x}) - \max_{x \in \partial B_r(\overline{x})} \varphi(x) - \delta r > 0, \quad \forall p \in B_\delta(0).$$

Thus, for such a $\delta > 0$ and any $p \in B_\delta(0)$, all the maximum points of $\varphi_p(\cdot)$ on $B_r(\overline{x})$ are in the interior of $B_r(\overline{x})$, namely, $\mathcal{K}$ lies in the interior of $B_r(\overline{x})$. Therefore, at any $x \in \mathcal{K}$ that is a maximum point of φ_p, we have

$$D\varphi + p = D\varphi_p = 0, \qquad D^2\varphi = D^2\varphi_p \leq 0. \tag{6.64}$$

From the first relation in (6.64), $p = -D\varphi(x)$ for some $x \in \mathcal{K}$, i.e.,

$$(D\varphi)(\mathcal{K}) \supseteq B_\delta(0). \tag{6.65}$$

Since φ is semiconvex, there exists a $\lambda > 0$ such that $\varphi(x) + \frac{\lambda}{2}|x|^2$ is convex. Then

$$-\lambda I \leq D^2\varphi(x) \leq 0, \qquad \forall x \in \mathcal{K}. \tag{6.66}$$

By Lemma 6.8 (with $g(x) = \chi_{\mathcal{K}}(x)$ and $f(x) = D\varphi(x)$), we have

$$\begin{aligned} |B_\delta(0)| &\leq |(D\varphi)(\mathcal{K})| = \int_{\mathbf{R}^n} \chi_{(D\varphi)(\mathcal{K})}(y)dy \\ &\leq \int_{\mathbf{R}^n} \Big\{ \sum_{x \in (D\varphi)^{-1}(y)} \chi_{\mathcal{K}}(x) \Big\} dy \\ &= \int_{\mathbf{R}^n} \chi_{\mathcal{K}}(x) \big| \det D^2\varphi(x) \big| dx \leq \int_{\mathcal{K}} \lambda^n dx = \lambda^n |\mathcal{K}|. \end{aligned} \tag{6.67}$$

This proves the desired result for the case where φ is C^2. Let us note that the lower bound of $|\mathcal{K}|$ depends only on λ and δ.

Now, for the general case, where φ is not necessarily C^2, we can find a $\lambda > 0$ and a sequence φ^m of semiconvex C^2 functions such that

$$\begin{cases} \varphi^m \to \varphi, & \text{uniformaly on } B_r(\overline{x}), \\ \varphi^m(x) + \dfrac{\lambda}{2}|x|^2 \text{ is convex for all } m \geq 1. \end{cases} \tag{6.68}$$

Let

$$\begin{aligned} \mathcal{K}_m = \{x \in B_r(\overline{x}) \mid \exists |p| \leq \delta \text{ such that } \varphi_p^m(\cdot) \stackrel{\Delta}{=} \varphi^m(\cdot) + \langle p, \cdot \rangle \\ \text{attains a strict local maximum at } x\}. \end{aligned} \tag{6.69}$$

Then we can find a $\delta > 0$ such that $\mathcal{K}_m$ lies in the interior of $B_r(\overline{x})$. Thus, similar to (6.67), we must have

$$|\mathcal{K}_m| \geq \lambda^{-n} |B_\delta(0)|, \qquad \forall m \geq 1. \tag{6.70}$$

By the convergence of φ^m to φ, we have $\mathcal{K} \supseteq \bigcap_{k=1}^{\infty} \bigcup_{m=k}^{\infty} \mathcal{K}_m$. Therefore,

$$|\mathcal{K}| = \lim_{k\to\infty} \Big| \bigcup_{m=k}^{\infty} \mathcal{K}_m \Big| \geq \varlimsup_{m\to\infty} |\mathcal{K}_m| \geq \lambda^{-n} |B_\delta(0)|.$$

This completes our proof. □

7. Historical Remarks

Bellman's *Principle of optimality* stated in Bellman [5, p. 83] is as follows:

> "*An optimal policy has the property that whatever the initial state and initial decision are, the remaining decisions must constitute an optimal policy with regard to the state resulting from the first decision.*"

According to Bellman [5, p. 115], the basic idea of dynamic programming was initiated by himself in his research done during 1949–1951, mainly for multistage decision problems. The first paper on dynamic programming was published in 1952 by Bellman [1], which was soon expanded to a monograph [2] in 1953 published by the Rand Corporation. On the other hand, R. Isaacs [1] published a Rand Corporation Report on November 17, 1951, introducing the concept of *tenet of transition*, which is closely related to the principle of optimality.* A widely accessible book is Bellman [5], published in 1957. In 1954, Bellman ([3,4]) found that the technique was also applicable to the calculus of variations and to optimal control problems whose state equations are ordinary differential equations, which led to a nonlinear PDE now called the HJB equation. It seems to us that at that time Bellman did not realize that this equation is closely related to the well-known Hamilton–Jacobi equation arising in mechanics, as such a name was not mentioned in the papers and books by Bellman prior to 1960. In 1960, Kalman [3] pointed out such a relation, and probably was the first to use the name *Hamilton–Jacobi equation of the control problem* (see also Kalman–Falb–Arbib [1]).

As a matter of fact, the idea of the principle of optimality actually goes back to Jakob Bernoulli while he was solving the famous brachistochrone problem posed by his brother Johann in 1696 (see Goldstine [1, p. 45] and Pesch–Bulirsch [1]). Moreover, an equation identical to what

* Isaacs explained in [3] about his tenet of transition as follows: "Once I felt that there was the heart of the subject and cited it often in the early Rand seminars. Later I felt that it—like other mathematical hearts—was a mere truism. Thus, in DG (his book [2]) it is mentioned only by title. This I regret. I had no idea that Pontryagin's principle and Bellman's maximal principle (a special case of the tenet, appearing a little later in the Rand seminar) would enjoy such a widespread citation." Bellman informally cited the work of Isaacs in his paper [3].

was later called the "Bellman equation" was first derived by Carathéodory [1] in 1926 while he was studying the sufficient conditions of the calculus of variations problem (his approach was named *Carathéodory's royal road*, see Boerner [1,2], Pesch–Bulirsch [1], and Giaquinta–Hildebrandt [1, pp. 327–331; p. 396]). We should also mention the work of Wald on sequential analysis (Wald [1,2]) in the late 1940s, which contains some ideas very similar to that of dynamic programming.

Although the discrete-time stochastic version of dynamic programming was already discussed in the early works of Bellman (see Bellman [5, Chapter 11]), the continuous-time stochastic version of dynamic programming (involving the Itô-type SDEs as the state equations) probably was first studied by Kushner [1] in 1962. Since then, numerous people have contributed to the subject. We refer the reader to the books by Fleming–Rishel [1], Krylov [2], and Fleming–Soner [1] for further discussion as well as more literature cited therein.

For a long time, the dynamic programming theory of deterministic controlled systems remained nonrigorous. The main mathematical difficulty for a rigorous treatment is that the corresponding HJB equation is a first-order partial differential equation, which generally does not admit a classical (smooth) solution, or the value functions are not necessarily continuously differentiable. There have been many authors who made various efforts to introduce different notions of generalized or weak solutions, and tried to prove the value function to be the solution of the HJB equation in a certain sense. During the 1960s, in a series of papers, Kružkov [1–5] built a systematic theory for first-order Hamilton–Jacobi (HJ, for short) equations with smooth and convex Hamiltonians. In particular, the vanishing viscosity approximation was introduced in Kružkov [2,3]. At about the same time, Fleming [1] independently introduced the vanishing of viscosity, combining with the differential games technique, to study the HJ equations. See Fleming [3] also. On the other hand, in the early 1980s, Subbotin [1] studied the HJ equations with nonconvex Hamiltonians by introducing the so-called *minimax solution*. For details of this approach, see Subbotin [2]. Clarke–Vinter [1] employed Clarke's generalized gradients to introduce generalized solutions of the HJB equations. Under Clarke's framework, the HJB equation may have more than one solution, and the value function is one of them. On the other hand, generalized gradients may not be used to readily solve second-order HJB equations that correspond to stochastic problems. There have been many other works related to the study of HJ equations. See Bardi–Capuzzo-Dolcetta [1] for a survey on this aspect.

In the early 1980s, Crandall and Lions made a breakthrough by introducing the notion of a viscosity solution for first-order HJ equations (Crandall–Lions [1,2]). Lions [1] applied the theory of viscosity solutions to deterministic optimal control problems. At about the same time, Lions [2] also investigated the degenerate second-order HJ equation using a Feynman–Kac-type technique, representing the solutions of the second-order PDEs by the value functions of some stochastic optimal control problems. Jensen [1] was the first to find a PDE proof of the uniqueness of viscosity solution to second-order HJB equations, using the semiconvex/semiconcave approximation technique. Later, Ishii [2] provided another clever approach to prove the uniqueness of viscosity solutions to HJB equations. His main observation, in terms of our terminology, is essentially the inequality in (6.41) whenever $v - \widehat{v} - \varphi$ attains a local maximum at (t, x, s, y). In other words, the Hessian of the test function φ can be controlled from below by a diagonal matrix involving the Hessians of v and $-\widehat{v}$. These results provided a rigorous foundation for the dynamic programming method in optimal control theory. See Fleming–Soner [1] for a more detailed description.

As we know, for a deterministic control problem, a natural way of applying the dynamic programming method is as follows: First, let the initial time and state vary and define the value function. Second, establish *Bellman's optimality principle* (which has to be *proved*, as it is not a natural property; in some situations such as the so-called non-Markovian case it is *not* valid), together with some continuity and local boundedness of the value function. Third, show that the value function is a viscosity solution of the HJB equation, based on the optimality principle. Fourth, prove that the HJB equation admits at most one viscosity solution. Then some other steps should follow, like applying the verification theorem, synthesize, etc. This procedure is pretty clear. People naturally would like to follow this in solving stochastic optimal control problems, like Problem (S). One finds immediately that to keep a parallel procedure for the stochastic case is by no means trivial, since the deterministic and stochastic problems are quite different. To handle this, the weak formulation has to be considered as an auxiliary formulation, namely, an admissible control is taken to be a 5-tuple $(\Omega, \mathcal{F}, \mathbf{P}, W(\cdot), u(\cdot))$, and the state equation is understood to be in the weak sense. This, however, works only for the case of deterministic coefficients (i.e., all the functions b, σ, f, and h are not explicitly dependent on $\omega \in \Omega$). Otherwise, this approach does not work, and the general consideration on this aspect is left open so far.

In the mid-1970s Nisio [1,2] constructed what is now widely known as the *Nisio semigroup*, which is a nonlinear semigroup, based on Bellman's optimality principle (see also Nisio [3]). The main feature of this approach is that it provides an abstract framework for applying dynamic programming to controlled stochastic processes that may cover a wider range of Markovian models beyond diffusion processes.

The material in Section 2 is a summary of the standard theory of deterministic optimal control via dynamic programming. The proof of the uniqueness of viscosity solutions (Theorem 2.5, Step 3) is mainly based on the idea of Ishii [1]; also see Yong [6], Fleming–Soner [1], and Bardi–Capuzzo-Dolcetta [1]. Lemmas 2.7-(ii) and 2.8-(ii) are new here. In Section 3, we present a rigorous proof of the stochastic optimality principle directly, without using approximation and partial differential equation theory. Thus, our approach is parallel to the deterministic case, and it is different from that given in Fleming–Soner [1]. Section 4 discussed some properties of the value function. The vanishing of viscosity is presented via optimal control theory, which is different from the pure PDE approach. It should be noted that this approach does not necessarily work for general fully nonlinear degenerate second-order parabolic PDEs. The semiconcavity property is proved in the manner similar to that in Li–Yong [1] (which is for infinite-dimensional control problems), and it is different from that given in Fleming–Soner [1] (where a bit more smoothness conditions were assumed). Based on the optimality principle established in Section 3, it is proved in a natural way that the value function is a viscosity solution to the corresponding HJB equation in Section 5. Section 6 is devoted to the uniqueness problem. We have made an effort to avoid some unnecessary technicalities (like the introduction of the graphs of the sub-/superdifferentials, the so-called theorem of sums, etc.; see Crandall–Ishii–Lions [1]) and provide a simplified, fully self-contained proof compared with other existing ones. Although the stochastic case is quite different from the deterministic case, we still follow a procedure similar to the deterministic case to prove the uniqueness, with the following modifications: (1) The semiconvex/semiconcave approximations have been introduced; (2) the results of Alexandrov and Jensen (on semiconvex/semiconcave functions) have been used; (3) the idea of Ishii (of fully utilizing the information from the Hessian of a function that attains a local maximum) has been adopted. These modifications have taken care of the essential difference between the first-order and second-order cases. Our idea is a combination of those presented in Jensen [1], Ishii [2], Crandall–Ishii–Lions [1], and Fleming–Soner [1], where alternative proofs can be found. The inclusion of the proofs of Alexandrov's theorem (see Alexandrov [1]) and Jensen's lemma (see Jensen [1]) is just for completeness and the reader's convenience, and fortunately, they are not lengthy.

For viscosity solutions in infinite-dimensional spaces, we refer to Lions [3], Nisio [4], Li–Yong [2], Stojanovic–Yong [1,2], and Yong [2,5]. Other related works concerning the HJB equation and/or the dynamic programming method can be found, to mention a few, in Barbu [1,2], Barbu–Da Prato [1,2], Barbu–Da Prato–Popa [1], Benton [1], Di Blasio [1], Cannarsa [1], Cannarsa–Da Prato [1,2], Cannarsa–Frankowska [1,2], Dong–Bian [1], Peng–Yong [1], Tataru [1,2], Zhou [11], and Zhu [1,2].

Chapter 5

The Relationship Between the Maximum Principle and Dynamic Programming

1. Introduction

In Chapters 3 and 4 we studied Pontryagin's maximum principle (MP, for short) and Bellman's dynamic programming (DP, for short). These two approaches serve as two of the most important tools in solving optimal control problems. Both MP and DP can be regarded as some necessary conditions of optimal controls (under certain conditions, they become sufficient ones). An interesting phenomenon one can observe from the literature is that to a great extent these two approaches have been developed separately and independently. Hence, a natural question arises: Are there any relations between these two? In this chapter we are going to address this question.

The Hamiltonian systems associated with MP are ordinary differential equations (ODEs) in the deterministic case and stochastic differential equations (SDEs) in the stochastic case. The HJB equations associated with DP are partial differential equations (PDEs), of first order in the deterministic case and of second order in the stochastic case.* Therefore, the relationship between the MP and DP is essentially the relationship between the Hamiltonian systems and the HJB equations or, even more generally, the relationship among ODEs, SDEs, and PDEs. Interestingly, such relations have been studied for a long time in different context in other fields. Let us briefly recall them here.

In classical mechanics, one describes dynamics of a particle system either by a system of ODEs called Hamilton's canonical system (or the Hamiltonian system), or by a PDE called the Hamilton–Jacobi (HJ for short) equation. These two approaches are equivalent in the sense that the solutions of one can be represented by those of the other. To represent the solutions of Hamiltonian systems by those of HJ equations, one finds all the *complete integrals* of the HJ equations and uses the implicit function theorem. This is called the *Hamilton–Jacobi theorem.* To represent the solutions of HJ equations by those of Hamiltonian systems, or more generally, to solve a first-order PDE via solving ODEs, one uses the *method of characteristics.* To be precise, for a given first-order PDE (HJ equation, in particular), there is an associated family of ODEs for curves (the corresponding Hamiltonian systems, in particular), called *characteristic strips*, by which the solutions to the PDE can be constructed.

---

* Here, deterministic and stochastic cases refers to the optimal control problems that we have discussed in Chapters 3 and 4, not any other types of problems involving discrete time, state equations having jumps, etc.

There is a strong analogy between optimal control theory and analytic mechanics. This should be expected, since the classical calculus of variations, which is the foundation of analytic mechanics, is indeed the precursor of optimal control theory (see Chapter 2, Section 8). In the context of (deterministic) optimal control theory, the Hamiltonian systems involved in the maximum principle actually serve as the characteristics for the HJB equations involved in dynamic programming.

Remarkably, the stochastic version of the method of characteristics leads to nothing but the so-called *Feynman–Kac formula*, which represents the solutions to a *linear* second-order parabolic or elliptic PDE by those to some SDEs. Thus, one expects similar relations of MP and DP in the stochastic optimal control context as well.

Finally, we point out that the fundamental results of optimal control theory—MP and DP, have an interesting economic interpretation via the so-called *shadow price* of a resource, which, by definition, is the rate of change of the performance measure with respect to the change of the resource. It turns out that the shadow price, the performance measure, and the resource specialize the adjoint variable, the value function, and the state variable, respectively in optimal controls. Consequently, the relationship between MP and DP recovers the natural relation among these three entities in economic theory.

This chapter is the core of the entire book. The results presented in this chapter can be regarded as some further developments, in the context of (stochastic) optimal control theory, of the well-known classical theory on the relationship between the Hamiltonian systems and HJ equations. An important issue in studying the problem is that the derivatives of the value functions are unavoidably involved in these results and, as we have seen in Chapter 4, that the value functions are not necessarily smooth. Therefore, once again we need to employ the viscosity solutions and the associated super-/subdifferentials to handle them.

The rest of the chapter is organized as follows. In Section 2 we review the Hamilton–Jacobi theory in classical mechanics to demonstrate the origin of the problem. Section 3 presents the relationship between the MP and DP for deterministic control problems. The economic interpretation, the method of characteristics, and the Feynman–Kac formula are also discussed to show the analogy in areas other than optimal controls. In Section 4 the relationship between the MP and DP for stochastic systems within the framework of viscosity solutions is studied. Section 5 investigates the stochastic verification theorem in a nonsmooth form, which can be regarded as an extension of the relationship between the MP and DP from open-loop controls to feedback ones. Section 6 describes a way of constructing optimal feedback controls by virtue of the verification theorem. This can also be regarded as a continuation of Chapter 4 in applying dynamic programming to obtain optimal controls. Finally, Section 7 gives some historical remarks.

2. Classical Hamilton–Jacobi Theory

Let us begin with a typical problem in classical mechanics. Consider a system in $\mathbb{R}^3$ consisting of N particles. If these particles are free from constraints, we need $3N$ independent coordinates to describe their positions in space. Let us denote the position of the ith particle by $\mathbf{r}_i$ (a vector in $\mathbb{R}^3$). Suppose there are k (independent) constraints of the form

$$f_j(t, \mathbf{r}_1, \ldots, \mathbf{r}_N) = 0, \qquad 1 \leq j \leq k, \tag{2.1}$$

presented. Then the *degree of freedom* is reduced to $n \stackrel{\Delta}{=} 3N - k$. We refer to (2.1) as *holonomic constraints*. In such a case, one can introduce n independent variables $x_1, \ldots, x_n$ to describe the system. These variables are called the *generalized coordinates*. Note that in classical mechanics, people use q_j's instead of x_j's. We use x_j's to match our subsequent context. It should also be pointed out that the generalized coordinates x_j are not necessarily the Cartesian coordinates of the particles. However, we must have

$$\mathbf{r}_i = \mathbf{r}_i(t, x_1, \ldots, x_n), \qquad 1 \leq i \leq N, \tag{2.2}$$

for some functions $\mathbf{r}_i$. A simple example of this is to consider a single particle in $\mathbb{R}^3$ restricted to the unit circle in the xy-plane. In this case, we need only one generalized coordinate x_1, and the function $\mathbf{r}_1(t, x_1)$ in (2.2) can be chosen as

$$\mathbf{r}_1(t, x_1) = \begin{pmatrix} \cos x_1 \\ \sin x_1 \\ 0 \end{pmatrix}, \qquad 0 \leq x_1 < 2\pi.$$

In what follows, we define $x = (x_1, \cdots, x_n)$. Our purpose is to determine x as a function of time, which gives a complete description of the motion for the particle system. We assume that the motion of the system is determined by the so-called *kinetic energy* T and the *potential energy* V, which are known functions of the following types:

$$\begin{cases} T = T(t, x, \dot{x}) \equiv \displaystyle\sum_{i,j=1}^{n} P_{ij}(t, x)\dot{x}_i\dot{x}_j, \\ V = V(t, x), \end{cases} \qquad \forall (t, x, \dot{x}) \in [0, \infty) \times \mathbb{R}^{2n}. \tag{2.3}$$

Here, x and $\dot{x}$ are regarded as independent variables. Next, one defines the *Lagrangian* of the system as follows:

$$L(t, x, \dot{x}) = T(t, x, \dot{x}) - V(t, x), \qquad \forall (t, x, \dot{x}) \in [0, \infty) \times \mathbb{R}^{2n}. \tag{2.4}$$

Hamilton's principle asserts that if $t_1 < t_2$ are two instants, $x(t)$ is the motion of the system and $\dot{x}(t) = \frac{d}{dt}x(t)$, then the variation of the following integral is zero:

$$J(x(\cdot)) = \int_{t_1}^{t_2} L(t, x(t), \dot{x}(t))dt. \tag{2.5}$$

This implies that for any $\theta \in C^1([t_1,t_2];\mathbb{R}^n)$, with $\theta(t_1)=\theta(t_2)=0$,

$$\lim_{\delta\to 0}\frac{J(x(\cdot)+\delta\theta(\cdot))-J(x(\cdot))}{\delta}=0. \tag{2.6}$$

By a standard technique in the calculus of variations, from (2.6) we derive the following *Lagrange equation:*

$$\frac{d}{dt}\big(L_{\dot x}(t,x(t),\dot x(t))\big)-L_x(t,x(t),\dot x(t))=0,\qquad t\in(t_1,t_2). \tag{2.7}$$

Once we find the solution $x(\cdot)$ of equation (2.7) (with some initial conditions), the motion of the particle system is determined, and the original problem is solved.

It is seen that (2.7) is a system of second-order nonlinear ordinary differential equations. Most of the time, solving (2.7) is complicated. Legendre suggested a remarkable transformation, which reduces (2.7) to a system of first-order ordinary differential equations that has some "symmetric" structure and sometimes is easier to solve. We now explain such a transformation for our system. In what follows, (t,x) is fixed. The objective is to transform $(\dot x, L)$ to (p,H) with p being a new independent variable and H being a new function such that $(\dot x,L)$ and (p,H) have some duality relation. As a result, equation (2.7) will become more "symmetric." To this end, we define the *generalized momenta* as follows:

$$p\equiv\begin{pmatrix}p_1\\ \vdots\\ p_n\end{pmatrix}=L_{\dot x}(t,x,\dot x)\equiv\begin{pmatrix}L_{\dot x_1}(t,x,\dot x)\\ \vdots\\ L_{\dot x_n}(t,x,\dot x)\end{pmatrix}. \tag{2.8}$$

Let us suppose that from (2.8), one can solve for $\dot x$ (in terms of (t,x,p)):

$$\dot x=\varphi(t,x,p)\equiv(\varphi_1(t,x,p),\ldots,\varphi_n(t,x,p))^\top. \tag{2.9}$$

This is the case, for example, when $L(t,x,\dot x)$ is uniformly convex in $\dot x$. Next, we introduce (noting (2.9))

$$\begin{aligned}H(t,x,p)&\stackrel{\Delta}{=}\langle p,\varphi(t,x,p)\rangle-L(t,x,\varphi(t,x,p))\\&\equiv\langle p,\dot x\rangle-L(t,x,\dot x).\end{aligned} \tag{2.10}$$

The function $H(t,x,p)$ is called the *Hamiltonian*. By (2.10) and (2.8), one has

$$\begin{aligned}H_p(t,x,p)&=\varphi(t,x,p)+\varphi_p(t,x,p)p-\varphi_p(t,x,p)L_{\dot x}(t,x,\dot x)\\&=\dot x+\varphi_p(t,x,p)\big[p-L_{\dot x}(t,x,\dot x)\big]=\dot x.\end{aligned} \tag{2.11}$$

Putting (2.8) and (2.10)–(2.11) together, we obtain

$$\begin{cases}p=L_{\dot x}(t,x,\dot x), & \dot x=H_p(t,x,p),\\ L(t,x,\dot x)+H(t,x,p)=\langle p,\dot x\rangle.\end{cases} \tag{2.12}$$

This is called *Legendre's transformation* between $(\dot{x}, L)$ and (p, H). It is clear that $(\dot{x}, L)$ and (p, H) are completely symmetric (with (t, x) fixed).

Proposition 2.1. *Suppose for any fixed (t,x), Legendre's transformation (2.12) can be performed both from $(\dot{x}, L)$ to (p, H) and from (p, H) to $(\dot{x}, L)$, with $\varphi(t,x,p)$ in (2.9) being C^1 in x. Let $x(\cdot)$ be a C^1 function and let $p(t) \stackrel{\Delta}{=} L_{\dot{x}}(t, x(t), \dot{x}(t))$. Then $(x(\cdot), p(\cdot))$ is a solution of the system*

$$\begin{cases} \dot{x}(t) = H_p(t, x(t), p(t)), \\ \dot{p}(t) = -H_x(t, x(t), p(t)), \end{cases} \tag{2.13}$$

if and only if $x(\cdot)$ is a solution of (2.7).

Proof. First, let $x(\cdot)$ be a solution of (2.7) with the Lagrangian $L(t, x, \dot{x})$ given, and define $p(\cdot)$ as in the proposition. By (2.12), the first equation in (2.13) is already satisfied. Next, since (noting (2.12))

$$\begin{aligned} L(t,x,\dot{x}) &= \langle\, p, \dot{x}\,\rangle - H(t,x,p) \\ &= \langle\, L_{\dot{x}}(t,x,\dot{x}), \dot{x}\,\rangle - H(t,x,L_{\dot{x}}(t,x,\dot{x})), \end{aligned}$$

one has

$$\begin{aligned} L_x(t,x,\dot{x}) &= L_{\dot{x}x}(t,x,\dot{x})\dot{x} - H_x(t,x,L_{\dot{x}}(t,x,\dot{x})) \\ &\quad - L_{\dot{x}x}(t,x,\dot{x})H_p(t,x,L_{\dot{x}}(t,x,\dot{x})) \\ &= -H_x(t,x,p) + L_{\dot{x}x}(t,x,\dot{x})\big[\dot{x} - H_p(t,x,p)\big] \\ &= -H_x(t,x,p). \end{aligned}$$

Hence, by Lagrange's equation (2.7), we obtain

$$\begin{aligned} \dot{p}(t) &= \frac{d}{dt}\Big(L_{\dot{x}}(t,x(t),\dot{x}(t))\Big) = L_x(t,x(t),\dot{x}(t)) \\ &= -H_x(t,x(t),p(t)). \end{aligned}$$

This proves "if" part.

Now, let $(x(\cdot), p(\cdot))$ be a solution of (2.13) with the given Hamiltonian $H(t,x,p)$. Define the Lagrangian $L(t,x,\dot{x})$ via (2.12). Then (2.10) holds, and

$$\begin{aligned} H_x(t,x,p) &= \varphi_x(t,x,p)p - L_x(t,x,\varphi(t,x,p)) \\ &\quad - \varphi_x(t,x,\dot{x})L_{\dot{x}}(t,x,\dot{x}) = -L_x(t,x,\dot{x}). \end{aligned}$$

Thus, by (2.13) and (2.12), we have

$$\begin{aligned} -L_x(t,x(t),\dot{x}(t)) &= H_x(t,x(t),p(t)) = -\dot{p}(t) \\ &= -\frac{d}{dt}\big(L_{\dot{x}}(t,x(t),\dot{x}(t))\big). \end{aligned}$$

Hence, $x(\cdot)$ is a solution of (2.7), proving the "only if" part. □

System (2.13) is called a *Hamilton's canonical system of differential equations*, or simply a *Hamiltonian system*. The above result tells that

Lagrange's equation (2.7) is equivalent to the Hamiltonian system (2.13), under proper conditions.

From the above, we see that to solve a standard mechanics problem, one may take the following steps.

Step 1. Form the Lagrangian $L(t, x, \dot{x})$ by (2.3)–(2.4).

Step 2. Define the generalized momenta p by (2.8) and suppose from (2.8) that one can solve for $\dot{x}$ as in (2.9).

Step 3. Define the Hamiltonian $H(t, x, p)$ by (2.10).

Step 4. Solve the Hamiltonian system (2.13). The obtained $x(\cdot)$ gives the trajectory of the motion for the particles.

Sometimes, solving the system (2.13) directly is still complicated. Thus, let us supply a very interesting way of solving (2.13). This is done by solving a first-order partial differential equation, called the *Hamilton–Jacobi equation.* Once this is done, the solutions of (2.13) can be represented by the solutions of such a partial differential equation. We now introduce the following *Hamilton–Jacobi equation* associated with our mechanics problem:

$$v_t + H(t, x, v_x) = 0, \tag{2.14}$$

where $H(t, x, p)$ is the same as (2.10).

Definition 2.2. Suppose $v(t, x, a)$ is a solution of (2.14) with parameter $a \in \mathbb{R}^n$ such that

$$\det(v_{ax}) \neq 0. \tag{2.15}$$

Then, $v(t, x, a) + a_0$, with $a_0 \in \mathbb{R}$, is called a *complete integral* of (2.14).

We note that when $v(t, x, a) + a_0$ is a complete integral of (2.14), due to (2.15) and the implicit function theorem it is possible to determine a pair of functions $(x(\cdot), p(\cdot))$ from the following:

$$\begin{cases} v_a(t, x(t), a) = b, \\ v_x(t, x(t), a) = p(t), \end{cases} \tag{2.16}$$

at least locally in $(t, a, b) \in \mathbb{R}^{1+2n}$. The following result, called the *Hamilton–Jacobi theorem*, gives an interpretation of the pair $(x(\cdot), p(\cdot))$ found through (2.16).

Theorem 2.3. *Let $v(t, x, a) + a_0$ be a complete integral of (2.14) and let (2.16) determine a pair of functions $(x(t; a, b), p(t; a, b))$ defined for all $(t, a, b) \in \mathbb{R}^{1+2n}$. Then $(x(\cdot\,; a, b), p(\cdot\,; a, b))$ is a family of solutions to the Hamiltonian system (2.13), parametrized by $(a, b) \in \mathbb{R}^{2n}$.*

Proof. First of all, by the definition of $v(t, x, a)$, we have

$$v_t(t, x, a) + H(t, x, v_x(t, x, a)) = 0. \tag{2.17}$$

By differentiating (2.17) with respect to x and a, respectively, we obtain the following:

$$(2.18)\quad v_{xt}(t,x,a)+H_x(t,x,v_x(t,x,a))+v_{xx}(t,x,a)H_p(t,x,v_x(t,x,a))=0,$$

and

$$(2.19)\qquad v_{at}(t,x,a)+v_{ax}(t,x,a)H_p(t,x,v_x(t,x,a))=0.$$

On the other hand, differentiating both equations in (2.16) with respect to t, we have

$$(2.20)\qquad v_{at}(t,x(t),a)+v_{ax}(t,x(t),a)\dot{x}(t)=0,$$

and

$$(2.21)\qquad v_{xt}(t,x(t),a)+v_{xx}(t,x(t),a)\dot{x}(t)=\dot{p}(t).$$

Substituting $x=x(t)$ into (2.19) and combining it with (2.20), we get the following (noting the second relation in (2.16)):

$$(2.22)\qquad v_{ax}(t,x(t),a)\big[\dot{x}(t)-H_p(t,x(t),p(t))\big]=0.$$

Thus, by (2.15), we must have the first equation in (2.13). Next, we substitute $x=x(t)$ into (2.18) and combine it with (2.21) to get the second equation in (2.13), proving the theorem. $\square$

In solving mechanics problems, the parameters $a,b\in\mathbb{R}^n$ are usually determined by the given conditions. Thus, in principle, if one can find complete integrals of the corresponding Hamilton–Jacobi equation, the mechanics problem is solved. In some cases, the above procedure works quite well. Several interesting examples can be found in Courant–Hilbert [1].

The above result reveals that one can solve the Hamiltonian system (2.13) via solving the corresponding Hamilton–Jacobi equation (2.14). A natural question is whether we can do this the other way around. In the rest of this section we would like to discuss a method of solving a Hamilton–Jacobi equation via solving the corresponding Hamiltonian system. Actually, the method that we are going to present applies to more general first-order partial differential equations. However, we restrict ourselves to the Hamilton–Jacobi equation here. This approach is called the *method of characteristics*, and it is regarded as a part of the *Hamilton–Jacobi theory*.

We consider the following initial value problem for the Hamilton–Jacobi equation:

$$(2.23)\qquad \begin{cases} v_t+H(t,x,v_x)=0,\\ v\big|_{t=0}=\psi(x).\end{cases}$$

Here, the function H is a general nonlinear function and not necessarily the one determined through (2.10). We assume that H and ψ are smooth enough. No growth conditions or monotonicity conditions are assumed for

H. Thus, we can study only the *local solvability*, by which we mean that there exists a $t_0 > 0$ such that (2.23) admits a solution on $[0, t_0]$. We suggest the following procedure to solve (2.23) locally:

Step 1. For some $t_0 > 0$, solve the following Hamiltonian system:

$$\begin{cases} \dot{x}(t) = H_p(t, x(t), p(t)), & t \in [0, t_0], \\ \dot{p}(t) = -H_x(t, x(t), p(t)), & t \in [0, t_0], \\ x(0) = \xi \in \mathbb{R}^n, \\ p(0) = \psi_\xi(\xi). \end{cases} \tag{2.24}$$

In the present context, (2.24) is called the system of *characteristic equations* of (2.23). The solution of (2.24), denoted by $(x(t,\xi), p(t,\xi))$, is called a *characteristic strip* of (2.23).

Step 2. From $x = x(t,\xi)$, solve for $\xi = \xi(t,x)$.

Step 3. Define

$$\begin{aligned} v(t,\xi) = \psi(\xi) + \int_0^t \{ \langle H_p(r, x(r,\xi), p(r,\xi)), p(r,\xi) \rangle \\ - H(r, x(r,\xi), p(r,\xi)) \} dr, \end{aligned} \tag{2.25}$$

and substitute $\xi = \xi(t,x)$ (obtained in Step 2) into (2.25) to get

$$v(t,x) = v(t, \xi(t,x)). \tag{2.26}$$

Theorem 2.4. *The above Steps 1–3 can be processed, and the function $v(t,x)$ defined by (2.26) is a local solution of (2.23).*

Proof. First of all, from the assumed condition, there exists a unique solution $(x(t,\xi), p(t,\xi))$ of (2.24) for $t_0 > 0$ sufficiently small. Next, since

$$\frac{\partial x(t,\xi)}{\partial \xi}\Big|_{t=0} = I, \tag{2.27}$$

we can solve for $\xi = \xi(t,x)$ from $x = x(t,\xi)$ for all $t > 0$ small enough. Thus, Steps 1–3 can be processed.

We now prove that $v(t,x)$ is a local solution of (2.23). To this end, let us define

$$s(t,\xi) = -H(0, \xi, \psi_\xi(\xi)) - \int_0^t H_t(r, x(r,\xi), p(r,\xi)) dr. \tag{2.28}$$

Then it follows that (noting (2.24))

$$\begin{aligned} &\frac{d}{dt}\{ s(t,\xi) + H(t, x(t,\xi), p(t,\xi)) \} \\ &\quad = -H_t + H_t + \langle H_x, \dot{x} \rangle + \langle H_p, \dot{p} \rangle = 0. \end{aligned} \tag{2.29}$$

Setting $t = 0$ in (2.28), one has

$$s(0,\xi) + H(0, x(0,\xi), p(0,\xi)) = 0. \tag{2.30}$$

Combining (2.29)–(2.30) yields

$$s(t,\xi)+H(t,x(t,\xi),p(t,\xi))\equiv 0. \tag{2.31}$$

On the other hand, substituting the relation $\xi=\xi(t,x)$ (solved from $x=x(t,\xi)$ in Step 2) into the expressions $s(t,\xi)$ and $p(t,\xi)$, we obtain the functions $s(t,x)$ and $p(t,x)$. For these functions, by (2.31) we have

$$s(t,x)+H(t,x,p(t,x))\equiv 0. \tag{2.32}$$

Thus, comparing (2.32) with (2.23), we see that to prove $v(t,x)$ to be a (local) solution of (2.23), it suffices to show that

$$s(t,x)=v_t(t,x),\qquad p(t,x)=v_x(t,x). \tag{2.33}$$

We now prove (2.33). Differentiating (2.31) with respect to ξ_i gives

$$\begin{aligned}&s_{\xi_i}(t,\xi)+\langle H_x(t,x(t,\xi),p(t,\xi)),x_{\xi_i}(t,\xi)\rangle\\&\quad+\langle H_p(t,x(t,\xi),p(t,\xi)),p_{\xi_i}(t,\xi)\rangle=0.\end{aligned} \tag{2.34}$$

Next, let $U(t,\xi)\stackrel{\Delta}{=}(U^1(t,\xi),\ldots,U^n(t,\xi))^\top$ with

$$U^i(t,\xi)\stackrel{\Delta}{=}v_{\xi_i}(t,\xi)-\langle x_{\xi_i}(t,\xi),p(t,\xi)\rangle,\qquad 1\le i\le n, \tag{2.35}$$

or, equivalently,

$$U(t,\xi)\stackrel{\Delta}{=}v_\xi(t,\xi)-x_\xi(t,\xi)^\top p(t,\xi),$$

where

$$x_\xi=\begin{pmatrix}x_{1\xi_1}&\cdots&x_{1\xi_n}\\ \vdots&&\vdots\\ x_{n\xi_1}&\cdots&x_{n\xi_n}\end{pmatrix}\equiv\frac{\partial x(t,\xi)}{\partial\xi}.$$

We also let (noting (2.24), (2.25), and (2.31))

$$\begin{aligned}V(t,\xi)&\stackrel{\Delta}{=}v_t(t,\xi)-\langle x_t(t,\xi),p(t,\xi)\rangle-s(t,\xi)\\&=\langle H_p(t,x(t,\xi),p(t,\xi)),p(t,\xi)\rangle-H(t,x(t,\xi),p(t,\xi))\\&\quad-\langle H_p(t,x(t,\xi),p(t,\xi)),p(t,\xi)\rangle-s(t,\xi)\equiv 0.\end{aligned} \tag{2.36}$$

Consequently, by (2.24) and (2.34), we obtain (noting (2.35)–(2.36))

$$\begin{aligned}U_t^i(t,\xi)&=U_t^i(t,\xi)-V_{\xi_i}(t,\xi)\\&=v_{\xi_i t}(t,\xi)-\langle x_{\xi_i t}(t,\xi),p(t,\xi)\rangle-\langle x_{\xi_i}(t,\xi),p_t(t,\xi)\rangle\\&\quad-v_{\xi_i t}(t,\xi)+\langle x_{\xi_i t}(t,\xi),p(t,\xi)\rangle\\&\quad+\langle x_t(t,\xi),p_{\xi_i}(t,\xi)\rangle+s_{\xi_i}(t,\xi)\\&=\langle H_x(t,x(t,\xi),p(t,\xi)),x_{\xi_i}(t,\xi)\rangle\\&\quad+\langle H_p(t,x(t,\xi),p(t,\xi)),p_{\xi_i}(t,\xi)\rangle+s_{\xi_i}(t,\xi)=0.\end{aligned} \tag{2.37}$$

By (2.24), (2.25), and (2.27),

$$U(0,\xi) = \psi_\xi(\xi) - p(0,\xi) = 0, \tag{2.38}$$

which yields

$$0 \equiv U(t,\xi) = v_\xi(t,\xi) - x_\xi(t,\xi)^\top p(t,\xi). \tag{2.39}$$

On the other hand, by differentiating $v(t,\xi) = v(t,x(t,\xi))$ in ξ, we have

$$v_\xi(t,\xi) = x_\xi(t,\xi)^\top v_x(t,x(t,\xi)). \tag{2.40}$$

Hence, it follows from (2.39)–(2.40) that

$$x_\xi(t,\xi)^\top \{p(t,\xi) - v_x(t,x(t,\xi))\} = 0.$$

By the nondegeneracy of $x_\xi(t,\xi)$ (for $t > 0$ small), we obtain

$$p(t,\xi) \equiv v_x(t,x(t,\xi)). \tag{2.41}$$

The second relation in (2.33) is then proved.

Finally, differentiating $x \equiv x(t,\xi(t,x))$ with respect to t yields

$$0 = x_t(t,\xi(t,x)) + x_\xi(t,\xi(t,x))\xi_t(t,x). \tag{2.42}$$

By (2.36) and (2.40)–(2.42), we obtain

$$\begin{aligned}
\frac{\partial}{\partial t}[v(t,x)] &= \frac{\partial}{\partial t}[v(t,\xi(t,x))] \\
&= v_t(t,\xi(t,x)) + \langle\, v_\xi(t,\xi(t,x)), \xi_t(t,x)\,\rangle \\
&= s(t,\xi(t,x)) + \langle\, p(t,\xi(t,x)), x_t(t,\xi(t,x))\,\rangle \\
&\quad + \langle\, x_\xi(t,\xi(t,x))^\top v_x(t,x(t,\xi(t,x))), \xi_t(t,x)\,\rangle \\
&= s(t,\xi(t,x)) \\
&\quad + \langle\, p(t,\xi(t,x)), x_t(t,\xi(t,x)) + x_\xi(t,\xi(t,x))\xi_t(t,x)\,\rangle \\
&= s(t,\xi(t,x)) = s(t,x).
\end{aligned}$$

This proves the first relation in (2.33), and our theorem is proved. □

We see that (2.25)–(2.26) gives a representation of solutions v to Hamilton–Jacobi equation (2.23) by solutions to Hamiltonian system (2.24). We refer to (2.25)–(2.26) as a *deterministic Feynman–Kac formula*. Here, we borrow this name from stochastic analysis, where the Feynman–Kac formula gives a representation of solutions to linear *second-order* parabolic or elliptic PDEs by solutions to some stochastic differential equations. See Section 3.3 of this chapter and Chapter 7 for more details.

Observe that if the function H in (2.23) is determined by (2.10), then using (2.24), we can rewrite (2.25) as follows:

$$\begin{aligned} v(t,\xi) &= \psi(\xi) + \int_0^t \Big\{ \langle \dot{x}(r), p(r) \rangle - \langle p(r), \dot{x}(r) \rangle \\ &\qquad\qquad + L(r, x(r), \dot{x}(r)) \Big\} dr \\ &= \psi(\xi) + \int_0^t L(r, x(r), \dot{x}(r)) dr. \end{aligned} \tag{2.43}$$

Thus, in this case, we have (see (2.5))

$$J(x(\cdot)) \stackrel{\Delta}{=} \int_{t_1}^{t_2} L(t, x(t), \dot{x}(t)) dt = v(t_2, \xi) - v(t_1, \xi). \tag{2.44}$$

The function $v(t,\xi)$ is called *Hamilton's principal function.* It is seen that Bellman's principle of optimality proved in Chapter 4, Theorem 2.1, is very similar to the above. In fact, we can regard Bellman's principle as a generalization of (2.44).

Combining Theorems 2.3 and 2.4, we may roughly conclude that the Hamiltonian system (2.13) (or (2.24)) and the Hamilton–Jacobi equation (2.14) (or (2.23)) are equivalent. We point out that the crucial relation linking these two equations is the second relation in (2.16) or (2.33). We will see later that such a relation holds as well in the context of optimal controls (both deterministic and stochastic), and it plays an essential role in understanding the general relationship between MP and DP.

To conclude this section, let us make one more observation. Suppose $T > 0$ and we have a terminal value problem of following type:

$$\begin{cases} -v_t + H(t, x, -v_x) = 0, & (t,x) \in [0,T] \times \mathbb{R}^n, \\ v\big|_{t=T} = \psi(x), & x \in \mathbb{R}^n. \end{cases} \tag{2.45}$$

One can make a time-reverse transformation $s \mapsto T - t$ to transform (2.45) into a form similar to (2.23). Then the above local solvability result applies (on $[T - t_0, T]$ for small $t_0 > 0$). One may also directly develop the (local) solvability of (2.45) with exactly the same idea as before. We leave the details to the interested reader as an exercise.

3. Relationship for Deterministic Systems

Let us recall the deterministic optimal control problem formulated in Chapters 2–4. Consider the *deterministic controlled system*

$$\begin{cases} \dot{x}(t) = b(t, x(t), u(t)), & \text{a.e. } t \in [s,T], \\ x(s) = y, \end{cases} \tag{3.1}$$

with the *cost functional*

$$J(s, y; u(\cdot)) = \int_s^T f(t, x(t), u(t)) dt + h(x(T)), \tag{3.2}$$

where $(s,y) \in [0,T] \times \mathbb{R}^n$ is given representing the *initial time* and the *initial state* of the system, and the *control* $u(\cdot)$ is taken from the following set:

$$\mathcal{V}[s,T] \stackrel{\Delta}{=} \{u : [s,T] \to U \mid u(\cdot) \text{ measurable}\}.$$

The optimal control problem can be stated as follows.

Problem $(\mathbf{D}_{sy})$. For given $(s,y) \in [0,T) \times \mathbb{R}^n$, minimize (3.2) subject to (3.1) over $\mathcal{V}[s,T]$.

The *value function* is defined as

$$\begin{cases} V(s,y) = \inf_{u(\cdot)\in\mathcal{V}[s,T]} J(s,y;u(\cdot)), & (s,y) \in [0,T) \times \mathbb{R}^n, \\ V(T,y) = h(y), & y \in \mathbb{R}^n. \end{cases} \tag{3.3}$$

We recall assumptions (D1) from Chapter 3, Section 2, and (D2)′ from Chapter 4, Section 2.

(D1) (U,d) is a separable metric space, and $T > 0$.

(D2)′ The maps $b : [0,T] \times \mathbb{R}^n \times U \to \mathbb{R}^n$, $f : [0,T] \times \mathbb{R}^n \times U \to \mathbb{R}$, and $h : \mathbb{R}^n \to \mathbb{R}$ are uniformly continuous, and there exists a constant $L > 0$ such that for $\varphi(t,x,u) = b(t,x,u), f(t,x,u), h(x)$,

$$\begin{cases} |\varphi(t,x,u) - \varphi(t,\widehat{x},u)| \le L|x - \widehat{x}|, \\ \qquad\qquad \forall t \in [0,T],\ x,\widehat{x} \in \mathbb{R}^n,\ u \in U, \\ |\varphi(t,0,u)| \le L, \qquad \forall (t,u) \in [0,T] \times U. \end{cases} \tag{3.4}$$

Let us also introduce the following:

(D3)′ The maps b, f, and h are C^1 in x. Moreover, there exists a modulus of continuity $\overline{\omega} : [0,\infty) \to [0,\infty)$ such that for $\varphi(t,x,u) = b(t,x,u), f(t,x,u), h(x)$,

$$\begin{aligned} &|\varphi_x(t,x,u) - \varphi_x(t,\widehat{x},u)| \le \overline{\omega}(|x - \widehat{x}|), \\ &\qquad\qquad \forall t \in [0,T], x, \widehat{x} \in \mathbb{R}^n, u \in U. \end{aligned} \tag{3.5}$$

The difference between (D3)′ and (D3) (of Chapter 3, Section 2) is that no continuity in u has been assumed in the former.

Recall that the *Hamilton–Jacobi–Bellman* (HJB) *equation* associated with the optimal control problem is as follows (see Chapter 4, Section 2).

$$\begin{cases} -v_t(t,x) + \sup_{u\in U} H(t,x,u,-v_x(t,x)) = 0, & (t,x) \in [0,T) \times \mathbb{R}^n, \\ v\big|_{t=T} = h(x), \quad x \in \mathbb{R}^n, \end{cases} \tag{3.6}$$

where the *Hamiltonian* H is defined as

$$\begin{aligned} &H(t,x,u,p) \stackrel{\Delta}{=} \langle p, b(t,x,u) \rangle - f(t,x,u), \\ &\qquad\qquad (t,x,u,p) \in [0,T] \times \mathbb{R}^n \times U \times \mathbb{R}^n. \end{aligned} \tag{3.7}$$

On the other hand, the dynamics (3.1) and its *adjoint system* along with the maximum condition (as given by the maximum principle; see Chapter 3, Section 2) can be written as

$$
(3.8)\quad \begin{cases} \dot{x}(t) = H_p(t, x(t), u(t), p(t)), & \text{a.e. } t \in [s, T], \\ \dot{p}(t) = -H_x(t, x(t), u(t), p(t)), & \text{a.e. } t \in [s, T], \\ x(s) = y, \\ p(T) = -h_x(x(T)), \\ H(t, x(t), u(t), p(t)) = \max_{u \in U} H(t, x(t), u, p(t)), & \text{a.e. } t \in [s, T]. \end{cases}
$$

This system is a two-point boundary value problem (coupled through the maximum condition). System (3.8) is called an (*extended*) *Hamiltonian system.* It takes the same form as the Hamiltonian system (2.24) in mechanics (after a time reversal for p), in which case the maximum condition is satisfied automatically. Recall that (see Chapter 3, Section 2) when $(x(\cdot), u(\cdot))$ is an optimal pair of Problem (D_{sy}) and $p(\cdot)$ is the corresponding adjoint variable, we call $(x(\cdot), u(\cdot), p(\cdot))$ an *optimal triple* of Problem (D_{sy}). Sometimes, we also refer to (3.8) as a system of *forward-backward differential equations* (in order to be able to compare a similar system in the stochastic case).

We note that the two approaches, maximum principle and dynamic programming, are nothing but analogies of the Hamiltonian system and the Hamilton–Jacobi equation, respectively, in classical mechanics. Certainly, the present situation is more complicated, since unlike (2.24), the control $u(\cdot)$ appears in (3.8), which is subject to the maximum condition. In addition, unlike (2.23), $\sup_{u \in U}$ appears in (3.6), which makes the coefficient of the equation nonsmooth in v_x. These are all due to the underlying optimization problem. In the following subsections we shall systematically investigate the relationship between (3.6) and (3.8) from various aspects.

3.1. Adjoint variable and value function: Smooth case

The adjoint variable $p(\cdot)$ and the value function $V(\cdot\,,\cdot)$ play the key roles in the maximum principle and dynamic programming approaches, respectively. The following result tells that $p(\cdot)$ and $V(\cdot\,,\cdot)$ are actually closely related, at least formally.

Theorem 3.1. *Let (D1) and (D2)′–(D3)′ hold and $(s, y) \in [0, T) \times \mathbb{R}^n$ be fixed. Let $(\bar{x}(\cdot), \bar{u}(\cdot), p(\cdot))$ be an optimal triple of Problem (D_{sy}). Suppose that the value function $V \in C^{1,1}([0, T] \times \mathbb{R}^n)$. Then*

$$
(3.9)\quad \begin{aligned} V_t(t, \bar{x}(t)) &= H\big(t, \bar{x}(t), \bar{u}(t), -V_x(t, \bar{x}(t))\big) \\ &= \max_{u \in U} H\big(t, \bar{x}(t), u, -V_x(t, \bar{x}(t))\big), \quad \text{a.e. } t \in [s, T]. \end{aligned}
$$

Further, if $V \in C^{1,2}([0, T] \times \mathbb{R}^n)$ and V_{tx} is also continuous, then

$$
(3.10)\qquad V_x(t, \bar{x}(t)) = -p(t), \qquad \forall t \in [s, T].
$$

Proof. By the optimality of $(\bar{x}(\cdot), \bar{u}(\cdot))$, we have

$$V(t, \bar{x}(t)) = h(\bar{x}(T)) + \int_t^T f(r, \bar{x}(r), \bar{u}(r))dr, \quad \forall t \in [s, T]. \tag{3.11}$$

Differentiating both sides of (3.11) with respect to t, one has

$$V_t(t, \bar{x}(t)) + \langle V_x(t, \bar{x}(t)), b(t, \bar{x}(t), \bar{u}(t)) \rangle = -f(t, \bar{x}(t), \bar{u}(t)),$$

or (see (3.7))

$$V_t(t, \bar{x}(t)) = H\big(t, \bar{x}(t), \bar{u}(t), -V_x(t, \bar{x}(t))\big), \quad \text{a.e. } t \in [0, T]. \tag{3.12}$$

This gives the first equality in (3.9). On the other hand, since $V \in C^{1,1}([0, T] \times \mathbb{R}^n)$, this function satisfies the HJB equation (3.6). The second equality in (3.9) then follows. Next, combining (3.6) and (3.12), we obtain

$$\begin{aligned} &H\big(t, \bar{x}(t), \bar{u}(t), -V_x(t, \bar{x}(t))\big) - V_t(t, \bar{x}(t)) \\ &\qquad = 0 \geq H\big(t, x, \bar{u}(t), -V_x(t, x)\big) - V_t(t, x), \quad \forall x \in \mathbb{R}^n. \end{aligned} \tag{3.13}$$

Hence, if we further assume $V \in C^{1,2}([0, T] \times \mathbb{R}^n)$ with V_{tx} being also continuous, then (3.13) gives rise to

$$\frac{\partial}{\partial x}\big\{H(t, x, \bar{u}(t), -V_x(t, x)) - V_t(t, x)\big\}\Big|_{x=\bar{x}(t)} = 0.$$

A simple variant of the above equality is

$$\begin{aligned} -\frac{d}{dt}V_x(t, \bar{x}(t)) =& b_x(t, \bar{x}(t), \bar{u}(t))^\top V_x(t, \bar{x}(t)) \\ &+ f_x(t, \bar{x}(t), \bar{u}(t)), \qquad t \in [s, T]. \end{aligned}$$

Also, it is clear that $-V_x(T, \bar{x}(T)) = -h_x(\bar{x}(T))$. Hence $-V_x(t, \bar{x}(t))$ satisfies the adjoint equation (see (3.8)), and therefore (3.10) follows from the uniqueness of the solutions to the adjoint equation (for given $(\bar{x}(\cdot), \bar{u}(\cdot))$). $\square$

It is interesting to note that the second equality in (3.9) (combined with (3.10)) is nothing but the maximum condition appearing in the MP. Hence, we have actually shown in the above proof that the maximum principle (for unconstrained problems) can be derived *directly* from dynamic programming if V is smooth enough.

As we have seen in Chapter 4, the value function is not necessarily continuously differentiable. Thus, Theorem 3.1 is nothing more than a heuristic statement. We will discuss the general case in Section 3.4 below.

On the other hand, Theorem 3.1 can be regarded as a generalization of the relation presented in Section 2 between the Hamilton–Jacobi equations and the Hamiltonian systems. Relation (3.10) is essentially analogous to the second relation in (2.16) or (2.33) (after a time reversal). Furthermore, we point out that such a relation is both a key to understanding the economic interpretation of the optimal control theory and a bridge to connect the maximum principle and dynamic programming. We shall respectively elaborate these two points in the next two subsections.

3.2. Economic interpretation

Theorem 3.1 enables us to give an interesting interpretation of the adjoint variable $p(t)$, which plays a very important role in economic theory. Let us now explain this.

Consider a *decision-making problem* of a manufacturing firm that wishes to maximize its total profit over a period of time $[0, T]$. Denote by $x(t)$ the *resource* (such as manufacturing facility, manpower, capital) that the firm has at time t, and by $u(t)$ some *decisions* (such as production rate, personnel policy, marketing strategy) taken at t. For simplicity, we assume that all the variables are scalar-valued. The rate of change in the resource depends on the time, the present situation of the resource, and the decision taken. Therefore, we have

$$\begin{cases} \dot{x}(t) = r(t, x(t), u(t)), & t \in [0, T], \\ x(0) = x_0, \end{cases} \tag{3.14}$$

where r is some function representing the above dependence and x_0 is the initial resource. Let $n(t, x, u)$ denote the rate at which the net profit is being earned at time t as a result of having resource x and taking decision u. The problem is then to maximize the total profit earned during the time period $[0, T]$, which is $\int_0^T n(t, x(t), u(t))dt$, or equivalently to minimize

$$J = -\int_0^T n(t, x(t), u(t))dt. \tag{3.15}$$

We let $V(t, x)$ be the value function corresponding to the problem of minimizing (3.15), and $(\bar{x}(\cdot), \bar{u}(\cdot), p(\cdot))$ be an optimal triple.

Suppose at any time t the optimal state value $\bar{x}(t)$ is slightly changed to $\bar{x}(t) + \delta x(t)$; then by (3.10), the effect of this change on the minimum value of the cost is

$$V(t, \bar{x}(t) + \delta x(t)) - V(t, \bar{x}(t)) \approx -p(t)\delta x(t). \tag{3.16}$$

Intuitively, the more the resource, the better the performance (i.e., the smaller the value of V). Thus, we accept the fact that both sides of (3.16) are negative if $\delta x(t) > 0$, or positive if $\delta x(t) < 0$. Therefore, $p(t)$ is positive, and it measures the rate at which the best performance could be increased (decreased) by slightly increasing (decreasing) the amount of the resource. We call $p(t)$ the *marginal value* of the resource. Alternatively, if the increment in resource could be realized by purchasing the resource in the market outside the firm, then $p(t)$ represents the *maximum* price that would be worth paying. For this reason, we also call $p(t)$ the *shadow price* of the resource. This gives the adjoint variable $p(t)$ a clear economic meaning, which is suggested by the relationship (3.10).

Once this is understood, all the equations involved in the maximum principle make perfect sense in economic theory. For example, the adjoint equation

$$-\dot{p}(t) = n_x(t, \overline{x}(t), \overline{u}(t)) + r_x(t, \overline{x}(t), \overline{u}(t))p(t) \tag{3.17}$$

simply means that along the optimal path $\overline{x}(\cdot)$ of the resource, the *depreciation rate* $-\dot{p}(t)$ of the marginal value of the resource over a short period of time is the sum of its contribution $n_x(t, \overline{x}(t), \overline{u}(t))$ to the profit made in the period and its contribution $r_x(t, \overline{x}(t), \overline{u}(t))p(t)$ to the enhancement of the marginal value of the resource at the end of the period. Certainly, $p(T) = 0$ (noting (3.8) and (3.15)) means that the resource should be used up if the firm does not wish to leave any bequest behind. On the other hand, the Hamiltonian in this problem is

$$H(t, x, u) = n(t, x, u) + r(t, x, u)p(t), \tag{3.18}$$

which signifies the sum of the direct contribution rate $n(t, x, u)$ to the overall profit and the accumulation rate $r(t, x, u)p(t)$ of the marginal value of the resource. Understandably, the decision at any time should be chosen to make this sum as large as possible. This is the maximum principle. Finally, once an optimal path is followed, this sum is equal to V_t (see (3.9)), representing the rate of change for the maximum possible profit with respect to time.

3.3. Method of characteristics and the Feynman–Kac formula

One of the most important implications of Theorem 3.1 is that it reveals a deep relationship between MP and DP, in a way analogous to that between systems of ordinary differential equations and first-order partial differential equations via the classical *method of characteristics.* We have already seen such an example in Theorem 2.4. Now we proceed to explore it in the context of optimal controls.

We have seen from Section 2 that for a first-order partial differential equation there is an associated family of ordinary differential equations for curves, the characteristic strips, by which the solutions to the partial differential equation can be constructed. It happens that the family of adjoint equations along with the original system equations (which are ODEs) parametrized by the initial (s, y) serve as the characteristic strips for the HJB equation (which is a PDE). Specifically, we have the following result.

Theorem 3.2. *Let (D1) and (D2)′-(D3)′ hold. Suppose there exists a control $u(\cdot) \in \mathcal{V}[0,T]$ such that for any $(s, y) \in [0,T] \times \mathbb{R}^n$ there exists a pair $(x^{s,y}(\cdot), p^{s,y}(\cdot))$ satisfying*

$$\begin{cases} \dot{x}^{s,y}(t) = H_p(t, x^{s,y}(t), u(t), p^{s,y}(t)), & t \in [s,T], \\ \dot{p}^{s,y}(t) = -H_x(t, x^{s,y}(t), u(t), p^{s,y}(t)), & t \in [s,T], \\ x^{s,y}(s) = y, \\ p^{s,y}(T) = -h_x(x^{s,y}(T)), \end{cases} \tag{3.19}$$

and

$$H(s,y,u(s),p^{s,y}(s)) = \max_{u\in U} H(s,y,u,p^{s,y}(s)), \ \forall (s,y)\in[0,T]\times\mathbb{R}^n. \tag{3.20}$$

Then the function

$$J(s,y) \triangleq \int_s^T f(t,x^{s,y}(t),u(t))dt + h(x^{s,y}(T)) \tag{3.21}$$

solves the HJB equation (3.6).

Proof. First, by a standard result in ordinary differential equation theory $x^{s,y}(\cdot)$ is differentiable in (s,y) for any $y\in\mathbb{R}^n$ and almost all $s\in[0,T)$. Define

$$m(t) \triangleq \frac{\partial}{\partial y} x^{s,y}(t) \quad \text{and} \quad n(t) \triangleq \frac{\partial}{\partial s} x^{s,y}(t). \tag{3.22}$$

Then

$$\begin{cases} \dot m(t) = b_x(t,x^{s,y}(t),u(t))m(t), & t\in[s,T],\\ m(s) = I, \end{cases} \tag{3.23}$$

and

$$\begin{cases} \dot n(t) = b_x(t,x^{s,y}(t),u(t))n(t), & t\in[s,T],\\ n(s) = -b(s,y,u(s)). \end{cases} \tag{3.24}$$

Second, we directly compute that (making use of (3.22))

$$J_y(s,y) = \int_s^T m(t)^\top f_x(t,x^{s,y}(t),u(t))dt + m(T)^\top h_x(x^{s,y}(T)) \tag{3.25}$$

and

$$\begin{aligned} J_s(s,y) = -f(s,y,u(s)) &+ \int_s^T \langle f_x(t,x^{s,y}(t),u(t)), n(t)\rangle\, dt \\ &+ \langle h_x(x^{s,y}(T)), n(T)\rangle . \end{aligned} \tag{3.26}$$

By (3.23) and (3.19), one has

$$\frac{d}{dt}(m(t)^\top p^{s,y}(t)) = m(t)^\top f_x(t,x^{s,y}(t),u(t)).$$

Hence, integrating this equation from s to T, we get

$$-m(T)^\top h_x(x^{s,y}(T)) - p^{s,y}(s) = \int_s^T m(t)^\top f_x(t,x^{s,y}(t),u(t))dt. \tag{3.27}$$

Comparing (3.27) with (3.25), we obtain

$$J_y(s,y) = -p^{s,y}(s). \tag{3.28}$$

Similarly, by (3.24) and (3.19), we have

$$\frac{d}{dt}\langle p^{s,y}(t), n(t)\rangle = \langle f_x(t, x^{s,y}(t), u(t)), n(t)\rangle .$$

Thus,

$$\begin{aligned} & -\langle h_x(x^{s,y}(T)), n(T)\rangle + \langle p^{s,y}(s), b(s,y,u(s))\rangle \\ & \qquad = \int_s^T \langle f_x(t, x^{s,y}(t), u(t)), n(t)\rangle\, dt. \end{aligned} \tag{3.29}$$

It follows from (3.26) and (3.29) that

$$J_s(s,y) = H(s,y,u(s),p^{s,y}(s)). \tag{3.30}$$

Therefore, by the maximum condition (3.20), together with (3.28) and (3.30), we conclude that

$$\begin{aligned} & -J_s(s,y) + \sup_{u\in U} H(s,y,u,-J_y(s,y)) \\ & \qquad = -H(s,y,u(s),p^{s,y}(s)) + \max_{u\in U} H(s,y,u,p^{s,y}(s)) = 0. \end{aligned}$$

Further, it is clear that J satisfies the terminal condition in (3.6). The desired result is therefore proved. □

Note that Theorem 3.2 is very similar to Theorem 2.4. We have represented the solution of (3.6) by (3.21), which involves the solution of (3.19), under certain conditions. Again, the key of the proof is (3.28), which is analogous to (3.10) or the second relation in (2.33) and (2.16).

Note that the assumption (3.20), which is important in the above proof, is a rather strong condition. Indeed, (3.20) requires that the maximum of H be attained at $u(s)$ for *all* $(s,y) \in \mathbb{R}^{n+1}$. This may be valid only for very special cases.

To conclude this subsection, let us look at the case where U is a singleton. In this case, the HJB equation (3.6) is linear.

Corollary 3.3. *Let $x^{s,y}(\cdot)$ be the solution of*

$$\begin{cases} \dot{x}^{s,y}(t) = b(t, x^{s,y}(t)), & t \in [s,T], \\ x^{s,y}(s) = y. \end{cases} \tag{3.31}$$

Then the solution of the first-order PDE

$$\begin{cases} v_t(t,x) + \langle b(t,x), v_x(t,x)\rangle + f(t,x) = 0, & (t,x) \in [0,T) \times \mathbb{R}^n, \\ v\big|_{t=T} = h(x), \end{cases} \tag{3.32}$$

has the following representation:

$$v(s,y) = \int_s^T f(t, x^{s,y}(t))dt + h(x^{s,y}(T)). \tag{3.33}$$

Proof. Note that (3.20) is valid automatically in the present case. Thus, Theorem 3.2 applies. □

In probability theory there is the well-known Feynman–Kac formula, which gives a probabilistic representation of the solution to a linear (second-order) parabolic or elliptic partial differential equation by means of the solution to a stochastic differential equation (see Chapter 7, Theorem 4.1). Corollary 3.3 is actually a deterministic (namely, the diffusion term vanishes) version of the Feynman–Kac formula. Therefore, the relation between MP and DP presented in Theorem 3.1, as well as the method of characteristics, may be considered to be certain variants of the Feynman–Kac formula. Later, in Chapter 7, we shall study various Feynman–Kac-type formulae.

3.4. Adjoint variable and value function: Nonsmooth case

We have seen that Theorem 3.1 is of significant importance in connecting MP and DP. However, Theorem 3.1 is based on the assumption that the value function is smooth. This assumption does not necessarily hold, even in many simple cases (see Chapter 4). Therefore, we wish to know to what extent the relationship between MP and DP discussed in previous subsections remains valid, without assuming any differentiability of the value function. It turns out that the viscosity solution theory, once again, provides an excellent framework to deal with the problem.

In Chapter 4, Section 2, we have introduced the super- and subdifferentials $D^{1,\pm}_{t,x}v(t,x)$ for any $v \in C([0,T]\times \mathbb{R}^n)$. Also, we may define the partial super- and subdifferentials, with respect to either t or x. We denote them by $D^{1,\pm}_t v(t,x)$ and $D^{1,\pm}_x v(t,x)$, respectively.

From Chapter 4, Theorem 2.9, we know that $v \in C([0,T]\times \mathbb{R}^n)$ is a viscosity solution of the HJB equation (3.6) if and only if $v(T,x) = h(x)$, and for all $(t,x)\in [0,T)\times\mathbb{R}^n$,

$$
(3.34)\qquad \begin{cases} -q + \sup\limits_{u\in U} H(t,x,u,-p) \le 0, & \forall (q,p)\in D^{1,+}_{t,x}v(t,x); \\ -q + \sup\limits_{u\in U} H(t,x,u,-p) \ge 0, & \forall (q,p)\in D^{1,-}_{t,x}v(t,x). \end{cases}
$$

The following result is a *nonsmooth* version of Theorem 3.1.

Theorem 3.4. *Let (D1) and (D2)′–(D3)′ hold, and $(s,y)\in[0,T)\times\mathbb{R}^n$ be fixed. Let $(\bar{x}(\cdot),\bar{u}(\cdot),p(\cdot))$ be an optimal triple of Problem (D_{sy}). Then*

$$
(3.35)\qquad \begin{aligned} D^{1,-}_{t,x}V(t,\bar{x}(t)) &\subseteq \big\{\big(H(t,\bar{x}(t),\bar{u}(t),p(t)),-p(t)\big)\big\} \\ &\subseteq D^{1,+}_{t,x}V(t,\bar{x}(t)), \qquad \text{a.e. } t\in[s,T], \end{aligned}
$$

$$
(3.36)\qquad D^{1,-}_x V(t,\bar{x}(t)) \subseteq \{-p(t)\} \subseteq D^{1,+}_x V(t,\bar{x}(t)), \quad \forall t\in[s,T],
$$

and

$$
(3.37)\qquad \begin{aligned} &q = H(t,\bar{x}(t),\bar{u}(t),-p) = \max_{u\in U} H(t,\bar{x}(t),u,-p), \\ &\forall (q,p)\in D^{1,+}_{t,x}V(t,\bar{x}(t)) \cup D^{1,-}_{t,x}V(t,\bar{x}(t)), \quad \text{a.e. } t\in[s,T]. \end{aligned}
$$

We note that (3.36) holds for all $t \in [s,T]$, whereas (3.35) holds only for almost all $t \in [s,T]$. So (3.36) is not implied by (3.35). Combining (3.35) and (3.37), we obtain

$$H(t,\bar{x}(t),\bar{u}(t),p(t)) = \max_{u\in U} H(t,\bar{x}(t),u,p(t)), \quad \text{a.e. } t \in [s,T].$$

Thus, we recover the maximum principle from Theorem 3.4 *directly*. Note that we do not need the differentiability of the value function V here (note the remark following the proof of Theorem 3.1). In other words, the above shows that the maximum principle holds under (D1), (D2)′, and (D3)′, which coincides with the result in Chapter 3 (with (D2)–(D3) replaced by (D2)′–(D3)′).

In view of (3.36), the following inclusion always holds along any optimal path $\bar{x}(\cdot)$:

$$D_x^{1,-}V(t,\bar{x}(t)) \subseteq D_x^{1,+}V(t,\bar{x}(t)), \qquad \forall t \in [s,T].$$

Therefore, either $D_x^{1,-}V(t,\bar{x}(t))$ is empty or V is differentiable in x at $x = \bar{x}(t)$. In the latter case, the equalities in (3.36) hold, which recovers (3.10).

Similarly, from (3.35), we have

$$D_{t,x}^{1,-}V(t,\bar{x}(t)) \subseteq D_{t,x}^{1,+}V(t,\bar{x}(t)), \qquad \text{a.e. } t \in [s,T]. \tag{3.38}$$

Thus, along any optimal trajectory $\bar{x}(\cdot)$, for almost all $t \in [s,T]$, either $D_{t,x}^{1,-}V(t,\bar{x}(t))$ is empty or V is differentiable at $(t,\bar{x}(t))$. When the latter happens, one has set equalities in (3.35), which, along with (3.37), yields (3.9).

In particular, if V is differentiable (not necessarily *continuously* differentiable), all the inclusions in (3.35)–(3.36) become equalities. This implies (3.10) and (3.9). Thus, Theorem 3.1 is fully recovered with even *less* of a smoothness assumption on the value function V.

Proof of Theorem 3.4. First of all, we note that

$$\lim_{h\to 0} \frac{1}{h}\int_t^{t+h} \varphi(r)dr = \varphi(t), \qquad \text{a.e. } t \in (s,T), \tag{3.39}$$

for $\varphi(r) = b(r,\bar{x}(r),\bar{u}(r)), f(r,\bar{x}(r),\bar{u}(r))$. Fix a $t \in (s,T)$ such that (3.39) holds. For any $z \in \mathbb{R}^n$ and $\tau \in [s,T]$, denote by $x^{\tau,z}(\cdot)$ the solution of (3.1) starting from (τ,z) under the control $\bar{u}(\cdot)$. Note that τ is allowed to be smaller than t. Then, for any $r \in [\tau, T]$,

$$
\begin{aligned}
x^{\tau,z}(r) - \overline{x}(r) &= z - \overline{x}(t) - \int_t^\tau b(\theta, \overline{x}(\theta), \overline{u}(\theta))d\theta \\
&\quad + \int_\tau^r \big[b(\theta, x^{\tau,z}(\theta), \overline{u}(\theta)) - b(\theta, \overline{x}(\theta), \overline{u}(\theta))\big] d\theta \\
&= z - \overline{x}(t) - \int_t^\tau b(\theta, \overline{x}(\theta), \overline{u}(\theta))d\theta \\
&\quad + \int_\tau^r b_x(\theta, \overline{x}(\theta), \overline{u}(\theta))(x^{\tau,z}(\theta) - \overline{x}(\theta))d\theta \\
&\quad + \int_\tau^r \varepsilon_{\tau,z}(\theta)(x^{\tau,z}(\theta) - \overline{x}(\theta))d\theta,
\end{aligned} \tag{3.40}
$$

where

$$
\begin{cases}
\varepsilon_{\tau,z}(\theta) \overset{\Delta}{=} \displaystyle\int_0^1 \{b_x(\theta, \overline{x}(\theta) + \alpha(x^{\tau,z}(\theta) - \overline{x}(\theta)), \overline{u}(\theta)) \\
\qquad\qquad - b_x(\theta, \overline{x}(\theta), \overline{u}(\theta))\}d\alpha, \\
\displaystyle\lim_{\tau \to t, z \to \overline{x}(t)} \varepsilon_{\tau,z}(\theta) = 0, \qquad \forall \theta \in [0, T], \\
\displaystyle\sup_{\theta,\tau,z} |\varepsilon_{\tau,z}(\theta)| \le K.
\end{cases} \tag{3.41}
$$

Here, (3.5) has been used. Now, by the definition of $V(\tau, z)$ and the optimality of $(\overline{x}(\cdot), \overline{u}(\cdot))$, we have

$$
\begin{aligned}
&V(\tau, z) - V(t, \overline{x}(t)) \\
&\le \int_\tau^T \{f(r, x^{\tau,z}(r), \overline{u}(r)) - f(r, \overline{x}(r), \overline{u}(r))\}dr \\
&\quad + \{h(x^{\tau,z}(T)) - h(\overline{x}(T))\} - \int_t^\tau f(r, \overline{x}(r), \overline{u}(r))dr \\
&= \int_\tau^T \langle f_x(r, \overline{x}(r), \overline{u}(r)), x^{\tau,z}(r) - \overline{x}(r) \rangle \, dr \\
&\quad + \langle h_x(\overline{x}(T)), x^{\tau,z}(T) - \overline{x}(T) \rangle - \int_t^\tau f(r, \overline{x}(r), \overline{u}(r))dr \\
&\quad + \int_\tau^T \varepsilon^0_{\tau,z}(r)(x^{\tau,z}(r) - \overline{x}(r))dr + o(|x^{\tau,z}(T) - \overline{x}(T)|),
\end{aligned} \tag{3.42}
$$

where $\varepsilon^0_{\tau,z}(\cdot)$ is defined in a fashion similar to $\varepsilon_{\tau,z}(\cdot)$, with b_x replaced by f_x, and has the same properties as specified in (3.41). Recall that $p(\cdot)$ is the solution of the adjoint equation. Thus, by the duality relation between

$p(\cdot)$ and $x^{\tau,z}(\cdot)-\bar{x}(\cdot)$, one has

$$\begin{aligned}&\int_\tau^T \langle f_x(r,\bar{x}(r),\bar{u}(r)), x^{\tau,z}(r)-\bar{x}(r)\rangle\,dr\\&\quad+\langle h_x(\bar{x}(T)), x^{\tau,z}(T)-\bar{x}(T)\rangle\\&=-\langle p(\tau), z-\bar{x}(\tau)\rangle-\int_\tau^T \langle p(r),\varepsilon_{\tau,z}(r)[x^{\tau,z}(r)-\bar{x}(r)]\rangle\,dr\\&=\langle -p(t)+o(1), z-\bar{x}(t)-\int_t^\tau b(r,\bar{x}(r),\bar{u}(r))dr\rangle\\&\quad+o(|\tau-t|+|z-\bar{x}(t)|)\end{aligned}\tag{3.43}$$

Hence, (3.42) together with (3.43) yields (noting (3.39))

$$\begin{aligned}V(\tau,z)-V(t,\bar{x}(t))\leq&-\langle p(t),z-\bar{x}(t)\rangle+(\tau-t)H(t,\bar{x}(t),\bar{u}(t),p(t))\\&+o(|\tau-t|+|z-\bar{x}(t)|),\end{aligned}\tag{3.44}$$

which leads to the second inclusion in (3.35). This also implies that for such a t, $D^{1,+}_{t,x}V(t,\bar{x}(t))$ is nonempty. Now, for the other inclusion in (3.35) (still with a $t\in(s,T)$ satisfying (3.39)), if $D^{1,-}_{t,x}V(t,\bar{x}(t))=\phi$, then we are done. Otherwise, $V(t,x)$ must be differentiable at $(t,\bar{x}(t))$. In this case, equality in (3.35) holds due to its proved second relation. Either way, (3.35) must be true.

Taking $\tau=t$ in the above proof, we can show (3.36). In this case, we do not need t to satisfy (3.39). Therefore, (3.36) holds for *all* $t\in[s,T]$.

Finally, we prove (3.37). Again, take $t\in(s,T)$ such that (3.39) holds. If $(q,p)\in D^{1,+}_{t,x}V(t,\bar{x}(t))$, then by definition of superdifferential and Bellman's principle of optimality,

$$\begin{aligned}0&\geq\varlimsup_{r\downarrow t}\frac{V(r,\bar{x}(r))-V(t,\bar{x}(t))-q(r-t)-\langle p,\bar{x}(r)-\bar{x}(t)\rangle}{|r-t|+|\bar{x}(r)-\bar{x}(t)|}\\&=\Big\{\lim_{r\downarrow t}\frac{1}{|r-t|}\Big[-\int_t^r f(\theta,\bar{x}(\theta),\bar{u}(\theta))d\theta-q(r-t)\\&\quad-\int_t^r\langle p,b(\theta,\bar{x}(\theta),\bar{u}(\theta))\rangle\,d\theta\Big]\Big\}\cdot\varlimsup_{r\downarrow t}\frac{|r-t|}{|r-t|+|\bar{x}(r)-\bar{x}(t)|},\end{aligned}$$

where the second equality is valid because the limit of the first term on the right-hand side exists. Since $|\bar{x}(r)-\bar{x}(t)|\leq K|r-t|$ for some constant $K>0$, the above inequality yields

$$\begin{aligned}0&\geq -f(t,\bar{x}(t),\bar{u}(t))-q-\langle p,b(t,\bar{x}(t),\bar{u}(t))\rangle\\&=H(t,\bar{x}(t),\bar{u}(t),-p)-q.\end{aligned}$$

Similarly, by letting $r\uparrow t$, we can deduce the reverse inequality. Thus, the first equality in (3.37) follows. Next, as V is the viscosity solution of the HJB equation (3.6), by (3.34) we have

$$-q+\sup_{u\in U}H(t,\bar{x}(t),u,-p)\leq 0,$$

which yields the second equality of (3.37) by virtue of its proved first equality. Finally, due to (3.38) (which is implied by (3.35)) we obtain (3.37) for $(q,p) \in D^{1,-}_{t,x} V(t,\bar{x}(t))$. □

Although in a nonsmooth form, Theorem 3.4 still admits an economic interpretation. To see this, let us again consider the model in Section 3.2. Suppose at time t the optimal resource state $\bar{x}(t)$ is *increased* to $\bar{x}(t)+\delta x(t)$ ($\delta x(t) > 0$ small enough). Then the corresponding *improvement* on the performance is given by

$$V(t,\bar{x}(t)+\delta x(t)) - V(t,x(t)) \le -p(t)\delta x(t), \tag{3.45}$$

where the inequality is due to the second inclusion in (3.36), and we have neglected the term $o(|\delta x(t)|)$ on the right-hand side. As we noted in Section 3.2, both sides of (3.45) are negative (if $\delta x(t) > 0$). Thus,

$$|V(t,\bar{x}(t)+\delta x(t)) - V(t,x(t))| \ge p(t)|\delta x(t)|.$$

This implies that if the additional resource $\delta x(t)$ can be purchased in the market, then the resulting gain would justify the purchase as long as the unit price of the resource is no more expensive than $p(t)$. It follows that the marginal value of the resource is *at least* $p(t)$. On the other hand, suppose the optimal resource state $\bar{x}(t)$ is *decreased* to $\bar{x}(t)-\delta x(t)$ ($\delta x(t) > 0$), then the corresponding *loss* is given by

$$V(t,\bar{x}(t)-\delta x(t)) - V(t,x(t)) \le p(t)\delta x(t). \tag{3.46}$$

Since both sides (3.46) are positive, we conclude that

$$|V(t,\bar{x}(t)-\delta x(t)) - V(t,x(t))| \le p(t)|\delta x(t)|.$$

This basically says that if the company can sell the resource in the market, then the resulting additional income will compensate the loss in profit if the selling price is no cheaper than $p(t)$. So the marginal value of the resource is *at most* $p(t)$. In other words, $p(t)$ is exactly the marginal value or shadow price of the resource. In a similar fashion, we may also use (3.35) to interpret $H(t,\bar{x}(t),\bar{u}(t),p(t))$ as the rate of change for the maximum possible profit with respect to time (see Section 3.2).

We point out that Theorem 3.4 is a *true* extension of Theorem 3.1, by which we mean that it is possible to have strict set inclusions in (3.35)–(3.36). The following example gives such a situation.

Example 3.5. Consider the following control system: ($n = 1$)

$$\begin{cases} \dot{x}(t) = x(t)u(t), & t \in [s,T], \\ x(s) = y, \end{cases} \tag{3.47}$$

with the control domain being $U = [0,1]$ and the cost functional being

$$J(s,y;u(\cdot)) = -x(T). \tag{3.48}$$

The value function can be easily calculated as

$$(3.49)\qquad V(t,x)=\begin{cases}-x, & \text{if } x\le 0,\\ -xe^{T-t}, & \text{if } x>0.\end{cases}$$

Now consider Problem (D_{00}) (i.e., $s=0$ and $y=0$). Clearly, $(\bar{x}(\cdot),\bar{u}(\cdot))\equiv(0,0)$ is an optimal pair. One can show that

$$(3.50)\qquad \begin{cases}D_x^{1,-}V(t,\bar{x}(t))=\phi, & t\in[0,T),\\ D_x^{1,+}V(t,\bar{x}(t))=[-e^{T-t},-1], & t\in[0,T],\\ p(t)=1, & t\in[0,T],\end{cases}$$

and

$$(3.51)\qquad \begin{cases}D_{t,x}^{1,-}V(t,\bar{x}(t))=\phi, & t\in[0,T),\\ D_{t,x}^{1,+}V(t,\bar{x}(t))=\{0\}\times[-e^{T-t},-1], & t\in[0,T],\\ (H(t,\bar{x}(t),\bar{u}(t),p(t)),-p(t))=(0,-1), & t\in[0,T].\end{cases}$$

Thus, all the set inclusions in (3.35)–(3.36) are strict for $t\in[0,T)$.

Let us now discuss some easy but interesting consequences of Theorem 3.4. First of all, under (D1), (D2)$'$, and (D3)$'$, if V is convex in x, then V is differentiable in x at $\bar{x}(t)$ for all $t\in[s,T]$. The reason is that in this case, $D_x^{1,-}V(t,\bar{x}(t))$ coincides with the subgradient (in the sense of convex analysis; see Chapter 4, Proposition 2.6-(iv)) of the map $x\mapsto V(t,x)$ at $x=\bar{x}(t)$, which is *always* nonempty (see Chapter 3, Lemma 2.3-(i)). Then (3.36) implies that V has to be differentiable in x at $x=\bar{x}(t)$.

Second, from (3.37), we see that any point (q,p) in $D_{t,x}^{1,-}V(t,\bar{x}(t))$ or $D_{t,x}^{1,+}V(t,\bar{x}(t))$ is uniquely determined by the second entry p.

Third, if the map $p\mapsto \max_{u\in U}H(t,\bar{x}(t),u,-p)$ is *strictly* convex (note that this map is always convex), then $D_{t,x}^{1,+}V(t,\bar{x}(t))$ is a singleton, which has to be $\{(H(t,\bar{x}(t),\bar{u}(t),p(t)),-p(t))\}$ for almost all t. In fact, if there are two points $(q_i,p_i)\in D_{t,x}^{1,+}V(t,\bar{x}(t))$, $i=1,2$, with $p_1\ne p_2$, then by the convexity of $D_{t,x}^{1,+}V(t,\bar{x}(t))$, we have

$$\Big(\frac{q_1+q_2}{2},\frac{p_1+p_2}{2}\Big)\in D_{t,x}^{1,+}V(t,\bar{x}(t)).$$

Thus, by (3.37), we have

$$\begin{aligned}\frac{q_1+q_2}{2}&=\max_{u\in U}H(t,\bar{x}(t),u,-\frac{p_1+p_2}{2})\\ &<\frac{1}{2}\max_{u\in U}H(t,\bar{x}(t),u,-p_1)+\frac{1}{2}\max_{u\in U}H(t,\bar{x}(t),u,-p_2)\\ &=\frac{q_1+q_2}{2},\end{aligned}$$

which is a contradiction. Therefore, $p_1=p_2$, and then $q_1=q_2$.

Finally, note that both (3.35) and (3.36) serve as necessary conditions for $(\bar{x}(\cdot),\bar{u}(\cdot),p(\cdot))$ to be optimal. Thus, it is easy to state and prove some negative results concerning the optimality of given $x(\cdot)$ or $(x(\cdot), u(\cdot), p(\cdot))$. Below is one such result.

Corollary 3.6. *Let (D1) and (D2)′–(D3)′ hold. Let a state trajectory $x(\cdot)$ be such that the set*

$$\{t \in [0,T] \mid D_{t,x}^{1,-} V(t, x(t)) \text{ contains more than one point}\}$$

has a positive Lebesgue measure. Then $x(\cdot)$ is not optimal.

Proof. If $x(\cdot)$ is optimal, then for almost every $t \in [0,T]$, $D_{t,x}^{1,-} V(t, x(t))$ is either empty or a singleton by (3.35). This results in a contradiction. □

3.5. Verification theorems

Solving an optimal control problem requires finding an optimal control and the corresponding state trajectory. As we saw in Chapter 4, the main motivation of introducing dynamic programming is that one might be able to construct an optimal feedback control via the value function. The following result gives a way of testing whether a given admissible pair is optimal and, more importantly, suggests how to construct an optimal feedback control. Such a result is called a *verification theorem*.

Theorem 3.7. *Let (D1) and (D2)′ hold. Let $v \in C^{1,1}([0,T] \times \mathbb{R}^n)$ be a solution of the HJB equation (3.6). Then*

$$v(s,y) \le J(s,y;u(\cdot)), \quad \forall u(\cdot) \in \mathcal{V}[s,T], \ (s,y) \in [0,T) \times \mathbb{R}^n. \tag{3.52}$$

Furthermore, a given admissible pair $(\bar{x}(\cdot), \bar{u}(\cdot))$ is optimal for Problem (D_{sy}) if and only if

$$\begin{aligned} v_t(t, \bar{x}(t)) &\equiv \max_{u \in U} H\big(t, \bar{x}(t), u, -v_x(t, \bar{x}(t))\big) \\ &= H\big(t, \bar{x}(t), \bar{u}(t), -v_x(t, \bar{x}(t))\big), \quad \text{a.e. } t \in [s,T]. \end{aligned} \tag{3.53}$$

Proof. For any $u(\cdot) \in \mathcal{V}[s,T]$ with the corresponding state trajectory $x(\cdot)$, we have

$$\begin{aligned} \frac{d}{dt} v(t, x(t)) &= v_t(t, x(t)) + \langle\, v_x(t, x(t)), b(t, x(t), u(t))\,\rangle \\ &= -\, f(t, x(t), u(t)) \\ &\quad + \big\{ v_t(t, x(t)) - H\big(t, x(t), u(t), -v_x(t, x(t))\big)\big\} \\ &\ge -\, f(t, x(t), u(t)), \qquad \text{a.e. } t \in [s,T], \end{aligned} \tag{3.54}$$

where the last inequality is due to the HJB equation (3.6). Integrating both sides of (3.54) from s to T gives (3.52).

Next, applying the second equality of (3.54) to $(\overline{x}(\cdot), \overline{u}(\cdot))$, and integrating from s to T, we get

$$\begin{aligned} v(T,\overline{x}(T)) - v(s,y) = &- \int_s^T f(r,\overline{x}(r),\overline{u}(r))dr \\ &+ \int_s^T \Big\{v_t(r,\overline{x}(r)) - H\big(r,\overline{x}(r),\overline{u}(r),-v_x(r,\overline{x}(r))\big)\Big\}dr, \end{aligned}$$

or

$$\begin{aligned} v(s,y) &= J(s,y;\overline{u}(\cdot)) \\ &\quad - \int_s^T \Big\{v_t(r,\overline{x}(r)) - H\big(r,\overline{x}(r),\overline{u}(r),-v_x(r,\overline{x}(r))\big)\Big\}dr. \end{aligned} \tag{3.55}$$

The desired result follows immediately from the fact that

$$v_t(r,\overline{x}(r)) - H\big(r,\overline{x}(r),\overline{u}(r),-v_x(r,\overline{x}(r))\big) \geq 0, \quad \text{a.e. } r \in [s,T],$$

which is due to the HJB equation (3.6). □

In the above, v is called a *verification function*. When practically applying Theorem 3.7, one usually takes the verification function v to be the value function V if it is in $C^{1,1}([0,T] \times \mathbb{R}^n)$, since in this case V solves the HJB equation (3.6). Unfortunately, as we have already discussed before (see Chapter 4), it is very likely that the corresponding HJB equation (3.6) has no smooth solutions at all. Thus, the above result is very restrictive in applications. Once again, one tries to make use of viscosity solutions to overcome this pitfall. A natural question is, Does the verification theorem still hold, with the solutions of the HJB equation (3.6) in the classical sense replaced by the ones in the viscosity sense, and the derivatives involved replaced by the super- and/or subdifferentials?

We shall answer the above question by deriving verification theorems within the framework of viscosity solutions. First we need some technical preliminaries.

For $\hat{x} \in Q \subseteq \mathbb{R}^n$ and $\xi \in \mathbb{R}^n$, we denote by $v'(\hat{x};\xi)$ the (one-sided) *directional derivative* (in the direction ξ) of v at $\hat{x}$, namely

$$v'(\hat{x};\xi) \stackrel{\Delta}{=} \lim_{h\to 0+} \frac{v(\hat{x}+h\xi) - v(\hat{x})}{h}, \tag{3.56}$$

whenever the limit exists.

Lemma 3.8. *Let Q be an open subset of $\mathbb{R}^n$, and $v : \overline{Q} \to \mathbb{R}$ a continuous function. If $v'(\hat{x};\xi)$ exists for given $\hat{x} \in Q$ and $\xi \in \mathbb{R}^n$, then*

$$\sup_{p\in D_x^{1,-}v(\hat{x})} \langle p,\xi\rangle \leq v'(\hat{x};\xi) \leq \inf_{p\in D_x^{1,+}v(\hat{x})} \langle p,\xi\rangle, \tag{3.57}$$

where $\sup\phi \stackrel{\Delta}{=} -\infty$ and $\inf\phi \stackrel{\Delta}{=} +\infty$.

Proof. The result is clear if $\xi = 0$. So we assume $\xi \neq 0$. For any $p \in D_x^{1,+}v(\hat{x})$, by definition,

$$\varlimsup_{h\to 0+} \frac{v(\hat{x}+h\xi) - v(\hat{x}) - \langle p, h\xi \rangle}{h|\xi|} \leq 0.$$

Hence, noting (3.56), $v'(\hat{x};\xi) \leq \langle p, \xi \rangle$. This implies the second inequality in (3.57). Similarly, we can show the first one. $\square$

The following is a nonsmooth version of Theorem 3.7.

Theorem 3.9. *Let (D1) and (D2)′ hold. Let $v \in C([0,T]\times \mathbb{R}^n)$ be a viscosity solution of the HJB equation (3.6). Then (3.52) holds. Furthermore, let $(s,y) \in [0,T)\times\mathbb{R}^n$ be fixed and let $(\bar{x}(\cdot),\bar{u}(\cdot))$ be a given admissible pair of Problem (D_{sy}) such that there exists a measurable pair $(\bar{q}(t),\bar{p}(t))$ satisfying*

$$(3.58) \qquad (\bar{q}(t),\bar{p}(t)) \in D_{t,x}^{1,+}v(t,\bar{x}(t)) \bigcup D_{t,x}^{1,-}v(t,\bar{x}(t)), \quad \text{a.e. } t\in[s,T].$$

Then $(\bar{x}(\cdot),\bar{u}(\cdot))$ is optimal if and only if

$$(3.59) \qquad \int_s^T \bar{q}(t)dt \leq \int_s^T H(t,\bar{x}(t),\bar{u}(t),-\bar{p}(t))dt.$$

Proof. The first conclusion is clear, because v coincides with the value function V due to the uniqueness of the viscosity solutions to the HJB equation (3.6) (see Chapter 4, Theorem 2.5).

Now we prove the second assertion. The "only if" part comes immediately from (3.37) (we actually have a stronger conclusion, which leads to the equality in (3.59)). Now let us prove the "if" part. We shall set $\bar{b}(t) \stackrel{\Delta}{=} b(t,\bar{x}(t),\bar{u}(t))$, etc. to simplify the notation. Note that $v(\cdot\,,\cdot)$ has now been identified to be the value function. Thus, by Chapter 4, Theorem 2.5, $v(t,x)$ is Lipschitz continuous in (t,x). Further, by the Lipschitz continuity of $\bar{x}(\cdot)$, the map $t\mapsto v(t,\bar{x}(t))$ is also Lipschitz, which implies the almost everywhere differentiability of this map. Fix an $r\in(s,T)$ such that $\frac{d}{dt}v(t,\bar{x}(t))|_{t=r}$ exists, (3.58) holds, and

$$\lim_{h\to 0+}\frac{1}{h}\int_r^{r+h}\bar{b}(t)dt = \lim_{h\to 0+}\frac{1}{h}\int_{r-h}^{r}\bar{b}(t)dt = \bar{b}(r).$$

If $(\bar{q}(r),\bar{p}(r)) \in D_{t,x}^{1,+}v(r,\bar{x}(r))$, then

$$\begin{aligned}
\frac{d}{dt}v(t,\bar{x}(t))|_{t=r} &= \lim_{h\to 0+}\frac{v(r+h,\bar{x}(r+h)) - v(r,\bar{x}(r))}{h}\\
&= \lim_{h\to 0+}\frac{v(r+h,\bar{x}(r)+\int_r^{r+h}\bar{b}(\tau)d\tau) - v(r,\bar{x}(r))}{h}\\
&= \lim_{h\to 0+}\frac{v(r+h,\bar{x}(r)+h\bar{b}(r)+o(h)) - v(r,\bar{x}(r))}{h}.
\end{aligned}$$

Note that v is Lipschitz in x. Hence,

$$
(3.60)\qquad \begin{aligned}
\frac{d}{dt}v(t,\bar{x}(t))|_{t=r} &= \lim_{h\to 0+}\frac{v(r+h,\bar{x}(r)+h\bar{b}(r))-v(r,\bar{x}(r))}{h}\\
&= v'\big((r,\bar{x}(r));(1,\bar{b}(r))\big)\\
&\le \bar{q}(r)+\langle\,\bar{p}(r),\bar{b}(r)\,\rangle,
\end{aligned}
$$

where the last inequality is due to Lemma 3.8.

If $(\bar{q}(r),\bar{p}(r))\in D^{1,-}_{t,x}v(r,\bar{x}(r))$, then

$$
(3.61)\qquad \begin{aligned}
\frac{d}{dt}v(t,\bar{x}(t))|_{t=r} &= -\lim_{h\to 0+}\frac{v(r-h,\bar{x}(r-h))-v(r,\bar{x}(r))}{h}\\
&= -\lim_{h\to 0+}\frac{v(r-h,\bar{x}(r)-\int_{r-h}^{r}\bar{b}(\tau)d\tau)-v(r,\bar{x}(r))}{h}\\
&= -\lim_{h\to 0+}\frac{v(r-h,\bar{x}(r)-h\bar{b}(r)+o(h))-v(r,\bar{x}(r))}{h}\\
&= -v'\big((r,\bar{x}(r));(-1,-\bar{b}(r))\big)\\
&\le \bar{q}(r)+\langle\,\bar{p}(r),\bar{b}(r)\,\rangle,
\end{aligned}
$$

where the last inequality, once again, is due to Lemma 3.8. Thus (3.60) or (3.61) always holds. Integrating the inequality from s to T, we obtain (noting (3.59))

$$
(3.62)\qquad h(\bar{x}(T))-v(s,y)\le\int_s^T\{\bar{q}(r)+\langle\,\bar{p}(r),\bar{b}(r)\,\rangle\}dr\le-\int_s^T\bar{f}(r)dr,
$$

which yields

$$
(3.63)\qquad J(s,y;\bar{u}(\cdot))\le v(s,y).
$$

Combining (3.52) with (3.63) shows that $\bar{u}(\cdot)$ is an optimal control. □

We point out that the Lipschitz continuity of $v(s,y)$ in *both* s and y is very crucial in the above proof.

The following result shows that the integral condition (3.59) can be replaced by a pointwise condition that looks much stronger. Such a condition is comparable with (3.53).

Corollary 3.10. *Let (D1) and (D2)′ hold. Let $v\in C([0,T]\times\mathbb{R}^n)$ be the viscosity solution of the HJB equation (3.6). Let $(s,y)\in[0,T)\times\mathbb{R}^n$ be fixed and $(\bar{x}(\cdot),\bar{u}(\cdot))$ be an admissible pair of Problem (D_{sy}). Assume that*

$$
(3.64)\qquad D^{1,+}_{t,x}v(t,\bar{x}(t))\cup D^{1,-}_{t,x}v(t,\bar{x}(t))\ne\phi,\quad \text{a.e. } t\in[s,T].
$$

Then $(\bar{x}(\cdot),\bar{u}(\cdot))$ is optimal if and only if for almost every $t\in[s,T]$ there exists a pair $(q,p)\in D^{1,+}_{t,x}v(t,\bar{x}(t))\cup D^{1,-}_{t,x}v(t,\bar{x}(t))$ such that

$$
(3.65)\qquad q=H(t,\bar{x}(t),\bar{u}(t),-p)=\max_{u\in U}H(t,\bar{x}(t),u,-p).
$$

Proof. By (3.37), we have the necessity of condition (3.65) (in fact, the conclusion would be stronger, i.e., (3.65) holds for *any* point $(q,p) \in D_{t,x}^{1,+}v(t,\bar{x}(t)) \cup D_{t,x}^{1,-}v(t,\bar{x}(t))$). We now prove the sufficiency. Let $(q,p) \in D_{t,x}^{1,+}v(t,\bar{x}(t)) \cup D_{t,x}^{1,-}v(t,\bar{x}(t))$ be such that (3.65) holds. Then the same argument as (3.60) or (3.61) leads to

$$\frac{d}{dt}v(t,\bar{x}(t))|_{t=r} \leq q + \langle p, \bar{b}(r) \rangle = -\bar{f}(r),$$

where the last equality follows from the first equality of (3.65). This yields (3.62), which implies the optimality of $(\bar{x}(\cdot),\bar{u}(\cdot))$. □

Note that if (D1), (D2)′, and (D3)′ hold, and $(\bar{x}(\cdot),\bar{u}(\cdot),p(\cdot))$ is any optimal triple of Problem (D_{sy}), then (3.37) holds by Theorem 3.4. Consequently, (3.59) holds with

$$\begin{cases} \bar{q}(t) = H(t,\bar{x}(t),\bar{u}(t),-p(t)), \\ \bar{p}(t) = p(t), \end{cases}$$

and (3.65) clearly holds for $(q,p) = \big(H(t,\bar{x}(t),\bar{u}(t),-p(t)),p(t)\big)$.

Theorem 3.9 is an extension of the classical verification theorem (Theorem 3.7). On the other hand, the following example shows that the classical verification theorem may be not able to verify the optimality of a given admissible pair, whereas Theorem 3.9 or Corollary 3.10 can.

Example 3.11. Consider the same problem as in Example 3.5. Take an admissible pair $(\bar{x}(\cdot),\bar{u}(\cdot)) \equiv (0,0)$ for Problem (D_{00}). Theorem 3.7 cannot tell whether this pair is optimal, since $V_x(t,x)$ does not exist on the *whole* trajectory $\bar{x}(t), t \in [0,T]$. On the other hand, from (3.51) we have

$$\begin{cases} D_{t,x}^{1,+}V(t,\bar{x}(t)) = D_{t,x}^{1,+}V(t,0) = \{0\} \times [-e^{T-t},-1], \\ H(t,\bar{x}(t),u,-p) = 0, \quad \forall p \in \mathbb{R}, u \in [0,1], \end{cases} \qquad t \in [0,T].$$

Now if we take

$$(q,p) = (0,-1) \in D_{t,x}^{1,+}V(t,\bar{x}(t)), \qquad \forall t \in [0,T],$$

then (3.65) is satisfied. This implies that the pair $(\bar{x}(\cdot),\bar{u}(\cdot))$ is optimal by Corollary 3.10.

Sometimes, Theorem 3.9 or Corollary 3.10 can also be used to exclude some admissible pairs from being optimal. Below is such a situation.

Corollary 3.12. *Let $(\bar{x}(\cdot),\bar{u}(\cdot))$ be an admissible pair of Problem (D_{sy}). If there exists a set $T_0 \subseteq [s,T]$ with positive Lebesgue measure, such that for any $t \in T_0$, there is a pair $(q,p) \in D_{t,x}^{1,+}V(t,\bar{x}(t)) \bigcup D_{t,x}^{1,-}V(t,\bar{x}(t))$ satisfying*

$$q \neq H(t,\bar{x}(t),\bar{u}(t),-p),$$

then $(\bar{x}(\cdot),\bar{u}(\cdot))$ is not optimal.

Proof. The result follows immediately from Theorem 3.4. □

Now let us describe how to construct an optimal control by the verification theorems obtained above. First we introduce the following definition.

Definition 3.13. A measurable function $\mathbf{u} : [0,T] \times \mathbb{R}^n \to U$ is called an *admissible feedback control* if for any $(s, y) \in [0,T) \times \mathbb{R}^n$, there is a solution $x(\cdot\,; s, y)$ to the following equation:

$$\begin{cases} \dot{x}(t) = f\big(t, x(t), \mathbf{u}(t, x(t))\big), & \text{a.e. } t \in [s,T], \\ x(s) = y. \end{cases} \tag{3.66}$$

An admissible feedback control $\mathbf{u}^*$ is said to be optimal if for each $(s, y) \in [0,T) \times \mathbb{R}^n$, $\big(x^*(\cdot\,; s, y), \mathbf{u}^*(\cdot\,, x^*(\cdot\,; s, y))\big)$ is optimal for Problem (D_{sy}), where $x^*(\cdot\,; s, y)$ is the solution of (3.66) corresponding to $\mathbf{u}^*$.

Lemma 3.14. *Let $v \in C([0,T] \times \mathbb{R}^n)$ be the viscosity solution of the HJB equation (3.6). Then for each $(t, x) \in (0,T) \times \mathbb{R}^n$,*

$$\begin{cases} -q + \sup\limits_{u \in U} H(t, x, u, -p) \le 0, & \forall (q,p) \in D^{1,+}_{t,x} v(t,x), \\ -q + \sup\limits_{u \in U} H(t, x, u, -p) = 0, & \forall (q,p) \in D^{1,-}_{t,x} v(t,x). \end{cases} \tag{3.67}$$

Proof. The first relation follows from the definition of viscosity solutions (see (3.34)). For the second one, recall that $D^{1,-}_{t,x} v(t,x) \subseteq \partial v(t,x)$, the latter being Clarke's generalized gradient, which equals the convex hull of the set of all the limits $\lim(q_k, p_k) \equiv \lim(v_t(t_k, x_k), v_x(t_k, x_k))$, where $(t_k, x_k) \to (t, x)$ (see Chapter 4, Proposition 2.6-(iii) and Chapter 3, Lemma 2.3-(iv)). For any $(q_k, p_k) = (v_t(t_k, x_k), v_x(t_k, x_k))$, we have by the HJB equation

$$-q_k + \sup_{u \in U} H(t_k, x_k, u, -p_k) = 0.$$

But $\sup_{u \in U} H(t, x, u, -p)$ is continuous in (t, x) and convex in p, and so it follows that

$$-q + \sup_{u \in U} H(t, x, u, -p) \le 0, \quad \forall (q,p) \in D^{1,-}_{t,x}(t,x).$$

Since the opposite inequality always holds by (3.34), the proof is completed. □

An immediate consequence of Lemma 3.14 is the following result.

Corollary 3.15. *Let $v \in C([0,T] \times \mathbb{R}^n)$ be the viscosity solution of the HJB equation (3.6). Then for each $(t, x) \in (0,T) \times \mathbb{R}^n$,*

$$\inf_{(q,p,u) \in [D^{1,+}_{t,x} v(t,x) \cup D^{1,-}_{t,x} v(t,x)] \times U} [q - H(t, x, u, -p)] \ge 0. \tag{3.68}$$

Theorem 3.16. *Let* $\overline{\mathbf{u}}$ *be an admissible feedback control, and* $\overline{\mathbf{p}}$ *and* $\overline{\mathbf{q}}$ *two measurable functions satisfying*

$$(\overline{\mathbf{q}}(t,x), \overline{\mathbf{p}}(t,x)) \in D^{1,+}_{t,x}V(t,x) \bigcup D^{1,-}_{t,x}V(t,x), \qquad \forall (t,x) \in [0,T] \times \mathbb{R}^n. \tag{3.69}$$

Further, suppose that

$$\begin{aligned} &\overline{\mathbf{q}}(t,x) - H(t,x,\overline{\mathbf{u}}(t,x), -\overline{\mathbf{p}}(t,x)) \\ &\quad = \min_{(q,p,u)\in[D^{1,+}_{t,x}v(t,x)\cup D^{1,-}_{t,x}v(t,x)]\times U} [q - H(t,x,u,-p)] = 0, \\ &\qquad \forall (t,x) \in [0,T] \times \mathbb{R}^n. \end{aligned} \tag{3.70}$$

Then $\overline{\mathbf{u}}$ *is optimal.*

Proof. For any $(s,y) \in [0,T) \times \mathbb{R}^n$, let $\bar{x}(\cdot)$ be the corresponding trajectory under $\overline{\mathbf{u}}$ for Problem (D_{sy}). Put

$$\bar{u}(t) \stackrel{\Delta}{=} \overline{\mathbf{u}}(t,\bar{x}(t)), \quad \bar{q}(t) = \overline{\mathbf{q}}(t,\bar{x}(t)), \quad \bar{p}(t) = \overline{\mathbf{p}}(t,\bar{x}(t)). \tag{3.71}$$

By (3.70), $(\bar{x}(t), \bar{u}(t), \bar{q}(t), \bar{p}(t))$ satisfies (3.58)–(3.59). The desired result then follows from Corollary 3.10. □

By Theorem 3.16, one can formally obtain an optimal feedback control by minimizing $q - H(t,x,u,-p)$ over $[D^{1,+}_{t,x}V(t,x) \cup D^{1,-}_{t,x}V(t,x)] \times U$ for each (t,x). We said "formally" because there are some points that are not clear. First, although the infimum in (3.68) can be achieved (note that each $D^{1,+}_{t,x}V(t,x) \cup D^{1,-}_{t,x}V(t,x)$ is compact due to the Lipschitz property of V), we do not know in general whether the infimum is zero and whether there is a *measurable selection* $(\mathbf{q}^*(t,x), \mathbf{p}^*(t,x), \mathbf{u}^*(t,x))$ (the answers are positive if V is semiconvex or semiconcave). Second, even if there exists a measurable selection such that (3.70) holds, it is still generally difficult to verify whether equation (3.66) under $\mathbf{u}^*$ has a unique solution, because the right-hand side of (3.66) is measurable only in x. This is not clear even when V is smooth (cf. Fleming–Rishel [1, p. 99 and p. 170]). All these problems remain open.

To conclude this section, let us make one more observation. It is seen that Theorems 3.7, 3.9, and 3.16 are, one way or another, trying to derive an optimal pair via the (viscosity) solution $v(\cdot\,,\cdot)$ to the HJB equation (3.6). This in turn gives rise to a solution to the Hamiltonian system (3.8). Recall that this procedure is exactly the same as that of the Hamilton–Jacobi theorem (Theorem 2.3). Hence, the verification theorems along with the (formal) construction of optimal feedback controls may be regarded as a generalization of the Hamilton–Jacobi theorem presented in Section 2.

4. Relationship for Stochastic Systems

Starting from this section, we study the relationship between the maximum principle and dynamic programming for stochastic optimal control problems. Since dynamic programming is to be involved, we need to work under the weak formulation. Let us recall this from Chapter 4. Let $T > 0$ be given and let U be a metric space. For any $(s, y) \in [0, T) \times \mathbb{R}^n$, consider the controlled system

$$(4.1) \qquad \begin{cases} dx(t) = b(t, x(t), u(t))dt + \sigma(t, x(t), u(t))dW(t), & t \in [s, T], \\ x(s) = y, \end{cases}$$

along with the cost functional

$$(4.2) \qquad J(s, y; u(\cdot)) = E\Big\{ \int_s^T f(t, x(t), u(t))dt + h(x(T)) \Big\}.$$

Given $s \in [0, T)$, we denote by $\mathcal{U}^w[s, T]$ the set of all 5-tuples $(\Omega, \mathcal{F}, \mathbf{P}, W(\cdot), u(\cdot))$ satisfying the following:

(i) $(\Omega, \mathcal{F}, \mathbf{P})$ is a complete probability space.

(ii) $\{W(t)\}_{t \geq s}$ is an m-dimensional standard Brownian motion defined on $(\Omega, \mathcal{F}, \mathbf{P})$ over $[s, T]$ (with $W(s) = 0$ almost surely), and $\mathcal{F}_t^s = \sigma\{W(r) : s \leq r \leq t\}$ augmented by all the $\mathbf{P}$-null sets in $\mathcal{F}$.

(iii) $u : [s, T] \times \Omega \to U$ is an $\{\mathcal{F}_t^s\}_{t \geq s}$-adapted process on $(\Omega, \mathcal{F}, \mathbf{P})$.

(iv) Under $u(\cdot)$, for any $y \in \mathbb{R}^n$ equation (4.1) admits a unique solution (in the sense of Chapter 1, Definition 6.15) $x(\cdot)$ on $(\Omega, \mathcal{F}, \{\mathcal{F}_t^s\}_{t \geq 0}, \mathbf{P})$.

(v) $f(\cdot, x(\cdot), u(\cdot)) \in L^1_{\mathcal{F}}(0, T; \mathbb{R})$ and $h(x(T)) \in L^1_{\mathcal{F}_T}(\Omega; \mathbb{R})$. Here, the spaces $L^1_{\mathcal{F}}(0, T; \mathbb{R})$ and $L^1_{\mathcal{F}_T}(\Omega; \mathbb{R})$ are defined on the given filtered probability space $(\Omega, \mathcal{F}, \{\mathcal{F}_t^s\}_{t \geq s}, \mathbf{P})$ (associated with the given 5-tuple).

We write $(\Omega, \mathcal{F}, \mathbf{P}, W(\cdot), u(\cdot)) \in \mathcal{U}^w[s, T]$, but occasionally we will write only $u(\cdot) \in \mathcal{U}^w[s, T]$ if no ambiguity arises. The stochastic optimal control problem can be stated as follows:

Problem ($\mathbf{S}_{sy}$). For given $(s, y) \in [0, T) \times \mathbb{R}^n$, minimize (4.2) subject to (4.1) over $\mathcal{U}^w[s, T]$.

The value function is defined as follows (see Chapter 4):

$$(4.3) \qquad \begin{cases} V(s, y) = \inf_{u(\cdot) \in \mathcal{U}^w[s, T]} J(s, y; u(\cdot)), & \forall (s, y) \in [0, T) \times \mathbb{R}^n, \\ V(T, y) = h(y), & \forall y \in \mathbb{R}^n. \end{cases}$$

For convenience, let us recall (S1)′–(S2)′ from Chapter 4, Section 3, and (S3) from Chapter 3, Section 3:

(S1)′ (U, d) is a Polish space and $T > 0$.

(S2)′ The maps $b, \sigma, f : [0, T] \times \mathbb{R}^n \times U \to \mathbb{R}^n, \mathbb{R}^{n \times m}, \mathbb{R}$, respectively, and $h : \mathbb{R}^n \to \mathbb{R}$ are all uniformly continuous, and there exists a constant

$L>0$ such that for $\varphi(t,x,u)=b(t,x,u),\sigma(t,x,u),f(t,x,u),h(x)$,

$$
(4.4)\qquad \begin{cases} |\varphi(t,x,u)-\varphi(t,\widehat{x},u)|\le L|x-\widehat{x}|, \\ \qquad\qquad\qquad\qquad \forall t\in[0,T], x,\widehat{x}\in\mathbb{R}^n,\ u\in U, \\ |\varphi(t,0,u)|\le L, \qquad \forall (t,u)\in[0,T]\times U. \end{cases}
$$

(S3) The maps b,σ,f, and h are C^2 in x. Moreover, there exists a constant $L>0$ and a modulus of continuity $\overline{\omega}:[0,\infty)\to[0,\infty)$ such that for $\varphi=b,\sigma,f,h$,

$$
(4.5)\qquad \begin{cases} |\varphi_x(t,x,u)-\varphi_x(t,\widehat{x},\widehat{u})|\le L|x-\widehat{x}|+\overline{\omega}(d(u,\widehat{u})), \\ |\varphi_{xx}(t,x,u)-\varphi_{xx}(t,\widehat{x},\widehat{u})|\le \overline{\omega}(|x-\widehat{x}|+d(u,\widehat{u})), \\ \qquad\qquad \forall t\in[0,T],\ x,\widehat{x}\in\mathbb{R}^n,\ u,\widehat{u}\in U. \end{cases}
$$

Note that (S3) is a little stronger than (S3)$'$ in Chapter 4, Section 4. In the sequel, we shall denote by $b^i(t,x,u)$ the ith coordinate of the column vector $b(t,x,u)$, $i=1,2,\ldots,n$, by $\sigma^j(t,x,u)$ the jth column of the matrix $\sigma(t,x,u)$, and by $\sigma^{ij}(t,x,u)$ the ijth element of $\sigma(t,x,u)$.

Recall that the Hamilton–Jacobi–Bellman (HJB) equation associated with our optimal control problem is as follows (see Chapter 4, (3.24))

$$
(4.6)\qquad \begin{cases} -v_t(t,x)+\sup\limits_{u\in U} G(t,x,u,-v_x(t,x),-v_{xx}(t,x))=0, \\ \qquad\qquad\qquad\qquad (t,x)\in[0,T)\times\mathbb{R}^n, \\ v\big|_{t=T}=h(x), \quad x\in\mathbb{R}^n, \end{cases}
$$

where the *generalized Hamiltonian* G is defined by (see Chapter 3, (3.15), or Chapter 4, (3.25))

$$
(4.7)\qquad \begin{aligned} &G(t,x,u,p,P) \\ &\quad\triangleq \frac{1}{2}\mathrm{tr}\,\Big(\sigma(t,x,u)^{\top}P\sigma(t,x,u)\Big)+\langle p,b(t,x,u)\rangle-f(t,x,u), \\ &\qquad\qquad (t,x,u,p,P)\in[0,T]\times\mathbb{R}^n\times U\times\mathbb{R}^n\times\mathcal{S}^n. \end{aligned}
$$

On the other hand, associated with each admissible pair $(x(\cdot),u(\cdot))$ there are pairs of processes (called the *first-order* and *second-order adjoint processes*, respectively)

$$
(4.8)\qquad \begin{cases} (p(\cdot),q(\cdot))\in L^2_{\mathcal{F}}(0,T;\mathbb{R}^n)\times\big[L^2_{\mathcal{F}}(0,T;\mathbb{R}^n)\big]^m, \\ (P(\cdot),Q(\cdot))\in L^2_{\mathcal{F}}(0,T;\mathcal{S}^n)\times\big[L^2_{\mathcal{F}}(0,T;\mathcal{S}^n)\big]^m, \end{cases}
$$

where

$$
\begin{cases} q(\cdot)=(q_1(\cdot),\ldots,q_m(\cdot)), \quad Q(\cdot)=(Q_1(\cdot),\ldots,Q_m(\cdot)), \\ q_j(\cdot)\in L^2_{\mathcal{F}}(0,T;\mathbb{R}^n), \quad Q_j(\cdot)\in L^2_{\mathcal{F}}(0,T;\mathcal{S}^n), \quad 1\le j\le m, \end{cases}
$$

such that (see Chapter 3, (3.8)–(3.10))

$$
(4.9)\qquad \begin{cases} dp(t) = -\Big\{b_x(t,x(t),u(t))^\top p(t) + \sum_{j=1}^{m} \sigma_x^j(t,x(t),u(t))^\top q_j(t) \\ \qquad\qquad - f_x(t,x(t),u(t))\Big\}dt + q(t)dW(t), \quad t \in [s,T], \\ p(T) = -h_x(x(T)), \end{cases}
$$

and

$$
(4.10)\qquad \begin{cases} dP(t) = -\Big\{b_x(t,x(t),u(t))^\top P(t) + P(t)b_x(t,x(t),u(t)) \\ \qquad + \sum_{j=1}^{m} \sigma_x^j(t,x(t),u(t))^\top P(t)\sigma_x^j(t,x(t),u(t)) \\ \qquad + \sum_{j=1}^{m} \{\sigma_x^j(t,x(t),u(t))^\top Q_j(t) + Q_j(t)\sigma_x^j(t,x(t),u(t))\} \\ \qquad + H_{xx}(t,x(t),u(t),p(t),q(t))\Big\}dt \\ \qquad + \sum_{j=1}^{m} Q_j(t)dW^j(t), \quad t \in [s,T], \\ P(T) = -h_{xx}(x(T)), \end{cases}
$$

with

$$
(4.11)\qquad H(t,x,u,p,q) = \langle p, b(t,x,u) \rangle + \sum_{j=1}^{m} \langle q_j, \sigma^j(t,x,u) \rangle - f(t,x,u).
$$

Following Chapter 3, we call (4.9) and (4.10) the *first-order* and *second-order adjoint equations*, respectively. Under (S1)′, (S2)′, and (S3), both of them admit unique adapted solutions $(p(\cdot),q(\cdot))$ and $(P(\cdot),Q(\cdot))$ satisfying (4.8). The system consisting of (4.1), (4.9), (4.10), and the maximum condition (see Chapter 3, (3.20)) is called a *stochastic Hamiltonian system*. In what follows, if $(\bar{x}(\cdot),\bar{u}(\cdot))$ is an optimal pair of Problem (S_{sy}), and $(p(\cdot),q(\cdot))$ and $(P(\cdot),Q(\cdot))$ are adapted solutions of the corresponding (4.9) and (4.10), respectively, then $(\bar{x}(\cdot),\bar{u}(\cdot),p(\cdot),q(\cdot))$ is referred to as an *optimal 4-tuple* of Problem (S_{sy}) and $(\bar{x}(\cdot),\bar{u}(\cdot),p(\cdot),q(\cdot),P(\cdot),Q(\cdot))$ as an *optimal 6-tuple* of Problem (S_{sy}).

4.1. Smooth case

We first study the case where the value function $V(s,y)$ is sufficiently smooth. The following result is a stochastic analogue of Theorem 3.1.

Theorem 4.1. *Let (S1)′–(S2)′ hold and let $(s,y) \in [0,T) \times \mathbb{R}^n$ be fixed. Let $(\bar{x}(\cdot),\bar{u}(\cdot),p(\cdot),q(\cdot))$ be an optimal 4-tuple of Problem (S_{sy}). Suppose* *that the value function* $V \in C^{1,2}([0,T] \times \mathbb{R}^n)$. *Then*

$$
(4.12)\qquad \begin{aligned} V_t(t,\bar{x}(t)) &= G\big(t,\bar{x}(t),\bar{u}(t),-V_x(t,\bar{x}(t)),-V_{xx}(t,\bar{x}(t))\big) \\ &= \max_{u\in U} G\big(t,\bar{x}(t),u,-V_x(t,\bar{x}(t)),-V_{xx}(t,\bar{x}(t))\big), \\ &\qquad\qquad \text{a.e. } t \in [s,T], \ \mathbf{P}\text{-a.s.} \end{aligned}
$$

Further, if $V \in C^{1,3}([0,T] \times \mathbb{R}^n)$ *and* V_{tx} *is also continuous, then*

$$
(4.13)\qquad \begin{cases} V_x(t,\bar{x}(t)) = -p(t), \quad \forall t \in [s,T], \ \mathbf{P}\text{-a.s.}, \\ V_{xx}(t,\bar{x}(t))\sigma(t,\bar{x}(t),\bar{u}(t)) = -q(t), \quad \text{a.e. } t \in [s,T], \ \mathbf{P}\text{-a.s.} \end{cases}
$$

Proof. By Chapter 4, Theorem 3.4, we have

$$
(4.14)\qquad \begin{aligned} V(t,\bar{x}(t)) &= E\Big\{\int_t^T f(r,\bar{x}(r),\bar{u}(r))dr + h(\bar{x}(T))|\mathcal{F}_t^s\Big\}, \\ &\qquad \forall t \in [s,T], \ \mathbf{P}\text{-a.s.} \end{aligned}
$$

Define

$$
(4.15)\qquad m(t) \stackrel{\Delta}{=} E\Big\{\int_s^T f(r,\bar{x}(r),\bar{u}(r))dr + h(\bar{x}(T))|\mathcal{F}_t^s\Big\}, \qquad t \in [s,T].
$$

Clearly, $m(\cdot)$ is a square-integrable $\mathcal{F}_t^s$-martingale (recall that $s \in [0,T)$ is fixed). Thus, by the martingale representation theorem (Chapter 1, Theorem 5.7), we have

$$
(4.16)\qquad \begin{aligned} m(t) &= m(s) + \int_s^t M(r)dW(r) \\ &= V(s,y) + \int_s^t M(r)dW(r), \qquad t \in [s,T], \end{aligned}
$$

where $M \in (L^2_{\mathcal{F}}(s,T;\mathbb{R}^n))^m$. Then, by (4.14) and (4.16),

$$
(4.17)\qquad \begin{aligned} V(t,\bar{x}(t)) &= m(t) - \int_s^t f(r,\bar{x}(r),\bar{u}(r))dr \\ &= V(s,y) - \int_s^t f(r,\bar{x}(r),\bar{u}(r))dr + \int_s^t M(r)dW(r). \end{aligned}
$$

On the other hand, applying Itô's formula to $V(t,\bar{x}(t))$, we obtain

$$
(4.18)\qquad \begin{aligned} dV(t,\bar{x}(t)) &= \Big\{V_t(t,\bar{x}(t)) + \langle V_x(t,\bar{x}(t)), b(t,\bar{x}(t),\bar{u}(t))\rangle \\ &\quad + \frac{1}{2}\operatorname{tr}\big(\sigma(t,\bar{x}(t),\bar{u}(t))^\top V_{xx}(t,\bar{x}(t))\sigma(t,\bar{x}(t),\bar{u}(t))\big)\Big\}dt \\ &\quad + V_x(t,\bar{x}(t))^\top \sigma(t,\bar{x}(t),\bar{u}(t))dW(t). \end{aligned}
$$

Comparing (4.18) with (4.17), we conclude that

$$(4.19)\qquad \begin{cases} V_t(t,\overline{x}(t)) + \langle V_x(t,\overline{x}(t)), b(t,\overline{x}(t),\overline{u}(t))\rangle \\ \qquad + \frac{1}{2}\mathrm{tr}\left(\sigma(t,\overline{x}(t),\overline{u}(t))^\top V_{xx}(t,\overline{x}(t))\sigma(t,\overline{x}(t),\overline{u}(t))\right) \\ \qquad\qquad = -f(t,\overline{x}(t),\overline{u}(t)), \\ V_x(t,\overline{x}(t))^\top \sigma(t,\overline{x}(t),\overline{u}(t)) = M(t). \end{cases}$$

This proves the first equality in (4.12). Since $V \in C^{1,2}([0,1]\times\mathbb{R}^n)$, it satisfies the HJB equation (4.6), which implies the second equality in (4.12). Also, by (4.6) we have

$$(4.20)\qquad \begin{aligned} &G\big(t,\overline{x}(t),\overline{u}(t),-V_x(t,\overline{x}(t)),-V_{xx}(t,\overline{x}(t))\big) - V_t(t,\overline{x}(t)) = 0 \\ &\quad \geq G\big(t,x,\overline{u}(t),-V_x(t,x),-V_{xx}(t,x)\big) - V_t(t,x), \quad \forall x\in\mathbb{R}^n. \end{aligned}$$

Consequently, if $V \in C^{1,3}([0,T]\times\mathbb{R}^n)$ with V_{tx} being also continuous, then

$$(4.21)\qquad \frac{\partial}{\partial x}\Big\{G(t,x,\overline{u}(t),-V_x(t,x),-V_{xx}(t,x)) - V_t(t,x)\Big\}\Big|_{x=\overline{x}(t)} = 0.$$

This is equivalent to (recall (4.7))

$$(4.22)\qquad \begin{aligned} &V_{tx}(t,\overline{x}(t)) + V_{xx}(t,\overline{x}(t))b(t,\overline{x}(t),\overline{u}(t)) \\ &\quad + b_x(t,\overline{x}(t),\overline{u}(t))^\top V_x(t,\overline{x}(t)) \\ &\quad + \frac{1}{2}\mathrm{tr}\left(\sigma(t,\overline{x}(t),\overline{u}(t))^\top V_{xxx}(t,\overline{x}(t))\sigma(t,\overline{x}(t),\overline{u}(t))\right) \\ &\quad + \sum_{j=1}^m \left(\sigma_x^j(t,\overline{x}(t),\overline{u}(t))\right)^\top \left(V_{xx}(t,\overline{x}(t))\sigma(t,\overline{x}(t),\overline{u}(t))\right)^j \\ &\quad + f_x(t,\overline{x}(t),\overline{u}(t)) = 0, \end{aligned}$$

where

$$\mathrm{tr}\,(\sigma^\top V_{xxx}\sigma) \triangleq \Big(\mathrm{tr}\,(\sigma^\top((V_x)^1)_{xx}\sigma),\ldots,\mathrm{tr}\,(\sigma^\top((V_x)^n)_{xx}\sigma)\Big)^\top$$

with

$$((V_x)^1,\ldots,(V_x)^n)^\top \equiv V_x.$$

On the other hand, applying Itô's formula to $V_x(t,\overline{x}(t))$, we get (noting

(4.22))

$$
\begin{aligned}
& dV_x(t,\overline{x}(t)) \\
&= \Big\{ V_{tx}(t,\overline{x}(t)) + V_{xx}(t,\overline{x}(t)) b(t,\overline{x}(t),\overline{u}(t)) \\
&\qquad + \frac{1}{2}\operatorname{tr}\Big(\sigma(t,\overline{x}(t),\overline{u}(t))^\top V_{xxx}(t,\overline{x}(t))\sigma(t,\overline{x}(t),\overline{u}(t))\Big)\Big\} dt \\
&\quad + V_{xx}(t,\overline{x}(t))\sigma(t,\overline{x}(t),\overline{u}(t)) dW(t) \\
&= -\Big\{ b_x(t,\overline{x}(t),\overline{u}(t))^\top V_x(t,\overline{x}(t)) + f_x(t,\overline{x}(t),\overline{u}(t)) \\
&\qquad + \sum_{j=1}^m \Big(\sigma_x^j(t,\overline{x}(t),\overline{u}(t))\Big)^\top \Big(V_{xx}(t,\overline{x}(t))^\top \sigma(t,\overline{x}(t),\overline{u}(t))\Big)^j \Big\} dt \\
&\quad + V_{xx}(t,\overline{x}(t))\sigma(t,\overline{x}(t),\overline{u}(t)) dW(t).
\end{aligned} \tag{4.23}
$$

Note that

$$
-V_x(T,\overline{x}(T)) = -h_x(\overline{x}(T)).
$$

Hence, by the uniqueness of the solutions to (4.9), we obtain (4.13). □

The second equality in (4.12) may be regarded as a "maximum principle" in terms of the derivatives of the value function. Note that it is *different* from the stochastic maximum principle derived in Chapter 3, where no derivative of the value function is involved.

Corollary 4.2. *Let (S1)′–(S2)′ hold and $(s,y) \in [0,T) \times \mathbb{R}^n$ be fixed. Let $(\overline{x}(\cdot),\overline{u}(\cdot))$ be an optimal pair of Problem (S_{sy}). Assume further that $V \in C^{1,2}([0,T]\times\mathbb{R}^n)$. Then*

$$
\begin{aligned}
V(t,\overline{x}(t)) &= V(s,y) - \int_s^t f(r,\overline{x}(r),\overline{u}(r)) dr \\
&\quad + \int_s^t V_x(r,\overline{x}(r))^\top \sigma(r,\overline{x}(r),\overline{u}(r)) dW(r).
\end{aligned} \tag{4.24}
$$

Proof. This is immediate from (4.17) and (4.19). □

The above corollary gives an explicit expression of the Itô process $V(t,\overline{x}(t))$. In particular, it yields that along the optimal trajectory $\overline{x}(\cdot)$,

$$
t \mapsto V(t,\overline{x}(t)) + \int_s^t f(r,\overline{x}(r),\overline{u}(r)) dr, \qquad t \in [s,T],
$$

is a martingale.

Theorem 4.1 implies that the first-order adjoint variable $p(t)$ in the stochastic situation continues to serve as the *shadow price*, which now is certainly a random variable at each time instance. Moreover, the instantaneous standard deviation of the depreciation rate of this random shadow price process is $V_{xx}(t,\overline{x}(t))\sigma(t,\overline{x}(t),\overline{u}(t))$ (see (4.9) and (4.13)). Let us employ an example to explain the economic interpretation of Theorem 4.1.

Example 4.3. Consider the same manufacturing firm studied in Section 3.2. Suppose now there are random disturbances in the system, so that (3.14) is modified to

$$\begin{cases} dx(t) = r(t, x(t), u(t))dt + \sigma(t, x(t), u(t))dW(t), & t \in [0, T], \\ x(0) = x_0, \end{cases} \tag{4.25}$$

where σ represents the volatility of the appreciation rate of the resource. Accordingly, the objective (3.15) is modified to maximize the expected total profit earned during the time period $[0, T]$, or to minimize

$$J = -E \int_0^T n(t, x(t), u(t))dt. \tag{4.26}$$

The first-order adjoint equation of this problem takes the form (recall that all the variables are assumed to be one-dimensional)

$$\begin{aligned} -dp(t) = \{&n_x(t, \bar{x}(t), \bar{u}(t)) + r_x(t, \bar{x}(t), \bar{u}(t))p(t) \\ &+ \sigma_x(t, \bar{x}(t), \bar{u}(t))q(t)\}dt - q(t)dW(t). \end{aligned} \tag{4.27}$$

An interesting observation is, according to Theorem 4.1, that the diffusion coefficient of the depreciation rate of the marginal value $p(\cdot)$ of the resource $-q(t)$ is not identical to that of the resource $-\sigma(t)$; rather, it is $-\sigma(t)$ multiplied by a factor $R(t) \triangleq -V_{xx}(t, \bar{x}(t))$. In view of the first relation in (4.13), this factor can be written as

$$R(t) = \frac{p(t)V_{xx}(t, \bar{x}(t))}{V_x(t, \bar{x}(t))}. \tag{4.28}$$

This ratio is called the *relative risk aversion* (see Arrow [1]) with respect to the fluctuation in capital or resource value if V is regarded as the *utility function* (or, more accurately, *disutility function*, as V is the *minimized* cost) of the resource. The generalized Hamiltonian is now of the form (recall (4.7))

$$\begin{aligned} &G(t, x, u, -V_x(t, \bar{x}(t)), -V_{xx}(t, \bar{x}(t))) \\ &\quad = G(t, x, u, p(t), R(t)) \\ &\quad = n(t, x, u) + r(t, x, u)p(t) + \frac{1}{2}R(t)\sigma(t, x, u)^2. \end{aligned} \tag{4.29}$$

The first two terms in the above signify the sum of the expected direct contribution rate to the overall profit and the expected accumulation rate of the marginal value of the resource. The third term, which is the instantaneous variance σ^2 of the resource multiplied by $\frac{1}{2}R(t)$, reflects the reduction (or addition, depending on the sign of the factor $R(t)$) in expected marginal value of resource due to the decision-maker's uncertainty about the anticipated change in resource. This reduction/addition is called a *risk adjustment*. More specifically, when V is convex, meaning that the decision-maker is *risk-averse* (as there are higher costs at the extreme ends), then $R(t) \leq 0$, representing a reduction. A similar discussion can be applied to the case where V is concave. Once again, the decision at any time should be chosen to make this sum as large as possible. This is what the second equality in (4.12) means. Moreover, this sum, once an optimal path is followed, represents the rate of change for the maximum possible profit as time elapses, which justifies (4.12) in Theorem 4.1. Finally, let us go back to (4.27). The three terms of the drift part in (4.27) represent the mean depreciation rate of the marginal value of the resource along the optimal path of the resource over a short period of time. The first $[n_x(t,\bar{x}(t),\bar{u}(t))]$ and the second $[r_x(t,\bar{x}(t),\bar{u}(t))p(t)]$ terms have the same interpretations as those in the deterministic case (see Section 3.2). The additional third term,

$$\sigma_x(t,\bar{x}(t),\bar{u}(t))q(t) \equiv \frac{1}{2}R(t)\big[\sigma(t,x,\bar{u}(t))^2\big]_x\Big|_{x=\bar{x}(t)}, \tag{4.30}$$

reflects the enhancement of the risk adjustment in the expected marginal value of the resource at the end of the period. It is interesting to note that the mean depreciation rate of the marginal value (i.e., the drift term of (4.27)) equals $G_x(t,\bar{x}(t),\bar{u}(t),p(t),R(t))$.

4.2. Nonsmooth case: Differentials in the spatial variable

Now we proceed to handle the case where the value function is not necessarily smooth. Note that this is likely to occur if the underlying stochastic system is degenerate. Once again we shall use the viscosity solution theory for second-order PDEs to study the relationship between the maximum principle and dynamic programming. In this subsection, we first derive nonsmooth versions of (4.13), leaving that of (4.12) to the next subsection.

Let us first recall the second-order right parabolic super- and subdifferentials of a continuous function on $[0,T]\times\mathbb{R}^n$. For $v\in C([0,T]\times\mathbb{R}^n)$ and $(\hat{t},\hat{x})\in[0,T)\times\mathbb{R}^n$, the second-order *right* parabolic superdifferential of v at $(\hat{t},\hat{x})$ is the set

$$\begin{aligned} D^{1,2,+}_{t+,x}v(\hat{t},\hat{x}) \stackrel{\Delta}{=} \Big\{(q,p,P)\in\mathbb{R}\times\mathbb{R}^n\times\mathcal{S}^n \;\Big|\; &\varlimsup_{t\downarrow\hat{t},x\to\hat{x}} \frac{1}{|t-\hat{t}|+|x-\hat{x}|^2}\Big[v(t,x)-v(\hat{t},\hat{x}) \\ &-q(t-\hat{t})-\langle p,x-\hat{x}\rangle-\frac{1}{2}(x-\hat{x})^\top P(x-\hat{x})\Big]\leq 0\Big\}, \end{aligned} \tag{4.31}$$

and the second-order *right* subdifferential of v at $(\hat{t},\hat{x})$ is the set

$$\begin{aligned} D^{1,2,-}_{t+,x}v(\hat{t},\hat{x}) \stackrel{\Delta}{=} \Big\{(q,p,P)\in\mathbb{R}\times\mathbb{R}^n\times\mathcal{S}^n \;\Big|\; &\varliminf_{t\downarrow\hat{t},x\to\hat{x}} \frac{1}{|t-\hat{t}|+|x-\hat{x}|^2}\Big[v(t,x)-v(\hat{t},\hat{x}) \\ &-q(t-\hat{t})-\langle p,x-\hat{x}\rangle-\frac{1}{2}(x-\hat{x})^\top P(x-\hat{x})\Big]\geq 0\Big\}. \end{aligned} \tag{4.32}$$

From the above definitions, we see immediately that

$$
(4.33)\qquad \begin{cases} D_{t+,x}^{1,2,+}v(\widehat{t},\widehat{x}) + [0,\infty)\times\{0\}\times\mathcal{S}_+^n = D_{t+,x}^{1,2,+}v(\widehat{t},\widehat{x}),\\ D_{t+,x}^{1,2,-}v(\widehat{t},\widehat{x}) - [0,\infty)\times\{0\}\times\mathcal{S}_+^n = D_{t+,x}^{1,2,-}v(\widehat{t},\widehat{x}),\end{cases}
$$

where $\mathcal{S}_+^n \stackrel{\Delta}{=} \{S\in\mathcal{S}^n \mid S\geq 0\}$, and $A\pm B \stackrel{\Delta}{=} \{a\pm b \mid a\in A, b\in B\}$ for any subsets A and B in a same Euclidean space. Occasionally, we will also make use of the *partial* super- and subdifferentials in one of the variables t and x. Therefore, we need the following definitions:

$$
(4.34)\qquad \begin{cases} D_x^{2,+}v(\widehat{t},\widehat{x}) = \Big\{(p,P)\in\mathbb{R}^n\times\mathcal{S}^n \mid \\ \qquad \varlimsup\limits_{x\to\widehat{x}} \dfrac{v(\widehat{t},x)-v(\widehat{t},\widehat{x})-\langle p,x-\widehat{x}\rangle-\frac{1}{2}(x-\widehat{x})^\top P(x-\widehat{x})}{|x-\widehat{x}|^2}\leq 0\Big\},\\ D_x^{2,-}v(\widehat{t},\widehat{x}) = \Big\{(p,P)\in\mathbb{R}^n\times\mathcal{S}^n \mid \\ \qquad \varliminf\limits_{x\to\widehat{x}} \dfrac{v(\widehat{t},x)-v(\widehat{t},\widehat{x})-\langle p,x-\widehat{x}\rangle-\frac{1}{2}(x-\widehat{x})^\top P(x-\widehat{x})}{|x-\widehat{x}|^2}\geq 0\Big\},\end{cases}
$$

$$
(4.35)\qquad \begin{cases} D_{t+}^{1,+}v(\widehat{t},\widehat{x}) = \Big\{q\in\mathbb{R}\,\Big|\, \varlimsup\limits_{t\downarrow\widehat{t}} \dfrac{v(t,\widehat{x})-v(\widehat{t},\widehat{x})-q(t-\widehat{t})}{|t-\widehat{t}|}\leq 0\Big\},\\ D_{t+}^{1,-}v(\widehat{t},\widehat{x}) = \Big\{q\in\mathbb{R}\,\Big|\, \varliminf\limits_{t\downarrow\widehat{t}} \dfrac{v(t,\widehat{x})-v(\widehat{t},\widehat{x})-q(t-\widehat{t})}{|t-\widehat{t}|}\geq 0\Big\}.\end{cases}
$$

Under (S1)$'$ (S2)$'$, the value function $V\in C([0,T]\times\mathbb{R}^n)$ of Problem (S_{sy}) is the unique viscosity solution of HJB equation (4.6), which is equivalent to the following (see Chapter 4, Theorem 6.2): V is the only continuous function that satisfies (3.9)–(3.10) of Chapter 4 such that for all $(t,x)\in[0,T)\times\mathbb{R}^n$,

$$
(4.36)\qquad \begin{cases} -q+\sup\limits_{u\in U} G(t,x,u,-p,-P)\leq 0, \quad \forall (q,p,P)\in D_{t+,x}^{1,2,+}V(t,x);\\ -q+\sup\limits_{u\in U} G(t,x,u,-p,-P)\geq 0, \quad \forall (q,p,P)\in D_{t+,x}^{1,2,-}V(t,x);\\ V(T,x)=h(x).\end{cases}
$$

For convenience, we define $[S,\infty)\stackrel{\Delta}{=}\{\widehat{S}\in\mathcal{S}^n \mid \widehat{S}\geq S\}$ for any $S\in\mathcal{S}^n$, and similarly for $(-\infty,S]$. The following result is a nonsmooth version of (4.13) in Theorem 4.1.

Theorem 4.4. *Let (S1)$'$, (S2)$'$, and (S3) hold and let $(s,y)\in[0,T)\times\mathbb{R}^n$ be fixed. Let $(\overline{x}(\cdot),\overline{u}(\cdot),p(\cdot),q(\cdot),P(\cdot),Q(\cdot))$ be an optimal 6-tuple of Problem (S_{sy}). Then*

$$
(4.37)\qquad \{-p(t)\}\times[-P(t),\infty)\subseteq D_x^{2,+}V(t,\overline{x}(t)),\quad \forall t\in[s,T],\ \mathbf{P}\text{-a.s.},
$$

$$
(4.38)\qquad D_x^{2,-}V(t,\overline{x}(t))\subseteq\{-p(t)\}\times(-\infty,-P(t)],\quad \forall t\in[s,T],\ \mathbf{P}\text{-a.s.}
$$

Proof. We split the proof into several steps.

Step 1. Variational equations.

Fix a $t \in [s,T]$. For any $z \in \mathbb{R}^n$, denote by $x^z(\cdot)$ the solution of the following SDE on $[t,T]$:

$$x^z(r) = z + \int_t^r b(\alpha, x^z(\alpha), \overline{u}(\alpha))d\alpha + \int_t^r \sigma(\alpha, x^z(\alpha), \overline{u}(\alpha))dW(\alpha). \tag{4.39}$$

Set $\xi^z(r) = x^z(r) - \overline{x}(r)$. It is clear that (4.39) can be regarded as an SDE on $(\Omega, \mathcal{F}, \{\mathcal{F}_r^s\}_{r\geq s}, \mathbf{P}(\cdot|\mathcal{F}_t^s)(\omega)$ for $\mathbf{P}$-a.s. ω. Thus, the following estimate holds for any $k \geq 1$ (see Chapter 1, Theorem 6.3):

$$E\Big\{ \sup_{t\leq r\leq T} |\xi^z(r)|^{2k} |\mathcal{F}_t^s \Big\} \leq K|z - \overline{x}(t)|^{2k}, \quad \mathbf{P}\text{-a.s.} \tag{4.40}$$

Now we write the equation for $\xi^z(\cdot)$ in two different ways based on different orders of expansion:

$$\begin{cases} d\xi^z(r) = \overline{b}_x(r)\xi^z(r)dr + \sum_{j=1}^m \overline{\sigma}_x^j(r)\xi^z(r)dW^j(r) \\ \qquad\qquad + \varepsilon_{z1}(r)dr + \sum_{j=1}^m \varepsilon_{z2}^j(r)dW^j(r), \quad r \in [t,T], \\ \xi^z(t) = z - \overline{x}(t), \end{cases} \tag{4.41}$$

and

$$\begin{cases} d\xi^z(r) = \Big\{\overline{b}_x(r)\xi^z(r) + \frac{1}{2}\xi^z(r)^\top \overline{b}_{xx}(r)\xi^z(r)\Big\}dr \\ \qquad\qquad + \sum_{j=1}^m \Big\{\overline{\sigma}_x^j(r)\xi^z(r) + \frac{1}{2}\xi^z(r)^\top \overline{\sigma}_{xx}^j(r)\xi^z(r)\Big\}dW^j(r) \\ \qquad\qquad + \varepsilon_{z3}(r)dr + \sum_{j=1}^m \varepsilon_{z4}^j(r)dW^j(r), \quad s \in [t,T], \\ \xi^z(t) = z - \overline{x}(t), \end{cases} \tag{4.42}$$

where

$$\begin{cases} \overline{b}_x(r) \stackrel{\Delta}{=} b_x(r, \overline{x}(r), \overline{u}(r)), \quad \overline{\sigma}_x^j(r) \stackrel{\Delta}{=} \sigma_x^j(r, \overline{x}(r), \overline{u}(r)), \\ \xi^z(r)^\top \overline{b}_{xx}(r)\xi^z(r) \stackrel{\Delta}{=} (\xi^z(r)^\top \overline{b}_{xx}^1(r)\xi^z(r), \ldots, \xi^z(r)^\top \overline{b}_{xx}^n(r)\xi^z(r))^\top, \\ \xi^z(r)^\top \overline{\sigma}_{xx}^j(r)\xi^z(r) \stackrel{\Delta}{=} (\xi^z(r)^\top \overline{\sigma}_{xx}^{1j}(r)\xi^z(r), \ldots, \xi^z(r)^\top \overline{\sigma}_{xx}^{nj}(r)\xi^z(r))^\top, \\ \overline{b}_{xx}^i(r) \stackrel{\Delta}{=} b_{xx}^i(r, \overline{x}(r), \overline{u}(r)), \\ \overline{\sigma}_{xx}^{ij}(r) \stackrel{\Delta}{=} \sigma_{xx}^{ij}(r, \overline{x}(r), \overline{u}(r)), \end{cases}$$

and

$$\begin{cases} \varepsilon_{z1}(r) \stackrel{\Delta}{=} \displaystyle\int_0^1 \{b_x(r,\overline{x}(r)+\theta\xi^z(r),\overline{u}(r)) - \overline{b}_x(r)\}\xi^z(r)d\theta, \\ \varepsilon_{z2}^j(r) \stackrel{\Delta}{=} \displaystyle\int_0^1 \{\sigma_x^j(r,\overline{x}(r)+\theta\xi^z(r),\overline{u}(r)) - \overline{\sigma}_x^j(r)\}\xi^z(r)d\theta, \\ \varepsilon_{z3}(r) \stackrel{\Delta}{=} \displaystyle\int_0^1 (1-\theta)\xi^z(r)^\top\{b_{xx}(r,\overline{x}(r)+\theta\xi^z(r),\overline{u}(r)) - \overline{b}_{xx}(r)\}\xi^z(r)d\theta, \\ \varepsilon_{z4}^j(r) \stackrel{\Delta}{=} \displaystyle\int_0^1 (1-\theta)\xi^z(r)^\top\{\sigma_{xx}^j(r,\overline{x}(r)+\theta\xi^z(r),\overline{u}(r)) - \overline{\sigma}_{xx}^j(r)\}\xi^z(r)d\theta. \end{cases}$$

Step 2. Estimates of remainder terms.

We are going to show that for any $k \geq 1$, there exists a deterministic continuous and increasing function $\delta : [0,\infty) \to [0,\infty)$, independent of $z \in \mathbb{R}^n$, with $\frac{\delta(r)}{r} \to 0$ as $r \to 0$, such that

$$\begin{cases} E\Big[\displaystyle\int_t^T |\varepsilon_{z1}(r)|^{2k}dr|\mathcal{F}_t^s\Big](\omega) \leq \delta(|z-\overline{x}(t,\omega)|^{2k}), \quad \mathbf{P}\text{-a.s.}\,\omega, \\ E\Big[\displaystyle\int_t^T |\varepsilon_{z2}^j(r)|^{2k}dr|\mathcal{F}_t^s\Big](\omega) \leq \delta(|z-\overline{x}(t,\omega)|^{2k}), \quad \mathbf{P}\text{-a.s.}\,\omega, \end{cases} \tag{4.43}$$

and

$$\begin{cases} E\Big[\displaystyle\int_t^T |\varepsilon_{z3}(r)|^{k}dr|\mathcal{F}_t^s\Big](\omega) \leq \delta(|z-\overline{x}(t,\omega)|^{2k}), \quad \mathbf{P}\text{-a.s.}\,\omega, \\ E\Big[\displaystyle\int_t^T |\varepsilon_{z4}^j(r)|^{k}dr|\mathcal{F}_t^s\Big](\omega) \leq \delta(|z-\overline{x}(t,\omega)|^{2k}), \quad \mathbf{P}\text{-a.s.}\,\omega. \end{cases} \tag{4.44}$$

To this end, let us set $E^t \stackrel{\Delta}{=} E(\cdot\,|\mathcal{F}_t^s)(\omega)$ for a fixed $\omega \in \Omega$ such that (4.40) holds. Then, by setting $b_x(r,\theta) \stackrel{\Delta}{=} b_x(r,\overline{x}(r)+\theta\xi^z(r),\overline{u}(r))$, and using (4.5), we have

$$\begin{aligned} &E^t\int_t^T |\varepsilon_{z1}(r)|^{2k}dr \\ &\leq \int_t^T E^t\Big\{\int_0^1 |b_x(r,\theta)-\overline{b}_x(r)|^{2k}d\theta|\xi^z(r)|^{2k}\Big\}dr \\ &\leq K\int_t^T E^t|\xi^z(r)|^{4k}dr \leq K|z-\overline{x}(t,\omega)|^{4k}. \end{aligned} \tag{4.45}$$

Thus, the first equality in (4.43) follows for an obvious $\delta(\cdot)$. We can prove the second one similarly.

We proceed to show (4.44). Let $b_{xx}(r,\theta) \stackrel{\Delta}{=} b_{xx}(r,\overline{x}(r)+\theta\xi^z(r),\overline{u}(r))$.

A calculation similar to (4.45) shows that

$$
\begin{aligned}
&E^t \int_t^T |\varepsilon_{z3}(r)|^k dr \\
&\quad \le \int_t^T E^t\Big\{ \int_0^1 |b_{xx}(r,\theta) - \overline{b}_{xx}(r)|^k d\theta |\xi^z(r)|^{2k} \Big\} dr \\
&\quad \le \int_t^T \Big\{ E^t \big[\overline{\omega}(|\xi^z(r)|)^{2k}\big]\Big\}^{\frac{1}{2}} \Big\{ E^t |\xi^z(r)|^{4k}\Big\}^{\frac{1}{2}} dr \\
&\quad \le K|z - \overline{x}(t,\omega)|^{2k} \int_t^T \Big\{ E^t \big[\overline{\omega}(|\xi^z(r)|)^{2k}\big]\Big\}^{\frac{1}{2}} dr,
\end{aligned}
\tag{4.46}
$$

This yields the first equality in (4.44) for some $\delta(\cdot)$. We can prove the second one similarly. Finally, we may pick the largest $\delta(\cdot)$ obtained in the above four calculations, and (4.43)–(4.44) follows with a $\delta(\cdot)$ independent of $z \in \mathbb{R}^n$.

Step 3. Duality relation.

Applying the duality relation between $\xi^z(\cdot)$ and $p(\cdot)$, which satisfy (4.42) and (4.9), respectively (see Chapter 3, Section 4.3), and defining $\overline{f}_x(r) = f_x(r, \overline{x}(r), \overline{u}(r))$, we get

$$
\begin{aligned}
&E\Big\{ \int_t^T \langle \overline{f}_x(r), \xi^z(r) \rangle \, dr + \langle h_x(\overline{x}(T)), \xi^z(T) \rangle \, |\mathcal{F}_t^s \Big\} \\
&\quad = \langle -p(t), \xi^z(t) \rangle \\
&\qquad - E\Big\{ \frac{1}{2} \int_t^T \{ \langle p(r), \xi^z(r)^\top \overline{b}_{xx}(r) \xi^z(r) \rangle \\
&\qquad\qquad + \sum_{j=1}^m \langle q_j(r), \xi^z(r)^\top \overline{\sigma}^j_{xx}(r) \xi^z(r) \rangle \} dr \\
&\qquad - \int_t^T \{ \langle p(r), \varepsilon_{z3}(r) \rangle + \sum_{j=1}^m \langle q_j(r), \varepsilon^j_{z4}(r) \rangle \} dr | \mathcal{F}_t^s \Big\}, \quad \mathbf{P}\text{-a.s.}
\end{aligned}
\tag{4.47}
$$

On the other hand, setting $\Phi^z(r) \stackrel{\Delta}{=} \xi^z(r)\xi^z(r)^\top$, to which we apply Itô's formula, yields (compare with Chapter 3, (4.61))

$$
\left\{
\begin{aligned}
d\Phi^z(r) &= \Big\{ \overline{b}_x(r)\Phi^z(r) + \Phi^z(r)\overline{b}_x(r)^\top \\
&\qquad + \sum_{j=1}^m \overline{\sigma}^j_x(r)\Phi^z(r)\overline{\sigma}^j_x(r)^\top + \varepsilon_{z5}(r) \Big\} dr \\
&\quad + \sum_{j=1}^m \Big\{ \overline{\sigma}^j_x(r)\Phi^z(r) + \Phi^z(r)\overline{\sigma}^j_x(r)^\top \\
&\qquad\qquad + \varepsilon^j_{z6}(r) \Big\} dW^j(r), \qquad r \in [t,T], \\
\Phi^z(t) &= \xi(t)\xi(t)^\top,
\end{aligned}
\right.
\tag{4.48}
$$

where

$$\begin{cases} \varepsilon_{z5}(r) \stackrel{\Delta}{=} \varepsilon_{z1}(r)\xi^z(r)^\top + \xi^z(r)\varepsilon_{z1}(r)^\top \\ \qquad + \sum_{j=1}^m \{\bar\sigma_x^j(r)\xi^z(r)\varepsilon_{z2}^j(r)^\top + \varepsilon_{z2}^j(r)\xi^z(r)^\top\bar\sigma_x^j(r)^\top + \varepsilon_{z2}^j(r)\varepsilon_{z2}^j(r)^\top\}, \\ \varepsilon_{z6}^j(r) \stackrel{\Delta}{=} \varepsilon_{z2}^j(r)\xi^z(r)^\top + \xi^z(r)\varepsilon_{z2}^j(r)^\top. \end{cases}$$

Applying the duality relation again between $\Phi^z(\cdot)$ and $P(\cdot)$ using (4.48) and (4.10), we get

$$\begin{aligned} & -E\Big\{\int_t^T \xi^z(r)^\top H_{xx}(r)\xi^z(r)dr + \xi^z(T)^\top h_{xx}(\bar x(T))\xi^z(T)|\mathcal{F}_t^s\Big\} \\ & = -\xi^z(t)^\top P(t)\xi^z(t) \qquad\qquad (4.49) \\ & \quad - E\Big\{\int_t^T \operatorname{tr}\big[P(r)\varepsilon_{z5}(r) + \sum_{j=1}^m Q_j(r)\varepsilon_{z6}^j(r)\big]dr|\mathcal{F}_t^s\Big\}, \ \mathbf{P}\text{-a.s.} \end{aligned}$$

Step 4. Completion of the proof.

Let us call a $z \in \mathbb{R}^n$ *rational* if all its coordinates are rational numbers. Since the set of all rational $z \in \mathbb{R}^n$ is countable, we may find a subset $\Omega_0 \subseteq \Omega$ with $P(\Omega_0) = 1$ such that for any $\omega_0 \in \Omega_0$,

$$\begin{cases} V(t,\bar x(t,\omega_0)) = E\Big\{\int_t^T f(r,\bar x(r),\bar u(r))dr + h(\bar x(T))|\mathcal{F}_t^s\Big\}(\omega_0), \\ (4.40), (4.43), (4.44), (4.47), (4.49) \text{ are satisfied for any rational } z, \\ (\Omega,\mathcal{F},\mathbf{P}(\cdot|\mathcal{F}_t^s)(\omega_0), W(\cdot)-W(t);\bar u(\cdot)|_{[t,T]}) \in \mathcal{U}^w[t,T], \text{ and} \\ \sup_{s\le r\le T}(|p(r,\omega_0)| + |P(r,\omega_0)|) < +\infty. \end{cases}$$

The first equality of the above is due to Bellman's principle of optimality (Chapter 4, Theorem 3.4), while the last inequality holds because of the fact that $E\sup_{s\le r\le T}(|p(r)|^2 + |P(r)|^2) < +\infty$ (see Chapter 7, Theorem 2.2). Let $\omega_0 \in \Omega_0$ be fixed, and once again set $E^t \stackrel{\Delta}{=} E(\cdot\,|\mathcal{F}_t^s)(\omega_0)$. Then for any rational $z \in \mathbb{R}^n$, we have

$$\begin{aligned} & V(t,z) - V(t,\bar x(t,\omega_0)) \\ & \le E^t\Big\{\int_t^T \{f(r,x^z(r),\bar u(r)) - \bar f(r)\}dr + h(x^z(T)) - h(\bar x(T))\Big\} \\ & = E^t\Big\{\int_t^T \langle \bar f_x(r),\xi^z(r)\rangle dr + \langle h_x(\bar x(T)),\xi^z(T)\rangle\Big\} \qquad\qquad (4.50) \\ & \quad + \frac{1}{2}E^t\Big\{\int_t^T \xi^z(r)^\top \bar f_{xx}(r)\xi^z(r)dr + \xi^z(T)^\top h_{xx}(\bar x(T))\xi^z(T)\Big\} \\ & \quad + o(|z-\bar x(t,\omega_0)|^2). \end{aligned}$$

Appealing to (4.47) and (4.49), we have

$$
\begin{aligned}
&V(t,z)-V(t,\overline{x}(t,\omega_0))\\
&\quad\le -\langle p(t,\omega_0),\xi^z(t,\omega_0)\rangle\\
&\qquad - E^t\Big\{\frac{1}{2}\int_t^T \xi^z(r)^\top H_{xx}(r)\xi^z(r)dr+\frac{1}{2}\xi^z(T)^\top h_{xx}(\overline{x}(T))\xi^z(T)\Big\}\\
&\qquad + o(|z-\overline{x}(t,\omega_0)|^2)\\
&\quad= -\langle p(t,\omega_0),\xi^z(t,\omega_0)\rangle-\frac{1}{2}\xi^z(t,\omega_0)^\top P(t,\omega_0)\xi^z(t,\omega_0)\\
&\qquad + o(|z-\overline{x}(t,\omega_0)|^2)\\
&\quad= -\langle p(t,\omega_0),z-\overline{x}(t,\omega_0)\rangle-\tfrac{(z-\overline{x}(t,\omega_0))^\top P(t,\omega_0)(z-\overline{x}(t,\omega_0))}{2}\\
&\qquad + o(|z-\overline{x}(t,\omega_0)|^2).
\end{aligned}
\tag{4.51}
$$

Note that the term $o(|z-\overline{x}(t,\omega_0)|)$ in the above depends only on the size of $|z-\overline{x}(t,\omega_0)|$, and it is independent of z. Therefore, by the continuity of $V(t,\cdot)$, we see that (4.51) holds for all $z\in\mathbb{R}^n$, which proves

$$
(-p(t),-P(t))\in D_x^{2,+}V(t,\overline{x}(t)). \tag{4.52}
$$

Then, by (4.34), we obtain (4.37).

Let us now show (4.38). Fix an $\omega\in\Omega$ such that (4.51) holds for any $z\in\mathbb{R}^n$. For any $(p,P)\in D_x^{2,-}V(t,\overline{x}(t))$, by definition we have

$$
\begin{aligned}
0&\le \varliminf_{z\to\overline{x}(t)}\frac{V(t,z)-V(t,\overline{x}(t))-\langle p,z-\overline{x}(t)\rangle-\frac{1}{2}(z-\overline{x}(t))^\top P(z-\overline{x}(t))}{|z-\overline{x}(t)|^2}\\
&\le \varliminf_{z\to\overline{x}(t)}\frac{-\langle p(t)+p,z-\overline{x}(t)\rangle-\frac{1}{2}(z-\overline{x}(t))^\top(P(t)+P)(z-\overline{x}(t))}{|z-\overline{x}(t)|^2},
\end{aligned}
$$

where the last inequality is due to (4.51). Then, it is necessary that

$$
p=-p(t),\quad P\le -P(t).
$$

Thus, (4.38) follows. □

The following corollary is immediate.

Corollary 4.5. *Let the assumptions of Theorem 4.4 hold. Then*

$$
D_x^{1,-}V(t,\overline{x}(t))\subseteq\{-p(t)\}\subseteq D_x^{1,+}V(t,\overline{x}(t)),\quad \forall t\in[s,T],\ \mathbf{P}\text{-a.s.} \tag{4.53}
$$

Corollary 4.5 generalizes (3.36) from the deterministic case to the stochastic one. As a consequence, $p(t)$ can again be interpreted as the shadow price even when V is not smooth, based on some similar discussion as in Section 3.4. We omit the details here.

It is interesting to note that if $V\in C^{1,2}([0,T]\times\mathbb{R}^n)$, then (4.37)–(4.38) reduce to

$$
\begin{cases}
V_x(t,\overline{x}(t))=-p(t),\\
V_{xx}(t,\overline{x}(t))\le -P(t).
\end{cases}
\tag{4.54}
$$

We point out that the strict inequality $V_{xx}(t, \bar{x}(t)) < -P(t)$ may happen, as shown in the following example.

Example 4.6. Consider the following control system ($n = m = 1$):

$$\begin{cases} dx(t) = 2u(t)dt + \sqrt{2}dW(t), & t \in [0, T], \\ x(s) = y, \end{cases} \tag{4.55}$$

with the control domain being $U = [-1, 1]$ and the cost functional being

$$J(s, y; u(\cdot)) = E\Big\{ \int_s^T (u^2(t) + 1)dt - \log \text{ch}x(T) \Big\}, \tag{4.56}$$

where $\text{ch}x \triangleq \frac{1}{2}(e^x + e^{-x})$. For any fixed (s, y) and $u(\cdot) \in \mathcal{U}^w[s, T]$, applying Itô's formula to the process $\log \text{ch}x(t)$, then combining with (4.56), we get

$$\begin{aligned} &J(s, y; u(\cdot)) + \log \text{ch}y \\ &\quad = E \int_s^T \{u^2(t) + 1 - [\text{ch}x(t)]^{-2} - 2u(t)\text{th}x(t)\}dt \\ &\quad = E \int_s^T \{u(t) - \text{th}x(t)\}^2 dt \geq 0, \end{aligned} \tag{4.57}$$

where $x(\cdot)$ is the solution of (4.55), and $\text{th}x \triangleq \frac{e^x - e^{-x}}{e^x + e^{-x}}$. Equality holds if and only if $u(t) = \text{th}x(t)$. It follows then that

$$V(s, y) = -\log \text{ch}y \equiv -\log \Big(\frac{e^y + e^{-y}}{2} \Big), \tag{4.58}$$

which is analytic in $(s, y) \in [0, T] \times \mathbb{R}$. Now we consider Problem (S_{sy}) with $(s, y) = (0, 0)$. The above argument shows that $(\bar{x}(\cdot), \bar{u}(\cdot))$ is an optimal pair, where $\bar{u}(t) = \text{th}\bar{x}(t)$, and $\bar{x}(\cdot)$ satisfies

$$\begin{cases} d\bar{x}(t) = 2\text{th}\bar{x}(t)dt + \sqrt{2}dW(t), & t \in [0, T], \\ x(0) = 0. \end{cases} \tag{4.59}$$

Applying Itô's formula to $\text{th}\bar{x}(t)$, we obtain

$$\begin{cases} d[\text{th}\bar{x}(t)] = \sqrt{2}[\text{ch}\bar{x}(t)]^{-2}dW(t), & t \in [0, T], \\ \text{th}\bar{x}(T) = \text{th}\bar{x}(T). \end{cases} \tag{4.60}$$

The uniqueness of the adapted solution $(p(\cdot), q(\cdot))$ to the first-order adjoint equation (4.9) yields (compare (4.60) with (4.9))

$$\begin{cases} p(t) = \text{th}\bar{x}(t), \\ q(t) = \sqrt{2}[\text{ch}\bar{x}(t)]^{-2}, \end{cases} \quad t \in [0, T]. \tag{4.61}$$

Next, applying the martingale representation theorem to $E([\text{ch}\bar{x}(T)]^{-2} | \mathcal{F}_t^s)$, we get

$$\begin{cases} dE([\text{ch}\bar{x}(T)]^{-2} | \mathcal{F}_t^s) = Q(t)dW(t), & t \in [0, T], \\ E([\text{ch}\bar{x}(T)]^{-2} | \mathcal{F}_T^s) = [\text{ch}\bar{x}(T)]^{-2}, \end{cases} \tag{4.62}$$

for some $Q \in L^2_{\mathcal{F}}(0,T;\mathbb{R})$. Then the uniqueness of adapted solutions $(P(\cdot), Q(\cdot))$ of the second-order adjoint equation (4.10) implies (compare (4.62) with (4.10))

$$P(t) = E([\text{ch}\bar{x}(T)]^{-2}|\mathcal{F}_t^s). \tag{4.63}$$

It follows from Itô's formula again that

$$\begin{aligned}[\text{ch}\bar{x}(T)]^{-2} &= [\text{ch}\bar{x}(t)]^{-2} - 2\int_t^T [\text{ch}\bar{x}(r)]^{-4}dr \\ &\quad - 2\sqrt{2}\int_t^T \text{sh}\bar{x}(r)[\text{ch}\bar{x}(r)]^{-3}dW(r),\end{aligned} \tag{4.64}$$

which results in

$$P(t) \equiv E([\text{ch}\bar{x}(T)]^{-2}|\mathcal{F}_t^s) < [\text{ch}\bar{x}(t)]^{-2} \equiv -V_{xx}(t,\bar{x}(t)). \tag{4.65}$$

4.3. Nonsmooth case: Differentials in the time variable

In this subsection we proceed to study the super- and subdifferential of the value function in the time variable t along an optimal trajectory. What we are going to obtain is not simply some nonsmooth version of (4.12). Recall the result in the deterministic case, (3.35). We observe that the superdifferential of the value function in t contains $H(t,\bar{x}(t),\bar{u}(t),p(t))$, which is the maximum value of $H(t,\bar{x}(t),u,p(t))$ over $u \in U$ by the deterministic maximum principle. In the stochastic case, on the other hand, it is *not* the generalized Hamiltonian G that is to be maximized in the maximum principle (unless V is sufficiently smooth; see (4.12)). Instead, it is the following $\mathcal{H}$-function that appears in the stochastic maximum principle (see Chapter 3, (3.16)):

$$\begin{aligned}\mathcal{H}(t,x,u) &\triangleq G(t,x,u,p(t),P(t)) \\ &\quad + \text{tr}\Big(\sigma(t,x,u)^\top\big[q(t) - P(t)\sigma(t,\bar{x}(t),\bar{u}(t))\big]\Big),\end{aligned} \tag{4.66}$$

where $p(\cdot), q(\cdot)$, and $P(\cdot)$ are the solutions to (4.9) and (4.10) associated with the optimal pair $(\bar{x}(\cdot),\bar{u}(\cdot))$. Therefore, the following result appears rather natural.

Theorem 4.7. *Under the assumptions of Theorem 4.4, we have*

$$\mathcal{H}(t,\bar{x}(t),\bar{u}(t)) \in D^{1,+}_{t+}V(t,\bar{x}(t)), \quad \text{a.e. } t \in [s,T], \quad \mathbf{P}\text{-a.s.} \tag{4.67}$$

Proof. For any $t \in (s,T)$, take $\tau \in (t,T]$. Denote by $x_\tau(\cdot)$ the solution of the following SDE on $[\tau,T]$:

$$x_\tau(r) = \bar{x}(t) + \int_\tau^r b(\theta,x_\tau(\theta),\bar{u}(\theta))d\theta + \int_\tau^r \sigma(\theta,x_\tau(\theta),\bar{u}(\theta))dW(\theta). \tag{4.68}$$

Set $\xi_\tau(r) = x_\tau(r) - \bar{x}(r)$ for $r \in [\tau, T]$. Working under the new probability measure $\mathbf{P}(\cdot\,|\mathcal{F}_\tau^s)$, we get the following estimate for any $k \geq 1$:

$$E\{\sup_{\tau \leq r \leq T} |\xi_\tau(r)|^{2k} |\mathcal{F}_\tau^s\} \leq K|\bar{x}(\tau) - \bar{x}(t)|^{2k}, \quad \mathbf{P}\text{-a.s.} \tag{4.69}$$

Taking $E(\cdot\,|\mathcal{F}_t^s)$ on both sides and noting that $\mathcal{F}_t^s \subseteq \mathcal{F}_\tau^s$, we obtain, with a possibly new constant K:

$$E\{\sup_{\tau \leq r \leq T} |\xi_\tau(r)|^{2k} |\mathcal{F}_t^s\} \leq K|\tau - t|^k, \quad \mathbf{P}\text{-a.s.} \tag{4.70}$$

The process $\xi_\tau(\cdot)$ satisfies the following variational equations (compare with (4.41) and (4.42)):

$$\begin{cases} d\xi_\tau(r) = \bar{b}_x(r)\xi_\tau(r)dr + \displaystyle\sum_{j=1}^m \bar{\sigma}_x^j(r)\xi_\tau(r)dW^j(r) \\ \qquad\qquad + \varepsilon_{\tau 1}(r)dr + \displaystyle\sum_{j=1}^m \varepsilon_{\tau 2}^j(r)dW^j(r), \quad r \in [\tau, T], \\ \xi_\tau(\tau) = -\displaystyle\int_t^\tau \bar{b}(r)dr - \int_t^\tau \bar{\sigma}(r)dW(r), \end{cases} \tag{4.71}$$

and

$$\begin{cases} d\xi_\tau(r) = \Big\{\bar{b}_x(r)\xi_\tau(r) + \dfrac{1}{2}\xi_\tau(r)^\top \bar{b}_{xx}(r)\xi_\tau(r)\Big\}dr \\ \qquad + \displaystyle\sum_{j=1}^m \Big\{\bar{\sigma}_x^j(r)\xi_\tau(r) + \dfrac{1}{2}\xi_\tau(r)^\top \bar{\sigma}_{xx}^j(r)\xi_\tau(r)\Big\}dW^j(r) \\ \qquad + \varepsilon_{\tau 3}(r)dr + \displaystyle\sum_{j=1}^m \varepsilon_{\tau 4}^j(r)dW^j(r), \quad s \in [\tau, T], \\ \xi_\tau(\tau) = -\displaystyle\int_t^\tau \bar{b}(r)dr - \int_t^\tau \bar{\sigma}(r)dW(r), \end{cases} \tag{4.72}$$

where $\varepsilon_{\tau i}$ $(i = 1, 3)$ and $\varepsilon_{\tau i}^j$ $(i = 2, 4; j = 1, 2, \cdots, m)$ satisfy, for any $k \geq 1$,

$$\begin{cases} E\Big\{\displaystyle\int_\tau^T |\varepsilon_{\tau 1}(r)|^{2k} |\mathcal{F}_t^s\Big\}(\omega) \leq \delta(|\tau - t|^k), \quad \mathbf{P}\text{-a.s.}, \\ E\Big\{\displaystyle\int_\tau^T |\varepsilon_{\tau 2}^j(r)|^{2k} |\mathcal{F}_t^s\Big\}(\omega) \leq \delta(|\tau - t|^k), \quad \mathbf{P}\text{-a.s.}, \\ E\Big\{\displaystyle\int_\tau^T |\varepsilon_{\tau 3}(r)|^{k} |\mathcal{F}_t^s\Big\}(\omega) \leq \delta(|\tau - t|^k), \quad \mathbf{P}\text{-a.s.}, \\ E\Big\{\displaystyle\int_\tau^T |\varepsilon_{\tau 4}^j(r)|^{k} |\mathcal{F}_t^s\Big\}(\omega) \leq \delta(|\tau - t|^k), \quad \mathbf{P}\text{-a.s.}, \end{cases} \tag{4.73}$$

for some deterministic continuous increasing function $\delta : [0, \infty) \to [0, \infty)$ with $\frac{\delta(r)}{r} \to 0$ as $r \to 0$. The estimates (4.73) can be proved as in Step 2 of the proof of Theorem 4.4. The details are left to the reader. Note that

$$\big(\Omega,\mathcal{F},\mathbf{P}(\cdot\,|\mathcal{F}^s_\tau)(\omega),W(\cdot)-W(\tau),\bar{u}(\cdot)|_{[\tau,T]}\big)\in\mathcal{U}^w[\tau,T],\quad \mathbf{P}\text{-a.s.}$$

Thus, by the definition of the value function V,

$$V(\tau,\bar{x}(t))\le E\Big\{\int_\tau^T f(r,x_\tau(r),\bar{u}(r))dr+h(x_\tau(T))|\mathcal{F}^s_\tau\Big\},\quad \mathbf{P}\text{-a.s.} \tag{4.74}$$

Taking $E(\cdot\,|\mathcal{F}^s_t)$ on both sides of (4.74) and noting that $t<\tau$, we conclude that

$$V(\tau,\bar{x}(t))\le E\Big\{\int_\tau^T f(r,x_\tau(r),\bar{u}(r))dr+h(x_\tau(T))|\mathcal{F}^s_t\Big\},\quad \mathbf{P}\text{-a.s.} \tag{4.75}$$

Choose a subset $\Omega_0\subseteq\Omega$ with $\mathbf{P}(\Omega_0)=1$ such that for any $\omega_0\in\Omega_0$,

$$\begin{cases} V(t,\bar{x}(t,\omega_0))=E\Big\{\int_t^T f(r,\bar{x}(r),\bar{u}(r))dr+h(\bar{x}(T))|\mathcal{F}^s_t\Big\}(\omega_0),\\ (4.70),(4.73),(4.75)\text{ are satisfied for any rational }\tau>t,\\ \big(\Omega,\mathcal{F},\mathbf{P}(\cdot\,|\mathcal{F}^s_\tau)(\omega_0),W(\cdot)-W(\tau);\bar{u}(\cdot)|_{[\tau,T]}\big)\in\mathcal{U}^w[\tau,T],\\ \qquad\qquad\text{for any rational }\tau>t,\text{ and}\\ \sup\limits_{s\le r\le T}(|p(r,\omega_0)|+|P(r,\omega_0)|)<+\infty. \end{cases}$$

Let $\omega_0\in\Omega_0$ be fixed, and set $E^t\overset{\Delta}{=}E(\cdot\,|\mathcal{F}^s_t)(\omega_0)$. Then for any rational $\tau>t$, we have (noting (4.75))

$$\begin{aligned} &V(\tau,\bar{x}(t,\omega_0))-V(t,\bar{x}(t,\omega_0))\\ &\quad\le E^t\Big\{-\int_t^\tau \bar{f}(r)dr+\int_\tau^T[f(r,x_\tau(r),\bar{u}(r))-\bar{f}(r)]dr\\ &\qquad+h(x_\tau(T))-h(\bar{x}(T))\Big\}\\ &\quad=E^t\Big\{-\int_t^\tau \bar{f}(r)dr+\int_\tau^T\langle\bar{f}_x(r),\xi_\tau(r)\rangle\,dr\\ &\qquad+\langle h_x(\bar{x}(T)),\xi_\tau(T)\rangle+\frac{1}{2}\int_\tau^T\operatorname{tr}\Big(\bar{f}_{xx}(r)\xi_\tau(r)\xi_\tau(r)^\top\Big)dr\\ &\qquad+\frac{1}{2}\operatorname{tr}\Big(h_{xx}(\bar{x}(T))\xi_\tau(T)\xi_\tau(T)^\top\Big)\Big\}+o(|\tau-t|). \end{aligned} \tag{4.76}$$

As in (4.51), we have

$$\begin{aligned} &V(\tau,\bar{x}(t,\omega_0))-V(t,\bar{x}(t,\omega_0))\\ &\quad\le -E^t\int_t^\tau\bar{f}(r)dr-E^t\Big\{\langle p(\tau),\xi_\tau(\tau)\rangle+\frac{1}{2}\xi_\tau(\tau)^\top P(\tau)\xi_\tau(\tau)\Big\}\\ &\qquad+o(|\tau-t|). \end{aligned} \tag{4.77}$$

Now let us estimate the terms on the right-hand side of (4.77). To this end, we first note that for any $\varphi, \psi \in L^2_{\mathcal{F}}(0,T;\mathbb{R}^n)$,

$$
\begin{aligned}
&E^t \langle \int_t^\tau \varphi(r)dr, \int_t^\tau \psi(r)dr \rangle \\
&\leq \Big\{E^t \Big| \int_t^\tau \varphi(r)dr\Big|^2\Big\}^{\frac{1}{2}} \Big\{E^t \Big| \int_t^\tau \psi(r)dr\Big|^2\Big\}^{\frac{1}{2}} \\
&= (\tau - t)\Big\{ \int_t^\tau E^t|\varphi(r)|^2 dr \int_t^\tau E^t|\psi(r)|^2 dr\Big\}^{\frac{1}{2}} \\
&= o(|\tau - t|), \qquad \text{as } \tau \downarrow t, \quad \forall t \in [s,T), \quad \mathbf{P}\text{-a.s.},
\end{aligned}
\tag{4.78}
$$

and

$$
\begin{aligned}
&E^t \langle \int_t^\tau \varphi(r)dr, \int_t^\tau \psi(r)dW(r) \rangle \\
&\leq \Big\{E \Big| \int_t^\tau \varphi(r)dr\Big|^2\Big\}^{\frac{1}{2}} \Big\{E \Big| \int_t^\tau \psi(r)dW(r)\Big|^2\Big\}^{\frac{1}{2}} \\
&= (\tau - t)^{\frac{1}{2}}\Big\{ \int_t^\tau E^t|\varphi(r)|^2 dr \int_t^\tau E^t|\psi(r)|^2 dr\Big\}^{\frac{1}{2}} \\
&= o(|\tau - t|), \qquad \text{as } \tau \downarrow t, \quad \text{a.e. } t \in [s,T), \quad \mathbf{P}\text{-a.s.}
\end{aligned}
\tag{4.79}
$$

The last equality in (4.79) is due to the fact that the sets of Lebesgue points have full Lebesgue measures for integrable functions and $t \mapsto \mathcal{F}_t^s$ is continuous in t. Note that in (4.79), we only have a.e. $t \in [s,T)$, which is different from what we have in (4.78). Thus, by (4.71) and (4.9),

$$
\begin{aligned}
&E^t \langle p(\tau), \xi_\tau(\tau) \rangle \\
&= E^t\Big\{ \langle p(t), \xi_\tau(\tau)\rangle + \langle p(\tau) - p(t), \xi_\tau(\tau)\rangle \Big\} \\
&= E^t\Big\{\langle p(t), -\int_t^\tau \bar{b}(r)dr - \int_t^\tau \bar{\sigma}(r)dW(r)\rangle \\
&\quad + \langle -\int_t^\tau \big[\bar{b}_x(r)^\top p(r) + \sum_{j=1}^m \bar{\sigma}_x^j(r)^\top q_j(r) - \bar{f}_x(r)\big]dr \\
&\quad + \int_t^\tau q(r)dW(r), -\int_t^\tau \bar{b}(r)dr - \int_t^\tau \bar{\sigma}(r)dW(r)\rangle \Big\} \\
&= E^t\Big\{ -\langle p(t), \int_t^\tau \bar{b}(r)dr\rangle - \int_t^\tau \operatorname{tr}\big[q(r)^\top \bar{\sigma}(r)\big]dr\Big\} \\
&\quad + o(|\tau - t|).
\end{aligned}
\tag{4.80}
$$

Similarly,

$$
E^t \xi_\tau(\tau)^\top P(\tau)\xi_\tau(\tau) = E^t \int_t^\tau \operatorname{tr}\Big(\bar{\sigma}(r)^\top P(t)\bar{\sigma}(r)\Big)dr + o(|\tau - t|).
\tag{4.81}
$$

It follows from (4.77) and (4.80)–(4.81) that for any rational $\tau > t$ and at

$\omega = \omega_0$,

$$
\begin{aligned}
& V(\tau, \bar{x}(t)) - V(t, \bar{x}(t)) \\
& \leq E^t \Big\{ \langle p(t), \int_t^\tau \bar{b}(r)dr \rangle + \sum_{j=1}^m \int_t^\tau \langle q_j(r), \bar{\sigma}^j(r) \rangle \, dr \\
& \quad - \frac{1}{2} \int_t^\tau \mathrm{tr}\Big(\bar{\sigma}(r)^\top P(t) \bar{\sigma}(r)\Big) dr - \int_t^\tau \bar{f}(r)dr \Big\} + o(|\tau - t|) \\
& = (\tau - t)\mathcal{H}(t, \bar{x}(t), \bar{u}(t)) + o(|\tau - t|).
\end{aligned} \tag{4.82}
$$

By the same argument as in the paragraph following (4.51), we conclude that (4.82) holds for any (not only rational) $\tau > t$. This completes the proof. □

Note that it is vital in the above proof to have the sequence $\{\tau\}$ converge to t from the *right* side. This is essentially due to the adaptiveness requirement of all the processes involved. Therefore, our result, Theorem 4.7, involves the right superdifferential. In fact, the *right* superdifferential in (4.67) *cannot* be improved to the *two-sided* superdifferential $D_t^{1,+}V(t, \bar{x}(t))$, which is a smaller set. Below is a counterexample.

Example 4.8. Consider the optimal control problem (4.55)–(4.56) in Example 4.6. Since V is given by (4.58), which is independent of t, we have

$$
D_t^{1,+}V(t, \bar{x}(t)) = \{0\}, \qquad \forall t \in [0, T].
$$

However, for the optimal pair $(\bar{x}(\cdot), \bar{u}(\cdot))$ given there, we have (see (4.66))

$$
\mathcal{H}(t, \bar{x}(t), \bar{u}(t)) = -P(t) + [\mathrm{ch}\bar{x}(t)]^{-2} > 0, \tag{4.83}
$$

where the last inequality is due to (4.65). Therefore, $\mathcal{H}(t, \bar{x}(t), \bar{u}(t)) \notin D_t^{1,+}V(t, \bar{x}(t))$ in this case.

Now, let us combine Theorems 4.4 and 4.7 to get the following result.

Theorem 4.9. *Under the assumptions of Theorem 4.4, we have*

$$
\begin{aligned}
& [\mathcal{H}(t, \bar{x}(t), \bar{u}(t)), \infty) \times \{-p(t)\} \times [-P(t), \infty) \subseteq D_{t+,x}^{1,2,+}V(t, \bar{x}(t)), \\
& \qquad\qquad\qquad\qquad \text{a.e. } t \in [0, T], \ \mathbf{P}\text{-a.s.},
\end{aligned} \tag{4.84}
$$

and

$$
\begin{aligned}
D_{t+,x}^{1,2,-}V(t, \bar{x}(t)) & \subseteq (-\infty, \mathcal{H}(t, \bar{x}(t), \bar{u}(t))] \\
& \quad \times \{-p(t)\} \times (-\infty, -P(t)], \quad \text{a.e. } t \in [0, T], \ \mathbf{P}\text{-a.s.}
\end{aligned} \tag{4.85}
$$

Proof. The first conclusion can be proved by combining the proofs of (4.37) and (4.67) and making use of (4.33). For (4.85), it can be proved in a way similar to the proof of (4.38). The details are left to the reader. □

It has been shown by Theorem 4.1 that $q(t) = -V_{xx}(t, \bar{x}(t))\bar{\sigma}(t)$ (where $\bar{\sigma}(t) = \sigma(t, \bar{x}(t), \bar{u}(t))$) when V is smooth enough. It has also been shown by Example 4.6 that the nonsmooth version, i.e., $q(t) = P(t)\overline{\sigma}(t)$, does not hold in general (see (4.65) and (4.61)). We are therefore interested in knowing the general relationship among q, P, and $\overline{\sigma}$, which is given by the following result.

Proposition 4.10. *Under the assumptions of Theorem 4.4, we have*

$$\operatorname{tr}\Big(\sigma(t,\overline{x}(t),\overline{u}(t))^\top \big(q(t) - P(t)\sigma(t,\overline{x}(t),\overline{u}(t))\big)\Big) \geq 0, \qquad \text{a.e. } t \in [s,T],\ \mathbf{P}\text{-a.s.}, \tag{4.86}$$

or, equivalently,

$$\mathcal{H}(t,\overline{x}(t),\overline{u}(t)) \geq G(t,\overline{x}(t),\overline{u}(t),p(t),P(t)), \qquad \text{a.e. } t \in [s,T],\ \mathbf{P}\text{-a.s.} \tag{4.87}$$

Proof. By (4.84) and the fact that V is a viscosity solution of the HJB equation (4.6), we have

$$\begin{aligned} 0 \geq & -\mathcal{H}(t,\overline{x}(t),\overline{u}(t)) + \sup_{u\in U} G(t,\overline{x}(t),u,p(t),P(t)) \\ = & -G(t,\overline{x}(t),\overline{u}(t),p(t),P(t)) + \sup_{u\in U} G(t,\overline{x}(t),u,p(t),P(t)) \\ & -\operatorname{tr}\Big(\sigma(t,\overline{x}(t),\overline{u}(t))^\top \big(q(t) - P(t)\sigma(t,\overline{x}(t),\overline{u}(t))\big)\Big) \\ \geq & -\operatorname{tr}\Big(\sigma(t,\overline{x}(t),\overline{u}(t))^\top \big(q(t) - P(t)\sigma(t,\overline{x}(t),\overline{u}(t))\big)\Big). \end{aligned}$$

Then, (4.86) (or (4.87)) follows. □

5. Stochastic Verification Theorems

As in the deterministic case, we now investigate the *stochastic verification theorems*. Applying these results is a key step in utilizing the dynamic programming technique. Note, however, that we put them in this chapter (rather than in Chapter 4) because they also reveal part of the relationship problem between the maximum principle and dynamic programming.

5.1. Smooth case

We begin with the case where the value function is sufficiently smooth. The HJB equation (4.6) suggests that as far as the dynamic programming approach is concerned, the generalized Hamiltonian G in the stochastic case plays a role similar to that of the Hamiltonian H in the deterministic case. Thus, it is natural to expect that the following stochastic analogue of Theorem 3.7 holds.

Theorem 5.1. *Let (S1)′ and (S2)′ hold. Let $v \in C^{1,2}([0,T] \times \mathbb{R}^n)$ be a solution of the HJB equation (4.6). Then*

$$v(s,y) \leq J(s,y;u(\cdot)), \qquad \forall u(\cdot) \in \mathcal{U}^w[s,T],\ (s,y) \in [0,T] \times \mathbb{R}^n. \tag{5.1}$$

Furthermore, an admissible pair $(\bar{x}(\cdot),\bar{u}(\cdot))$ is optimal for Problem (S_{sy}) if and only if

$$
\begin{aligned}
v_t(t,\bar{x}(t)) &\equiv \max_{u\in U} G\big(t,\bar{x}(t),u,-v_x(t,\bar{x}(t)),-v_{xx}(t,\bar{x}(t))\big) \\
&= G\big(t,\bar{x}(t),\bar{u}(t),-v_x(t,\bar{x}(t)),-v_{xx}(t,\bar{x}(t))\big), \\
&\qquad\qquad\qquad \text{a.e. } t\in[s,T],\ \mathbf{P}\text{-a.s.}
\end{aligned}
\tag{5.2}
$$

Proof. For any $u(\cdot)\in\mathcal{U}^w[s,T]$ with the corresponding state trajectory $x(\cdot)$, applying Itô's formula to $v(t,x(t))$, we obtain

$$
\begin{aligned}
v(s,y) &= Eh(x(T)) \\
&\quad - E\int_s^T \Big\{ v_t(t,x(t)) + \langle\, v_x(t,x(t)), b(t,x(t),u(t))\,\rangle \\
&\quad + \frac{1}{2}\operatorname{tr}\Big(\sigma(t,x(t),u(t))^T v_{xx}(t,x(t))\sigma(t,x(t),u(t))\Big)\Big\}dt \\
&= J(s,y;u(\cdot)) + E\int_s^T \Big\{ - v_t(t,x(t)) \\
&\quad + G\big(t,x(t),u(t),-v_x(t,x(t)),-v_{xx}(t,x(t))\big)\Big\}dt \\
&\le J(s,y;u(\cdot)) + E\int_s^T \Big\{ - v_t(t,x(t)) \\
&\quad + \sup_{u\in U} G\big(t,x(t),u,-v_x(t,x(t)),-v_{xx}(t,x(t))\big)\Big\}dt \\
&= J(s,y;u(\cdot)).
\end{aligned}
\tag{5.3}
$$

Then (5.1) follows.

Next, applying (5.3) to $(\bar{x}(\cdot),\bar{u}(\cdot))$, we have

$$
\begin{aligned}
v(s,y) =& J(s,y;\bar{u}(\cdot)) + E\int_s^T \{ - v_t(t,\bar{x}(t)) \\
&+ G\big(t,\bar{x}(t),\bar{u}(t),-v_x(t,\bar{x}(t)),-v_{xx}(t,\bar{x}(t))\big)\}dt.
\end{aligned}
\tag{5.4}
$$

The desired result follows immediately from the fact that

$$
-v_t(t,\bar{x}(t)) + G\big(t,\bar{x}(t),\bar{u}(t),-v_x(t,\bar{x}(t)),-v_{xx}(t,\bar{x}(t))\big) \le 0,
\tag{5.5}
$$

which is due to the HJB equation (4.6). □

Theorem 5.1 above is referred to as the *classical stochastic verification theorem.*

5.2. Nonsmooth case

Note that Theorem 5.1 requires the value function V to be smooth enough. We now look at the case where V is *not* necessarily smooth. Recall that the proof of Theorem 3.9 for the deterministic case heavily relies on the value function being Lipschitz continuous in *both* the time and spatial variables, which is not necessarily true for the stochastic systems under consideration. Indeed, since $\int_0^t \sigma dW$ is only of order $t^{1/2}$, the value function in the stochastic case is at most Hölder continuous of order $\frac{1}{2}$. This prevents us from using the same idea as in the proof of Theorem 3.9. A key to overcoming the difficulty is the following lemma.

Lemma 5.2. *Let $g \in C[0,T]$. Extend g to $(-\infty,+\infty)$ with $g(t)=g(T)$ for $t>T$, and $g(t)=g(0)$ for $t<0$. Suppose there is a $\rho \in L^1(0,T)$ such that*

$$\varlimsup_{h\to 0} \frac{g(t+h)-g(t)}{h} \le \rho(t), \qquad \text{a.e. } t\in[0,T]. \tag{5.6}$$

Then

$$g(\beta)-g(\alpha) \le \int_\alpha^\beta \varlimsup_{h\to 0+} \frac{g(r+h)-g(r)}{h}dr, \quad \forall 0\le\alpha\le\beta\le T. \tag{5.7}$$

Proof. In view of (5.6), we can apply Fatou's lemma to get

$$\begin{aligned}\int_\alpha^\beta \varlimsup_{h\to 0+} \frac{g(r+h)-g(r)}{h}dr &\ge \varlimsup_{h\to 0+}\int_\alpha^\beta \frac{g(r+h)-g(r)}{h}dr\\ &= \varlimsup_{h\to 0+}\frac{\int_{\alpha+h}^{\beta+h} g(r)dr - \int_\alpha^\beta g(r)dr}{h}\\ &= \varlimsup_{h\to 0+}\frac{\int_{\beta}^{\beta+h} g(r)dr - \int_\alpha^{\alpha+h} g(r)dr}{h}\\ &= g(\beta)-g(\alpha).\end{aligned}$$

This proves (5.7). □

It is known that if g is a real-valued absolutely continuous function (in particular, if g is Lipschitz), then g is almost everywhere differentiable with the derivative $\dot g \in L^1(0,T)$, and we have the Newton-Leibniz formula

$$g(\beta)-g(\alpha) = \int_\alpha^\beta \dot g(r)dr.$$

Lemma 5.2 is a generalization of the above formula for continuous (not necessarily absolutely continuous) functions.

Theorem 5.3. *Let (S1)′–(S2)′ hold. Let $v \in C([0,T]\times\mathbb{R}^n)$, satisfying (3.9)–(3.10) of Chapter 4, be a viscosity solution of the HJB equation (4.6). Then (5.1) holds. Furthermore, let $(s,y)\in[0,T)\times\mathbb{R}^n$ be fixed and let $(\bar x(\cdot),\bar u(\cdot))$ be an admissible pair for Problem (S_{sy}) such that there exists a triple*

$$(\bar q,\bar p,\bar P)\in L^2_{\mathcal F}(s,T;\mathbb{R})\times L^2_{\mathcal F}(s,T;\mathbb{R}^n)\times L^2_{\mathcal F}(s,T;\mathcal S^n)$$

satisfying

$$(\bar q(t),\bar p(t),\bar P(t))\in D^{1,2,+}_{t+,x}v(t,\bar x(t)), \quad \text{a.e. } t\in[s,T],\ \mathbf{P}\text{-a.s.} \tag{5.8}$$

and

$$E\int_s^T \overline{q}(t)dt \le E\int_s^T G(t,\overline{x}(t),\overline{u}(t),-\overline{p}(t),-\overline{P}(t))dt. \tag{5.9}$$

Then $(\overline{x}(\cdot),\overline{u}(\cdot))$ *is optimal.*

Proof. First of all, (5.1) follows from the uniqueness of solutions to the HJB equation (4.6). To prove the second assertion, note that the proof of Chapter 4, Lemma 5.4-(ii) gives the following: For any $v(\cdot\,,\cdot) \in C([0,T]\times \mathbb{R}^n)$, there exists a $\varphi(\cdot\,,\cdot) \equiv \varphi(\cdot\,,\cdot\,;t,x,q,p,P) \in C^{1,2}([0,T]\times\mathbb{R}^n)$ with $(t,x)\in[0,T)\times\mathbb{R}^n$ and $(q,p,P)\in\mathbb{R}\times\mathbb{R}^n\times\mathcal{S}^n$ such that whenever $(t,x)\in[0,T)\times\mathbb{R}^n$ and $(q,p,P)\in D^{1,2,+}_{t+,x}v(t,x)$, one has

$$(\varphi_t(t,x),\varphi_x(t,x),\varphi_{xx}(t,x)) = (q,p,P), \tag{5.10}$$

and $v-\varphi$ attains a strict maximum over $[t,T]\times\mathbb{R}^n$ at (t,x). Now, for almost all $\omega\in\Omega$, let

$$\phi(r,z)\equiv\phi(r,z;\omega)\stackrel{\Delta}{=}\varphi(r,z;t,\overline{x}(t;\omega),\overline{q}(t;\omega),\overline{p}(t;\omega),\overline{P}(t;\omega)), \tag{5.11}$$

where $(\overline{q}(\cdot),\overline{p}(\cdot),\overline{P}(\cdot))$ are the processes satisfying (5.8) and (5.9). Then, on the probability space $(\Omega,\mathcal{F},\mathbf{P}(\cdot\,|\mathcal{F}_t^s)(\omega))$, applying Itô's formula to $\phi(r,\overline{x}(r))$, we have

$$\begin{aligned}
&E\Big\{v(t+h,\overline{x}(t+h))-v(t,\overline{x}(t))\mid\mathcal{F}_t^s\Big\}\\
&\quad\le E\Big\{\phi(t+h,\overline{x}(t+h))-\phi(t,\overline{x}(t))\mid\mathcal{F}_t^s\Big\}\\
&\quad= E\Big\{\int_t^{t+h}\Big[\phi_t(r,\overline{x}(r))+\langle\,\phi_x(r,\overline{x}(r)),\overline{b}(r)\,\rangle\\
&\qquad\qquad+\frac{1}{2}\mathrm{tr}\,\Big(\overline{\sigma}(r)^\top\phi_{xx}(r,\overline{x}(r))\overline{\sigma}(r)\Big)\Big]dr\;\Big|\;\mathcal{F}_t^s\Big\},\quad \mathbf{P}\text{-a.s.}
\end{aligned}$$

Taking the expectation in the above and letting t be the Lebesgue point of the integrand, we have

$$\begin{aligned}
&E\{v(t+h,\overline{x}(t+h))-v(t,\overline{x}(t))\}\\
&\quad= hE\{\phi_t(t,\overline{x}(t))+\langle\,\phi_x(t,\overline{x}(t)),\overline{b}(t)\,\rangle\\
&\qquad+\frac{1}{2}\mathrm{tr}\,[\overline{\sigma}(t)^\top\phi_{xx}(t,\overline{x}(t))\overline{\sigma}(t)]\}+o(h)\\
&\quad= hE\{\overline{q}(t)+\langle\,\overline{p}(t),\overline{b}(t)\,\rangle+\frac{1}{2}\mathrm{tr}\,[\overline{\sigma}(t)^\top\overline{P}(t)\overline{\sigma}(t)]\}+o(h).
\end{aligned} \tag{5.12}$$

Consequently,

$$\begin{aligned}
&\varlimsup_{h\to0+}\frac{Ev(t+h,\overline{x}(t+h))-Ev(t,\overline{x}(t))}{h}\\
&\quad\le E\Big\{\overline{q}(t)+\langle\,\overline{p}(t),\overline{b}(t)\,\rangle+\frac{1}{2}\mathrm{tr}\,(\overline{\sigma}(t)^\top\overline{p}(t)\overline{\sigma}(t))\Big\}.
\end{aligned} \tag{5.13}$$

Now, applying Lemma 5.2 to $g(t) \equiv Ev(t, \bar{x}(t))$, we arrive at

$$\begin{aligned}
&Ev(T, \bar{x}(T)) - v(s, y) \\
&\quad \leq E \int_s^T \Big\{ \bar{q}(t) + \langle \bar{p}(t), \bar{b}(t) \rangle + \frac{1}{2} \mathrm{tr}\left(\bar{\sigma}(t)^\top \overline{P}(t) \bar{\sigma}(t) \right) \Big\} dt \\
&\quad \leq -E \int_s^T \bar{f}(t) dt,
\end{aligned}$$

where the last inequality is due to (5.9). This leads to

$$v(s, y) \geq J(s, y; \bar{u}(\cdot)).$$

Then, combining the above with the first assertion (i.e., (5.1)), we obtain the optimality of $(\bar{x}(\cdot), \bar{u}(\cdot))$. □

The sufficient condition (5.9) can be replaced by an equivalent condition, which, however, looks much stronger and is of a form analogous to (5.2). We state this condition below.

Proposition 5.4. *Condition (5.9) in Theorem 5.3 is equivalent to the following:*

$$\begin{aligned}
\bar{q}(t) &= G(t, \bar{x}(t), \bar{u}(t), -\bar{p}(t), -\overline{P}(t)) \\
&= \max_{u \in U} G(t, \bar{x}(t), u, -\bar{p}(t), -\overline{P}(t)), \quad \text{a.e. } t \in [s, T], \ \mathbf{P}\text{-a.s.}
\end{aligned} \tag{5.14}$$

Proof. It is clear that (5.14) implies (5.9). Suppose now (5.9) holds. Since v is the viscosity solution of the HJB (4.6), by definition we have

$$-\bar{q}(t) + \sup_{u \in U} G(t, \bar{x}(t), u, -\bar{p}(t), -\overline{P}(t)) \leq 0.$$

The above inequality along with (5.9) easily yields (5.14). □

It is clear that the sufficiency part of the classical result Theorem 5.1 is recovered from Theorem 5.3 (along with Proposition 5.4) if v is smooth. Moreover, we do have a simple example showing that Theorem 5.1 may not be able to verify the optimality of a given admissible pair, whereas Theorem 5.3 can.

Example 5.5. Consider the following control system ($n = m = 1$):

$$\begin{cases} dx(t) = x(t)u(t)dt + x(t)dW(t), & t \in [s, T], \\ x(s) = y, \end{cases} \tag{5.15}$$

with the control domain being $U = [0, 1]$ and the cost functional being

$$J(s, y; u(\cdot)) = -Ex(T). \tag{5.16}$$

The corresponding HJB equation reads

$$\begin{cases} -v_t(t, x) + \sup\limits_{0 \leq u \leq 1} [-v_x(t, x)xu] - \dfrac{1}{2} x^2 v_{xx}(t, x) = 0, \\ v(T, x) = -x. \end{cases} \tag{5.17}$$

It is not difficult to directly verify that the following function is a viscosity solution of (5.17):

$$V(t,x) = \begin{cases} -x, & \text{if } x \le 0, \\ -xe^{T-t}, & \text{if } x > 0, \end{cases} \tag{5.18}$$

which clearly satisfies (3.9)–(3.10) of Chapter 4. Thus, by the uniqueness of the viscosity solutions, V coincides with the value function of the control problem. Let us consider an admissible control $\overline{u}(\cdot) \equiv 0$ for initial time $s = 0$ and initial state $y = 0$. The trajectory under $\overline{u}(\cdot)$ is easily seen to be $\overline{x}(t) \equiv 0$. Now we want to see whether the pair $(\overline{x}(\cdot), \overline{u}(\cdot))$ is optimal. Theorem 5.1 cannot tell us anything, because $V_x(t,x)$ does not exist along the *whole* trajectory $\overline{x}(t)), t \in [s,T]$. However, we have

$$D_{t+,x}^{1,2,+}V(t,\overline{x}(t)) = [0,+\infty) \times [-e^{T-t}, -1] \times [0,+\infty).$$

If we take

$$(\overline{q}(t), \overline{p}(t), \overline{P}(t)) = (0,-1,0) \in D_{t+,x}^{1,2,+}V(t,\overline{x}(t)), \qquad t \in [0,T],$$

then (5.9) or (5.14) is satisfied. This implies that the pair $(\overline{x}(\cdot), \overline{u}(\cdot))$ is indeed optimal by Theorem 5.3 (or Proposition 5.4).

Theorem 5.3 is expressed in terms of superdifferential. One naturally asks whether a similar result holds for *sub*differential. The answer was yes for the *deterministic* case (in terms of the first-order subdifferential; see Theorem 3.9). Unfortunately, the answer is no for the stochastic case, namely, the statement of Theorem 5.3 no longer holds true if the superdifferential in (5.8) is replaced by the subdifferential. Here is a counterexample.

Example 5.6. Consider the problem in Example 4.6. The unique optimal feedback control has been computed to be $\mathbf{u}(t,x) = \text{th } x$. The admissible pair $(\overline{x}(t), \overline{u}(t)) = (\sqrt{2}W(t), 0)$ is therefore *not* optimal for Problem (S$_{00}$). However, we shall show that there does exist $(\overline{q}(t), \overline{p}(t), \overline{P}(t)) \in D_{t+,x}^{1,2,-}V(t,\overline{x}(t))$ such that (5.9) holds. To this end, note that $V(s,y)$ is given by (4.58). Hence, we can compute (noting (4.33))

$$\begin{aligned} D_{t+,x}^{1,2,-}V(t,\overline{x}(t)) = \{(q,p,P) | q \le 0, p = -\text{th}(\sqrt{2}W(t)), \\ P \le -[\text{ch}(\sqrt{2}W(t))]^{-2}\}. \end{aligned}$$

Take

$$(\overline{q}(t), \overline{p}(t), \overline{P}(t)) = (0, -\text{th}(\sqrt{2}W(t)), -1) \in D_{t+,x}^{1,2-}V(t,\overline{x}(t)).$$

Then it is easy to verify that

$$\overline{q}(t) = G(t, \overline{x}(t), \overline{u}(t), -\overline{p}(t), -\overline{P}(t)) = 0,$$

which implies (5.9).

As a matter of fact, for the deterministic case we have the inclusion (3.35). However, for the stochastic case we have only the inclusions (4.84)–(4.85), which do not guarantee the following:

$$D^{1,2,-}_{t+,x}V(t,\bar{x}(t)) \subseteq D^{1,2,+}_{t+,x}V(t,\bar{x}(t)), \quad \text{a.e. } t \in [0,T], \ \mathbf{P}\text{-a.s.}$$

To conclude this subsection, let us present a nonsmooth version of the necessity part of Theorem 5.1.

Theorem 5.7. *Let (S1)′–(S2)′ hold and let V be the value function. Let $(s,y) \in [0,T) \times \mathbb{R}^n$ be fixed and let $(\bar{x}(\cdot),\bar{u}(\cdot))$ be an optimal pair for Problem (S_{sy}). Then for any $(\bar{q},\bar{p},\overline{P}) \in L^2_{\mathcal{F}}(s,T;\mathbb{R}^n) \times L^2_{\mathcal{F}}(s,T;\mathbb{R}^n) \times L^2_{\mathcal{F}}(s,T;\mathcal{S}^n)$ satisfying*

$$(\bar{q}(t),\bar{p}(t),\overline{P}(t)) \in D^{1,2,-}_{t+,x}V(t,\bar{x}(t)), \quad \text{a.e. } t \in [s,T], \ \mathbf{P}\text{-a.s.}, \tag{5.19}$$

it must hold that

$$E\bar{q}(t) \le EG(t,\bar{x}(t),\bar{u}(t),-\bar{p}(t),-\overline{P}(t)), \quad \text{a.e. } t \in [s,T]. \tag{5.20}$$

Proof. Fix a $t \in [s,T)$ and an $\omega \in \Omega$ such that $(\bar{q}(t),\bar{p}(t),\overline{P}(t)) \in D^{1,2,-}_{t+,x}V(t,\bar{x}(t))$. As in the proof of Theorem 5.3, we have (using Chapter 4, Lemma 5.5-(ii)) a test function $\phi(\cdot\,,\cdot) \equiv \varphi(\cdot\,,\cdot\,;t,x,q,p,P) \in C^{1,2}([0,T]\times\mathbb{R}^n)$ with $(t,x) \in [0,T) \times \mathbb{R}^n$ and $(q,p,P) \in \mathbb{R} \times \mathbb{R}^n \times \mathcal{S}^n$ such that $V - \phi$ achieves its minimum at $(t,\bar{x}(t))$ and (5.10) holds. Then for sufficiently small $h > 0$, and a.e. $t \in [0,T)$,

$$\begin{aligned}
&EV(t+h,\bar{x}(t+h)) - EV(t,\bar{x}(t)) \\
&\quad \ge E\phi(t+h,\bar{x}(t+h)) - E\phi(t,\bar{x}(t)) \\
&\quad = E\int_t^{t+h} \Big\{\phi_t(r,\bar{x}(r)) + \langle\, \phi_x(r,\bar{x}(r)),\bar{b}(r)\,\rangle \\
&\qquad\qquad + \frac{1}{2}\operatorname{tr}\Big(\bar{\sigma}(r)^\top \phi_{xx}(r,\bar{x}(r))\bar{\sigma}(r)\Big)\Big\}dr. \\
&\quad = hE\big\{\bar{q}(t) + \langle\, \bar{p}(t),\bar{b}(t)\,\rangle + \frac{1}{2}\operatorname{tr}\big[\bar{\sigma}(t)^\top \overline{P}(t)\bar{\sigma}(t)\big]\big\} + o(h).
\end{aligned} \tag{5.21}$$

However, since $(\bar{x}(\cdot),\bar{u}(\cdot))$ is optimal, by Bellman's principle of optimality, we have

$$V(r,\bar{x}(r)) = E\Big\{\int_r^T \bar{f}(\alpha)d\alpha + h(\bar{x}(T))|\mathcal{F}^s_r\Big\}, \quad \forall r \in [s,T], \ \mathbf{P}\text{-a.s.},$$

which implies

$$EV(t+h,\bar{x}(t+h)) - EV(t,\bar{x}(t)) = -E\int_t^{t+h} \bar{f}(r)dr.$$

Then, by (5.21), we obtain

$$E\bar{q}(t) \le EG(t,\bar{x}(t),\bar{u}(t),-\bar{p}(t),-\overline{P}(t)), \quad \text{a.e. } t \in [s,T].$$

The proof is completed. □

Theorem 5.7 reduces to the "only if" part of Theorem 5.1 if V is smooth. Indeed, in the smooth case, we conclude from Theorem 5.7 that

$$EV_t(t,\bar{x}(t)) \leq EG(t,\bar{x}(t),\bar{u}(t),-V_x(t,\bar{x}(t)),-V_{xx}(t,\bar{x}(t))).$$

On the other hand, since V is the solution of the HJB equation (4.6), we have

$$\begin{aligned}&G\big(t,\bar{x}(t),\bar{u}(t),-V_x(t,\bar{x}(t)),-V_{xx}(t,\bar{x}(t))\big)\\ &\quad\leq \sup_{u\in U} G\big(t,\bar{x}(t),u,-V_x(t,\bar{x}(t)),-V_{xx}(t,\bar{x}(t))\big) = V_t(t,\bar{x}(t)).\end{aligned}\tag{5.22}$$

Thus the equality in (5.22) holds, which yields (5.2).

6. Optimal Feedback Controls

This section describes how to construct optimal feedback controls by the results presented in the previous section. First we introduce the following definition (compare with Definition 3.13).

Definition 6.1. A measurable function $\mathbf{u} : [0,T]\times \mathbb{R}^n \to U$ is called an *admissible feedback control* if for any $(s,y)\in[0,T)\times\mathbb{R}^n$ there is a weak solution $x(\cdot)\equiv x(\cdot\,;s,y)$ of the following equation:

$$\begin{cases} dx(t) = b\big(t,x(t),\mathbf{u}(t,x(t))\big)dt + \sigma\big(t,x(t),\mathbf{u}(t,x(t))\big)dW(t),\\ x(s)=y.\end{cases}\tag{6.1}$$

An admissible feedback control $\bar{\mathbf{u}}$ is said to be *optimal* if for each $(s,y)\in[0,T)\times\mathbb{R}^n$, $\big(\bar{x}(\cdot\,;s,y),\bar{\mathbf{u}}(\cdot\,,\bar{x}(\cdot\,;s,y))\big)$ is optimal for Problem (S_{sy}) (under the weak formulation), where $\bar{x}(\cdot\,;s,y)$ is a weak solution of (6.1) corresponding to $\bar{\mathbf{u}}$.

Theorem 6.2. *Let (S1)′–(S2)′ hold and let $v\in C([0,T]\times\mathbb{R}^n)$ be the viscosity solution of the HJB equation (4.6). Then for each $(t,x)\in(0,T)\times\mathbb{R}^n$,*

$$\inf_{(q,p,P,u)\in[D^{1,2,+}_{t+,x}v(t,x)]\times U}[q-G(t,x,u,-p,-P)]\geq 0.\tag{6.2}$$

Further, if $\bar{\mathbf{u}}$ is an admissible feedback control, and $\bar{\mathbf{q}}$, $\bar{\mathbf{p}}$, and $\bar{\mathbf{P}}$ are measurable functions satisfying $(\bar{\mathbf{q}}(t,x),\bar{\mathbf{p}}(t,x),\bar{\mathbf{P}}(t,x))\in D^{1,2,+}_{t+,x}V(t,x)$ for all (t,x), and

$$\begin{aligned}&\bar{\mathbf{q}}(t,x)-G(t,x,\bar{\mathbf{u}}(t,x),-\bar{\mathbf{p}}(t,x),-\bar{\mathbf{P}}(t,x))\\ &\quad= \min_{(q,p,P,u)\in[D^{1,2,+}_{t+,x}v(t,x)]\times U}\{q-G(t,x,u,-p,-P)\}=0\end{aligned}\tag{6.3}$$

for all $(t,x)\in[0,T]\times\mathbb{R}^n$, then $\bar{\mathbf{u}}$ is optimal.

Proof. By the uniqueness of viscosity solutions to the HJB equation (4.6), v is the value function V of Problem (S_{sy}). Then, by Chapter 4, Theorem 6.2, we obtain (6.2).

Next, for any $(s,y) \in [0,T) \times \mathbb{R}^n$, let $\bar{x}(\cdot)$ be the trajectory of (6.1) corresponding to $\overline{\mathbf{u}}$. Put

$$\begin{cases} \bar{u}(t) \stackrel{\Delta}{=} \overline{\mathbf{u}}(t,\bar{x}(t)), & \bar{q}(t) \stackrel{\Delta}{=} \overline{\mathbf{q}}(t,\bar{x}(t)), \\ \bar{p}(t) \stackrel{\Delta}{=} \overline{\mathbf{p}}(t,\bar{x}(t)), & \overline{P}(t) \stackrel{\Delta}{=} \overline{\mathbf{P}}(t,\bar{x}(t)). \end{cases}$$

By (6.3), $(\bar{x}(t),\bar{u}(t),\bar{q}(t),\bar{p}(t),\overline{P}(t))$ satisfies (5.8)–(5.9). The desired result then follows readily from Theorem 5.3. □

By Theorem 6.2, we see that under proper conditions, one can obtain an optimal feedback control by minimizing $q - G(t,x,u,-p,-P)$ over $(q,p,P,u) \in [D^{1,2,+}_{t+,x} v(t,x)] \times U$ for each (t,x). Now let us investigate the conditions imposed in Theorem 6.2. First of all, (6.3) requires that

$$\min_{(q,p,P,u)\in[D^{1,2,+}_{t+,x} V(t,x)]\times U} \{q - G(t,x,u,-p,-P)\} = 0. \tag{6.4}$$

This condition in fact partially characterizes the existence of an optimal feedback control, although rather implicitly in the sense that the value function (or the viscosity solution to the HJB equation) is involved. In particular, this condition is satisfied *automatically* if v is smooth, as seen from the HJB equation (4.6). Next, in order to apply Filippov's lemma to obtain a *measurable selection* $(\overline{\mathbf{q}}(t,x),\overline{\mathbf{p}}(t,x),\overline{\mathbf{P}}(t,x),\overline{\mathbf{u}}(t,x))$ of $D^{1,2,+}_{t+,x} v(t,x)$ that achieves the minimum in (6.4), we need to study the measurability of the *multifunction* $(t,x) \mapsto D^{1,2,+}_{t+,x} v(t,x)$. To do this, let us first recall the measurability of multifunctions.

Definition 6.3. Let $X \subseteq \mathbb{R}^n$ a Lebesgue measurable set, Y a metric space, and $\Gamma : X \to 2^Y$ a multifunction. We say that Γ is measurable if for any closed set $F \subseteq Y$ the set

$$\Gamma^{-1}(F) \stackrel{\Delta}{=} \{x \in X \mid \Gamma(x) \bigcap F \neq \phi\} \tag{6.5}$$

is Lebesgue measurable.

Note that in the above definition, Γ need not be closed-set-valued. It is known that when Y is a Polish space (i.e., a separable complete metric space), the closed set F in the above definition can be replaced by any open set. Consequently, we have the following simple result.

Lemma 6.4. *Let $X \subseteq \mathbb{R}^n$ be a Lebesgue measurable set, Y a Polish space, and $\Gamma : X \to 2^Y$ be a multifunction. Then, Γ is measurable if and only if the multifunction $x \mapsto \overline{\Gamma}(x) \stackrel{\Delta}{=} \overline{\Gamma(x)}$ is measurable.*

Proof. Note that for any open set $U \subseteq Y$ and $x \in X$,

$$\Gamma(x) \bigcap U \neq \phi \iff \overline{\Gamma(x)} \bigcap U \neq \phi.$$

Hence, by the definition of Γ^{-1} (see (6.5))

$$\Gamma^{-1}(U) = \overline{\Gamma}^{-1}(U), \qquad \forall \text{ open set } U \subseteq Y.$$

Then, by the above observation, we obtain the conclusion. □

Proposition 6.5. *Both of the multifunctions* $(t,x)\mapsto D^{1,2,+}_{t+,x}v(t,x)$ *and* $(t,x)\mapsto \overline{D^{1,2,+}_{t+,x}v(t,x)}$ *are convex-set-valued and are measurable.*

Proof. For any $(s,y)\in(0,T]\times\mathbb{R}^n$, define

$$W(t,x,q,p,P;s,y) \triangleq \begin{cases} \dfrac{1}{|s-t|+|y-x|^2}\Big\{v(s,y)-v(t,x)-q(s-t)-\langle p,y-x\rangle \\ \qquad\qquad -\dfrac{1}{2}(y-x)^T P(y-x)\Big\}, & \text{if } t\in[0,s), \\ 0, & \text{if } t\in[s,T]. \end{cases}$$

Clearly, the function $(t,x,q,p,P)\mapsto W(t,x,q,p,P;s,y)$ is Borel measurable (with (s,y) as parameters). Hence, the function

$$\varlimsup_{s\to t+,y\to x} W(t,x,q,p,P;s,y)\triangleq \widetilde W(t,x,q,p,P)$$

is also Borel measurable. This implies that the multifunction

$$D^{1,2,+}_{t+,x}v(t,x)\equiv\{(q,p,P)\mid \widetilde W(t,x,q,p,P)\le 0\}$$

is measurable (see Li–Yong [1, p. 99, Theorem 2.2]). By Lemma 6.4, we obtain the measurability of the multifunction $(t,x)\mapsto\overline{D^{1,2,+}_{t+,x}v(t,x)}$. The convexity of these two multifunctions is obvious. □

Filippov's lemma (see, e.g., Li–Yong [1, p. 102, Corollary 2.26]) says that if Γ is a measurable multifunction defined on some Lebesgue measurable set taking closed-set values in a Polish space, then it admits a measurable selection. Therefore, if we assume that $D^{1,2,+}_{t+,x}V(t,x)$ is closed and that the minimum in (6.3) exists, then, by Proposition 6.5 and Filippov's lemma, we can find a measurable selection $(\overline{\mathbf{q}}(t,x),\overline{\mathbf{p}}(t,x),\overline{\mathbf{P}}(t,x),\overline{\mathbf{u}}(t,x))\in D^{1,2,+}_{t+,x}v(t,x)$ that minimizes $q-G(t,x,u,-p,-P)$.

Suppose now we have selected a measurable function $\overline{\mathbf{u}}(t,x)$. It may not be an admissible feedback control. The reason is that the coefficients $\tilde b(t,x)\triangleq b(t,x,\overline{\mathbf{u}}(t,x))$ and $\tilde\sigma(t,x)\triangleq\sigma(t,x,\overline{\mathbf{u}}(t,x))$ of the SDE (6.1) are measurable only in (t,x), which does not guarantee the existence of a solution in general. This difficulty occurs in the deterministic case as well. However, for the stochastic case, there are some elegant existence results for SDEs with measurable coefficients. Let us briefly discuss two situations.

Case 1. Assume that $b(t,x,u)$ and $\sigma(t,x,u)$ are uniformly bounded and $\sigma(t,x,u)$ is an $n\times n$ matrix and is uniformly elliptic, i.e.,

$$\langle\sigma(t,x,u)\lambda,\lambda\rangle\ge\delta|\lambda|^2,\qquad\forall(t,x,u)\in[0,T]\times\mathbb{R}^n\times U, \tag{6.6}$$

for some constant $\delta>0$. Then by Chapter 1, Theorem 6.13, there exists a weak solution to SDE (6.1) under $\overline{\mathbf{u}}(t,x)$. By Theorem 6.2, $\overline{\mathbf{u}}(t,x)$ is an optimal feedback control.

Case 2. Assume that $\sigma(t,x,u) \equiv \sigma(t,x)$ is a nonsingular $n \times n$ matrix independent of u. Moreover, $\sigma(t,x)$ is uniformly bounded, Lipschitz in x, and

$$\sup_{t,x} |\sigma(t,x)^{-1}| < +\infty. \tag{6.7}$$

Then under $\overline{\mathbf{u}}(t,x)$, the existence and uniqueness of the weak solutions to (6.1) are obtained by Girsanov's transformation (see Chapter 1, Theorems 6.11 and 6.12). Therefore, $\overline{\mathbf{u}}(t,x)$ is an optimal feedback control by Theorem 6.2.

To summarize the above discussion, we have the following theorem.

Theorem 6.6. *Assume that $D^{1,2,+}_{t+,x}V(t,x)$ is closed and (6.4) holds. Then in either of the above two cases, there is a measurable selection $(\overline{\mathbf{q}}(t,x), \overline{\mathbf{p}}(t,x), \overline{\mathbf{P}}(t,x), \overline{\mathbf{u}}(t,x))$ of $\left[D^{1,2,+}_{t+,x}v(t,x)\right] \times U$ that minimizes $q - G(t,x,u,-p,-P)$. Moreover, the fourth component $\overline{\mathbf{u}}(t,x)$ is an optimal feedback control.*

Basically, the verification theorem derived in this section reduces the original stochastic control problem to a two-phase problem. In the first phase, one has to solve the HJB equations, which are fully nonlinear second-order partial differential equations. In most cases one has to rely on numerical methods to solve the equations (see Kushner–Dupuis [1]), whereas only in some exceptional cases may one obtain analytical solutions (like the one in Example 5.5). In the second phase, one finds the optimal feedback $\overline{\mathbf{u}}$ by minimizing $q - G(t,x,u,-p,-P)$ over *both* the superdifferential of v and the control region. The second phase is relatively easy because the superdifferential of v is explicitly known once v is known. However, if v is approximated by numerical solutions v_n, then a natural question is what conditions ensure that the feedback controls obtained by applying our verification theorem to v_n are good enough. To answer the question requires a study of the asymptotic behavior of the super-/subdifferentials of the approximating solutions v_n. These remain challenging open problems.

7. Historical Remarks

Both the Hamiltonian system and the Hamilton–Jacobi equation were introduced by William R. Hamilton in 1834 and 1835 for studying the same class of problems in mechanics (Jacobi's name was associated due to his fundamental improvement of Hamilton's original work). Thus, the equivalence between these two has intrinsically existed from the very moment when they were given birth. The equivalence between general first-order partial differential equations and certain systems of ordinary differential equations was first noticed by Constantin Carathéodory [1] in the 1920s. The material presented in Section 2 is a modification of the presentations in Goldstein [1] and Courant–Hilbert [1].

The relationship between the adjoint function and the value function (or, equivalently, that between the maximum principle and dynamic programming) in the smooth case (as given in Theorem 3.1) was understood as early as when Pontryagin and his colleagues proved the maximum principle; see Pontryagin–Boltyanski–Gamkrelidze–Mischenko [1]. The relationship has been elaborated by many others (including a proof of deriving the maximum principle directly from dynamic programming) since then, see, e.g., Citron [1], Berkovitz [1], and Fleming–Rishel [1]. The economic interpretation of the relationship presented in Section 3.2 is based on the paper of Dorfman [1]. On the other hand, the method of characteristics is a classical approach (prior to the Sobolev space method) in solving linear partial differential equations; see, for example, Courant–Hilbert [1]. Theorem 3.2, which is basically the method of characteristics in the context of optimal control, is based on Fleming–Rishel [1]. Corollary 3.3 is a deterministic version of the Feynman–Kac formula (Feynman [1] and Kac [1]). See Chapter 7 and Yong [9] for some more results related to the Feynman–Kac formula.

An attempt to relate the maximum principle and dynamic programming without assuming the smoothness of the value function was first made by Barron–Jensen [1] in 1986, where the viscosity solution was used to derive the maximum principle directly from the dynamic programming equation. The relationships (3.35) and (3.36), which are nonsmooth versions of (3.9) and (3.10), respectively, within the framework of the viscosity solution, were obtained by Zhou in 1988 for his Ph.D. thesis and published in 1990 (Zhou [1]). Similar results were derived by Subbotina [1] independently, under assumptions that the control region is compact and the Roxin condition holds. More research along this line were carried out in Cannarsa–Frankowska [1], Frankowska [1], and Li–Yong [2]. On the other hand, the verification theorem using the viscosity solution was first obtained by Zhou [9] in 1993, based on which optimal feedback control may be constructed without involving derivatives of the value function. It is interesting to note that this "nonsmooth" verification theorem can be regarded as an extension of the relationship between the maximum principle and dynamic programming from open-loop controls to feedback controls. The discussions in Sections 3.4 and 3.5 follow those in Zhou [1,9].

Research on the relationship between the maximum principle and dynamic programming using a different type of nonsmooth analysis, namely, Clarke's generalized gradient (Clarke [2]), was carried out a bit earlier than the research using the viscosity solution mentioned above. Clarke–Vinter [2] obtained a nonsmooth version of (3.10). Later, Vinter [1] derived a nonsmooth version of (3.9) using the generalized gradient. A nonsmooth verification theorem within this framework was given even earlier in Clarke–Vinter [1]. It should be noted that the viscosity solution and generalized gradient are two quite different types of nonsmooth analysis (although the latter is where the term *nonsmooth analysis* originates). Roughly speaking, if the super-/subdifferential is a nonsmooth version of *differentiability*, then the generalized gradient can be regarded as a nonsmooth version of *continuous differentiability*. The advantage of the generalized gradient theory is that it can treat optimal control problems with state constraints, while its disadvantage is that it may not work well for *stochastic* control problems, since an extension to the second-order generalized Hessian is not straightforward and obvious. Indeed, there have been a few different notions of *second-order generalized gradient* in the literature; see, for example, Chaney [1], Rockafellar [2], and Cominetti–Correa [1]. Haussmann [6,7] introduced a notion called the *generalized Hessian* based on Brownian motion, and then applied to the stochastic optimal control problems in a way analogous to the generalized gradient applied to the deterministic problems. Some connections between the viscosity solution and the generalized gradient were discussed in Frankowska [1] and Zhou [9].

For stochastic optimal control problems, the relationship between the maximum principle and dynamic programming was intuitively discussed by Bismut [3] and Bensoussan [1]. The proof of Theorem 4.1 is more related to Bensoussan [1]. The nonsmooth version of the first relation in (4.13) was first obtained in Zhou [2]. Further relations involving the second-order differential in x and the first-order differential in t were developed by Zhou in his Ph.D. thesis (see also Zhou [4]) using the viscosity solution and second-order adjoint equation. Section 4 is based on Zhou [4]. It is worth mentioning that Example 4.6 was provided by a referee in his/her review report for an earlier version of Zhou [4], and we include it in this book with high appreciation to this anonymous referee for supplying this very nice example. The relationship in the context of optimal control of stochastic partial differential equations was explored in Zhou [3].

The classical verification theorem, Theorem 5.1, is standard and may be found in many books on stochastic controls, such as Fleming–Rishel [1] and Fleming–Soner [1]. The nonsmooth verification theorem, Theorem 5.3, was obtained by Zhou–Yong–Li [1] four years after the publication of its deterministic counterpart (Zhou [9]). Optimal feedback controls for deterministic problems (without assuming smoothness of the value function) were investigated by Berkovitz [2], Cannarsa–Frankowska [2], and Li–Yong [2]. The materials in Sections 5 and 6 are mainly taken from Zhou [9] and Zhou–Yong–Li [1].

Chapter 6

Linear Quadratic Optimal Control Problems

1. Introduction

We have studied general nonlinear optimal control problems for both the deterministic and stochastic cases in previous chapters. In this chapter we are going to investigate a special case of optimal control problems where the state equations are linear in both the state and control with nonhomogeneous terms, and the cost functionals are quadratic. Such a control problem is called a linear quadratic optimal control problem (LQ problem, for short). The LQ problems constitute an extremely important class of optimal control problems, since they can model many problems in applications, and more importantly, many nonlinear control problems can be reasonably approximated by the LQ problems. On the other hand, solutions of LQ problems exhibit elegant properties due to their simple and nice structures.

Since the state equation in an LQ problem is linear, by the *variation of constants formula* the state process can be explicitly expressed in terms of the initial state, the control, and the nonhomogeneous term in a linear form. Substituting this relation into the cost functional (which is quadratic in the state and control), we obtain a functional quadratic in the triple of the initial state, the control, and the nonhomogeneous term. Thus, the original LQ problem can be transformed to a quadratic optimization (minimizing) problem (parametrized by the initial state and the nonhomogeneous term) in the (Hilbert) space consisting of all square-integrable control processes. This leads to some necessary and sufficient conditions for the existence of optimal controls. However, this functional analysis approach can determine optimal controls only in a very abstract form. On the other hand, employing the elementary method of *completion of squares*, one can obtain an optimal control in a linear state feedback form via the so-called *Riccati equation*. Along this line, the solvability of the Riccati equation leads to that of the LQ problem. It is interesting to note that both the maximum principle and dynamic programming can lead to the Riccati equation, by which one can see more clearly the relationship between MP and DP (actually, these two approaches are *equivalent* in the LQ case).

At first glance, one might consider the stochastic LQ problems to be just some routine extension of their deterministic counterpart. However, the presence of the control in the diffusion term makes a stochastic LQ problem significantly different from the deterministic ones. In a deterministic LQ problem (formulated in Chapter 2, Section 2), the control *weighting matrix* $R(t)$ in the cost functional (see Chapter 2, (2.13)) has to be nonnegative definite for almost all $t \in [0,T]$; otherwise, the optimization problem would be ill-posed. In fact, roughly speaking, if the control weighting matrix $R(t)$ is indefinite (on a set $F \subseteq [0,T]$ of positive Lebesgue measure), i.e., some of its eigenvalues are negative (which means that the control components corresponding to those eigenvalues are *beneficial* rather than *costly*), then an optimal control $u(\cdot)$ is simply such that those control components are infinitely large, resulting in a negative infinite cost. In other words, *the larger those control components, the better.* Surprisingly, some stochastic LQ problems could be well-posed when the weighting matrix $R(t)$ is negative definite, i.e., the "the-larger-the-better" policy no longer applies. Let us look at a very simple example in one dimension. The deterministic LQ problem

$$(1.1)\qquad \begin{aligned} &\text{Minimize} && J = \frac{1}{2}\int_0^1 [x(t)^2 + r(t)u(t)^2]dt + \frac{1}{2}x(1)^2, \\ &\text{Subject to} && dx(t) = 0, \quad x(0) = 0, \end{aligned}$$

where $r(t) < 0$, is not well-posed. In fact,

$$J = \int_0^1 r(t)u(t)^2 dt \to -\infty, \qquad \text{as } |u(t)| \to +\infty.$$

Now consider a stochastic version of (1.1):

$$(1.2)\qquad \begin{aligned} &\text{Minimize} && J = E\Big\{\frac{1}{2}\int_0^1 [x(t)^2 + r(t)u(t)^2]dt + \frac{1}{2}x(1)^2\Big\}, \\ &\text{Subject to} && dx(t) = u(t)dW(t), \quad x(0) = 0. \end{aligned}$$

Substituting $x(t) = \int_0^t u(s)dW(s)$ into the cost function, we obtain

$$(1.3)\qquad J = \frac{1}{2}E\int_0^1 [r(t) + (2-t)]u(t)^2 dt.$$

Hence, when $r(t)$ is a deterministic function with $r(t) > t - 2$, the optimization problem is well-posed (with the optimal control $\overline{u}(t) = 0$). In this case, the control weighting $r(t)$ could be *negative* as long as, say, $r(t) > -1$. Certainly, $r(t)$ should not be *too* negative. For example, the problem would obviously become ill-posed if $r(t) < -2$.

The above seemingly surprising observation indeed makes perfect sense when we think a little deeper: The gain due to a larger control size may *not* outweigh the loss due to a greater uncertainty (because the diffusion depends on the control). It is emphasized that the fact that the control enters into the diffusion term, which means that the controller could control the volatility or the decision made would unavoidably affect the scale of the uncertainty in the system, plays a key role here. This kind of situation happens in real-world systems. In a stock market, for example, the trading made by the so-called *large investors* is going to influence the fluctuations of stock prices. If the control is absent in the diffusion, then the control weighting matrix has to be nonnegative definite in order for the LQ problem to be meaningful; for such a case, the stochastic LQ problem is almost completely parallel to the deterministic one. However, if the control is present in the diffusion, then the control weighting matrix, even being negative, could be compensated by a quadratic term related to the degree to which the control can affect the underlying uncertainty. This observation reveals a fundamental difference between deterministic and stochastic systems.

Let us take one more concrete example to illustrate the above idea. Suppose an oil company is investing in an oil prospecting project. This project will cause a certain degree of pollution, and suppose the pollution level $x(t)$ during a period of time $[0, T]$ is described by

$$\begin{cases} dx(t) = [\alpha x(t) + \beta u(t)]dt + \delta u(t)dW(t), \\ x(0) = x_0, \end{cases} \tag{1.4}$$

where $u(t)$ represents the investment level of the company at time t; x_0 is the initial pollution level; and α, β, and δ are given constants. Suppose the investment is expected to be very profitable, and the return in the time period $[t, t + \Delta t]$ is $r|u(t)|^2 \Delta t$ with a constant $r > 0$, and the company has sufficient funds to make the investment so that $u(t) \in (0, +\infty)$. On the other hand, the environmental impact of the project is supervised and monitored by the government, so that the pollution level $x(t)$ cannot deviate from an allowable level $x^*(t)$ by too much at any time. The objective of the company is on one hand to maximize the total expected return, $E \int_0^T r|u(t)|^2 dt$, and on the other hand to minimize the expected negative environmental impact, measured by $E \int_0^T |x(t) - x^*(t)|^2 dt$. This is a multiobjective optimization problem, and it may be converted into a single-objective problem by putting weights on the different objectives. Thus the following is to be minimized:

$$J = E \int_0^T (\lambda_1 |x(t) - x^*(t)|^2 - \lambda_2 r |u(t)|^2) dt, \tag{1.5}$$

where $\lambda_1, \lambda_2 \in (0, 1)$ with $\lambda_1 + \lambda_2 = 1$ represent the weights. This is a stochastic (minimizing) LQ problem with a negative control cost. If the problem were deterministic (i.e., there were no risk) where a positive return is guaranteed, then by the deterministic LQ theory, the control cost $-\lambda_2 r|u(t)|^2$ would be overwhelming when $\lambda_2 > 0$, so the optimal policy would be $u(t) = +\infty$ (i.e., the larger the investment size the better) and the problem would become trivial. However, the problem is actually stochastic, where the diffusion coefficient depends on the control (i.e., the risk in pollution increases as the investment level increases). Thus there is a trade-off (no matter how small λ_1 is) between the return (or the investment size) and the risk, which makes the optimization problem sensible.

More generally, such a phenomenon can happen in the following situation. Suppose in a deterministic (minimizing) optimization problem, the cost *decreases* as the level of activity the decision-maker carries out *increases* (a typical example of such situations is an investment that would be "guaranteed" to be profitable if the risk were to be excluded from consideration). Then it is not really an optimization problem because there is no *trade-off* in it, and the optimal decision is simply to take the maximum possible activity level. So the problem is trivial (or ill-posed). But in a stochastic environment, suppose the uncertainty *increases* with *increasing* magnitude of the activity level and the uncertainty would result in certain additional cost. Then there is a trade-off between the activity level and the uncertainty, and the decision-maker has to carefully balance the two to achieve an optimal solution. The problem therefore becomes interesting. Needless to say, such phenomena may occur in a much wider class of optimization problems that can go beyond linear systems and optimal control problems.

For stochastic LQ problems, the method of completion of squares, the stochastic maximum principle, and dynamic programming all give rise to a *stochastic Riccati equation.* This equation is quite different from the conventional Riccati equation arising in the deterministic LQ problems. A stochastic LQ problem is well-posed if there are solutions to the stochastic Riccati equation, and an optimal feedback control can then be obtained via these solutions. Therefore, the stochastic LQ problem can be reduced to that of solving the stochastic Riccati equation. However, the existence and uniqueness of the solutions to the stochastic Riccati equation are available only for certain special cases so far.

The rest of this chapter is organized as follows. Section 2 revisits some important aspects in the deterministic LQ theory. In Section 3 the formulation of the stochastic LQ problems is given. In Section 4 the method of functional analysis is used to obtain the existence of optimal controls. Next, based on the results of Section 4, we derive the stochastic Hamiltonian system along with the optimal pair of the LQ problems in Section 5. This is the maximum principle for the optimal control. In Section 6 the stochastic Riccati equation is introduced, and the optimal control is represented in a state feedback form, through the solution to the stochastic Riccati equation. It is demonstrated that the three approaches, namely, the stochastic maximum principle, dynamic programming, and the completion of squares technique, all lead to the stochastic Riccati equation. In Section 7 the solvability of the stochastic Riccati equation is discussed for several special cases. Section 8 is devoted to a continuous-time mean-variance portfolio selection problem that can be solved via the stochastic LQ theory developed in the preceding sections. Finally, some historical remarks will be given in Section 9.

2. Deterministic LQ Problems Revisited

2.1. Formulation

We consider the following state equation:

$$
\begin{cases} \dot{x}(t) = A(t)x(t) + B(t)u(t) + b(t), & t \in [s,T], \\ x(s) = y, \end{cases} \tag{2.1}
$$

where $0 \leq s \leq T$, $A(\cdot) \in L^\infty(0,T;\mathbb{R}^{n\times n})$, $B(\cdot) \in L^\infty(0,T;\mathbb{R}^{n\times k})$, and $b(\cdot) \in L^2(0,T;\mathbb{R}^n)$. The cost functional takes the form

$$
(2.2)\qquad \begin{aligned} J(s,y;u(\cdot)) = \frac{1}{2}\int_s^T &\{\langle Q(t)x(t),x(t)\rangle + 2\langle S(t)x(t),u(t)\rangle \\ &+\langle R(t)u(t),u(t)\rangle\}dt + \frac{1}{2}\langle Gx(T),x(T)\rangle, \end{aligned}
$$

with $Q(\cdot) \in L^\infty(0,T;\mathcal{S}^n)$, $S \in L^\infty(0,T;\mathbb{R}^{k\times n})$, $R \in L^\infty(0,T;\mathcal{S}^k)$, and $G \in \mathcal{S}^n$. Note that while all the coefficients are dependent on the time t, in what follows the variable t will usually be suppressed if no confusion would occur. Further, we introduce the following notation: For any $F \in L^\infty(0,T;\mathcal{S}^n)$,

$$
\begin{cases} F \geq 0, & \iff \quad F(t) \geq 0, \quad \text{a.e. } t \in [0,T], \\ F > 0, & \iff \quad F(t) > 0, \quad \text{a.e. } t \in [0,T], \\ F \gg 0, & \iff \quad F(t) \geq \delta I, \quad \text{a.e. } t \in [0,T], \text{ for some } \delta > 0. \end{cases}
$$

For any $s \in [0,T)$, we let

$$
\mathcal{V}[s,T] \stackrel{\Delta}{=} L^2(s,T;\mathbb{R}^k), \qquad \mathcal{Y}[s,T] \stackrel{\Delta}{=} L^2(s,T;\mathbb{R}^n).
$$

A $u(\cdot) \in \mathcal{V}[s,T]$ is called an admissible control (over $[s,T]$). Our optimal control problem is the following.

Problem (DLQ). For each $(s,y) \in [0,T]\times\mathbb{R}^n$, find a control $\bar{u}(\cdot) \in \mathcal{V}[s,T]$ such that

$$
(2.3)\qquad J(s,y;\bar{u}(\cdot)) = \inf_{u(\cdot)\in\mathcal{V}[s,T]} J(s,y;u(\cdot)) \stackrel{\Delta}{=} V(s,y).
$$

The above problem is referred to as a (deterministic) *linear quadratic optimal control problem.* Note that we do not assume the nonnegativity of the matrices Q, G, and R, in general. Let us now introduce the following definition.

Definition 2.1. Problem (DLQ) is said to be

(i) *finite* at $(s,y) \in [0,T]\times\mathbb{R}^n$ if the right-hand side of (2.3) is finite;

(ii) (*uniquely*) *solvable* at $(s,y) \in [0,T]\times\mathbb{R}^n$ if there exists a (unique) $\bar{u}(\cdot) \in \mathcal{V}[s,T]$ satisfying (2.3). In this case, $\bar{u}(\cdot)$ is called an (the) *optimal control,* and $\bar{x}(\cdot)$ and $(\bar{x}(\cdot),\bar{u}(\cdot))$ are called an (the) *optimal trajectory* and an (the) *optimal pair,* respectively;

(iii) *finite* (resp. (*uniquely*) *solvable*) at $s \in [0,T]$ if it is finite (resp. (uniquely) solvable) at (s,y) for any $y \in \mathbb{R}^n$; Problem (DLQ) is said to be finite (resp. (uniquely) solvable) if it is finite (resp. (uniquely) solvable) at any $s \in [0,T]$.

We note that Problem (DLQ) is trivially solvable at $s = T$. Also, it is clear that solvability implies finiteness.

2.2. A minimization problem of a quadratic functional

To study Problem (DLQ), we introduce the following operators/functions:

$$
(2.4)\qquad \begin{cases} (L_s u(\cdot))(\cdot) \stackrel{\Delta}{=} \int_s^{\cdot} \Phi(\cdot, r)Bu(r)dr, \quad \widehat{L}_s u(\cdot) \stackrel{\Delta}{=} (L_s u(\cdot))(T), \\ \qquad\qquad\qquad\qquad\qquad\qquad \forall u(\cdot) \in \mathcal{V}[s,T], \\ (\Gamma_s y)(\cdot) \stackrel{\Delta}{=} \Phi(\cdot, s)y, \quad \widehat{\Gamma}_s y \stackrel{\Delta}{=} (\Gamma_s y)(T), \qquad \forall y \in \mathbb{R}^n, \\ f^s(t) \stackrel{\Delta}{=} \int_s^t \Phi(t,r)b(r)dr, \quad \widehat{f}^s \stackrel{\Delta}{=} f^s(T), \end{cases}
$$

where $\Phi(\cdot,\cdot)$ is the fundamental solution matrix of $A(t)$ on $[s,T]$. If we denote by $x(\cdot) \stackrel{\Delta}{=} x(\cdot\,; s, y, u(\cdot))$ the solution of (2.1) corresponding to $(s,y) \in [0,T] \times \mathbb{R}^n$ and $u(\cdot) \in \mathcal{V}[s,T]$, then by the variation of constants formula, we have

$$
(2.5)\qquad \begin{cases} x(\cdot) = (\Gamma_s y)(\cdot) + (L_s u(\cdot))(\cdot) + f^s(\cdot), \\ x(T) = \widehat{\Gamma}_s y + \widehat{L}_s u(\cdot) + \widehat{f}^s. \end{cases}
$$

It is clear that $L_s \in \mathcal{L}(\mathcal{V}[s,T]; \mathcal{Y}[s,T])$ and $\widehat{L}_s \in \mathcal{L}(\mathcal{V}[s,T]; \mathbb{R}^n)$, and their adjoint operators $L_s^* \in \mathcal{L}(\mathcal{Y}[s,T]; \mathcal{V}[s,T])$ and $\widehat{L}_s^* \in \mathcal{L}(\mathbb{R}^n; \mathcal{V}[s,T])$ are given by

$$
(2.6)\qquad \begin{cases} (L_s^* x(\cdot))(\cdot) = \int_{\cdot}^T B^\top \Phi(r,\cdot)^\top x(r)dr, \ \forall x(\cdot) \in \mathcal{Y}[s,T], \\ (\widehat{L}_s^* y)(\cdot) = \Phi(T,\cdot)^\top y, \ \forall y \in \mathbb{R}^n. \end{cases}
$$

Further, in what follows, with an abuse of notation, we use the convention

$$
(2.7)\qquad \begin{cases} (Qx(\cdot))(\cdot) = Q(\cdot)x(\cdot), \qquad \forall x(\cdot) \in \mathcal{Y}[s,T], \\ (Sx(\cdot))(\cdot) = S(\cdot)x(\cdot), \qquad \forall x(\cdot) \in \mathcal{Y}[s,T], \\ (Ru(\cdot))(\cdot) = R(\cdot)u(\cdot), \qquad \forall u(\cdot) \in \mathcal{V}[s,T]. \end{cases}
$$

Then by (2.2) and (2.4)–(2.7), for any $(y, u(\cdot)) \in \mathbb{R}^n \times \mathcal{V}[s,T]$, we have the following representation of the cost functional:

$$
(2.8)\qquad \begin{aligned} &J(s,y;u(\cdot)) \\ &= \frac{1}{2}\Big\{ \langle (L_s^* Q L_s + S L_s + L_s^* S^\top + \widehat{L}_s^* G \widehat{L}_s + R)u(\cdot), u(\cdot) \rangle \\ &\qquad +2\,\langle (L_s^* Q + S)((\Gamma_s y)(\cdot) + f^s(\cdot)) + \widehat{L}_s^* G(\widehat{\Gamma}_s y + \widehat{f}^s), u(\cdot) \rangle \\ &\qquad + \langle Q((\Gamma_s y)(\cdot) + f^s(\cdot)), (\Gamma_s y)(\cdot) + f^s(\cdot) \rangle \\ &\qquad + \langle G(\widehat{\Gamma}_s y + \widehat{f}^s), \widehat{\Gamma}_s y + \widehat{f}^s \rangle \Big\} \\ &\equiv \frac{1}{2}\Big\{ \langle N_s u(\cdot), u(\cdot) \rangle + 2\,\langle H_s(y), u(\cdot) \rangle + M_s(y) \Big\}, \end{aligned}
$$

where $\langle \cdot, \cdot \rangle$ is the inner product on $\mathcal{V}[s,T]$, and

$$(2.9)\qquad \begin{cases} N_s \stackrel{\Delta}{=} R + L_s^* Q L_s + S L_s + L_s^* S^\top + \widehat{L}_s^* G \widehat{L}_s, \\ H_s(y) \stackrel{\Delta}{=} (L_s^* Q + S)((\Gamma_s y)(\cdot) + f^s(\cdot)) + \widehat{L}_s^* G(\widehat{\Gamma}_s y + \widehat{f}^s), \\ M_s(y) \stackrel{\Delta}{=} \langle Q((\Gamma_s y)(\cdot) + f^s(\cdot)), (\Gamma_s y)(\cdot) + f^s(\cdot) \rangle \\ \qquad\qquad + \langle G(\widehat{\Gamma}_s y + \widehat{f}^s), \widehat{\Gamma}_s y + \widehat{f}^s \rangle. \end{cases}$$

Hence, for any $(s,y) \in [0,T] \times \mathbb{R}^n$, $u(\cdot) \mapsto J(s,y;u(\cdot))$ is a (continuous) quadratic functional on the Hilbert space $\mathcal{V}[s,T]$, and the original Problem (DLQ) is transformed to a minimization problem of such a functional over $\mathcal{V}[s,T]$, with $(s,y) \in [0,T] \times \mathbb{R}^n$ and $b(\cdot) \in L^2(0,T;\mathbb{R}^n)$ being parameters.

Theorem 2.2. *We have the following results.*

(i) *If Problem (DLQ) is finite at some $(s,y) \in [0,T) \times \mathbb{R}^n$, then*

$$(2.10)\qquad N_r \geq 0, \qquad \forall r \in [s,T].$$

(ii) *Problem (DLQ) is (uniquely) solvable at $(s,y) \in [0,T) \times \mathbb{R}^n$ if and only if $N_s \geq 0$ and there exists a (unique) $\overline{u}(\cdot) \in \mathcal{V}[s,T]$ such that*

$$(2.11)\qquad N_s \overline{u}(\cdot) + H_s(y) = 0.$$

In this case, $\overline{u}(\cdot)$ is an (the) optimal control.

(iii) *If $N_s \gg 0$ (i.e., N_s^{-1} exists and $N_s^{-1} \in \mathcal{L}(\mathcal{V}[s,T];\mathcal{V}[s,T])$) for some $s \in [0,T)$, then for any $y \in \mathbb{R}^n$, $J(s,y;\cdot)$ admits a unique minimizer $\overline{u}(\cdot)$ given by*

$$(2.12)\qquad \overline{u}(\cdot) = -N_s^{-1} H_s(y).$$

In this case,

$$(2.13)\qquad \begin{aligned} V(s,y) &= \inf_{u(\cdot) \in \mathcal{V}[s,T]} J(s,y;u(\cdot)) = J(s,y;\overline{u}(\cdot)) \\ &= \frac{1}{2}\{M_s(y) - \langle N_s^{-1} H_s(y), H_s(y) \rangle\}, \\ &\qquad\qquad \forall (s,y) \in [0,T] \times \mathbb{R}^n. \end{aligned}$$

(iv) *If Problem (DLQ) is uniquely solvable at $s \in [0,T)$, so is Problem (DLQ) with $b(\cdot) = 0$.*

Proof. (i) Suppose (2.10) fails. Then, for some $r \in [s,T]$ and some $u_0(\cdot) \in \mathcal{V}[r,T]$, we have

$$(2.14)\qquad \langle N_r u_0(\cdot), u_0(\cdot) \rangle < 0.$$

Extend $u_0(\cdot)$ to an element, still denoted by $u_0(\cdot)$, in $\mathcal{V}[s,T]$ such that

$u_0(t) = 0$ for $t \in [s, r)$. Then, by (2.8), for any $k > 0$,

$$\begin{aligned} J(s, y; ku_0(\cdot)) &= J(r, x(r); ku_0(\cdot)) + \frac{1}{2}\int_s^r \langle Qx(t), x(t) \rangle dt \\ &= \frac{1}{2}k^2 \Big\{ \langle N_r u_0(\cdot), u_0(\cdot) \rangle + \frac{2\langle H_r(x(r)), u_0(\cdot) \rangle}{k} \\ &\qquad + \frac{M_r(x(r)) + \int_s^r \langle Qx(t), x(t) \rangle dt}{k^2} \Big\} \\ &\le k^2 \langle N_r u_0(\cdot), u_0(\cdot) \rangle, \end{aligned} \tag{2.15}$$

provided that k is large enough. The above, together with (2.14), implies that

$$\inf_{u(\cdot) \in \mathcal{V}[s,T]} J(s, y; u(\cdot)) = -\infty,$$

contradicting the finiteness of Problem (DLQ) at (s, y).

(ii) First, let $\overline{u}(\cdot) \in \mathcal{V}[s, T]$ be an optimal control of Problem (DLQ) for $(s, y) \in [0, T) \times \mathbb{R}^n$. Then, for any $u(\cdot) \in \mathcal{V}[s, T]$, we have

$$0 \le \lim_{\lambda \to 0} \frac{J(s, y; \overline{u}(\cdot) + \lambda u(\cdot)) - J(s, y; \overline{u}(\cdot))}{\lambda} = \langle N_s \overline{u}(\cdot) + H_s(y), u(\cdot) \rangle .$$

Since $\mathcal{V}[s, T]$ is a linear space, we must have equality in the above for any $u(\cdot) \in \mathcal{V}[s, T]$. Hence, (2.11) follows.

Conversely, let $(y, \overline{u}(\cdot)) \in \mathbb{R}^n \times \mathcal{V}[s, T]$ satisfy (2.11). Then, for any $u(\cdot) \in \mathcal{V}[s, T]$, we have (noting (2.10))

$$\begin{aligned} &J(s, y; u(\cdot)) - J(s, y; \overline{u}(\cdot)) \\ &\quad = J(s, y; \overline{u}(\cdot) + u(\cdot) - \overline{u}(\cdot)) - J(s, y; \overline{u}(\cdot)) \\ &\quad = \langle N_s \overline{u}(\cdot) + H_s(y), u(\cdot) - \overline{u}(\cdot) \rangle + \frac{1}{2} \langle N_s(u(\cdot) - \overline{u}(\cdot)), u(\cdot) - \overline{u}(\cdot) \rangle \ge 0. \end{aligned}$$

Thus, $\overline{u}(\cdot)$ is optimal.

(iii) By combining (ii) with $N_s \gg 0$, we obtain (2.12). Also, (2.13) follows easily.

(iv) Suppose Problem (DLQ) (with general $b(\cdot)$) is uniquely solvable at $s \in [0, T)$. Let

$$\widetilde{H}_s(y) = [(L_s^* Q + S)\Gamma_s + \widehat{L}_s^* G \widehat{\Gamma}_s] y.$$

Then (2.11) is equivalent to

$$N_s \overline{u}(\cdot) + \widetilde{H}_s(y) + H_s(0) = 0. \tag{2.16}$$

By (i) and (ii), we know that for any $y \in \mathbb{R}^n$, (2.16) admits a unique solution $\overline{u}(\cdot) \stackrel{\Delta}{=} \overline{u}(\cdot\,; y, b(\cdot))$. Then for any $y \in \mathbb{R}^n$, $\overline{u}(\cdot\,; y, b(\cdot))$ and $\overline{u}(\cdot\,; 0, b(\cdot))$ uniquely exist, and the difference $\widetilde{u}(\cdot) \stackrel{\Delta}{=} \overline{u}(\cdot\,; y, b(\cdot)) - \overline{u}(\cdot\,; 0, b(\cdot))$ satisfies

$$N_s \widetilde{u}(\cdot) + \widetilde{H}_s(y) = 0. \tag{2.17}$$

Thus, by (ii), Problem (DLQ) with $b(\cdot)=0$ is solvable at s. The uniqueness follows easily. □

If

$$R \gg 0, \quad Q - SR^{-1}S^{\top} \geq 0, \quad G \geq 0, \tag{2.18}$$

then (2.2) can be rewritten as

$$\begin{aligned} J(s,y;u(\cdot)) &= \frac{1}{2}\int_s^T \{\langle (Q - S^{\top}R^{-1}S)x(t), x(t)\rangle \\ &\quad + \left|R^{\frac{1}{2}}\left[u(t) + R^{-1}Sx(t)\right]\right|^2\}dt + \frac{1}{2}\langle Gx(T), x(T)\rangle. \end{aligned} \tag{2.19}$$

In this case, $N_s \gg 0$ holds. We refer to Problem (DLQ) as a *standard LQ problem* if (2.18) holds. Hence, a standard LQ problem is uniquely solvable by Theorem 2.2-(iii).

2.3. A linear Hamiltonian system

We note that the optimal control $\overline{u}(\cdot)$ determined by (2.12) is not easy to compute, since N_s^{-1} is in an abstract form and very complicated. Thus, we would like to find some more explicit form of the optimal control. In this subsection we are going to relate the solvability of Problem (DLQ) to that of a linear Hamiltonian system by using the maximum principle presented in Chapter 3, Section 2. Recall that the maximum principle is applicable to our Problem (DLQ) here (see the remark in Chapter 3, Section 2).

Theorem 2.3. *Let Problem (DLQ) be solvable at some $(s,y)\in[0,T)\times\mathbb{R}^n$ with an optimal pair $(\overline{x}(\cdot),\overline{u}(\cdot))$. Then there exists a solution $\overline{p}(\cdot)$ to the equation*

$$\begin{cases} \dot{\overline{p}}(t) = -A^{\top}\overline{p}(t) + Q\overline{x}(t) + S^{\top}\overline{u}(t), \\ \overline{p}(T) = -G\overline{x}(T), \end{cases} \tag{2.20}$$

on $[s,T]$ such that

$$R\overline{u}(t) - B^{\top}\overline{p}(t) + S\overline{x}(t) = 0, \quad \text{a.e. } t\in[s,T], \tag{2.21}$$

and

$$R(t) \geq 0, \quad \text{a.e. } t\in[s,T]. \tag{2.22}$$

Proof. By Chapter 3, Theorem 2.1, equation (2.20) is the adjoint equation associated with the optimal pair $(\overline{x}(\cdot),\overline{u}(\cdot))$. Thus, the maximum principle yields the following maximum condition for Problem (DLQ):

$$\begin{aligned} &\max_{u\in\mathbb{R}^k}\left\{\langle B^{\top}\overline{p}(t) - S\overline{x}(t), u\rangle - \frac{1}{2}\langle Ru, u\rangle\right\} \\ &= \langle B^{\top}\overline{p}(t) - S\overline{x}(t), \overline{u}(t)\rangle - \frac{1}{2}\langle R\overline{u}(t), \overline{u}(t)\rangle, \quad \text{a.e. } t\in[s,T]. \end{aligned} \tag{2.23}$$

Consequently, conditions in (2.21)–(2.22) are nothing but the first-order and second-order necessary conditions of maximizing the quadratic function (in u) in (2.23). □

The following shows that the nonnegativity condition (2.22) is necessary not only for the solvability of Problem (DLQ), but also for the finiteness of the problem.

Proposition 2.4. *Suppose Problem (DLQ) is finite at some $(s,y) \in [0,T) \times \mathbb{R}^n$. Then (2.22) holds.*

Proof. It suffices to show that $R(t) \geq 0$ at any Lebesgue point $t \in [s,T)$ of $R(\cdot)$. Suppose this is not true. Then there are a Lebesgue point $t \in [s,T)$ and a $u_0 \in \mathbb{R}^k$ such that $\langle R(t)u_0, u_0 \rangle < 0$. Now, for any $\lambda, \varepsilon > 0$, let

$$u_\varepsilon^\lambda(r) = \lambda u_0 \chi_{[t,t+\varepsilon]}(r), \qquad r \in [s,T).$$

Under this control, we have

$$\begin{aligned} x_\varepsilon^\lambda(r) &= \Phi(r,s)y + \int_s^r \Phi(r,r')[Bu_\varepsilon^\lambda(r') + b(r')]dr' \\ &= \Phi(r,s)y + \lambda \int_t^{r\wedge(t+\varepsilon)} \Phi(r,r')Bu_0 dr' + \int_s^r \Phi(r,r')b(r')dr', \\ &\qquad\qquad r \in [s,T). \end{aligned}$$

Thus,

$$|x_\varepsilon^\lambda(r)| \leq K(1+\lambda\varepsilon), \qquad \forall r \in [s,T],\ \lambda,\varepsilon > 0,$$

for some constant K independent of t, λ, ε. Consequently,

$$\begin{aligned} J(s,y;u_\varepsilon^\lambda(\cdot)) &= \frac{1}{2}\int_s^T \{ \langle Q(r)x_\varepsilon^\lambda(r), x_\varepsilon^\lambda(r) \rangle + 2\langle S(r)x_\varepsilon^\lambda(r), u_\varepsilon^\lambda(r) \rangle \\ &\qquad + \langle R(r)u_\varepsilon^\lambda(r), u_\varepsilon^\lambda(r) \rangle \}dr + \frac{1}{2}\langle Gx_\varepsilon^\lambda(T), x_\varepsilon^\lambda(T) \rangle \\ &\leq \frac{1}{2}\lambda^2\varepsilon\{ \langle R(t)u_0, u_0 \rangle + K\varepsilon \} + K, \end{aligned}$$

for some constant $K > 0$ depending on $A, B, Q, S, R, G, y, u_0, T, s$ and independent of $\lambda, \varepsilon > 0$. Thus, by choosing $\varepsilon > 0$ small enough so that

$$\langle R(t)u_0, u_0 \rangle + K\varepsilon < 0,$$

and then sending $\lambda \to \infty$, we conclude that Problem (DLQ) is not finite at (s,y), which is a contradiction. □

The following example shows that condition (2.22) is *not* sufficient for the solvability of Problem (DLQ). It also shows that the finiteness is strictly weaker than the solvability.

Example 2.5. Consider the one-dimensional control system

$$\begin{cases} \dot{x}(t) = u(t), & t \in [s,T], \\ x(s) = y \in \mathbb{R}, \end{cases} \tag{2.24}$$

with $s \in [0,T)$ and the cost functional

$$J(s,y;u(\cdot)) = \frac{1}{2}\int_s^T x(t)^2 dt. \tag{2.25}$$

In this case, $R = 0$. By taking

$$u_\varepsilon(t) = -\frac{y}{\varepsilon}\chi_{[s,s+\varepsilon]}(t), \qquad \forall t \in [s,T],$$

with $0 < \varepsilon < T - s$, we have

$$0 \le J(s,y;u_\varepsilon(\cdot)) = \frac{1}{2}\int_s^{s+\varepsilon} y^2\Big(1 - \frac{t-s}{\varepsilon}\Big)^2 dt = \frac{y^2\varepsilon}{6} \to 0, \qquad \text{as } \varepsilon \to 0.$$

Thus, the corresponding value function satisfies

$$V(s,y) = 0, \qquad \forall (s,y) \in [0,T) \times \mathbb{R}. \tag{2.26}$$

Consequently, the corresponding LQ problem is finite. On the other hand, for any $y \neq 0$, if there is an optimal pair $(\overline{x}(\cdot), \overline{u}(\cdot))$ achieving the minimum value 0 of the cost functional, then $\overline{x}(t) \equiv 0$, which is contradictory to the fact that $\overline{x}(0) = y \neq 0$. So the problem is *not* solvable.

We refer to Problem (DLQ) as a *singular* LQ problem if $R(t) \geq 0$ is degenerate (i.e., some of the eigenvalues of $R(t)$ are zero) on some set of positive Lebesgue measure. Thus, the above example is a singular LQ problem.

The following example further shows that condition (2.22) is not sufficient for the finiteness of Problem (DLQ) if Q and G are allowed to be indefinite.

Example 2.6. Consider system (2.24) with cost functional

$$J(s,y;u(\cdot)) = -\frac{1}{2}x(T)^2. \tag{2.27}$$

Take $u_\ell(t) \equiv \ell$. Then

$$J(s,y;u_\ell(\cdot)) = -\frac{1}{2}\big[\ell(T-s) + y\big]^2 \to -\infty \quad (\ell \to \infty).$$

Thus, the corresponding LQ problem is not finite.

Below, we are going to consider the case

$$R \gg 0. \tag{2.28}$$

The following result is a refinement of Theorem 2.3. It gives a relation between Problem (DLQ) and a two-point boundary value problem for a linear Hamiltonian system.

Theorem 2.7. *We have the following results.*

(i) *Let $N_s \geq 0$ for some $s \in [0,T)$ and (2.28) hold. Then Problem (DLQ) is (uniquely) solvable at $(s,y) \in [0,T) \times \mathbb{R}^n$ if and only if the two-point boundary value problem*

$$
(2.29)\quad \begin{cases} \dot{\overline{x}}(t) = (A - BR^{-1}S)\overline{x}(t) + BR^{-1}B^\top \overline{p}(t) + b(t), & t \in [s,T], \\ \dot{\overline{p}}(t) = -(A - BR^{-1}S)^\top \overline{p}(t) + (Q - S^\top R^{-1} S)\overline{x}(t), & t \in [s,T], \\ \overline{x}(s) = y, \qquad \overline{p}(T) = -G\overline{x}(T), \end{cases}
$$

admits a (unique) solution $(\overline{x}(\cdot), \overline{p}(\cdot)) \in \mathcal{Y}[s,T] \times \mathcal{Y}[s,T]$. In this case,

$$
(2.30)\qquad \overline{u}(t) = R^{-1}[B^\top \overline{p}(t) - S\overline{x}(t)], \qquad t \in [s,T],
$$

gives an (the) optimal control, and the function $\overline{x}(\cdot)$ obtained by solving (2.29) is the corresponding optimal state trajectory.

(ii) *Let $N_s \gg 0$ for some $s \in [0,T)$ and suppose (2.28) holds. Then (2.29) admits a unique solution, and Problem (DLQ) is uniquely solvable at s with the optimal feedback control given by (2.30).*

Proof. (i) If Problem (DLQ) is solvable at (s,y) with an optimal control $\overline{u}(\cdot)$, then by Theorem 2.2-(ii),

$$
(2.31)\quad \begin{aligned} 0 &= N_s \overline{u}(\cdot) + H_s(y) \\ &= (R + L_s^* Q L_s + S L_s + L_s^* S^\top + \widehat{L}_s^* G \widehat{L}_s)\overline{u}(\cdot) \\ &\quad + (L_s^* Q + S)((\Gamma_s y)(\cdot) + f^s(\cdot)) + \widehat{L}^* G(\widehat{\Gamma}_s y + \widehat{f}^s) \\ &= R\overline{u}(\cdot) + S\overline{x}(\cdot) + L_s^*(Q\overline{x}(\cdot) + S^\top \overline{u}(\cdot)) + \widehat{L}_s^* G \overline{x}(T), \end{aligned}
$$

where $\overline{x}(\cdot) \stackrel{\Delta}{=} (\Gamma_s y)(\cdot) + L_s \overline{u}(\cdot) + f^s(\cdot)$ is the corresponding optimal trajectory. Set

$$
\overline{p}(t) = -\Phi(T,t)^\top G \overline{x}(T) - \int_t^T \Phi(r,t)^\top [Q\overline{x}(r) + S^\top \overline{u}(r)]dr, \quad t \in [s,T].
$$

Then (noting (2.4) and (2.31))

$$
B^\top \overline{p}(\cdot) = -\widehat{L}_s^* G \overline{x}(T) - L_s^*[Q\overline{x}(\cdot) + S^\top \overline{u}(\cdot)] = R\overline{u}(\cdot) + S\overline{x}(\cdot).
$$

This yields (2.30). Clearly, the pair $(\overline{x}(\cdot), \overline{p}(\cdot)) \in \mathcal{Y}[s,T] \times \mathcal{Y}[s,T]$ is a solution of the two-point boundary value problem (2.29).

Conversely, if (2.29) admits a solution $(\overline{x}(\cdot), \overline{p}(\cdot))$, then by defining $\overline{u}(\cdot)$ by (2.30), reversing the procedure of (2.31), we obtain (2.11), giving the solvability of Problem (DLQ). The uniqueness part follows easily.

(ii) The conclusion follows from (i) and Theorem 2.2-(iii). ☐

Note that (2.29) and (2.30) constitute the Hamiltonian system introduced in Chapter 3, (2.9) (where the maximum condition is exactly (2.30) under (2.28)). Therefore, the above theorem implies that if $N_s \geq 0$ for some $s \in [0,T)$ and $R \gg 0$, then the Hamiltonian system completely characterizes the optimality of the LQ problem (on $[s,T]$). In particular, a standard LQ problem (see (2.18)) can be uniquely solved by the control (2.30).

It is seen that the function $\bar{p}(\cdot)$ is an auxiliary function via which the optimal control and the corresponding optimal state are related indirectly. Our next goal is to get rid of this auxiliary function and link the optimal control and the state directly. This will be carried out via the so-called Riccati equation.

2.4. The Riccati equation and feedback optimal control

By (2.30), if

$$\bar{p}(t) = -P(t)\bar{x}(t) - \varphi(t), \qquad t \in [s,T], \tag{2.32}$$

for some $P(\cdot)$ and $\varphi(\cdot)$, then the optimal control $\bar{u}(\cdot)$ will take a *state feedback* form:

$$\bar{u}(t) = -R^{-1}[(B^\top P(t) + S)\bar{x}(t) + B^\top \varphi(t)], \qquad t \in [s,T]. \tag{2.33}$$

Such a representation is very useful in applications.

To formally derive equations that $P(\cdot)$ and $\varphi(\cdot)$ should satisfy, let $(\bar{x}(\cdot), \bar{p}(\cdot))$ be a solution of (2.29) such that (2.32) holds for some $P(\cdot) \in C^1([s,T]; \mathbb{R}^{n\times n})$ and $\varphi(\cdot) \in C^1([s,T]; \mathbb{R}^n)$. Differentiating (2.32), using (2.29), (2.30), and (2.32), we obtain (t will be suppressed)

$$\begin{aligned}
(Q - S^\top R^{-1}S)\bar{x} &- (A^\top - S^\top R^{-1}B^\top)[-P\bar{x} - \varphi] = \dot{\bar{p}} \\
&= -\dot{P}\bar{x} - P\big[A\bar{x} + B\bar{u} + b\big] + \dot{\varphi} \\
&= -\dot{P}\bar{x} - P\big\{A\bar{x} - BR^{-1}\big[(B^\top P + S)\bar{x} + B^\top\varphi\big] + b\big\} + \dot{\varphi}.
\end{aligned}$$

This yields

$$\begin{aligned}
0 &= \big[\dot{P} + P(A - BR^{-1}S) + (A - BR^{-1}S)^\top P \\
&\qquad - PBR^{-1}B^\top P + Q - S^\top R^{-1}S\big]\bar{x} \\
&\qquad + \dot{\varphi} + \big[A^\top - S^\top R^{-1}B^\top - PBR^{-1}B^\top\big]\varphi + Pb \\
&= \big\{\dot{P} + PA + A^\top P - (PB + S^\top)R^{-1}(B^\top P + S) + Q\big\}\bar{x} \\
&\qquad + \dot{\varphi} + [(A - BR^{-1}S)^\top - PBR^{-1}B^\top]\varphi + Pb.
\end{aligned}$$

Thus, if we choose $P(\cdot)$ and $\varphi(\cdot)$ to satisfy the equations

$$\begin{cases}
\dot{P}(t) + P(t)A + A^\top P(t) + Q \\
\qquad - [B^\top P(t) + S]^\top R^{-1}[B^\top P(t) + S] = 0, \quad \text{a.e. } t \in [s,T], \\
P(T) = G,
\end{cases} \tag{2.34}$$

or equivalently,

$$\begin{cases}
\dot{P}(t) + P(t)(A - BR^{-1}S) + (A - BR^{-1}S)^\top P(t) \\
\qquad - P(t)BR^{-1}B^\top P(t) + Q - S^\top R^{-1}S = 0, \quad \text{a.e. } t \in [s,T], \\
P(T) = G,
\end{cases} \tag{2.34$'$}$$

and

$$
(2.35)\qquad \begin{cases} \dot\varphi(t) + \big[(A - BR^{-1}S)^\top - P(t)BR^{-1}B^\top\big]\varphi(t) + P(t)b(t) = 0, \\ \hfill \text{a.e. } t \in [s,T], \\ \varphi(T) = 0, \end{cases}
$$

then (2.32) will hold, and the feedback control (2.33) can be taken. We call (2.34) the *Riccati equation* associated with Problem (DLQ). Let us make a few observations about this equation. First, since it is a nonlinear ordinary differential equation with the velocity field (the left-hand side of (2.34) except the term $\dot P(t)$) being locally Lipschitz in the unknown $P(\cdot)$, it is *locally solvable*, namely, one can always find an $s < T$ such that (2.34) admits a solution on $[s,T]$. However, due to the presence of the quadratic term in the Riccati equation, the global existence of a solution is not guaranteed, unless there are some additional conditions imposed. Second, if R is assumed to be uniformly positive definite (i.e., $R \gg 0$), then the solution to (2.34), if any, must be unique. This can be easily shown by the boundedness of the coefficients along with the Gronwall inequality. Third, if the uniqueness of solutions holds, then any solution to (2.34) must be symmetric-matrix-valued. This is due to the fact that $P^\top$ solves (2.34) if P does. On the other hand, once (2.34) is solved, (2.35) can always be uniquely solved due to the boundedness of its coefficients.

From the above analysis, we see that provided one can solve (2.34), by setting $\overline{u}(\cdot)$ via (2.33) with $\varphi(\cdot)$ being the solution of (2.35), one obtains an optimal control. However, the above derivation is rather formal. The following gives a precise statement and a rigorous proof.

Theorem 2.8. *Assume (2.28). Let $P(\cdot) \in C([s,T];\mathcal{S}^n)$ be the solution of the Riccati equation (2.34) and $\varphi(\cdot)$ the solution of (2.35) on some interval $[s,T]$. Then Problem (DLQ) is uniquely solvable at s, and the optimal control has the state feedback form (2.33). Moreover,*

$$
(2.36)\qquad \begin{aligned} V(s,y) &= \frac{1}{2}\langle P(s)y, y\rangle + \langle \varphi(s), y\rangle \\ &\quad + \frac{1}{2}\int_s^T \big[2\langle \varphi(t), b(t)\rangle - |R^{-\frac{1}{2}}B^\top\varphi(t)|^2\big]dt, \quad \forall y \in \mathbb{R}^n. \end{aligned}
$$

Proof. Let $\overline{x}(\cdot)$ be the solution of the following system:

$$
(2.37)\qquad \begin{cases} \dot{\overline{x}}(t) = \big\{A - BR^{-1}\big[B^\top P(t) + S\big]\big\}\overline{x}(t) \\ \qquad\qquad - BR^{-1}B^\top\varphi(t) + b(t), \quad t \in [s,T], \\ \overline{x}(s) = y. \end{cases}
$$

Define $\overline{u}(\cdot)$ by (2.33). Then $\overline{x}(\cdot)$ is the state trajectory corresponding to $(s,y,\overline{u}(\cdot))$. Now, for any $u(\cdot) \in \mathcal{V}[s,T]$, let $x(\cdot) = x(\cdot\,; s,y,u(\cdot))$. By applying the chain rule to $\langle P(t)x(t), x(t)\rangle$ and $\langle \varphi(t), x(t)\rangle$, respectively, we obtain

$$\begin{aligned}
&\langle P(T)x(T), x(T)\rangle - \langle P(s)y, y\rangle \\
&= \int_s^T \{ \langle [(P(t)B + S^\top)R^{-1}(B^\top P(t) + S) - Q]x(t), x(t)\rangle \\
&\qquad + 2\langle B^\top P(t)x(t), u(t)\rangle + 2\langle P(t)b(t), x(t)\rangle \}dt,
\end{aligned}$$

and

$$\begin{aligned}
-\langle \varphi(s), y\rangle &= \langle \varphi(T), x(T)\rangle - \langle \varphi(s), y\rangle \\
&= \int_s^T \{ \langle [P(t)B + S^\top)R^{-1}B^\top \varphi(t) - P(t)b(t), x(t)\rangle \\
&\qquad + \langle \varphi(t), Bu(t) + b(t)\rangle \}dt.
\end{aligned}$$

Hence, it follows that (we suppress t in the integrand)

$$\begin{aligned}
&J(s, y; u(\cdot)) - \frac{1}{2}\langle P(s)y, y\rangle - \langle \varphi(s), y\rangle \\
&= \frac{1}{2}\int_s^T \Big\{ \langle Ru, u\rangle + \langle (PB + S^\top)R^{-1}(B^\top P + S)x, x\rangle \\
&\qquad + 2\langle (B^\top P + S)x, u\rangle + 2\langle R^{-1}(B^\top P + S)x, B^\top \varphi\rangle \\
&\qquad + 2\langle B^\top \varphi, u\rangle + 2\langle \varphi, b\rangle \Big\}dt \\
&= \frac{1}{2}\int_s^T \Big\{ \big|R^{-\frac{1}{2}}[Ru + (B^\top P + S)x + B^\top \varphi]\big|^2 \\
&\qquad - |R^{-\frac{1}{2}}B^\top \varphi|^2 + 2\langle \varphi, b\rangle \Big\}dt.
\end{aligned} \tag{2.38}$$

This yields

$$\begin{aligned}
J(s, y, \bar{u}(\cdot)) &= \frac{1}{2}\langle P(s)y, y\rangle + \langle \varphi(s), y\rangle \\
&\quad + \frac{1}{2}\int_s^T [2\langle \varphi(t), b(t)\rangle - |R^{-\frac{1}{2}}B^\top \varphi(t)|^2]dt \\
&\leq J(s, y; u(\cdot)),
\end{aligned} \tag{2.39}$$

which implies both the optimality of $\bar{u}(\cdot)$ and the validity of (2.36). □

The above proof (in particular, (2.38)) is based on the *completion of squares* technique. In fact, we can carry out the proof without knowing a priori the forms of the Riccati equation (2.34) as well as (2.35). The idea is first to set $dP(t) = \Gamma(t)dt$ and $d\varphi(t) = \alpha(t)dt$, then rewrite $J(s, y; u(\cdot))$ similarly to the above proof, and finally get the expression of $\Gamma(t)$ and $\alpha(t)$ in order to complete a perfect square in the integrand involved. This provides another way of deriving the Riccati equation (2.34) and the associated (2.35) without applying the maximum principle. Details are left to the interested reader.

The following result gives an equivalence between the solvability of the Riccati equation (2.34) and that of Problem (DLQ).

Theorem 2.9. *Assume (2.28). Then Problem (DLQ) is uniquely solvable at each $r \in [s,T]$ if and only if the Riccati equation (2.34) is uniquely solvable on $[s,T]$.*

Proof. Sufficiency. If (2.34) is uniquely solvable on $[s,T]$, then by Theorem 2.8, Problem (DLQ) is solvable at any $r \in [s,T]$. From (2.38), the optimal control has to be a state feedback form (2.33). Then the desired uniqueness follows from that of the solutions to (2.37).

Necessity. If Problem (DLQ) is uniquely solvable at each $r \in [s,T]$, then by Theorem 2.2-(iv), for any $y \in \mathbb{R}^n$, Problem (DLQ) with $b(\cdot)=0$ is uniquely solvable at (s,y) as well. By Theorem 2.7, system (2.29) (with $b(\cdot)=0$) admits a unique solution $(\overline{x}(\cdot\,;s,y),\overline{p}(\cdot\,;s,y))$, which is linear in y. Hence, we my define $\mathbb{R}^{n\times n}$-valued functions $X(\cdot)$ and $Y(\cdot)$ such that

$$
(2.40)\qquad \begin{cases} \overline{x}(t) \equiv \overline{x}(t;s,y) = X(t)y, \\ \overline{p}(t) \equiv \overline{p}(t;s,y) = Y(t)y, \end{cases} \qquad t \in [s,T].
$$

Clearly, $X(\cdot)$ and $Y(\cdot)$ are C^1, and they satisfy

$$
(2.41)\qquad \begin{cases} \dot{X}(t) = \widehat{A}X(t) - BR^{-1}B^\top Y(t), \quad t \in [s,T], \\ \dot{Y}(t) = \widehat{Q}X(t) - \widehat{A}^\top Y(t), \quad t \in [s,T], \\ X(s) = I, \qquad Y(T) = -GX(T), \end{cases}
$$

where

$$
\begin{cases} \widehat{A} = A - BR^{-1}S, \\ \widehat{Q} = Q - S^\top R^{-1}S. \end{cases}
$$

We claim that

$$
(2.42)\qquad \det X(t) \neq 0, \qquad t \in [s,T].
$$

Suppose this is not the case. Then we can find an $r \in [s,T]$ and a $y \in \mathbb{R}^n$, $y \neq 0$, such that

$$
\overline{x}(r) \equiv X(r)y = 0.
$$

Note that $\overline{x}(\cdot)$ is the optimal trajectory (corresponding to (s,y)), and the corresponding optimal control is $\overline{u}(\cdot) = R^{-1}[B^\top \overline{p}(\cdot) - S\overline{x}(\cdot)]$ (see (2.30)). By (2.8) and the fact that $\overline{x}(r)=0$ (noting $b(\cdot)=0$),

$$
J(r,\overline{x}(r);u(\cdot)) = \frac{1}{2}\langle N_r u(\cdot), u(\cdot)\rangle, \quad \forall u(\cdot) \in \mathcal{V}[r,T].
$$

Since Problem (DLQ) is uniquely solvable, $N_r \geq 0$ by Theorem 2.2-(ii). Hence by the unique solvability again, the only optimal control for the LQ problem starting with time r and state $\bar{x}(r) = 0$ is the one that is 0 almost everywhere, i.e.,

$$\bar{u}(t) = 0, \qquad \text{a.e. } t \in [r, T],$$

which yields $\bar{x}(t) = 0$ for all $t \in [r, T]$ by the state equation (2.1). Then one further has $\bar{p}(T) = -G\bar{x}(T) = 0$. As a result, $(\bar{x}(\cdot), \bar{p}(\cdot))$ satisfies the equations in (2.29), which vanishes at $t = T$, and we then must have $(\bar{x}(t), \bar{p}(t))$ identically zero over $[s, T]$, contradicting $\bar{x}(s) = y \neq 0$. This proves the claim (2.42).

Since $X(\cdot)$ is C^1, so is $X(\cdot)^{-1}$ by (2.42). Define

$$P(t) \stackrel{\Delta}{=} -Y(t)X(t)^{-1}, \qquad t \in [s, T]. \tag{2.43}$$

We are going to prove that $P(\cdot)$ is the solution of the Riccati equation (2.34). First, $P(t)$ is symmetric. In fact, a direct computation using (2.41) gives us (t is suppressed in $X(t)$ and $Y(t)$ for simplicity)

$$\begin{cases} \dfrac{d}{dt}[X^\top Y - Y^\top X] = 0, & t \in [s, T], \\ [X^\top Y - Y^\top X]\big|_{t=T} = 0. \end{cases}$$

Hence, $X^\top Y - Y^\top X \equiv 0$. The symmetry of $P(t)$ follows.

Next, note that

$$(X^{-1})' = -X^{-1}\hat{A} + X^{-1}BR^{-1}B^\top YX^{-1}.$$

Therefore,

$$\begin{aligned} \dot{P} &= -[\hat{Q}X - \hat{A}^\top Y]X^{-1} + Y[X^{-1}\hat{A} - X^{-1}BR^{-1}B^\top YX^{-1}] \\ &= -\hat{Q} - \hat{A}^\top P - P\hat{A} + PBR^{-1}B^\top P, \end{aligned}$$

together with $P(T) = G$, proving that $P(\cdot)$ is the solution of (2.34)$'$, which is equivalent to (2.34). □

Corollary 2.10. *For a standard LQ problem (i.e., (2.18) holds), the Riccati equation (2.34) admits a unique solution $P(\cdot)$ over $[0, T]$, and Problem (DLQ) is uniquely solvable with the optimal control $\bar{u}(\cdot)$ given by (2.30). Moreover,*

$$P(t) \geq 0, \qquad \forall t \in [0, T]. \tag{2.44}$$

Proof. Under (2.18), all the conclusions of this corollary are clear from Theorems 2.2-(iii) and 2.9, except the nonnegativity of $P(\cdot)$, which we now prove. By Theorem 2.2-(iv), we know that in the current case, Problem (DLQ) with $b(\cdot) = 0$ is also uniquely solvable, for which $\varphi(t)$, the solution to (2.35), is identically 0. Thus, by considering this problem, using (2.19) and (2.36), we have

$$\langle P(s)y, y \rangle = 2V(s, y) = 2J(s, y; \bar{u}(\cdot)) \geq 0, \quad \forall (s, y) \in [0, T] \times \mathbb{R}^n,$$

which implies the nonnegativity of $P(\cdot)$. □

The above result shows that a standard LQ problem can be solved completely via the Riccati equation. However, the unique solvability of the LQ problem does not necessarily imply that the underlying problem is standard. Here is an example.

Example 2.11. Consider a control system (with both the state and control being one-dimensional)

$$\begin{cases} \dot{x}(t) = u(t), & t \in [s,T], \\ x(s) = y, \end{cases} \tag{2.45}$$

with the cost functional

$$J(s,y;u(\cdot)) = \frac{1}{2}\int_s^T u(t)^2 dt - \frac{1}{2}x(T)^2. \tag{2.46}$$

This is not a standard LQ problem, as the weight on the square of the terminal state is negative. Let $T > 1$ and $s \in (T-1,T]$. For any $u(\cdot) \in L^2(s,T;\mathbb{R})$, let $x(\cdot)$ be the corresponding state trajectory. Applying the Newton-Leibniz formula to $\frac{x(t)^2}{t+1-T}$ over $[s,T]$, we have

$$x(T)^2 = \frac{y^2}{s+1-T} + \int_s^T \Big\{ \frac{2x(t)u(t)}{t+1-T} - \frac{x(t)^2}{(t+1-T)^2} \Big\} dt.$$

Thus,

$$J(s,y;u(\cdot)) = \frac{-y^2}{2(s+1-T)} + \frac{1}{2}\int_s^T \Big\{ u(t) - \frac{x(t)}{t+1-T} \Big\}^2 dt.$$

Consequently, the optimal control is given by

$$u(t) = \frac{x(t)}{t+1-T}, \qquad t \in [s,T], \tag{2.47}$$

and the corresponding value function is given by

$$V(s,y) = \frac{-y^2}{2(s+1-T)}, \qquad \forall (s,y) \in (T-1,T] \times \mathbb{R}. \tag{2.48}$$

So the LQ problem is uniquely solvable at least on the interval $(T-1,T]$.

We have seen that both the maximum principle and the completion of squares can lead to the Riccati equation. We point out here that the dynamic programming equation can also result in the same Riccati equation (2.34) and the associated (2.35), and the corresponding verification theorem will give rise to the same optimal feedback control (2.33). We are not going to discuss this in detail here, as we will do it for the stochastic case in Section 6. Interested readers may work out the deterministic case on their own.

To conclude this section, let us present some results on the Riccati equation (2.34) in the standard case with $S = 0$. These results are not only important for the deterministic LQ problems, but also useful for the stochastic LQ problems in the sequel.

Define

$$\begin{cases} \mathcal{S}_+^k \stackrel{\Delta}{=} \{\Theta \in \mathcal{S}^k \mid \Theta \geq 0\}, \\ \widehat{\mathcal{S}}_+^k \stackrel{\Delta}{=} \{\Theta \in \mathcal{S}_+^k \mid \Theta^{-1} \text{ exists }\}, \\ \widehat{\mathcal{R}} \stackrel{\Delta}{=} \{\widehat{R} \in L^\infty(0,T;\widehat{\mathcal{S}}_+^k) \mid \widehat{R}^{-1} \in L^\infty(0,T;\widehat{\mathcal{S}}_+^k)\}. \end{cases}$$

Clearly, $C([0,T];\widehat{\mathcal{S}}_+^k) \subseteq \widehat{\mathcal{R}}$. Fix $Q \in C([0,T];\mathcal{S}_+^n)$ and $G \in \mathcal{S}_+^n$. For each $R \in \widehat{\mathcal{R}}$, by Corollary 2.10, the Riccati equation (2.34) with $S = 0$ admits a unique solution $P \in C(0,T;\mathcal{S}_+^n)$. Thus we can define a mapping $\Gamma : \widehat{\mathcal{R}} \to C([0,T];\mathcal{S}_+^n)$ as $\Gamma(R) \stackrel{\Delta}{=} P$.

Proposition 2.12. *The operator Γ is continuous and monotonically increasing. Moreover, there is a $\widehat{K} \in L^\infty(0,T;\mathcal{S}_+^n)$ such that*

$$\sup_{R \in \mathcal{R}} \langle \Gamma(R)y, y \rangle \leq \langle \widehat{K}y, y \rangle, \quad \forall y \in \mathbb{R}^n. \tag{2.49}$$

Proof. Let $R_i \in \widehat{\mathcal{R}}$ and $P_i = \Gamma(R_i)$ $(i = 1,2)$. Define $\widehat{P} = P_1 - P_2$. Then $\widehat{P}$ satisfies

$$\begin{cases} \dot{\widehat{P}} = -\widehat{P}\widehat{A} - \widehat{A}^\top \widehat{P} - \widehat{Q} + \widehat{P}B\widehat{R}_2^{-1}B^\top \widehat{P}, \\ \widehat{P}(T) = 0, \end{cases} \tag{2.50}$$

with $\widehat{A} = A - BR_2^{-1}B^\top P_2$ and $\widehat{Q} = P_1B(R_2^{-1} - R_1^{-1})B^\top P_1$. Thus, if $R_2 \to R_1$, then $\widehat{Q} \to 0$. Hence by Gronwall's inequality, $P_1 - P_2 = \widehat{P} \to 0$, proving the continuity. On the other hand, if $R_1 \geq R_2 (> 0)$, then $R_2^{-1} \geq R_1^{-1} > 0$, which results in $\widehat{Q} \geq 0$. Consequently, by Corollary 2.10, the solution to the auxiliary Riccati equation (2.50) is nonnegative, namely, $\widehat{P} \geq 0$. This proves the monotonicity. Finally, for any $R \in \widehat{\mathcal{R}}$ and $(s,y) \in [0,T] \times \mathbb{R}^n$, consider Problem (DLQ) with $b(\cdot) \equiv 0$ (in which case equation (2.35) has only the solution $\varphi(\cdot) \equiv 0$). By Theorem 2.8,

$$\frac{1}{2} \langle P(s)y, y \rangle = V(s,y) \leq J(s,y;0) = M_s(y),$$

where $M_s(y)$ is given in (2.9) and the last equality above is due to (2.8). Noting the definition of $M_s(y)$ and the fact that $b(\cdot) \equiv 0$, we have

$$M_s(y) = \langle \widehat{K}(s)y, y \rangle$$

for some $\widehat{K} \in L^\infty(0,T;\mathcal{S}_+^n)$. This completes the proof. □

3. Formulation of Stochastic LQ Problems

Now we start to discuss stochastic LQ problems. We emphasize that the stochastic LQ problem is *not* a routine generalization of its deterministic counterpart. Some significant differences will be revealed below.

3.1. Statement of the problems

Since we are going to involve the dynamic programming approach in treating the LQ problems, we shall adopt the weak formulation hereafter. Let $T > 0$ be given. For any $(s, y) \in [0, T) \times \mathbb{R}^n$, consider the following (nonhomogeneous) linear equation:

$$
\begin{cases} dx(t) = [A(t)x(t) + B(t)u(t) + b(t)]dt \\ \qquad\qquad + \sum_{j=1}^{m} [C_j(t)x(t) + D_j(t)u(t) + \sigma_j(t)]dW^j(t), \quad t \in [s, T], \\ x(s) = y, \end{cases} \tag{3.1}
$$

where $A, B, C_j, D_j, b, \sigma_j$ are deterministic matrix-valued functions of suitable sizes. In addition, we are given a quadratic cost functional,

$$
\begin{aligned} J(s, y; u(\cdot)) = E\Big\{\frac{1}{2}\int_s^T & [\langle Q(t)x(t), x(t) \rangle + 2\langle Sx(t), u(t) \rangle \\ & + \langle R(t)u(t), u(t) \rangle]dt + \frac{1}{2}\langle Gx(T), x(T) \rangle \Big\}, \end{aligned} \tag{3.2}
$$

where Q, S, and R are $\mathcal{S}^n$-, $\mathbb{R}^{k\times n}$-, and $\mathcal{S}^k$-valued functions, respectively, and $G \in \mathcal{S}^n$.

For any $s \in [0, T)$, we denote by $\mathcal{U}^w[s, T]$ the set of all 5-tuples $(\Omega, \mathcal{F}, \mathbf{P}, W(\cdot), u(\cdot))$ satisfying the following:

(i) $(\Omega, \mathcal{F}, \mathbf{P})$ is a complete probability space.

(ii) $\{W(t)\}_{t\geq s}$ is an m-dimensional standard Brownian motion defined on $(\Omega, \mathcal{F}, \mathbf{P})$ over $[s, T]$ (with $W(s) = 0$ almost surely), and $\mathcal{F}_t^s = \sigma\{W(r) : s \leq r \leq t\}$ augmented by all the $\mathbf{P}$-null sets in $\mathcal{F}$.

(iii) $u(\cdot) \in L^2_{\mathcal{F}}(s, T; \mathbb{R}^k)$.

(iv) Under $u(\cdot)$, for any $y \in \mathbb{R}^n$ equation (3.1) admits a unique solution (in the sense of Chapter 1, Definition 6.15) $x(\cdot)$ on $(\Omega, \mathcal{F}, \{\mathcal{F}_t^s\}_{t\geq 0}, \mathbf{P})$.

(v) The right-hand side of (3.2) is well-defined under $u(\cdot)$.

As before, for simplicity, we usually write $u(\cdot) \in \mathcal{U}^w[s, T]$ instead of $(\Omega, \mathcal{F}, \mathbf{P}, W(\cdot), u(\cdot)) \in \mathcal{U}^w[s, T]$ if the probability space $(\Omega, \mathcal{F}, \mathbf{P})$ and the Brownian motion $W(t)$ are clear from the context.

Under certain conditions on the coefficients (see Assumption (L1) below), (3.1) has a unique solution $x(\cdot) \in L^2_{\mathcal{F}}(s, T; \mathbb{R}^n) \stackrel{\Delta}{=} \mathcal{X}[s, T]$ and (3.2) is well-defined. Let us now state our stochastic linear quadratic optimal control problems (stochastic LQ problems, for short) as follows:

Problem (SLQ). For each $(s,y)\in[0,T]\times\mathbb{R}^n$, find a 5-tuple $(\overline{\Omega},\overline{\mathcal{F}},\overline{\mathbf{P}},\overline{W}(\cdot),\overline{u}(\cdot))\in\mathcal{U}^w[s,T]$ such that

$$J(s,y;\overline{u}(\cdot))=\inf_{u(\cdot)\in\mathcal{U}^w[s,T]}J(s,y;u(\cdot))\overset{\Delta}{=}V(s,y). \tag{3.3}$$

We call V the *value function* of Problem (SLQ). Note that

$$V(T,y)=\frac{1}{2}\langle Gy,y\rangle,\quad\forall y\in\mathbb{R}^n. \tag{3.4}$$

Let us introduce the following assumption.

(L1) The functions appearing in Problem (SLQ) satisfy

$$\begin{cases}A,C_j\in L^\infty(0,T;\mathbb{R}^{n\times n}), & B,D_j,S^\top\in L^\infty(0,T;\mathbb{R}^{n\times k}),\\ Q\in L^\infty(0,T;\mathcal{S}^n), \quad R\in L^\infty(0,T;\mathcal{S}^k), & G\in\mathcal{S}^n,\\ b,\sigma_j\in L^2(0,T;\mathbb{R}^n), & j=1,2,\ldots,m.\end{cases} \tag{3.5}$$

We remark that all the given data of the problem are assumed to be deterministic. We are not going to discuss the case of random coefficients here, which is quite different.

We introduce the following definitions (compare with Definition 2.1).

Definition 3.1. Problem (SLQ) is said to be

(i) *finite* at $(s,y)\in[0,T]\times\mathbb{R}^n$ if

$$V(s,y)>-\infty; \tag{3.6}$$

(ii) *solvable* at $(s,y)\in[0,T]\times\mathbb{R}^n$ if there exists a control $(\overline{\Omega},\overline{\mathcal{F}},\overline{\mathbf{P}},\overline{W}(\cdot),\overline{u}(\cdot))\in\mathcal{U}^w[s,T]$ such that

$$J(s,y;\overline{u}(\cdot))=V(s,y). \tag{3.7}$$

In this case, $\overline{u}(\cdot)$ is called an *optimal control*; the corresponding $\overline{x}(\cdot)$ and $(\overline{x}(\cdot),\overline{u}(\cdot))$ are called an *optimal state process* and an *optimal pair*, respectively;

(iii) *pathwise uniquely solvable* at $(s,y)\in[0,T]\times\mathbb{R}^n$ if it is solvable at $(s,y)\in[0,T]\times\mathbb{R}^n$ and, whenever $(\overline{\Omega},\overline{\mathcal{F}},\overline{\mathbf{P}},\overline{W}(\cdot),\overline{u}(\cdot))$ and $(\overline{\Omega},\overline{\mathcal{F}},\overline{\mathbf{P}},\overline{W}(\cdot),\tilde{u}(\cdot))$ are two optimal controls, it must hold that

$$\mathbf{P}\big(\overline{u}(t)=\tilde{u}(t),\quad \text{a.e. } t\in[s,T]\big)=1;$$

(iv) *finite* (resp. *(pathwise uniquely) solvable*) at $s\in[0,T]$ if it is finite (resp. (path uniquely) solvable) at (s,y) with all $y\in\mathbb{R}^n$; Problem (SLQ) is said to be *finite* (resp. *(pathwise uniquely) solvable*) if it is finite (resp. (pathwise uniquely) solvable) at all $s\in[0,T]$.

3.2. Examples

We know from Proposition 2.4 that in order for a deterministic Problem (DLQ) to be *finite* at some $(s,y)\in[0,T)\times\mathbb{R}^n$, it is necessary that $R(t)\geq 0$

for a.e. $t \in [s,T]$. Moreover, if $R(t) \geq 0$ and $R(t)$ is singular on a set of positive measure, then Problem (DLQ) may have no optimal control in general (see Example 2.5). For stochastic LQ problems, we will see that the condition $R(t) \geq 0$ is not necessary for the finiteness nor for the solvability (see an example below). This (among others) makes the stochastic LQ problems essentially different from the deterministic ones. In this section we present several examples to demonstrate these differences.

Example 3.2. Consider a control system (where both the state and control are one-dimensional)

$$
\begin{cases} \dot{x}(t) = u(t), & t \in [s,T], \\ x(s) = y, \end{cases} \tag{3.8}
$$

with the cost functional (which is the same as (2.25))

$$
J_1(s,y;u(\cdot)) = \frac{1}{2}\int_s^T x(t)^2 dt. \tag{3.9}
$$

In Example 2.5, we have shown that this problem is finite at any $s \in [0,T]$, but not solvable at any (s,y) with $y \neq 0$.

Now we consider a stochastic version of the problem:

$$
\begin{cases} dx(t) = u(t)dt + \delta u(t)dW(t), & t \in [s,T], \\ x(s) = y, \end{cases} \tag{3.10}
$$

for some $\delta \neq 0$, with the cost functionals

$$
J(s,y;u(\cdot)) = E\Big\{\frac{1}{2}\int_s^T x(t)^2 dt\Big\}. \tag{3.11}
$$

Define

$$
p(t) = \delta^2(1 - e^{\frac{t-T}{\delta^2}}), \qquad t \in [0,T]. \tag{3.12}
$$

For any $u(\cdot) \in \mathcal{U}^w[s,T]$, by Itô's formula we have

$$
\begin{aligned}
0 &= E\{p(T)x(T)^2\} \\
&= p(s)y^2 \\
&\quad + E\Big\{\int_s^T \Big[\Big(\frac{1}{\delta^2}p(t) - 1\Big)x(t)^2 + 2p(t)x(t)u(t) + \delta^2 p(t)u(t)^2\Big]dt\Big\}.
\end{aligned}
$$

Thus,

$$
\begin{aligned}
J(s,y;u(\cdot)) &= E\Big\{\frac{1}{2}\int_s^T x(t)^2 dt\Big\} \\
&= \frac{1}{2}p(s)y^2 + E\Big\{\frac{1}{2}\int_s^T \delta^2 p(t)\big[u(t) + \frac{x(t)}{\delta^2}\big]^2 dt\Big\}.
\end{aligned}
$$

This implies that Problem (SLQ) admits an optimal feedback control

$$u(t) = -\frac{x(t)}{\delta^2}, \qquad t \in [s, T], \tag{3.13}$$

and the value function is

$$V(s, y) = \frac{1}{2}\delta^2(1 - e^{\frac{s-T}{\delta^2}})y^2, \quad \forall (s, y) \in [0, T] \times \mathbb{R}. \tag{3.14}$$

This example shows that a stochastic LQ problem with $R = 0$ may still have optimal controls. As a matter of fact, the following example tells us that we can go even further.

Example 3.3. Consider the control system (3.8) with the cost functional

$$J_2(s, y; u(\cdot)) = -\frac{1}{2}\int_s^T u(t)^2 dt + \frac{1}{2}x(T)^2. \tag{3.15}$$

We see that (3.15) has a negative weight $R = -1$ on the control term. By Proposition 2.4, the corresponding LQ problem is not finite at any $(s, y) \in [0, T) \times \mathbb{R}$. As a matter of fact, by taking

$$u_\varepsilon(t) = \frac{1}{\varepsilon}\chi_{[T-\varepsilon, T]}(t), \qquad t \in [s, T],$$

with $0 < \varepsilon < T - s$, one has

$$J_2(s, y; u_\varepsilon(\cdot)) = -\frac{1}{2\varepsilon} + \frac{1}{2}(y + 1)^2. \tag{3.16}$$

Letting $\varepsilon \to 0$ in the above, we conclude that the corresponding value function

$$V_2(s, y) = -\infty, \qquad \forall (s, y) \in [0, T] \times \mathbb{R},$$

and, of course, no optimal control exists.

Consider now the stochastic control system (3.10) with the cost functional

$$J(s, y; u(\cdot)) = E\Big\{-\frac{1}{2}\int_s^T u(t)^2 dt + \frac{1}{2}x(T)^2\Big\}. \tag{3.17}$$

It is seen from (3.10) that the control affects the size of the noise in the system. We will prove (in Section 6) that for any $|\delta| > 1$ with

$$\delta^2(2\ln|\delta| - 1) > T - 1, \tag{3.18}$$

the problem (3.10) and (3.17) admits an optimal control for any $s \in [0, T]$, and in particular, the value function is finite.

This example shows that a stochastic LQ problem with a *negative* control weighting matrix in the cost may still be meaningful. The following example shows another interesting point.

Example 3.4. Consider control system (3.8) with the cost functional

$$J_3(s,y;u(\cdot)) = \frac{1}{2}\int_s^T u(t)^2 dt - \frac{1}{2}x(T)^2. \tag{3.19}$$

This is an LQ problem with the weight on the square of the terminal state being negative, and we have shown in Example 2.11 that if $T > 1$ and $s \in (T-1, T]$, an optimal control is given by

$$u(t) = \frac{x(t)}{t+1-T}, \qquad t \in [s,T], \tag{3.20}$$

and the corresponding value function is

$$V_3(s,y) = \frac{-y^2}{2(s+1-T)}, \qquad \forall (s,y) \in (T-1,T] \times \mathbb{R}. \tag{3.21}$$

Now we consider the stochastic control system (3.10) with $|\delta| > 1$ and with the cost functional

$$J(s,y;u(\cdot)) = E\Big\{\frac{1}{2}\int_s^T u(t)^2 dt - \frac{1}{2}x(T)^2\Big\}. \tag{3.22}$$

For any $\ell > 0$, we take

$$u_\ell(t) = \ell \chi_{[T-\frac{1}{\ell},T]}(t), \qquad t \in [s,T].$$

Let $x_\ell(\cdot)$ be the corresponding trajectory. Then

$$\begin{aligned} E|x_\ell(T)|^2 &= E\big|y + 1 + \delta\ell\big[W(T) - W(T-\frac{1}{\ell})\big]\big|^2 \\ &= (y+1)^2 + \delta^2\ell. \end{aligned} \tag{3.23}$$

Thus,

$$J(s,y;u_\ell(\cdot)) \le \frac{-(y+1)^2}{2} - \frac{\ell(\delta^2-1)}{2} \to -\infty, \text{ as } \ell \to +\infty,$$

which yields that $V(s,y)$ is not finite if $|\delta| > 1$ and $s < T$. This example shows that a well-posed deterministic LQ problem may become "ill-posed" if noise gets into the control system.

4. Finiteness and Solvability

Starting from this section, we consider only the case where $m = 1$ (i.e., the Brownian motion is one-dimensional) for simplicity of presentation. There is no essential difficulty with the multidimensional Brownian motions. All the indices j will then be dropped.

In this section we are going to study the existence of optimal controls for Problem (SLQ). The method is a combination of functional analysis and backward stochastic differential equations (BSDEs, for short).

Let us first introduce the following equation for matrix-valued processes:

$$\begin{cases} d\Phi(t) = A(t)\Phi(t)dt + C(t)\Phi(t)dW(t), & t \geq 0, \\ \Phi(0) = I. \end{cases} \tag{4.1}$$

From Chapter 1, Theorem 6.14, we know that $\Phi(t)^{-1}$ exists for all $t \geq 0$ and the following holds:

$$\begin{cases} d\Phi(t)^{-1} = -\Phi(t)^{-1}[A(t) - C(t)^2]dt - \Phi(t)^{-1}C(t)dW(t), & t \geq 0, \\ \Phi(0)^{-1} = I. \end{cases} \tag{4.2}$$

The solution $x(\cdot)$ of (3.1) can be written as follows:

$$\begin{aligned} x(t) &= \Phi(t)\Phi(s)^{-1}y \\ &\quad +\Phi(t)\int_s^t \Phi(r)^{-1}[(B(r)-C(r)D(r))u(r)+b(r)-C(r)\sigma(r)]dr \\ &\quad +\Phi(t)\int_s^t \Phi(r)^{-1}[D(r)u(r)+\sigma(r)]dW(r), \quad t \in [s,T]. \end{aligned} \tag{4.3}$$

By the Burkholder-Davis-Gundy inequality (see Chapter 1, Theorem 5.4), we have

$$\begin{aligned} &E\{\sup_{s\leq r\leq t}|x(r)|^2\} \\ &\quad \leq KE\{|y|^2+\int_s^t [|u(r)|^2+|b(r)|^2+|\sigma(r)|^2]dr\}, \quad \forall t \in [s,T]. \end{aligned} \tag{4.4}$$

Now we define the following operators: $\forall s \in [0,T]$, $y \in \mathbb{R}^n$, $u(\cdot) \in \mathcal{U}^w[s,T]$,

$$\begin{cases} (L_s u(\cdot))(\cdot) \stackrel{\Delta}{=} \Phi(\cdot)\Big\{\int_s^\cdot \Phi(r)^{-1}[B(r)-C(r)D(r)]u(r)dr \\ \qquad\qquad + \int_s^\cdot \Phi(r)^{-1}D(r)u(r)dW(r)\Big\}, \\ \widehat{L}_s u(\cdot) \stackrel{\Delta}{=} (L_s u(\cdot))(T), \qquad \forall u(\cdot) \in \mathcal{U}^w[s,T], \\ (\Gamma_s y)(\cdot) \stackrel{\Delta}{=} \Phi(\cdot)\Phi(s)^{-1}y, \quad \widehat{\Gamma}_s y \stackrel{\Delta}{=} (\Gamma_s y)(T), \quad \forall y \in \mathbb{R}^n, \\ f^s(\cdot) \stackrel{\Delta}{=} \Phi(\cdot)\Big\{\int_s^\cdot \Phi(r)^{-1}[b(r)-C(r)\sigma(r)]dr + \int_s^\cdot \Phi(r)^{-1}\sigma(r)dW(r)\Big\}, \\ \widehat{f}^s \stackrel{\Delta}{=} f^s(T). \end{cases} \tag{4.5}$$

Then, for any $(s,y) \in [0,T) \times \mathbb{R}^n$ and $u(\cdot) \in \mathcal{U}^w[s,T]$, the corresponding state process $x(\cdot)$ and its terminal value $x(T)$ are given by

$$\begin{cases} x(\cdot) = (\Gamma_s y)(\cdot) + (L_s u(\cdot))(\cdot) + f^s(\cdot), \\ x(T) = \widehat{\Gamma}_s y + \widehat{L}_s u(\cdot) + \widehat{f}^s. \end{cases} \tag{4.6}$$

The above (4.5)–(4.6) are similar to (2.4)–(2.5). Our next goal is to find a representation of the cost functional in terms of the control (like (2.8)). To this end, we first note that for any $s \in [0,T]$, the operators

$$\begin{cases} L_s : \mathcal{U}^w[s,T] \to \mathcal{X}[s,T], & \widehat{L}_s : \mathcal{U}^w[s,T] \to \mathcal{X}_T, \\ \Gamma_s : \mathbb{R}^n \to \mathcal{X}[s,T], & \widehat{\Gamma}_s : \mathbb{R}^n \to \mathcal{X}_T, \end{cases} \tag{4.7}$$

are all bounded linear operators (by (4.4)), where $\mathcal{X}[s,T] \stackrel{\Delta}{=} L^2_{\mathcal{F}}(s,T;\mathbb{R}^n)$ and $\mathcal{X}_T \stackrel{\Delta}{=} L^2_{\mathcal{F}}(\Omega;\mathbb{R}^n)$. We now want to find bounded linear operators

$$\begin{cases} L_s^* : \mathcal{X}[s,T] \to \mathcal{U}^w[s,T], & \widehat{L}_s^* : \mathcal{X}_T \to \mathcal{U}^w[s,T], \\ \Gamma_s^* : \mathcal{X}[s,T] \to \mathbb{R}^n, & \widehat{\Gamma}_s^* : \mathcal{X}_T \to \mathbb{R}^n, \end{cases} \tag{4.8}$$

such that

$$\begin{cases} E\displaystyle\int_s^T \langle (L_s u(\cdot))(t), \xi(t) \rangle \, dt = E\int_s^T \langle u(t), (L_s^*\xi(\cdot))(t) \rangle \, dt, \\ E\displaystyle\int_s^T \langle (\Gamma_s y)(t), \xi(t) \rangle \, dt = E \langle y, \Gamma_s^* \xi(\cdot) \rangle, \\ \qquad\qquad \forall u(\cdot) \in \mathcal{U}^w[s,T], \ \xi(\cdot) \in \mathcal{X}[s,T], \ y \in \mathbb{R}^n, \end{cases} \tag{4.9}$$

and

$$\begin{cases} E \langle \widehat{L}_s u(\cdot), \eta \rangle = E\displaystyle\int_s^T \langle u(t), (\widehat{L}_s^* \eta)(t) \rangle \, dt, \\ E \langle \widehat{\Gamma}_s y, \eta \rangle = E \langle y, \widehat{\Gamma}_s^* \eta \rangle, \\ \qquad\qquad \forall u(\cdot) \in \mathcal{U}^w[s,T], \ y \in \mathbb{R}^n, \ \eta \in \mathcal{X}_T. \end{cases} \tag{4.10}$$

In the above, we have used $\langle \cdot\,, \cdot \rangle$ as inner products in different Euclidean spaces, which can be identified from the context.

To find the operators (4.8) satisfying (4.9)–(4.10), let us introduce the following BSDE:

$$\begin{cases} dp(t) = -\big[A^\top p(t) + C^\top q(t) + \xi(t)\big]dt + q(t)dW(t), \quad t \in [s,T], \\ p(T) = \eta \in \mathcal{X}_T. \end{cases} \tag{4.11}$$

Proposition 4.1. *We have the following results.*

(i) *For any $\xi(\cdot) \in \mathcal{X}[s,T]$, let $(p_0(\cdot), q_0(\cdot)) \in \mathcal{X}[s,T] \times \mathcal{X}[s,T]$ be the adapted solution of (4.11) with $\eta = 0$. Define*

$$\begin{cases} (L_s^*\xi)(t) \stackrel{\Delta}{=} B^\top p_0(t) + D^\top q_0(t), \qquad t \in [s,T], \\ \Gamma_s^* y \stackrel{\Delta}{=} p_0(s). \end{cases} \tag{4.12}$$

Then $L_s^* : \mathcal{X}[s,T] \to \mathcal{U}^w[s,T]$ and $\Gamma_s^* : \mathcal{X}[s,T] \to \mathbb{R}^n$ are bounded operators satisfying (4.9).

(ii) *For any $\eta \in \mathcal{X}_T$, let $(p_1(\cdot), q_1(\cdot)) \in \mathcal{X}[s,T] \times \mathcal{X}[s,T]$ be the solution of (4.11) with $\xi(\cdot) = 0$. Define*

$$\begin{cases} (\widehat{L}_s^* \eta)(t) \stackrel{\Delta}{=} B^\top p_1(t) + D^\top q_1(t), & t \in [s,T], \\ \widehat{\Gamma}_s^* \eta \stackrel{\Delta}{=} p_1(s). \end{cases} \tag{4.13}$$

Then $\widehat{L}_s^* : \mathcal{X}_T \to \mathcal{U}^w[s,T]$ and $\widehat{\Gamma}_s^* : \mathcal{X}_T \to \mathbb{R}^n$ are bounded operators satisfying (4.10).

Proof. For any $\eta \in \mathcal{X}_T$ and $\xi(\cdot) \in \mathcal{X}[s,T]$, by Chapter 7, Theorem 2.2, there exists a unique adapted solution $(p(\cdot), q(\cdot)) \in \mathcal{X}[s,T] \times \mathcal{X}[s,T]$ to (4.11) satisfying

$$E\Big(\sup_{s \le t \le T} |p(t)|^2 + \int_s^T |q(t)|^2 dt\Big) \le KE\Big(|\eta|^2 + \int_s^T |\xi(t)|^2 dt\Big), \tag{4.14}$$

for some constant $K > 0$. It follows that all the operators defined in (4.12) and (4.13) are bounded. Next, for any $y \in \mathbb{R}^n$ and $u(\cdot) \in \mathcal{U}^w[s,T]$, let $x(\cdot)$ be the solution of (3.1). Applying Itô's formula to $\langle x(\cdot), p(\cdot) \rangle$, we obtain

$$\begin{aligned} &\langle x(T), \eta \rangle - \langle y, p(s) \rangle \\ &\quad = \int_s^T [\langle u(t), B^\top p(t) + D^\top q(t) \rangle - \langle x(t), \xi(t) \rangle] dt \\ &\qquad + \int_s^T [\cdots] dW(t). \end{aligned} \tag{4.15}$$

Consequently, using (4.3) and (4.5), we have

$$\begin{aligned} &E[\langle \widehat{\Gamma}_s y + \widehat{L}_s u(\cdot), \eta \rangle - \langle y, p(s) \rangle \\ &\quad = E \int_s^T \{ \langle u(t), B^\top p(t) + D^\top q(t) \rangle \\ &\qquad - \langle (\Gamma_s y)(t) + (L_s u(\cdot))(t), \xi(t) \rangle \} dt. \end{aligned} \tag{4.16}$$

Here, we have used $\langle \cdot, \cdot \rangle$ as inner products in different Hilbert spaces. Then, in (4.16), by taking $y = 0$ and $\eta = 0$, we obtain the first relations in (4.9); by taking $u(\cdot) = 0$ and $\eta = 0$, we obtain the second relations in (4.9); by taking $y = 0$ and $\xi(\cdot) = 0$, we obtain the first relations in (4.10); and by taking $u(\cdot) = 0$ and $\xi(\cdot) = 0$, we obtain the second relations in (4.10). □

Once we have the above result, similar to (2.8), we can obtain the following representation for the cost functional (3.2):

$$J(s, y; u(\cdot)) = \frac{1}{2}\{ \langle N_s u(\cdot), u(\cdot) \rangle + 2 \langle H_s(y), u(\cdot) \rangle + M_s(y) \}, \tag{4.17}$$

where

$$
(4.18)\qquad \begin{cases} N_s = R + L_s^* Q L_s + S L_s + L_s^* S^\top + \widehat{L}_s^* G \widehat{L}_s, \\ H_s(y) = (L_s^* Q + S)\big[(\Gamma_s y)(\cdot) + f^s(\cdot)\big] + \widehat{L}_s^* G(\widehat{\Gamma}_s y + \widehat{f}^s), \\ M_s(y) = \langle Q\big[(\Gamma_s y)(\cdot) + f^s(\cdot)\big], (\Gamma_s y)(\cdot) + f^s(\cdot)\rangle \\ \qquad\qquad + \langle G(\widehat{\Gamma}_s y + \widehat{f}^s), \widehat{\Gamma}_s y + \widehat{f}^s \rangle, \end{cases}
$$

and $\langle \cdot, \cdot \rangle$ represents inner products in different spaces. It is clear that the operator $N_s : \mathcal{U}^w[s,T] \to \mathcal{U}^w[s,T]$ is bounded.

Parallel to Theorem 2.2, we have the following result for the finiteness and solvability of our Problem (SLQ), whose proof is the same as that of Theorem 2.2.

Theorem 4.2. *Let (L1) hold.*

(i) *If Problem (SLQ) is finite at some $(s,y) \in [0,T] \times \mathbb{R}^n$, then*

$$
(4.19)\qquad N_r \geq 0, \qquad \forall r \in [s,T].
$$

(ii) *Problem (SLQ) is (pathwise uniquely) solvable at $(s,y) \in [0,T] \times \mathbb{R}^n$ if and only if $N_s \geq 0$ and there exists a (pathwise unique) $\bar{u}(\cdot) \in \mathcal{U}^w[s,T]$ such that*

$$
(4.20)\qquad N_s \bar{u}(\cdot) + H_s(y) = 0.
$$

In this case, $\bar{u}(\cdot)$ is an (the pathwise unique) optimal control.

(iii) *If $N_s \gg 0$ for some $s \in [0,T)$, then for any $y \in \mathbb{R}^n$, $J(s,y;\cdot)$ admits a pathwise unique minimizer $\bar{u}(\cdot)$ given by*

$$
(4.21)\qquad \bar{u}(\cdot) = -N_s^{-1} H_s(y).
$$

In this case,

(4.22)

$$
\begin{aligned} V(s,y) &\stackrel{\Delta}{=} \inf_{u(\cdot)\in\mathcal{V}[s,T]} J(s,y;u(\cdot)) = J(s,y;\bar{u}(\cdot)) \\ &= \frac{1}{2}\{M_s(y) - \langle N_s^{-1} H_s(y), H_s(y) \rangle\}, \ \forall (s,y) \in [0,T] \times \mathbb{R}^n. \end{aligned}
$$

(iv) *If Problem (SLQ) is pathwise uniquely solvable at $s \in [0,T)$, so is Problem (SLQ) with $b(\cdot) = \sigma(\cdot) = 0$.*

As in the deterministic case, when

$$
(4.23)\qquad R \gg 0, \quad Q - S R^{-1} S^\top \geq 0, \quad G \geq 0,
$$

Problem (SLQ) is referred to as a *standard stochastic LQ problem*. In such a case, $N_s \gg 0$, and thus by the above result the corresponding LQ problem is uniquely solvable.

5. A Necessary Condition and a Hamiltonian System

In this section we are going to apply the stochastic maximum principle obtained in Chapter 3 to derive a necessary condition of optimality for

Problem (SLQ). This will lead to the solvability of a Hamiltonian system. Note that by the remark at the end of Chapter 3, Section 4, the stochastic maximum principle is applicable to Problem (SLQ).

Theorem 5.1. *Let (L1) hold. Let Problem (SLQ) be solvable at* $(s,y) \in [0,T)\times\mathbb{R}^n$ *with* $(\bar{x}(\cdot),\bar{u}(\cdot))$ *being an optimal pair. Then there exist adapted processes* $(\bar{p}(\cdot),\bar{q}(\cdot))$ *and* $(\overline{P}(\cdot),\overline{\Lambda}(\cdot))$ *satisfying the following:*

$$(5.1)\qquad \begin{cases} d\bar{p}(t) = -[A(t)^\top\bar{p}(t) + C(t)^\top\bar{q}(t) - Q(t)\bar{x}(t) - S(t)^\top\bar{u}(t)]dt \\ \qquad\qquad + \bar{q}(t)dW(t), \quad t\in[s,T], \\ \bar{p}(T) = -G\bar{x}(T), \end{cases}$$

$$(5.2)\qquad \begin{cases} d\overline{P}(t) = -[A(t)^\top\overline{P}(t) + \overline{P}(t)A(t) + C(t)^\top\overline{P}(t)C(t) \\ \qquad\qquad +\overline{\Lambda}(t)C(t)+C(t)^\top\overline{\Lambda}(t)-Q(t)]dt+\overline{\Lambda}(t)dW(t), \ t\in[s,T], \\ \overline{P}(T) = -G, \end{cases}$$

such that

$$(5.3)\qquad \begin{cases} R(t)\bar{u}(t) - B(t)^\top\bar{p}(t) - D(t)^\top\bar{q}(t) + S(t)\bar{x}(t) = 0, \\ R(t) - D(t)^\top\overline{P}(t)D(t) \ge 0, \quad \text{a.e. } t\in[s,T], \ \mathbf{P}\text{-a.s.} \end{cases}$$

Proof. It is clear that the first and second adjoint equations corresponding to the given optimal pair $(\bar{x}(\cdot),\bar{u}(\cdot))$ are (5.1) and (5.2), respectively, for the present case. Moreover, the $\mathcal{H}$-function $\mathcal{H}(t,\bar{x}(t),u)$ is quadratic in u, which attains its maximum at $\bar{u}(t)$, a.e. $t\in[s,T]$, $\mathbf{P}$-a.s., by the stochastic maximum principle (Chapter 3, Theorem 3.2). It is easy to verify that (5.3) consists of nothing but the first-order and second-order necessary conditions of the maximum point $\bar{u}(t)$ for the quadratic function $\mathcal{H}(t,\bar{x}(t),\cdot)$. □

In view of (5.3), as long as the term $-D^\top P(t)D$ is sufficiently positive definite, the necessary condition might still be satisfied even if R is negative. We have seen such a case in Example 3.3 (the solvability of that LQ problem under condition (3.18) will be given in Section 6). On the other hand, if the term $-D^\top P(t)D$ is negative definite on an interval $[s,T]$, then the necessary condition might be violated even if $R(t)$ is positive definite. This explains Example 3.4. Therefore, condition (5.3) is essentially different from (2.21)–(2.22) for the deterministic case due to the presence of D.

An interesting corollary of the above result is the following.

Corollary 5.2. *Let (L1) hold with T being a Lebesgue point of $R(\cdot)$ and $D(\cdot)$. Suppose Problem (SLQ) is finite at some* $(s,y)\in[0,T)\times\mathbb{R}^n$. *Then*

$$(5.4)\qquad R(T) + D(T)^\top GD(T) \ge 0.$$

Proof. It follows from Theorem 4.2 that (4.19) holds under our assumption. Now, for any $\varepsilon > 0$, we consider an LQ problem that is the same as Problem (SLQ) except that $R(t)$ is replaced by $R(t)+\varepsilon I$. Then the corresponding operator N_s is replaced by $N_s+\varepsilon I$, which is invertible now. Thus, Theorem 4.2-(iii) implies that this LQ problem admits a (pathwise unique) optimal control. Consequently, by Theorem 5.1, one has

$$R(T)+\varepsilon I+D(T)^\top GD(T)\geq 0. \tag{5.5}$$

Since $\varepsilon>0$ is arbitrary, (5.4) follows. $\square$

Note that in Example 3.4, condition (5.4) is violated.

However, the following example shows that condition (5.4) is not sufficient for the finiteness of Problem (SLQ).

Example 5.3. Consider the one-dimensional control system

$$\begin{cases} dx(t)=\frac{1}{2}x(t)dt+u(t)dW(t), & t\in[s,T],\\ x(s)=y, \end{cases} \tag{5.6}$$

with the cost functional

$$J(s,y;u(\cdot))=E\Big\{\frac{1}{2}\int_s^T u(t)^2dt-\frac{1}{2}x(T)^2\Big\}. \tag{5.7}$$

In this case, $A=\frac{1}{2}$, $B=C=Q=0$, $D=R=1$, and $G=-1$. Thus,

$$R+D^\top GD=0.$$

Applying Itô's formula to $-e^{T-t}x(t)^2$, we have

$$\begin{aligned} J(s,y;u(\cdot))=&E\Big\{\int_s^T u(t)^2(1-e^{T-t})dt-e^{T-s}y^2\Big\},\\ &\forall u(\cdot)\in\mathcal{U}^w[s,T]. \end{aligned} \tag{5.8}$$

It follows that Problem (SLQ) is not finite at any $(s,y)\in[0,T)\times\mathbb{R}^n$.

Next, we introduce the following system of equations:

$$\begin{cases} dx(t)=[Ax(t)+Bu(t)+b(t)]dt\\ \qquad\qquad+[Cx(t)+Du(t)+\sigma(t)]dW(t),\\ dp(t)=-[A^\top p(t)-Qx(t)-S^\top u(t)+C^\top q(t)]dt\\ \qquad\qquad+q(t)dW(t),\\ x(s)=y,\qquad p(T)=-Gx(T). \end{cases} \tag{5.9}$$

This is a *decoupled* forward–backward stochastic differential equation (FBSDE). Given $y\in\mathbb{R}^n$ and $u(\cdot)\in\mathcal{U}^w[s,T]$, we may first solve the forward equation for $x(\cdot)$ and then solve the backward equation for $(p(\cdot),q(\cdot))$. Hence, (5.9) admits a unique adapted solution $(x(\cdot),p(\cdot),q(\cdot))$ corresponding to y and $u(\cdot)$. The following result is a further consequence of Proposition 4.1.

Lemma 5.4. *For any $(y, u(\cdot)) \in \mathbb{R}^n \times \mathcal{U}^w[s,T]$, let $(x(\cdot), p(\cdot), q(\cdot))$ be the adapted solution of (5.9). Then*

$$(N_s u(\cdot) + H_s(y))(t) = Ru(t) + Sx(t) - B^\top p(t) - D^\top q(t), \quad t \in [s,T], \ \mathbf{P}\text{-a.s.} \tag{5.10}$$

In particular, if $(x_0(\cdot), p_0(\cdot), q_0(\cdot))$ is the adapted solution of (5.9) with $y = 0$ and $b(\cdot) = \sigma(\cdot) = 0$, then

$$(N_s u(\cdot))(t) = Ru(t) + Sx_0(t) - B^\top p_0(t) - D^\top q_0(t), \quad t \in [s,T], \ \mathbf{P}\text{-a.s.} \tag{5.11}$$

Proof. Let $(x(\cdot), p(\cdot), q(\cdot))$ be the adapted solution of (5.9). By Proposition 4.1 and (4.18), we have (noting (4.6))

$$\begin{aligned} &N_s u(\cdot) + H_s(y) \\ &\quad = (R + L_s^* Q L_s + S L_s + L_s^* S^\top + \widehat{L}_s^* G \widehat{L}_s) u(\cdot) \\ &\qquad + (L_s^* Q + S)((\Gamma_s y)(\cdot) + f^s(\cdot)) + \widehat{L}_s^* G(\widehat{\Gamma}_s y + \widehat{f}^s) \\ &\quad = Ru(\cdot) + S((\Gamma_s y)(\cdot) + L_s u(\cdot) + f^s(\cdot)) \\ &\qquad + L_s^* Q((\Gamma_s y)(\cdot) + L_s u(\cdot) + f^s(\cdot)) \\ &\qquad + L_s^* S^\top u(\cdot) + \widehat{L}_s^* G(\widehat{\Gamma}_s y + \widehat{L}_s u(\cdot) + \widehat{f}^s) \\ &\quad = Ru(\cdot) + Sx(\cdot) + L_s^*\big[Qx(\cdot) + S^\top u(\cdot)\big] + \widehat{L}_s^* G x(T) \\ &\quad = Ru(\cdot) + Sx(\cdot) - B^\top p(\cdot) - D^\top q(\cdot). \end{aligned} \tag{5.12}$$

This gives (5.10). The proof of (5.11) is clear. □

The following result is a variant of Theorem 4.2 in which the conditions are given in terms of an FBSDE.

Proposition 5.5. *Let (L1) hold. Then Problem (SLQ) is (pathwise uniquely) solvable at $(s, y) \in [0,T] \times \mathbb{R}^n$ with an (the) optimal pair $(\overline{x}(\cdot), \overline{u}(\cdot))$ if and only if there (uniquely) exists a 4-tuple $(\overline{x}(\cdot), \overline{u}(\cdot), \overline{p}(\cdot), \overline{q}(\cdot))$ satisfying the FBSDE*

$$\begin{cases} d\overline{x}(t) = [A\overline{x}(t) + B\overline{u}(t) + b(t)]dt + [C\overline{x}(t) + D\overline{u}(t) + \sigma(t)]dW(t), \\ d\overline{p}(t) = -[A^\top \overline{p}(t) + C^\top \overline{q}(t) - Q\overline{x}(t) - S^\top \overline{u}(t)]dt + \overline{q}(t)dW(t), \\ \overline{x}(s) = y, \qquad \overline{p}(T) = -G\overline{x}(T), \end{cases} \tag{5.13}$$

with the condition

$$R\overline{u}(t) + S\overline{x}(t) - B^\top \overline{p}(t) - D^\top \overline{q}(t) = 0, \quad t \in [s,T], \ \mathbf{P}\text{-a.s.}, \tag{5.14}$$

and for any $u(\cdot) \in \mathcal{U}^w[s,T]$, the unique adapted solution $(x_0(\cdot), p_0(\cdot), q_0(\cdot))$ of (5.9) with $y = 0$ and $b(\cdot) = \sigma(\cdot) = 0$ satisfies

$$E\Big\{\int_s^T \langle Ru(t) + Sx_0(t) - B^\top p_0(t) - D^\top q_0(t), u(t)\rangle\, dt\Big\} \geq 0. \tag{5.15}$$

Proof. Necessity. The condition (5.14) follows from Theorem 5.1, while (5.15) follows from Theorem 4.2-(i) along with (5.11).

Sufficiency. From Lemma 5.4, condition (5.15) is equivalent to $N_s \geq 0$. Now let $(\bar{x}(\cdot), \bar{p}(\cdot), \bar{q}(\cdot))$ be an adapted solution of (5.13) such that (5.14) holds. Then, by Lemma 5.4, we see that (5.14) is the same as (4.20). Hence by Theorem 4.2-(ii), Problem (SLQ) is solvable.

The uniqueness part is clear. □

Now let us look at the case where $R(t)$ is invertible for all t and

$$R(\cdot)^{-1} \in L^\infty_{\mathcal{F}}(0,T;\mathcal{S}^m). \tag{5.16}$$

Note that in this case $R(t)$ is still not necessarily positive definite. When (5.16) holds, (5.13)–(5.14) are equivalent to the following (t is suppressed in the coefficients) by (5.14):

$$\begin{cases} d\bar{x}(t) = [\widehat{A}\bar{x}(t) + BR^{-1}B^\top \bar{p}(t) + BR^{-1}D^\top \bar{q}(t) + b(t)]dt \\ \qquad\qquad + [\widehat{C}\bar{x}(t) + DR^{-1}B^\top \bar{p}(t) + DR^{-1}D^\top \bar{q}(t) + \sigma(t)]dW(t), \\ d\bar{p}(t) = -[\widehat{Q}\bar{x}(t) - \widehat{A}^\top \bar{p}(t) - \widehat{C}^\top \bar{q}(t)]dt + \bar{q}(t)dW(s), \\ \bar{x}(s) = y, \qquad \bar{p}(T) = -G\bar{x}(T), \end{cases} \tag{5.17}$$

where

$$\widehat{A} = A - BR^{-1}S,\ \widehat{C} = C - DR^{-1}S,\ \widehat{Q} = Q - S^\top R^{-1}S.$$

Clearly, (5.17) is a *coupled* linear FBSDE. By Proposition 5.5, we have the following result.

Corollary 5.6. *Let (L1) and (5.16) hold. Let $N_s \geq 0$. Then Problem (SLQ) is (pathwise uniquely) solvable at $(s,y) \in [0,T) \times \mathbb{R}^n$ if and only if the FBSDE (5.17) admits a (unique) adapted solution $(\bar{x}(\cdot), \bar{p}(\cdot), \bar{q}(\cdot))$. In this case, the following is an (the) optimal control:*

$$\bar{u}(t) = -R(t)^{-1}\big[S\bar{x}(t) - B^\top \bar{p}(t) - D^\top \bar{q}(t)\big], \quad \forall t \in [s,T]. \tag{5.18}$$

The above result is comparable with Theorem 2.7. It also gives an intrinsic relation between the solvability of Problem (SLQ) and FBSDE (5.17), which is nothing but the stochastic Hamiltonian system associated with Problem (SLQ).

Corollary 5.7. *Let (L1) and (4.23) hold. Then FBSDE (5.17) admits a unique adapted solution $(\bar{x}(\cdot), \bar{p}(\cdot), \bar{q}(\cdot))$, and Problem (SLQ) is pathwise uniquely solvable at s with the optimal control having the representation (5.18).*

It is clear that if (4.23) holds, (4.19) or equivalently (5.15) automatically holds. Thus, for a standard stochastic LQ problem, Corollary 5.7 implies that (5.13)–(5.14) completely characterize the optimal pair $(\bar{x}(\cdot), \bar{u}(\cdot))$.

6. Stochastic Riccati Equations

Although we have obtained optimal controls for the stochastic LQ problems by (5.18), it is not an implementable control policy, since it involves $\overline{p}(\cdot)$ and $\overline{q}(\cdot)$. In this section we are going to obtain a state feedback representation of the optimal control for Problem (SLQ) via a Riccati-like equation, in the case where the problem is uniquely solvable at some $s \in [0,T)$.

As we have seen in the deterministic case, the optimal feedback control can be represented via the Riccati equation. What should then be an appropriate "Riccati equation" corresponding to our stochastic LQ problems? Are there any differences between the Riccati equations for Problems (DLQ) and (SLQ)? In this section we are going to derive an appropriate Riccati equation (called a *stochastic Riccati equation*) associated with Problem (SLQ), based on three different approaches: stochastic maximum principle/Hamiltonian systems, dynamic programming/HJB equations, and a purely completion of squares technique.

Let us start with the stochastic maximum principle, which is a continuation of the study in the previous section. Assume that Problem (SLQ) is uniquely solvable at $s \in [0,T)$. Then for any $y \in \mathbb{R}^n$, the corresponding stochastic Hamiltonian system is solvable, namely, (5.13) admits a unique adapted solution $(\overline{x}(\cdot), \overline{p}(\cdot), \overline{q}(\cdot))$ with $\overline{u}(\cdot) \in \mathcal{U}^w[s,T]$ satisfying

$$R\overline{u}(t) + S\overline{x}(t) - B^\top \overline{p}(t) - D^\top \overline{q}(t) = 0, \qquad \forall t \in [s,T], \ \mathbf{P}\text{-a.s.} \tag{6.1}$$

Now we conjecture that $\overline{x}(\cdot)$ and $\overline{p}(\cdot)$ are related by the following:

$$\overline{p}(t) = -P(t)\overline{x}(t) - \varphi(t), \qquad t \in [s,T], \tag{6.2}$$

with $P(\cdot) \in C^1([s,T];\mathcal{S}^n)$ and $\varphi(\cdot) \in C^1([s,T];\mathbb{R}^n)$. Applying Itô's formula to (6.2) and using (5.13), we have (suppressing t in the functions)

$$\begin{aligned} &[-\dot{P}\overline{x} - P(A\overline{x} + B\overline{u} + b) - \dot{\varphi}]dt - P(C\overline{x} + D\overline{u} + \sigma)dW(t) \\ &\quad = d\overline{p} \\ &\quad = -(A^\top \overline{p} + C^\top \overline{q} - Q\overline{x} - S^\top \overline{u})dt + \overline{q}dW(t). \end{aligned} \tag{6.3}$$

Thus,

$$\overline{q} = -PC\overline{x} - PD\overline{u} - P\sigma, \tag{6.4}$$

and (noting (6.2) and (6.4))

$$\begin{aligned} 0 &= \dot{P}\overline{x} + PA\overline{x} + PB\overline{u} + Pb + \dot{\varphi} + Q\overline{x} + S^\top \overline{u} - A^\top \overline{p} - C^\top \overline{q} \\ &= (\dot{P} + PA + Q)\overline{x} + (PB + S^\top)\overline{u} + A^\top(P\overline{x} + \varphi) \\ &\quad + C^\top(PC\overline{x} + PD\overline{u} + P\sigma) + Pb + \dot{\varphi} \\ &= (\dot{P} + PA + A^\top P + C^\top PC + Q)\overline{x} + (PB + S^\top + C^\top PD)\overline{u} \\ &\quad + A^\top \varphi + C^\top P\sigma + Pb + \dot{\varphi}. \end{aligned} \tag{6.5}$$

On the other hand, by (6.1)–(6.2) and (6.4), we have

$$\begin{aligned}0 &= R\bar{u} + S\bar{x} + B^\top(P\bar{x} + \varphi) + D^\top(PC\bar{x} + PD\bar{u} + P\sigma)\\&= (R + D^\top PD)\bar{u} + (B^\top P + S + D^\top PC)\bar{x} + B^\top\varphi + D^\top P\sigma.\end{aligned}$$

Thus, by assuming the existence of $(R + D^\top PD)^{-1}$, we obtain from the above that

$$\begin{aligned}0 = &[\dot{P} + PA + A^\top P + C^\top PC + Q\\&- (PB + S^\top + C^\top PD)(R + D^\top PD)^{-1}(B^\top P + S + D^\top PC)]\bar{x}\\&+ [A^\top - (PB + S^\top + C^\top PD)(R + D^\top PD)^{-1}B^\top]\varphi\\&+ [C^\top - (PB + S^\top + C^\top PD)(R + D^\top PD)^{-1}D^\top]P\sigma + Pb + \dot{\varphi}.\end{aligned}$$

Hence, $P(\cdot)$ should be a solution of

$$\text{(6.6)}\quad \begin{cases}\dot{P} + PA + A^\top P + C^\top PC + Q\\ \quad - \big(B^\top P+S+D^\top PC\big)^\top\big(R+D^\top PD\big)^{-1}\big(B^\top P+S+D^\top PC\big) = 0,\\ \hfill \text{a.e. } t \in [s,T],\\ P(T) = G,\\ R(t) + D(t)^\top P(t)D(t) > 0, \qquad \text{a.e. } t \in [s,T],\end{cases}$$

and $\varphi(\cdot)$ should be the solution of the following BSDE:

$$\text{(6.7)}\quad \begin{cases}\dot{\varphi} + [A - B(R + D^\top PD)^{-1}(B^\top P+S+D^\top PC)]^\top\varphi\\ \quad + [C-D(R + D^\top PD)^{-1}(B^\top P+S+D^\top PC)]^\top P\sigma + Pb = 0,\\ \hfill \text{a.e. } t \in [s,T],\\ \varphi(T) = 0.\end{cases}$$

We call (6.6) the *stochastic Riccati equation* associated with Problem (SLQ). This is a nonlinear ordinary differential equation. Note that we call it "stochastic" even though the equation itself is deterministic, because it arises from a stochastic LQ problem, and moreover, it would be a backward stochastic differential equation in the general case where all the coefficient matrices are random (cf. Chen–Li–Zhou [1]). On the other hand, one may note that the derivation above only requires $R(t) + D(t)^\top P(t)D(t)$ to have an inverse. Nevertheless, the constraint $R(t) + D(t)^\top P(t)D(t) > 0$ (as in (6.6)) is indeed needed in order for the original stochastic LQ problem to have a (minimizing) optimal control; see the proof of Theorem 6.1 below.

It should be noted that if T is a Lebesgue point of $R(\cdot)$ and $D(\cdot)$, and the condition

$$\text{(6.8)}\qquad R(T) + D(T)^\top GD(T) > 0$$

holds, then there exists an $s \in [0,T)$ such that (6.6) admits a solution $P(\cdot)$ on $[s,T]$. As a consequence, since (6.7) is a linear ordinary differential equation with bounded coefficients, it admits a solution $\varphi(\cdot)$ in the same interval $[s,T]$. Namely, both (6.6) and (6.7) are at least locally solvable.

The following result justifies the above heuristic derivation.

Theorem 6.1. *Let (L1) hold. Let $P(\cdot) \in C([s,T];\mathcal{S}^n)$ and $\varphi(\cdot) \in C([s,T];\mathbb{R}^n)$ be the solutions of (6.6) and (6.7), respectively, for some $s \in [0,T)$, such that*

$$
\left\{
\begin{array}{l}
B\Psi, D\Psi \in L^\infty(s,T;\mathbb{R}^{n\times n}), \text{ where} \\
\Psi \stackrel{\Delta}{=} (R + D^\top P D)^{-1}[B^\top P + S + D^\top P C],
\end{array}
\right.
\tag{6.9}
$$

and

$$
\left\{
\begin{array}{l}
B\psi, D\psi \in L^2(s,T;\mathbb{R}^n), \text{ where} \\
\psi \stackrel{\Delta}{=} (R + D^\top P D)^{-1}[B^\top \varphi + D^\top P \sigma].
\end{array}
\right.
\tag{6.10}
$$

Then Problem (SLQ) is solvable at s with the optimal control $\overline{u}(\cdot)$ being of a state feedback form,

$$
\overline{u}(t) = -\Psi(t)x(t) - \psi(t), \qquad t \in [s,T],
\tag{6.11}
$$

and

$$
\begin{aligned}
V(s,y) &= \frac{1}{2}\langle P(s)y, y\rangle + \langle \varphi(s), y\rangle \\
&\quad + \frac{1}{2}E\int_s^T \{2\langle \varphi, b\rangle + \langle P\sigma, \sigma\rangle - |(R + D^\top P D)^{\frac{1}{2}}\psi|^2\}dt, \\
&\quad \forall y \in \mathbb{R}^n.
\end{aligned}
\tag{6.12}
$$

Proof. By (L1) and (6.9)–(6.10), the following SDE admits a unique strong solution $\overline{x}(\cdot)$:

$$
\left\{
\begin{aligned}
d\overline{x}(t) &= \big\{[A(t) - B(t)\Psi(t)]\overline{x}(t) - B(t)\psi(t) + b(t)\big\}dt \\
&\qquad + \big\{[C(t) - D(t)\Psi(t)]\overline{x}(t) - D(t)\psi(t) + \sigma(t)\big\}dW(t), \\
\overline{x}(s) &= y.
\end{aligned}
\right.
\tag{6.13}
$$

Moreover, the following estimate holds:

$$
E \sup_{s\le t\le T} |\overline{x}(t)|^2 \le K(1 + |y|^2).
$$

So the control $\overline{u}(\cdot)$ defined by (6.11) is admissible, and by taking such a control on any probability space $(\Omega, \mathcal{F}, \mathbf{P})$ along with a Brownian motion $W(\cdot)$, $\overline{x}(\cdot)$ is the corresponding state process. We claim that $\overline{u}(\cdot)$ is an optimal control. To show this, take any $u(\cdot) \in \mathcal{U}^w[s,T]$ and let $x(\cdot)$ be the corresponding state process. Applying Itô's formula to $\langle P(\cdot)x(\cdot), x(\cdot)\rangle$, we have

$$
\begin{aligned}
&E\langle Gx(T),x(T)\rangle-\langle P(s)y,y\rangle\\
&\quad=E\int_s^T\{\langle \dot{P}x,x\rangle+\langle P(Ax+Bu+b),x\rangle+\langle Px,Ax+Bu+b\rangle\\
&\qquad\qquad+\langle P(Cx+Du+\sigma),Cx+Du+\sigma\rangle\}dt\\
&\quad=E\int_s^T\{\langle(\dot{P}+PA+A^\top P+C^\top PC)x,x\rangle\\
&\qquad\qquad+2\langle(B^\top P+D^\top PC)x+D^\top P\sigma,u\rangle+\langle D^\top PDu,u\rangle\\
&\qquad\qquad+2\langle Pb+C^\top P\sigma,x\rangle+\langle P\sigma,\sigma\rangle\}dt.
\end{aligned}
$$

Also, applying Itô's formula to $\langle\varphi(\cdot),x(\cdot)\rangle$, we obtain

$$
\begin{aligned}
&-\langle\varphi(s),y\rangle\\
&\quad=E[\langle\varphi(T),x(T)\rangle-\langle\varphi(s),y\rangle]\\
&\quad=E\int_s^T\{-\langle[A^\top-(PB+S^\top+C^\top PD)(R+D^\top PD)^{-1}B^\top]\varphi,x\rangle\\
&\qquad\qquad-\langle[C^\top-(PB+S^\top+C^\top PD)(R+D^\top PD)^{-1}D^\top]P\sigma+Pb,x\rangle\\
&\qquad\qquad+\langle\varphi,Ax+Bu+b\rangle\}dt\\
&\quad=E\int_s^T\{\langle(PB+S^\top+C^\top PD)(R+D^\top PD)^{-1}[B^\top\varphi+D^\top P\sigma],x\rangle\\
&\qquad\qquad-\langle C^\top P\sigma+Pb,x\rangle+\langle B^\top\varphi,u\rangle+\langle\varphi,b\rangle\}dt.
\end{aligned}
$$

Thus, by setting $\widehat{R}\triangleq R+D^\top PD$ and $\widehat{S}\triangleq B^\top P+S+D^\top PC$, we obtain the following:

$$
\begin{aligned}
&J(s,y;u(\cdot))-\frac{1}{2}\langle P(s)y,y\rangle-\langle\varphi(s),y\rangle\\
&\quad=\frac{1}{2}E\int_s^T\Big\{\langle\widehat{S}^\top\widehat{R}^{-1}\widehat{S}x,x\rangle+2\langle\widehat{S}x+B^\top\varphi+D^\top P\sigma,u\rangle\\
&\qquad\qquad+2\langle\widehat{S}^\top\widehat{R}^{-1}[B^\top\varphi+D^\top P\sigma],x\rangle+\langle\widehat{R}u,u\rangle+2\langle\varphi,b\rangle+\langle P\sigma,\sigma\rangle\Big\}dt\\
&\quad=\frac{1}{2}E\int_s^T\{|\widehat{R}^{-\frac{1}{2}}[\widehat{R}u+\widehat{S}x+B^\top\varphi+D^\top P\sigma]|^2\\
&\qquad\qquad-|\widehat{R}^{-\frac{1}{2}}[B^\top\varphi+D^\top P\sigma]|^2+2\langle\varphi,b\rangle+\langle P\sigma,\sigma\rangle\}dt\\
&\quad=\frac{1}{2}E\int_s^T\{|\widehat{R}^{\frac{1}{2}}[u+\Psi x+\psi]|^2-|\widehat{R}^{\frac{1}{2}}\psi|^2+2\langle\varphi,b\rangle+\langle P\sigma,\sigma\rangle\}dt.
\end{aligned}
$$

Noting that

$$
\widehat{R}(t)\equiv R(t)+D(t)^\top P(t)D(t)>0,\quad \text{a.e. } t\in[s,T],
$$

we conclude from the above that

$$\begin{aligned}J(s,y;\bar{u}(\cdot)) &= \frac{1}{2}\langle P(s)y,y\rangle + \langle \varphi(s),y\rangle \\ &\quad + \frac{1}{2}E\int_s^T \{2\langle \varphi,b\rangle + \langle P\sigma,\sigma\rangle - |(R+D^\top PD)^{\frac{1}{2}}\psi|^2\}dt \\ &\leq J(s,y;u(\cdot)),\end{aligned}$$

which implies that $\bar{u}(\cdot)$ is an optimal control. It also leads to (6.12). □

We have seen that the stochastic maximum principle/Hamiltonian system can formally lead to the stochastic Riccati equation, and the completion of squares technique further shows that the stochastic LQ problem indeed can be solved via the stochastic Riccati equation. It should be noted that in the proof of Theorem 6.1, even if one does not have the forms of the equations (6.6) and (6.7) a priori, one can set $dP(t) = \Gamma(t)dt$ and $d\varphi(t) = \alpha(t)dt$, and derive $\Gamma(t)$ and $\alpha(t)$ by a constructive technique purely of completion of squares. Namely, we can derive the stochastic Riccati equation simply by the completion of squares technique without involving the stochastic maximum principle. The reader is encouraged to work through the details of this derivation.

Next, we are going to show that there is a third way of deriving the stochastic Riccati equation along with the optimal feedback control, namely, the dynamic programming approach.

By the dynamic programming principle, the value function V should satisfy the HJB equation

$$\text{(6.14)}\qquad \begin{cases} -V_t + \sup\limits_{u\in\mathbf{R}^k} G(t,x,u,V_x,V_{xx}) = 0, \\ V(T,x) = \dfrac{1}{2}\langle Gx,x\rangle, \end{cases}$$

where the generalized Hamiltonian (t is suppressed below) in this case is

$$\text{(6.15)}\qquad \begin{aligned} &G(t,x,u,V_x,V_{xx}) \\ &= -\frac{1}{2}\langle Ru,u\rangle - \langle Sx,u\rangle - \frac{1}{2}\langle Qx,x\rangle - \langle V_x, Ax+Bu+b\rangle \\ &\quad - \frac{1}{2}\langle V_{xx}(Cx+Du+\sigma), Cx+Du+\sigma\rangle. \end{aligned}$$

We conjecture that $V(t,x)$ is quadratic in x, namely,

$$\text{(6.16)}\qquad V(t,x) = \frac{1}{2}\langle P(t)x,x\rangle + \langle \varphi(t),x\rangle + f(t),$$

for some suitable $P(\cdot), \varphi(\cdot)$ and $f(\cdot)$ (where $P(t)$ is symmetric) with

$$\text{(6.17)}\qquad P(T) = G, \quad \varphi(T) = 0, \quad f(T) = 0.$$

Substituting (6.16) into (6.15) and using completion of squares repeatedly, we get

$$
\begin{aligned}
&G(t,x,u,V_x,V_{xx})\\
&\quad = -\frac{1}{2}|\widehat{R}^{\frac{1}{2}}[u+\Psi x+\psi]|^2+\frac{1}{2}\langle\, \widehat{S}^\top \widehat{R}^{-1}\widehat{S}x,x\,\rangle\\
&\quad +\frac{1}{2}\langle(-Q-C^\top PC-P^\top A-A^\top P)x,x\,\rangle\\
&\quad -\langle\, A^\top\varphi+C^\top P\sigma-\Psi^\top\widehat{R}^{-1}\psi+Pb,x\,\rangle\\
&\quad -\frac{1}{2}|\widehat{R}^{\frac{1}{2}}\psi|^2-\langle\,\varphi,b\,\rangle-\frac{1}{2}\langle\, P\sigma,\sigma\,\rangle,
\end{aligned}
\tag{6.18}
$$

where

$$
\begin{cases}
\widehat{R}\triangleq R+D^\top PD, & \widehat{S}\triangleq B^\top P+S+D^\top PC,\\
\Psi\triangleq\widehat{R}^{-1}\widehat{S}, & \psi\triangleq\widehat{R}^{-1}(B^\top\varphi+D^\top P\sigma).
\end{cases}
$$

The stochastic verification theorem (Theorem 5.1 of Chapter 5; note that in the present case V is smooth) says that the optimal control $\overline{u}(\cdot)$ has the property that for a.e. $t\in[0,T]$, $\mathbf{P}$-a.s., $\overline{u}(t)$ should maximize

$$
u\mapsto G(t,\overline{x}(t),u,-V_x(t,\overline{x}(t)),-V_{xx}(t,\overline{x}(t))).
$$

Thus, by (6.18), we see that the optimal (feedback) control is given by

$$
\overline{u}(t)=-\Psi(t)x(t)-\psi(t),
\tag{6.19}
$$

provided that $\widehat{R}(t)\equiv R(t)+D(t)^\top P(t)D(t)>0$ for a.e. t. In addition, the HJB equation (6.14) now reads

$$
\begin{aligned}
&\frac{1}{2}\langle\, \dot{P}x,x\,\rangle+\langle\,\dot{\varphi},x\,\rangle+\dot{f}\\
&\quad=\frac{1}{2}\langle\, \widehat{S}^\top\widehat{R}^{-1}\widehat{S}x,x\,\rangle+\frac{1}{2}\langle(-Q-C^\top PC-P^\top A-A^\top P)x,x\,\rangle\\
&\quad-\langle\, A^\top\varphi+C^\top P\sigma-\Psi^\top\widehat{R}^{-1}\psi+Pb,x\,\rangle\\
&\quad+\frac{1}{2}|\widehat{R}^{\frac{1}{2}}\psi|^2-\langle\,\varphi,b\,\rangle-\frac{1}{2}\langle\, P\sigma,\sigma\,\rangle.
\end{aligned}
$$

Then we recover the stochastic Riccati equation (6.6) and the equation (6.7) by comparing the quadratic terms and linear terms in x, respectively, and noting (6.17). Finally, by

$$
\dot{f}=\frac{1}{2}|(R+D^\top PD)^{\frac{1}{2}}\psi|^2-\langle\,\varphi,b\,\rangle-\frac{1}{2}\langle\, P\sigma,\sigma\,\rangle,
$$

along with (6.17) and (6.16), we recover the representation of the value function (6.12).

We point out that the argument above appears rather formal. In order to derive equations (6.6) and (6.7), a rigorous argument can be made by going through a "reverse" direction. More precisely, if equations (6.6) and

(6.7) are given a priori and their solvability is known, then the function given by (6.16) or (6.12) is easily verified to satisfy the HJB equation (6.14), and therefore it must be the value function by the uniqueness of solutions to the HJB equation. Moreover, the verification theorem gives rise to the optimal feedback control (6.19).

One can also see clearly the relationship between the stochastic maximum principle and dynamic programming in the present LQ case. Indeed, (6.2) and (6.4) are exactly the relationship given in Chapter 5, (4.13).

Let us now complete Example 3.3 presented in Section 3.

Example 3.3. (continued) For the LQ problem with the state equation (3.10) and the cost functionals (3.15), the Riccati equation reads (noting $|\delta| > 1$)

$$\begin{cases} \dot P(t) = \dfrac{P(t)^2}{\delta^2 P(t) - 1}, & t \le T, \\ P(T) = 1, \\ -1 + \delta^2 P(t) > 1. \end{cases}$$

A direct computation shows that the above is equivalent to

$$\begin{cases} \delta^2 \ln P(t) + \dfrac{1}{P(t)} = t + 1 - T, \\ \delta^2 P(t) > 1, \end{cases} \qquad t \in (\bar s, T],$$

where $\bar s$ is such that $\delta^2 P(\bar s) = 1$. Thus

$$\delta^2 \big(1 - 2 \ln |\delta|\big) = \bar s + 1 - T,$$

or

$$\bar s = T - 1 - \delta^2 \big(2 \ln |\delta| - 1\big).$$

Hence, (3.18) ensures that $\bar s < 0$, which guarantees the solvability of Problem (SLQ) at all $s \in [0, T]$ in view of Theorem 6.1.

7. Global Solvability of Stochastic Riccati Equations

We have seen that the solvability of the stochastic Riccati equations is vital for the solution of the stochastic LQ problems. In this section we investigate the (global) existence and uniqueness of solutions to the equation for various cases.

Let us first prove the uniqueness of solutions to the stochastic Riccati equation (6.6) as well as (6.7). For this, we need an additional assumption.

(L2) $D(\cdot)$ and $R(\cdot)$ satisfy the following:

$$D \in C([0,T]; \mathbb{R}^{n\times k}), \quad R \in C([0,T]; \mathcal{S}^k). \tag{7.1}$$

Proposition 7.1. *Let (L1) and (L2) hold. If $P \in C([0,T]; \mathbb{R}^{n\times n})$ is a solution to the stochastic Riccati equation (6.6), then P is the only solution.*

As a consequence, $P \in C([0,T];\mathcal{S}^n)$. Moreover, in this case (6.7) admits only one solution $\varphi \in C([0,T];\mathbb{R}^n)$.

Proof. Suppose $\widetilde{P} \in C([0,T];\mathbb{R}^{n\times n})$ is another solution of (6.6). Set $\widehat{P} \triangleq P - \widetilde{P}$. Then $\widehat{P}$ satisfies

$$\begin{cases} \dot{\widehat{P}} + \widehat{P}A + A^\top \widehat{P} + C^\top \widehat{P}C \\ \qquad - (\widehat{P}B + C^\top \widehat{P}D)\widehat{R}^{-1}(B^\top P + S + D^\top PC) \\ \qquad + (\widetilde{P}B + S^\top + C^\top \widetilde{P}D)\widehat{R}^{-1}D^\top \widehat{P}D\widetilde{R}^{-1}(B^\top P + S + D^\top PC) \\ \qquad - (\widetilde{P}B + S^\top + C^\top \widetilde{P}D)\widetilde{R}^{-1}(B^\top \widehat{P} + D^\top \widehat{P}C) = 0, \\ \widehat{P}(T) = 0, \end{cases}$$

where $\widehat{R} \triangleq R + D^\top PD > 0$ and $\widetilde{R} \triangleq R + D^\top \widetilde{P}D > 0$. Since $|\widehat{R}(t)^{-1}|$ and $|\widetilde{R}(t)^{-1}|$ are uniformly bounded due to their continuity, we can apply Gronwall's inequality to get $\widehat{P}(t) \equiv 0$. This proves the uniqueness for the equation (6.6). Moreover, if P solves (6.6), so does $P^\top$. Hence $P = P^\top$ by uniqueness, implying $P \in C([0,T];\mathcal{S}^n)$. Finally, the existence and uniqueness of solutions to (6.7) is straightforward due to the uniform boundedness of the coefficients. □

We now focus on the existence of solutions to the stochastic Riccati equation (6.6). We mentioned that it is always locally solvable (on a small interval $[s,T]$) if (6.8) holds, but the local solvability is not very useful in view of solving the LQ problems where the initial times s are usually given a priori. Hence we are interested in global solvability over a *given* interval. We shall consider three special cases in the following three subsections, respectively.

7.1. Existence: The standard case

In this subsection we investigate the (global) solvability of (6.6) under condition (4.23), namely, the standard case. It is very natural to expect that (6.6) is solvable in this case, because due to Corollary 5.7, Problem (SLQ) is pathwise uniquely solvable at any $s \in [0,T]$ in the standard case.

Theorem 7.2. *Let (L1), (L2), and (4.23) hold. Then the stochastic Riccati equation (6.6) admits a unique solution over $[0,T]$.*

To prove this theorem, we need the following lemma.

Lemma 7.3. *The linear matrix-valued differential equation*

$$(7.2)\qquad \begin{cases} \dot{P}(t) + P(t)\widehat{A}(t) + \widehat{A}(t)^\top P(t) + \widehat{C}(t)^\top P(t)\widehat{C}(t) + \widehat{Q}(t) = 0, \\ \qquad\qquad\qquad t \in [0,T], \\ P(T) = \widehat{G}, \end{cases}$$

where $\widehat{A}(t), \widehat{C}(t), \widehat{Q}(t)$, and $\widehat{G}$ are measurable and uniformly bounded, admits a unique solution $P \in C([0,T];\mathcal{S}^n)$. Moreover, if $\widehat{Q} \in L^\infty(0,T;\mathcal{S}^n_+)$

and $\widehat{G} \in \mathcal{S}_+^n$, *then*

$$P \in C([0,T]; \mathcal{S}_+^n). \tag{7.3}$$

Proof. Since equation (7.2) is linear with bounded coefficients, the existence and uniqueness of its solution $P \in C([0,T]; \mathcal{S}^n)$ are clear. Now let $\Phi(\cdot)$ be the solution of the following SDE on a filtered probability space:

$$\begin{cases} d\Phi(t) = \widehat{A}(t)\Phi(t)dt + \widehat{C}(t)\Phi(t)dW(t), & t \in [0,T], \\ \Phi(0) = I. \end{cases} \tag{7.4}$$

Clearly, this equation admits a unique solution. Moreover, $\Phi(t)$ is invertible for all $t \in [0,T]$, $\mathbf{P}$-a.s. (see Chapter 1, Theorem 6.14). Now applying Itô's formula, we obtain

$$\begin{aligned} d(\Phi(t)^\top P(t)\Phi(t)) = & -\Phi(t)^\top \widehat{Q}(t)\Phi(t)dt \\ & + \Phi(t)^\top \big[\widehat{C}(t)^\top P(t) + P(t)\widehat{C}(t)\big]\Phi(t)dW(t). \end{aligned}$$

Consequently,

$$\begin{aligned} P(t) = & E\big\{[\Phi(t)^{-1}]^\top \Phi(T)^\top \widehat{G}\Phi(T)\Phi(t)^{-1} \\ & + \int_t^T [\Phi(t)^{-1}]^\top \Phi(r)^\top \widehat{Q}(r)\Phi(r)\Phi(t)^{-1} dr\big\} \geq 0, \end{aligned}$$

if $\widehat{Q} \geq 0$ and $\widehat{G} \geq 0$. This proves (7.3). □

Proof of Theorem 7.2. First of all, we claim that equation (6.6) is equivalent to the following:

$$\begin{cases} \dot{P} + P\widehat{A} + \widehat{A}^\top P + \widehat{C}^\top P\widehat{C} + \widehat{Q} = 0, \\ P(T) = G, \end{cases} \tag{7.5}$$

where

$$\begin{cases} \widehat{A} = A - B\Psi, \quad \widehat{C} = C - D\Psi, \\ \widehat{Q} = (\Psi - R^{-1}S)^\top R(\Psi - R^{-1}S) + Q - S^\top R^{-1}S, \\ \Psi = (R + D^\top PD)^{-1}(B^\top P + S + D^\top PC). \end{cases} \tag{7.6}$$

In fact, by defining $\widehat{A}, \widehat{C}, \widehat{Q}$, and Ψ as in (7.6), we have

$$(R + D^\top PD)\Psi = B^\top P + S + D^\top PC, \tag{7.7}$$

and

$$
\begin{aligned}
-\dot{P} &= P(\widehat{A}+B\Psi)+(\widehat{A}+B\Psi)^\top P+(\widehat{C}+D\Psi)^\top P(\widehat{C}+D\Psi)\\
&\quad +Q-\Psi^\top(R+D^\top PD)\Psi\\
&= P\widehat{A}+\widehat{A}^\top P+\widehat{C}^\top P\widehat{C}+PB\Psi+\Psi^\top B^\top P+\Psi^\top D^\top PD\Psi\\
&\quad +\widehat{C}^\top PD\Psi+\Psi^\top D^\top P\widehat{C}+Q-\Psi^\top R\Psi-\Psi^\top D^\top PD\Psi\\
&= P\widehat{A}+\widehat{A}^\top P+\widehat{C}^\top P\widehat{C}+(PB+\widehat{C}^\top PD)\Psi\\
&\quad +\Psi^\top(B^\top P+D^\top P\widehat{C})+Q-\Psi^\top R\Psi.
\end{aligned}
\tag{7.8}
$$

Note that

$$
\begin{aligned}
(PB+\widehat{C}^\top PD)\Psi &= (PB+C^\top PD+\Psi^\top D^\top PD)\Psi\\
&= \big[\Psi^\top(R+D^\top PD)-S^\top-\Psi^\top D^\top PD\big]\Psi\\
&= \Psi^\top R\Psi - S^\top\Psi.
\end{aligned}
$$

Plugging this into (7.8), we obtain (7.5). Since the procedure is reversible, we obtain the equivalence.

Next, we construct the following iterative scheme. For $i=0,1,2,\ldots$, set

$$
\begin{cases}
P_0 \equiv G,\\
\Psi_i = (R+D^\top P_iD)^{-1}(B^\top P_i+S+D^\top P_iC),\\
\widehat{A}_i = A-B\Psi_i, \quad \widehat{C}_i = C-D\Psi_i,\\
\widehat{Q}_i = (\Psi_i-R^{-1}S)^\top R(\Psi_i-R^{-1}S)+Q-S^\top R^{-1}S,
\end{cases}
\tag{7.9}
$$

and let P_{i+1} be the solution of

$$
\begin{cases}
\dot{P}_{i+1}+P_{i+1}\widehat{A}_i+\widehat{A}_i^\top P_{i+1}+\widehat{C}_i^\top P_{i+1}\widehat{C}_i+\widehat{Q}_i=0,\\
P_{i+1}(T)=G.
\end{cases}
\tag{7.10}
$$

By Lemma 7.3 and condition (4.23), we see that

$$
P_i(t)\ge 0, \qquad \forall t\in[0,T], \quad i\ge 1.
\tag{7.11}
$$

Hence, the above iterative scheme can be processed. We now claim that

$$
P_i(t)\ge P_{i+1}(t), \qquad \forall t\in[0,T], \quad i\ge 0.
\tag{7.12}
$$

To show this, define $\Delta_i \stackrel{\Delta}{=} P_i - P_{i+1}$ and $\Lambda_i \stackrel{\Delta}{=} \Psi_i - \Psi_{i-1}$. Subtracting the $(i+1)$th equation from the ith equation for (7.10), we get

$$
\begin{aligned}
-\dot{\Delta}_i &= \Delta_i\widehat{A}_i+\widehat{A}_i^\top\Delta_i+\widehat{C}_i^\top\Delta_i\widehat{C}_i+P_i(\widehat{A}_{i-1}-\widehat{A}_i)\\
&\quad +(\widehat{A}_{i-1}-\widehat{A}_i)^\top P_i+\widehat{C}_{i-1}^\top P_i\widehat{C}_{i-1}+\widehat{Q}_{i-1}-\widehat{Q}_i.
\end{aligned}
\tag{7.13}
$$

By (7.9), we have the following:

$$\begin{aligned}&\widehat{A}_{i-1}-\widehat{A}_i=-B\Lambda_i,\qquad \widehat{C}_{i-1}-\widehat{C}_i=-D\Lambda_i,\\&\widehat{C}_{i-1}^{\top}P_i\widehat{C}_{i-1}-\widehat{C}_i^{\top}P_i\widehat{C}_i=\Lambda_i^{\top}D^{\top}P_iD\Lambda_i-\widehat{C}_i^{\top}P_iD\Lambda_i-\Lambda_i^{\top}D^{\top}P_i\widehat{C}_i,\\&\widehat{Q}_{i-1}-\widehat{Q}_i=\Lambda_i^{\top}R\Lambda_i+(\Psi_i-R^{-1}S)^{\top}R\Lambda_i+\Lambda_i^{\top}R(\Psi_i-R^{-1}S).\end{aligned}$$

Thus, plugging the above into (7.13) yields

$$\begin{aligned}&-[\dot{\Delta}_i+\Delta_i\widehat{A}_i+\widehat{A}_i^{\top}\Delta_i+\widehat{C}_i^{\top}\Delta_i\widehat{C}_i]\\&=-P_iB\Lambda_i-\Lambda_i^{\top}B^{\top}P_i+\Lambda_i^{\top}D^{\top}P_iD\Lambda_i-\widehat{C}_i^{\top}P_iD\Lambda_i-\Lambda_i^{\top}D^{\top}P_i\widehat{C}_i\\&\quad+\Lambda_i^{\top}R\Lambda_i+(\Psi_i-R^{-1}S)^{\top}R\Lambda_i+\Lambda_i^{\top}R(\Psi_i-R^{-1}S)\\&=\Lambda_i^{\top}(R+D^{\top}P_iD)\Lambda_i-\Lambda_i^{\top}(B^{\top}P_i+D^{\top}P_i\widehat{C}_i+S-S\Psi_i)\\&\quad-(P_iB+\widehat{C}_i^{\top}P_i+S^{\top}-\Psi_i^{\top}R)\Psi_i\\&=\Lambda_i^{\top}(R+D^{\top}P_iD)\Lambda_i\geq 0.\end{aligned}\tag{7.14}$$

Noting that $\Delta_i(T)=0$, we can apply Lemma 7.3 again to obtain our claim (7.12). Thus $\{P_i\}$ is a decreasing sequence in $C([0,T];\mathcal{S}_+^n)$ and therefore has a limit (with respect to the max norm of $C([0,T];\mathcal{S}_+^n)$), denoted by P. Clearly, P is the solution to (7.5), hence (6.6). $\square$

Equations (7.9)–(7.10) actually give a numerical algorithm to compute the solution of the stochastic Riccati equation (6.6). The following proposition gives an estimate for the convergence speed of this algorithm.

Proposition 7.4. *Under the same assumptions of Theorem 7.2, let $\{P_i\}\subseteq C([0,T];\mathcal{S}_+^n)$ be constructed by the algorithms (7.9)–(7.10) and let P be the solution to the stochastic Riccati equation (6.6). Then*

$$|P_i(t)-P(t)|\leq K\sum_{j=i}^{\infty}\frac{c^{j-2}}{(j-2)!}(T-t)^{j-2},\quad i=2,3,\ldots,\tag{7.15}$$

where $K,c>0$ are constants that depend only on the coefficients of (6.6).

Proof. Note that $\Delta_i\equiv P_i-P_{i+1}$ satisfies (7.14). Set $\widehat{R}_i\triangleq R+D^{\top}P_iD$. Then

$$\begin{aligned}\Lambda_i&=\Psi_i-\Psi_{i-1}\\&=\widehat{R}_{i-1}^{-1}D^{\top}\Delta_{i-1}D\widehat{R}_i^{-1}(B^{\top}P_i+S+D^{\top}P_iC)\\&\quad-\widehat{R}_{i-1}^{-1}(B^{\top}\Delta_{i-1}+D^{\top}\Delta_{i-1}C).\end{aligned}\tag{7.16}$$

Equation (7.14), together with $\Delta_i(T)=0$, implies that

$$\Delta_i(t)=\int_t^T[\Delta_i\widehat{A}_i+\widehat{A}_i^{\top}\Delta_i+\widehat{C}_i^{\top}\Delta_i\widehat{C}_i+\Lambda_i^{\top}\widehat{R}_i\Lambda_i](s)ds.\tag{7.17}$$

Putting (7.16) into (7.17) and noting the uniform boundedness of the sequences $\{|P_i|\}, \{|\widehat{R}_i|\}$, and $\{|\widehat{R}_i^{-1}|\}$ (by the proof of Theorem 7.2), we get

$$|\Delta_i(t)| \leq K \int_t^T [|\Delta_{i-1}(s)| + |\Delta_i(s)|]ds. \tag{7.18}$$

Define $v_i(t) = \int_t^T |\Delta_i(s)|ds$. Then (7.18) reduces to

$$\dot{v}_i(t) + Kv_i(t) + Kv_{i-1}(t) \geq 0,$$

which implies

$$v_i(t) \leq Ke^{KT} \int_t^T v_{i-1}(s)ds \equiv c \int_t^T v_{i-1}(s)ds.$$

By induction, we deduce that

$$v_{i+1}(t) \leq \frac{c^i}{i!}(T-t)^i v_1(0).$$

It then follows from (7.18) that

$$|\Delta_i(t)| \leq K \Big\{ \frac{c^{i-1}}{(i-1)!}(T-t)^{i-1} + \frac{c^{i-2}}{(i-2)!}(T-t)^{i-2} \Big\} v_1(0).$$

This easily yields (7.15). $\square$

7.2. Existence: The case $C = 0$, $S = 0$, and $Q, G \geq 0$

In this subsection we show the global existence of the stochastic Riccati equation (6.6) for the case $C = 0$, $S = 0$, and $Q, G \geq 0$. Note that in this case R still could be indefinite!

In the present case, (6.6) takes the following form:

$$\begin{cases} \dot{P} + PA + A^\top P + Q - PB(R + D^\top PD)^{-1}B^\top P = 0, \quad \text{a.e. } t \in [0,T], \\ P(T) = G, \\ R(t) + D(t)^\top P(t)D(t) \geq 0, \quad \text{a.e. } t \in [0,T]. \end{cases} \tag{7.19}$$

To study this equation, we consider the following (conventional deterministic) Riccati equation (see (2.34) with $S = 0$):

$$\begin{cases} \dot{P} + PA + A^\top P + Q - PB\widehat{R}^{-1}B^\top P = 0, \quad \text{a.e. } t \in [0,T], \\ P(T) = G, \end{cases} \tag{7.20}$$

where $\widehat{R}(\cdot)$ is fixed. Recall that the mapping

$$\Gamma : \widehat{\mathcal{R}} \equiv \{\widehat{R} \in L^\infty(0,T;\widehat{\mathcal{S}}^k_+) \mid \widehat{R}^{-1} \in L^\infty(0,T;\widehat{\mathcal{S}}^k_+)\} \to C([0,T];\mathcal{S}^n_+)$$

is the solution operator of (7.20), namely, $\Gamma(\widehat{R})$ is the solution to (7.20) corresponding to $\widehat{R}$.

The solvability of the stochastic Riccati equation (6.6) can be studied via Γ, as shown in the following theorem.

Theorem 7.5. *Let (L1) and (L2) hold. Then the following are equivalent:*

(i) *The stochastic Riccati equation (7.19) admits a solution P.*

(ii) *There exist $\widehat{R}^+, \widehat{R}^- \in C([0,T]; \widehat{\mathcal{S}}^k_+)$ such that*

$$\widehat{R}^+ \geq R + D^\top \Gamma(\widehat{R}^+) D \geq R + D^\top \Gamma(\widehat{R}^-) D \geq \widehat{R}^-. \tag{7.21}$$

(iii) *There exists an $\widehat{R} \in C([0,T]; \widehat{\mathcal{S}}^k_+)$ such that*

$$R = \widehat{R} - D^\top \Gamma(\widehat{R}) D. \tag{7.22}$$

Proof. (i) $\Rightarrow$ (ii). When (7.19) admits a solution P, by taking

$$\widehat{R}^+ \equiv \widehat{R}^- \triangleq R + D^\top P D,$$

we obtain (7.21) (with all inequalities being equalities).

(ii) $\Rightarrow$ (iii). Let $\widehat{R}^+, \widehat{R}^-$ be given with (7.21) satisfied. Define sequences $\{\widehat{R}_i^+\}_{i\geq 0}$, $\{\widehat{R}_i^-\}_{i\geq 0}$, $\{P_i^+\}_{i\geq 0}$, and $\{P_i^-\}_{i\geq 0}$ iteratively as follows

$$\begin{cases} \widehat{R}_0^+ = \widehat{R}^+, \quad \widehat{R}_0^- = \widehat{R}^-, \quad P_0^+ = \Gamma(\widehat{R}_0^+), \quad P_0^- = \Gamma(\widehat{R}_0^-); \\ \widehat{R}_{i+1}^+ = R + D^\top P_i^+ D, \quad \widehat{R}_{i+1}^- = R + D^\top P_i^- D, \\ P_{i+1}^+ = \Gamma(\widehat{R}_{i+1}^+), \quad P_{i+1}^- = \Gamma(\widehat{R}_{i+1}^-), \qquad i = 0,1,2,\dots . \end{cases} \tag{7.23}$$

By (7.21), we have

$$\widehat{R}_0^+ \geq \widehat{R}_1^+ \geq \widehat{R}_1^- \geq \widehat{R}_0^- > 0.$$

Since Γ is increasing (Proposition 2.12), we also have

$$P_0^+ \geq P_1^+ \geq P_1^- \geq P_0^- \geq 0.$$

By (7.23), we easily obtain

$$\begin{cases} P_0^+ \geq P_i^+ \geq P_{i+1}^+ \geq P_{i+1}^- \geq P_i^- \geq P_0^- \geq 0, \\ \widehat{R}_0^+ \geq \widehat{R}_i^+ \geq \widehat{R}_{i+1}^+ \geq \widehat{R}_{i+1}^- \geq \widehat{R}_i^- \geq \widehat{R}_0^- > 0, \end{cases} \qquad i = 1,2,\dots . \tag{7.24}$$

From the above, we see that $\widehat{R}_i^+ \in \widehat{\mathcal{R}}$ and there exist $\widehat{R}^+ \in \widehat{\mathcal{R}}$ and P^+ such that

$$\lim_{i\to\infty} \widehat{R}_i^+ = \widehat{R}^+, \qquad \lim_{i\to\infty} P_i^+ = P^+. \tag{7.25}$$

By the continuity of Γ (Proposition 2.12), we have

$$P^+ = \lim_{i\to\infty} P_i^+ = \lim_{i\to\infty} \Gamma(\widehat{R}_i^+) = \Gamma(\lim_{i\to\infty} \widehat{R}_i^+) = \Gamma(\widehat{R}^+).$$

Taking limits in (7.23), we obtain $\widehat{R}^+ = R + D^\top P^+ D = R + D^\top \Gamma(R^+) D$.

(iii) $\Rightarrow$ (i). If such $\widehat{R}$ exists, then $P \stackrel{\Delta}{=} \Gamma(\widehat{R})$ is the solution of the stochastic Riccati equation (7.19). □

By the same argument as above, the limits

$$P^- \stackrel{\Delta}{=} \lim_{i\to\infty} P_i^-, \quad \widehat{R}^- \stackrel{\Delta}{=} \lim_{i\to\infty} \widehat{R}_i^-$$

exist as well, and P^- is also a solution of (7.19). In view of the uniqueness of the solutions, $P^- = P^+$.

The above result suggests an algorithm for computing the solution of the stochastic Riccati equation (7.19).

Proposition 7.6. *Let (L1), (L2), and (7.21) hold. Let the sequence $\{P_i\} \subseteq C([0,T]; \mathcal{S}^n)$ be constructed by the algorithms*

$$\widehat{R}_0 = \widehat{R}^+, \quad P_i = \Gamma(\widehat{R}_i), \quad \widehat{R}_{i+1} = R + D^\top P_i D, \quad i = 0,1,2,\ldots, \tag{7.26}$$

or

$$\widehat{R}_0 = \widehat{R}^-, \quad P_i = \Gamma(\widehat{R}_i), \quad \widehat{R}_{i+1} = R + D^\top P_i D, \quad i = 0,1,2,\ldots. \tag{7.27}$$

Let P be the solution to the stochastic Riccati equation (7.19). Then

$$|P_i(t) - P(t)| \le K \sum_{j=i}^{\infty} \frac{c^{j-2}}{(j-2)!}(T-t)^{j-2}, \quad i = 2,3,\ldots, \tag{7.28}$$

where $K, c > 0$ are constants that depend only on the coefficients of (7.19).

Proof. We prove only the estimate (7.28) for the algorithm (7.26). The other one is the same. By definition,

$$P_i(t) = G - \int_t^T [P_i A + A^\top P_i - P_i B \widehat{R}_i^{-1} B^\top P_i + Q](s)ds.$$

Set $\widehat{P}_i \stackrel{\Delta}{=} P_{i+1} - P_i$. Then

$$\begin{aligned}\widehat{P}_i(t) = -\int_t^T & [\widehat{P}_i(A - B\widehat{R}_{i+1}^{-1}B^\top P_{i+1}) + (A - B\widehat{R}_i^{-1}B^\top P_i)^T \widehat{P}_i \\ & + P_i B \widehat{R}_{i+1}^{-1} D^\top \widehat{P}_{i-1} D \widehat{R}_i^{-1} B^\top P_{i+1}](s)ds.\end{aligned}$$

In view of (7.24), the sequences $\{|P_i|\}, \{|\widehat{R}_i|\}$ and $\{|\widehat{R}_i^{-1}|\}$ are bounded. Hence,

$$|\widehat{P}_i(t)| \le K \int_t^T [|\widehat{P}_{i-1}(s)| + |\widehat{P}_i(s)|]ds. \tag{7.29}$$

So we have obtained an inequality similar to (7.18), and the estimate (7.28) thus follows from the same argument in proving (7.15). □

Let us examine the key condition (7.21) again. The existence of $\widehat{R}^+$ is in fact guaranteed. To see this, put $\widehat{R}^+_\lambda = \lambda I$ with $\lambda > 0$. In view of Proposition 2.12, $\Gamma(\widehat{R}^+_\lambda)$ is bounded uniformly in $\lambda > 0$. Therefore,

$$\widehat{R}^+_\lambda \geq R + D^T\Gamma(\widehat{R}^+_\lambda)D$$

for λ sufficiently large. Hence we have the following result.

Theorem 7.7. *Let (L1) and (L2) hold. Then the stochastic Riccati equation (7.19) admits a solution if and only if there exist an $\widehat{R} \in C([0,T]; \widehat{\mathcal{S}}^k_+)$ such that*

$$R + D^\top\Gamma(\widehat{R})D \geq \widehat{R}. \tag{7.30}$$

This theorem says that while R can be indefinite (or negative definite) for the stochastic Riccati equation to have solutions, it cannot be *too* negative. Indeed, in any case, R cannot be smaller than $\inf_{\widehat{R}\in C([0,T];\hat{\mathcal{S}}^k_+)} \big[\widehat{R} - D^T\Gamma(\widehat{R})D\big]$!

In the special case where R is positive definite, i.e., $R \in C([0,T]; \widehat{\mathcal{S}}^k_+)$, by Corollary 2.10, $\Gamma(\widehat{R}) \geq 0$ for all $\widehat{R} \in C([0,T]; \widehat{\mathcal{S}}^k_+)$. By choosing $\widehat{R} = \frac{1}{2}R$, we have (7.30). This means that if $R \in C([0,T]; \widehat{\mathcal{S}}^k_+)$, then the stochastic Riccati equation admits a unique solution $P \in C([0,T]; \mathcal{S}^n_+)$. So for the case $R > 0$, $C = 0$, $S = 0$, and $Q, G \geq 0$, Theorem 7.7 matches Theorem 7.2.

Theorem 7.7 is very useful in estimating intervals on which (7.19) is solvable without directly dealing with (7.19) itself. Let us give an example to demonstrate.

Example 7.8. Consider the control system

$$\begin{cases} dx(t) = u(t)dt + u(t)dW(t), & t \in [s,1], \\ x(s) = y, \end{cases} \tag{7.31}$$

with the cost functional

$$J(s,y;u(\cdot)) = E\Big\{ -\frac{1}{2}\int_s^1 ru(t)^2dt + \frac{1}{2}x(1)^2\Big\}, \tag{7.32}$$

where $r > 0$. By Corollary 5.2, we know that in order for the LQ problem to be finite, it is necessary that $r \leq 1$. Let us assume $0 < r < 1$ below. The corresponding stochastic Riccati equation reads

$$\begin{cases} \dot{P}(t) = \dfrac{P(t)^2}{P(t) - r}, \\ P(1) = 1, \\ P(t) - r > 0. \end{cases} \tag{7.33}$$

By Theorem 7.7, (7.33) admits a solution on some interval $[s,1]$ $(0 \le s < 1)$ if and only if there is a positive continuous function $\widehat{R}(\cdot)$ such that

$$-r + \Gamma(\widehat{R})(t) \ge \widehat{R}(t), \quad \forall t \in [s,1], \tag{7.34}$$

where $\Gamma(\widehat{R})$ is the solution of the conventional Riccati equation

$$\dot{p}(t) = \frac{p(t)^2}{\widehat{R}(t)}, \quad p(1) = 1.$$

So $\Gamma(\widehat{R})(t) = \left(1 + \int_t^1 \frac{1}{\widehat{R}(\tau)} d\tau\right)^{-1}$. Setting $f(t) \stackrel{\Delta}{=} \frac{1}{\widehat{R}(1-t)}$, one can rewrite (7.34) as

$$1 + \int_0^{1-t} f(\tau)d\tau \le \frac{f(1-t)}{1+rf(1-t)}, \quad \forall t \in [s,1],$$

or

$$\int_0^t f(\tau)d\tau \le \frac{(1-r)f(t)-1}{1+rf(t)}, \quad \forall t \in [0,1-s]. \tag{7.35}$$

Since $f(t) > 0$ and $r \in (0,1)$, it is then necessary that $f(t) > \frac{1}{1-r}$ for $t \in [0,1-s]$. Putting

$$f_1(t) \stackrel{\Delta}{=} f(t) - \frac{1}{1-r} > 0, \quad \forall t \in [0,1-s],$$

and substituting into (7.35), we obtain

$$\int_0^t f_1(\tau)d\tau \le \frac{(1-r)^2 f_1(t)}{1+r(1-r)f_1(t)} - \frac{t}{1-r}, \quad \forall t \in [0,1-s].$$

Hence f_1 must satisfy

$$(1-r)[(1-r)^2 - rt]f_1(t) > t, \quad \forall t \in [0,1-s].$$

It follows that

$$(1-r)^2 - rt > 0, \quad \forall t \in [0,1-s]. \tag{7.36}$$

The above inequality gives an estimate on the interval where the stochastic Riccati equation (7.33) admits a solution, namely,

$$1 - s < \left(\frac{1}{\sqrt{r}} - \sqrt{r}\right)^2.$$

In particular, in order for (7.23) to be solvable on the whole interval $[0,1]$, it must hold that

$$1 < \left(\frac{1}{\sqrt{r}} - \sqrt{r}\right)^2, \text{ or } r < \frac{3-\sqrt{5}}{2}.$$

Therefore, the weight of the control in the cost functional is allowed to be negative, but it should not be *too* negative.

7.3. Existence: One-dimensional case

As evident from the previous discussions, the solvability of (6.6) is in general very complicated. In fact, we have treated only some very special cases. In this section, we look at the case where both the state and control variables are one-dimensional with constant coefficients. For such a case, we can solve the stochastic Riccati equation (6.6) completely.

Note that the case $D = 0$ reduces (6.6) to the conventional Riccati equation (for deterministic LQ problems). Thus, let us assume that $D = 1$, after scaling. Then we can write the stochastic Riccati equation (6.6) as follows:

$$\begin{cases} \dot{P} = -(2A + C^2)P - Q + \dfrac{[S + (B + C)P]^2}{P + R}, \\ P(T) = G, \\ P(t) + R > 0. \end{cases} \tag{7.37}$$

By a change of variable,

$$y(t) \stackrel{\Delta}{=} P(T - t) + R, \qquad t \le T, \tag{7.38}$$

equation (7.37) further reduces to

$$\begin{cases} \dot{y} = \dfrac{\alpha y^2 + \beta y + \gamma}{y}, & t \ge 0, \\ y(0) = g, \\ y(t) > 0, \qquad t \ge 0, \end{cases} \tag{7.39}$$

where

$$\begin{cases} \alpha \stackrel{\Delta}{=} (2A + C^2) - (B + C)^2, \\ \beta \stackrel{\Delta}{=} Q - R\big[(2A + C^2) - 2(B + C)^2\big] - 2S(B + C), \\ \gamma \stackrel{\Delta}{=} -[R(B + C) - S]^2 \le 0, \\ g \stackrel{\Delta}{=} G + R > 0. \end{cases} \tag{7.40}$$

The last (strict) inequality in (7.40), which is necessary for any solution to (7.37) and is an analogue of (6.8), will be assumed in the sequel. It is easy to see that (7.37) admits a solution $P(\cdot)$ on $[s, T]$ (with $s \in [0, T]$) if and only if (7.39) admits a solution $y(\cdot)$ on $[0, T - s]$.

In what follows, we let

$$\theta \stackrel{\Delta}{=} \sup\{t > 0 \mid (7.39) \text{ admits a solution on } [0, t]\}. \tag{7.41}$$

Then θ will determine the maximal interval in which (7.37) is solvable (backward from T), and Problem (SLQ) is solvable *through the stochastic Riccati equation.* The following result gives the explicit description of θ in terms of α, β, γ, and g defined by (7.40).

Theorem 7.9. *Let all the coefficients A, B, C, D, Q, S, R be time-invariant with $n = k = 1$, and let α, β, γ, g be defined by (7.40). Then the following hold:*

(i) *If $\beta = \gamma = 0$, then $\theta = +\infty$.*

(ii) *If $\beta = 0$ and $\gamma \neq 0$, then*

$$\theta < +\infty \quad \Longleftrightarrow \quad \gamma < -\alpha^+ g^2. \tag{7.42}$$

In this case,

$$\theta = \begin{cases} \dfrac{g^2}{2|\gamma|}, & \alpha = 0, \\ \dfrac{1}{2\alpha} \ln \dfrac{\gamma}{\alpha g^2 + \gamma}, & \alpha \neq 0. \end{cases} \tag{7.43}$$

(iii) *If $\beta \neq 0$ and $\gamma = 0$, then*

$$\theta < +\infty \quad \Longleftrightarrow \quad \beta < -\alpha^+ g. \tag{7.44}$$

In this case,

$$\theta = \begin{cases} \dfrac{g}{|\beta|}, & \alpha = 0, \\ \dfrac{1}{\alpha} \ln \dfrac{\beta}{\alpha g + \beta}, & \alpha \neq 0. \end{cases} \tag{7.45}$$

(iv) *If $\alpha = 0$ and $\beta\gamma \neq 0$, then*

$$\theta < +\infty \quad \Longleftrightarrow \quad \beta < 0. \tag{7.46}$$

In this case,

$$\theta = \frac{|\gamma|}{\beta^2} \Big\{ \frac{\beta g}{\gamma} - \ln\Big(1 + \frac{\beta g}{\gamma}\Big) \Big\}. \tag{7.47}$$

(v) *If $\alpha\beta\gamma \neq 0$, then let $\Delta = \beta^2 - 4\alpha\gamma$, and one of the following three situations holds:*

(a) *if $\Delta > 0$, then $\theta < +\infty$ if and only if one of the following holds,*

$$\begin{cases} \alpha > 0, \ \beta < \sqrt{\Delta} - 2\alpha g, \\ \alpha < 0, \ \beta \notin \big[0, (2|\alpha| g - \sqrt{\Delta})^+\big], \end{cases} \tag{7.48}$$

and in either of the above two cases,

$$\theta = \frac{1}{\sqrt{\Delta}} \Big\{ y_+ \ln \frac{y_+}{y_+ - g} - y_- \ln \frac{y_-}{y_- - g} \Big\}, \tag{7.49}$$

where

$$y_\pm = \frac{-\beta \pm \sqrt{\Delta}}{2\alpha};$$

(b) *if* $\Delta = 0$, *then*

$$\theta < +\infty \quad \Longleftrightarrow \quad \beta \notin [0, 2|\alpha| g], \tag{7.50}$$

and in this case,

$$\theta = \frac{1}{\alpha}\Big\{ \ln \frac{\beta}{2\alpha g + \beta} + \frac{2\alpha g}{2\alpha g + \beta} \Big\}. \tag{7.51}$$

(c) *if* $\Delta < 0$, *then one always has* $\theta < +\infty$ *and*

$$\begin{aligned} \theta = \frac{1}{2\alpha}\Big\{ &\ln \frac{\gamma}{\alpha g^2 + \beta g + \gamma} \\ &- \frac{2\beta}{\sqrt{|\Delta|}}\Big[\tan^{-1} \frac{\beta}{\sqrt{|\Delta|}} - \tan^{-1} \frac{2\alpha g + \beta}{\sqrt{|\Delta|}} \Big] \Big\}. \end{aligned} \tag{7.52}$$

Proof. The central idea for all the cases is to find the first time θ at which the solution $y(\cdot)$ changes its sign from positive to negative, i.e., the first time θ when $y(\theta) = 0$.

(i) When $\beta = \gamma = 0$, (7.39) becomes

$$\dot y = \alpha y, \quad y(0) = g, \quad y(t) > 0,$$

whose solution is $y(t) = e^{\alpha t} g$. Thus, $\theta = +\infty$.

(ii) When $\beta = 0$ and $\gamma \neq 0$, (7.39) becomes

$$\frac{d}{dt}[y^2] = 2\alpha y^2 + 2\gamma, \quad y(0)^2 = g^2. \tag{7.53}$$

Thus,

$$y(t)^2 = \begin{cases} 2\gamma t + g^2, & \text{if } \alpha = 0, \\ e^{2\alpha t}\Big(g^2 + \dfrac{\gamma}{\alpha}\Big) - \dfrac{\gamma}{\alpha}, & \text{if } \alpha \neq 0. \end{cases} \tag{7.54}$$

Note that $\gamma \le 0$. Hence, (7.42) holds, and (7.43) follows from (7.54) (by setting $t = \theta$ and $y(\theta) = 0$ and then solving for θ).

(iii) When $\beta \neq 0$ and $\gamma = 0$, (7.39) becomes

$$\dot y = \alpha y + \beta, \quad y(0) = g,$$

which is similar to (7.53). Thus as in the proof of (ii), we obtain (7.44) and (7.45).

(iv) When $\alpha = 0$ and $\beta\gamma \neq 0$, (7.39) becomes

$$\dot y = \frac{\beta y + \gamma}{y}, \quad y(0) = g. \tag{7.55}$$

Thus,

$$y(t) - g - \frac{\gamma}{\beta} \ln \Big| \frac{\beta y(t) + \gamma}{\beta g + \gamma} \Big| = \beta t, \qquad t \in [0, \theta]. \tag{7.56}$$

From (7.55), we can easily show that

$$\begin{cases} \beta g+\gamma\ge 0 \quad\Rightarrow\quad y(t)\ge g>0, \qquad \forall t\ge 0,\\ \beta g+\gamma<0 \quad\Rightarrow\quad y(t)\downarrow, \quad \text{as } t\uparrow.\end{cases}$$

Thus, in order to have $\theta<+\infty$, we first need $\beta g+\gamma<0$. Then, in (7.56), setting $t=\theta$ and $y(\theta)=0$, we have

$$\theta=-\frac{1}{\beta}\Big[g+\frac{\gamma}{\beta}\ln\frac{\gamma}{\beta g+\gamma}\Big]=-\frac{\gamma}{\beta^2}\Big[\frac{\beta g}{\gamma}-\ln\Big(1+\frac{\beta g}{\gamma}\Big)\Big]. \tag{7.57}$$

Since $\gamma<0$, θ is positive if and only if

$$\frac{\beta g}{\gamma}-\ln\Big(1+\frac{\beta g}{\gamma}\Big)>0,$$

which is equivalent to $\frac{\beta g}{\gamma}>0$, namely $\beta<0$. This together with (7.57) yields (7.46)–(7.47).

(v) We now consider the case where $\alpha\beta\gamma\ne 0$.

(a) If $\Delta>0$, then $\alpha y^2+\beta y+\gamma=0$ has two distinct real roots $y_\pm$. In this case, we can rewrite the equation in (7.39) as

$$\dot y=\frac{\alpha}{y}(y-y_-)(y-y_+). \tag{7.58}$$

Now, if $\alpha>0$, we have $y_-y_+=\frac{\gamma}{\alpha}<0$, which leads to

$$y_-<0<y_+.$$

Thus, from (7.58), it follows that $\theta<+\infty$ if and only if (recall $g>0$)

$$g<y_+=\frac{-\beta+\sqrt{\Delta}}{2\alpha},$$

which gives the first case in (7.48). Next, for $\alpha<0$, we have $y_-y_+=\frac{\gamma}{\alpha}>0$. Note that $y_-+y_+=\frac{\beta}{-\alpha}=\frac{\beta}{|\alpha|}$, Thus, if $\beta<0$, we have

$$y_-<y_+<0.$$

By (7.58) again, noting $\alpha<0$, we obtain that $\theta<+\infty$. On the other hand, if $\beta>0$, then

$$0<y_-<y_+.$$

Hence, $\theta<+\infty$ if and only if

$$g<y_-=\frac{-\beta-\sqrt{\Delta}}{2\alpha}.$$

Combining the above, we obtain the second case in (7.48). Now we return to (7.58). In the present case, a direct computation shows that the solution $y(\cdot)$ satisfies

$$\alpha(y_+ - y_-)t = -y_- \ln\Big(\frac{y(t) - y_-}{g - y_-}\Big) + y_+ \ln\Big(\frac{y(t) - y_+}{g - y_+}\Big).$$

Then, setting $t = \theta$ and $y(\theta) = 0$ and noting $\alpha(y_+ - y_-) = \sqrt{\Delta}$, we obtain (7.49).

(b) If $\Delta = 0$, then $\alpha = \frac{\beta^2}{4\gamma} < 0$ and $\alpha y^2 + \beta y + \gamma = 0$ has only one real root $y_0 = \frac{\beta}{-2\alpha} = \frac{\beta}{2|\alpha|}$ of multiplicity 2. Thus, if $\beta < 0$, then $y_0 < 0$ and one always has $\theta < +\infty$. On the other hand, when $\beta > 0$, $y_0 > 0$, and in order to have $\theta < +\infty$, we need $g < y_0 \equiv \frac{\beta}{2|\alpha|}$, which yields (7.50). In this case, solving (7.39) gives

$$\ln\frac{y(t) - y_0}{g - y_0} + \frac{y_0(y(t) - g)}{(y(t) - y_0)(g - y_0)} = \alpha t.$$

Then (7.51) can be obtained by setting $t = \theta$ and $y(\theta) = 0$ in the above.

(c) Finally, if $\Delta < 0$, then $\alpha < 0$ and $\alpha y^2 + \beta y + \gamma < 0$ for any y. Thus, we always have $\theta < +\infty$. By solving (7.39) in the present case, we have

$$\begin{aligned} 2\alpha t = &\ln\Big(\frac{\alpha y(t)^2 + \beta y(t) + \gamma}{\alpha g^2 + \beta g + \gamma}\Big) \\ &- \frac{2\beta}{\sqrt{|\Delta|}}\Big\{\tan^{-1}\frac{2\alpha y(t) + \beta}{\sqrt{|\Delta|}} - \tan^{-1}\frac{2\alpha g + \beta}{\sqrt{|\Delta|}}\Big\}. \end{aligned}$$

Hence, we can obtain (7.52) similarly as before. $\square$

Let

(7.59) $\quad \bar{s} \stackrel{\Delta}{=} \inf\{s \in [0, T] \mid (7.37) \text{ admits a unique solution on } [s, T]\}.$

Then $\bar{s} = (T - \theta)^+$, and for any $s \in (\bar{s}, T]$, the stochastic Riccati equation (7.37) is solvable on $[s, T]$, which leads to the unique solvability of Problem (SLQ) at s.

So far we have discussed the stochastic LQ problems and the associated stochastic Riccati equations for the case where the Brownian motion involved in (3.1) is one-dimensional. As we mentioned, there is no essential difficulty in treating the multidimensional Brownian motion case. For the reader's convenience, we state the most important results for the multidimensional case below.

First of all, associated with the dynamics (3.1), the corresponding stochastic Riccati equation (6.6) as well as (6.7) become

$$
(7.60)\quad \begin{cases}
\dot P + PA + A^\top P + \sum_{j=1}^m C_j^\top P C_j + Q - (B^\top P + S + \sum_{j=1}^m D_j^\top P C_j)^\top \\
\qquad \cdot (R + \sum_{j=1}^m D_j^\top P D_j)^{-1} (B^\top P + S + \sum_{j=1}^m D_j^\top P C_j) = 0, \\
\qquad\qquad\qquad\qquad \text{a.e. } t \in [s,T], \\
P(T) = G, \\
R(t) + \sum_{j=1}^m D_j(t)^\top P(t) D_j(t) > 0, \qquad \text{a.e. } t \in [s,T],
\end{cases}
$$

and

$$
(7.61)\quad \begin{cases}
\dot\varphi + [A - B(R + \sum_{j=1}^m D_j^\top P D_j)^{-1}(B^\top P + S + \sum_{j=1}^m D_j^\top P C_j)]^\top \varphi \\
\quad + \sum_{j=1}^m \Big[C_j - D_j (R + \sum_{i=1}^m D_i^\top P D_i)^{-1} (B^\top P + S + \sum_{i=1}^m D_i^\top P C_i)\Big]^\top P\sigma_j \\
\quad + Pb = 0, \qquad \text{a.e. } t \in [s,T], \\
\varphi(T) = 0.
\end{cases}
$$

The following main result is an analogue of Theorem 6.1.

Theorem 7.10. *Let (L1) hold. Let $P(\cdot) \in C([s,T]; \mathcal{S}^n)$ and $\varphi(\cdot) \in C([s,T]; \mathbb{R}^n)$ be the solutions of (7.60) and (7.61), respectively, for some $s \in [0,T)$ such that*

$$
(7.62)\quad \begin{cases}
B\Psi, D\Psi \in L^\infty(s,T; \mathbb{R}^{n\times n}), \ \text{where} \\
\Psi \stackrel{\Delta}{=} (R + \sum_{j=1}^m D_j^\top P D_j)^{-1} [B^\top P + S + \sum_{j=1}^m D_j^\top P C_j],
\end{cases}
$$

and

$$
(7.63)\quad \begin{cases}
B\psi, D\psi \in L^2(s,T; \mathbb{R}^n), \ \text{where} \\
\psi \stackrel{\Delta}{=} (R + \sum_{j=1}^m D_j^\top P D_j)^{-1} [B^\top \varphi + \sum_{j=1}^m D_j^\top P \sigma_j].
\end{cases}
$$

Then Problem (SLQ) is solvable at s with the optimal control $\bar u(\cdot)$ being of a state feedback form:

$$
(7.64)\qquad \bar u(t) = -\Psi(t) x(t) - \psi(t), \qquad t \in [s,T],
$$

and

$$V(s,y) = \frac{1}{2}\langle P(s)y, y\rangle + \langle \varphi(s), y\rangle + \frac{1}{2}E\int_s^T \{ -|(R+\sum_{j=1}^m D_j^\top P D_j)^{\frac{1}{2}}\psi|^2 + 2\langle \varphi, b\rangle + \sum_{j=1}^m \langle P\sigma_j, \sigma_j\rangle \}dt, \quad \forall y \in \mathbb{R}^n. \tag{7.65}$$

The multidimensional versions of the other results of Sections 6–7 can be stated similarly. The details are left to the reader.

8. A Mean–Variance Portfolio Selection Problem

In this section we discuss a portfolio optimization problem that can be eventually formulated as a stochastic LQ problem and solved by the results presented in the preceding sections.

Portfolio selection seeks a best allocation (associated with the goal of an investor) of wealth among a basket of securities. Let us start with the model introduced in Chapter 2, Section 3.2, without consumption. Suppose there is a market in which $m+1$ assets (or securities) are traded continuously. One of the assets is the bond whose price process $P_0(t)$ is subject to the following (deterministic) ordinary differential equation:

$$\begin{cases} dP_0(t) = r(t)P_0(t)dt, & t \in [0,T], \\ P_0(0) = p_0 > 0, \end{cases} \tag{8.1}$$

where $r(t) > 0$ is the interest rate of the bond. The other m assets are stocks whose price processes $P_1(t), \ldots, P_m(t)$ satisfy the stochastic differential equation

$$\begin{cases} dP_i(t) = P_i(t)\{b_i(t)dt + \sum_{j=1}^m \sigma_{ij}(t)dW^j(t)\}, & t \in [0,T], \\ P_i(0) = p_i > 0, \end{cases} \tag{8.2}$$

where the deterministic function $b_i(t) > 0$ is the appreciation rate, and $\sigma_i(t) \stackrel{\Delta}{=} (\sigma_{i1}(t), \ldots, \sigma_{im}(t)) : [0,T] \to \mathbb{R}^m$ is the volatility or the dispersion of the stocks. Here, $W(t) \equiv (W^1(t), \ldots, W^m(t))^\top$ is a standard m-dimensional Brownian motion defined on some complete probability space $(\Omega, \mathcal{F}, P)$, and $\mathcal{F}_t = \sigma\{W(s) : s \le t\}$. Define the *variance matrix*

$$\sigma(t) \stackrel{\Delta}{=} \begin{pmatrix} \sigma_1(t) \\ \vdots \\ \sigma_m(t) \end{pmatrix} \equiv (\sigma_{ij}(t))_{m\times m}.$$

The basic assumptions throughout this section are

$$\sigma(t)\sigma(t)^\top \ge \delta I, \qquad \forall t \in [0,T], \tag{8.3}$$

for some $\delta > 0$, and

$$b_i(t) > r(t) > 0, \qquad \forall t \in [0,T], \quad i = 1,2,\ldots,m. \tag{8.4}$$

The first assumption is the so-called *nondegeneracy*, and the second one is a very natural assumption, since otherwise nobody is willing to invest in the risky stocks. We also assume that all the functions are measurable and uniformly bounded in t.

Consider an investor whose total wealth at time $t \geq 0$ is denoted by $x(t)$. Suppose he/she decides to hold $N_i(t)$ shares of the ith asset ($i = 0, 1, \ldots, m$) at time t. Then

$$x(t) = \sum_{i=0}^{m} N_i(t)P_i(t), \qquad t \geq 0. \tag{8.5}$$

Assume that the trading of shares takes place continuously and transaction cost and consumptions are not considered. Then one has (compare with Chapter 2, (3.12))

$$\begin{cases} dx(t) = \Big\{r(t)x(t) + \sum_{i=1}^{m} \big[b_i(t) - r(t)\big]u_i(t)\Big\}dt \\ \qquad\qquad + \sum_{j=1}^{m}\sum_{i=1}^{m} \sigma_{ij}(t)u_i(t)dW^j(t), \\ x(0) = x_0, \end{cases} \tag{8.6}$$

where

$$u_i(t) \stackrel{\Delta}{=} N_i(t)P_i(t), \qquad i = 0,1,2,\ldots,m, \tag{8.7}$$

denotes the total market value of the investor's wealth in the ith bond/stock. Note here that we allow short-selling, so that $u_i(t) < 0$ would be allowed. We call $u(t) = (u_1(t), \ldots, u_m(t))^\top$ a *portfolio* of the investor. Notice that we exclude the allocation to the bond, $u_0(t)$, from the portfolio, as it will be determined completely by the allocations to the stocks. The objective of the investor is to maximize the mean terminal wealth, $Ex(T)$, and at the same time to minimize the variance of the terminal wealth

$$\operatorname{Var} x(T) \equiv E\big[x(T) - Ex(T)\big]^2 \equiv Ex(T)^2 - \big[Ex(T)\big]^2. \tag{8.8}$$

This is a *multiobjective optimization* problem with two criteria in conflict.

Definition 8.1. A portfolio $u(\cdot)$ is said to be *admissible* if $u(\cdot) \in L^2_{\mathcal{F}}(0,T;\mathbb{R}^m)$. The *mean-variance portfolio optimization problem* is defined as

$$\begin{array}{ll} \text{Minimize} & \Big(J_1(u(\cdot)), J_2(u(\cdot))\Big) \stackrel{\Delta}{=} \Big(-Ex(T), \operatorname{Var} x(T)\Big), \\ \text{Subject to} & \begin{cases} u(\cdot) \in L^2_{\mathcal{F}}(0,T;\mathbb{R}^m), \\ (x(\cdot), u(\cdot)) \text{ satisfy equation (8.6).} \end{cases} \end{array} \tag{8.9}$$

Moreover, an admissible portfolio $\overline{u}(\cdot)$ is called an *efficient portfolio* if there exists no admissible portfolio $u(\cdot)$ such that

$$(8.10) \qquad J_1(u(\cdot)) \leq J_1(\overline{u}(\cdot)), \qquad J_2(u(\cdot)) \leq J_2(\overline{u}(\cdot)),$$

and at least one of the inequalities holds strictly. In this case, we call $(J_1(\overline{u}(\cdot)), J_2(\overline{u}(\cdot))) \in \mathbb{R}^2$ an *efficient point.* The set of all efficient points is called the *efficient frontier.*

In other words, an efficient portfolio is one for which there exists no better portfolio with respect to both the mean and variance criteria. The problem then is to identify all the efficient portfolios along with the efficient frontier.

By standard multiobjective convex optimization theory (see Zeleny [1]), an efficient portfolio can be found by solving a single-objective optimization problem where the objective is a weighted average of the two original criteria. The efficient frontier can then be generated by varying the weights. Therefore, the original problem can be solved via the following optimal control problem

$$(8.11) \qquad \begin{array}{ll} \text{Minimize} & J_1(u(\cdot)) + \mu J_2(u(\cdot)) \equiv -Ex(T) + \mu \text{Var}\, x(T), \\ \text{Subject to} & \begin{cases} u(\cdot) \in L^2_{\mathcal{F}}(0,T;\mathbb{R}^m), \\ (x(\cdot), u(\cdot)) \text{ satisfy equation (8.6)}, \end{cases} \end{array}$$

where the parameter (representing the weight) μ is positive. Denote the above problem by $P(\mu)$. Define

$$(8.12) \qquad \Pi_{P(\mu)} \stackrel{\Delta}{=} \{u(\cdot) | u(\cdot) \text{ is an optimal control of } P(\mu)\}.$$

Note that Problem $P(\mu)$ is *not* a standard stochastic optimal control problem treated in this book due to the presence of the term $[Ex(T)]^2$ in its cost function (see (8.8) and (8.11)). We now propose to embed the problem into a tractable auxiliary problem that turns out to be a stochastic LQ problem. To do this, set

$$(8.13) \qquad \begin{array}{ll} \text{Minimize} & J(u(\cdot); \mu, \lambda) \stackrel{\Delta}{=} E\{\mu x(T)^2 - \lambda x(T)\}, \\ \text{Subject to} & \begin{cases} u(\cdot) \in L^2_{\mathcal{F}}(0,T;\mathbb{R}^m), \\ (x(\cdot), u(\cdot)) \text{ satisfy equation (8.6)}, \end{cases} \end{array}$$

where the parameters μ and λ satisfy $\mu > 0$ and $-\infty < \lambda < +\infty$. Let us call the above Problem $A(\mu, \lambda)$. Define

$$(8.14) \qquad \Pi_{A(\mu,\lambda)} \stackrel{\Delta}{=} \{u(\cdot) | u(\cdot) \text{ is an optimal control of } A(\mu, \lambda)\}.$$

The following result tells the relationship between the problems $P(\mu)$ and $A(\mu, \lambda)$.

Theorem 8.2. *For any $\mu > 0$, one has*

$$\Pi_{P(\mu)} \subseteq \bigcup_{-\infty<\lambda<+\infty} \Pi_{A(\mu,\lambda)}. \tag{8.15}$$

Moreover, if $\overline{u}(\cdot) \in \Pi_{P(\mu)}$, then $\overline{u}(\cdot) \in \Pi_{A(\mu,\overline{\lambda})}$ with $\overline{\lambda} = 1 + 2\mu E\overline{x}(T)$.

Proof. We need to prove only the second assertion, as the first one is a direct consequence of the second. Let $\overline{u}(\cdot) \in \Pi_{P(\mu)}$. If $\overline{u}(\cdot) \notin \Pi_{A(\mu,\overline{\lambda})}$, then there exists $u(\cdot)$ such that

$$\mu\Big(Ex(T)^2 - E\overline{x}(T)^2\Big) - \overline{\lambda}\Big(Ex(T) - E\overline{x}(T)\Big) < 0. \tag{8.16}$$

Set a function

$$\pi(x,y) = \mu x - \mu y^2 - y. \tag{8.17}$$

It is a concave function in (x,y) and (noting (8.8))

$$\pi(Ex(T)^2, Ex(T)) = -Ex(T) + \mu \mathrm{Var}\, x(T), \tag{8.18}$$

which is the objective function of the problem $P(\mu)$. The concavity of π implies (noting $\pi_x(x,y) = \mu$ and $\pi_y(x,y) = -(1+2\mu y)$)

$$\begin{aligned} &\pi(Ex(T)^2, Ex(T)) \\ &\quad \le \pi(E\overline{x}(T)^2, E\overline{x}(T)) + \mu\Big(Ex(T)^2 - E\overline{x}(T)^2\Big) \\ &\qquad - (1 + 2\mu E\overline{x}(T))\Big(Ex(T) - E\overline{x}(T)\Big) \\ &\quad < \pi(E\overline{x}(T)^2, E\overline{x}(T)), \end{aligned} \tag{8.19}$$

where the last inequality is due to (8.16). By (8.19), $\overline{u}(\cdot)$ is not optimal for the problem $P(\mu)$, leading to a contradiction. □

The implication of Theorem 8.2 is that any optimal solution of the problem $P(\mu)$ (as long as it exists) can be found via solving Problem $A(\mu,\lambda)$.

Now let us solve Problem $A(\mu,\lambda)$. Putting

$$\gamma \stackrel{\Delta}{=} \frac{\lambda}{2\mu} \quad \text{and} \quad y(t) \stackrel{\Delta}{=} x(t) - \gamma, \tag{8.20}$$

Problem $A(\mu,\lambda)$ is equivalent to minimizing

$$E\Big[\frac{1}{2}\mu y(T)^2\Big], \tag{8.21}$$

subject to

$$\begin{cases} dy(t) = \big\{A(t)y(t) + B(t)u(t) + b(t)\big\}dt \\ \qquad\qquad + \sum_{j=1}^{m} D_j(t)u(t)dW^j(t), \\ y(0) = x_0 - \gamma, \end{cases} \tag{8.22}$$

where

$$
(8.23) \qquad \begin{cases} A(t) \stackrel{\Delta}{=} r(t), \quad B(t) \stackrel{\Delta}{=} (b_1(t) - r(t), \ldots, b_m(t) - r(t)), \\ b(t) \stackrel{\Delta}{=} \gamma r(t), \quad D_j(t) \stackrel{\Delta}{=} (\sigma_{1j}(t), \ldots, \sigma_{mj}(t)), \\ (s, y) \stackrel{\Delta}{=} (0, x_0 - \gamma). \end{cases}
$$

Thus, the above problem specializes Problem (SLQ) formulated in Section 3 with

$$
(8.24) \qquad \begin{cases} (Q(t), S(t), R(t)) = (0, 0, 0), \quad G = \mu, \\ C_j(t) = 0, \quad \sigma_j(t) = 0, \end{cases}
$$

and $A(t), B(t), b(t), D_j(t)$ given by (8.23).

Note that $R(t) = 0$ in this problem. Also it is a problem with a multidimensional Brownian motion. A special but important property of this problem is that the unknown $P(t)$ of the corresponding stochastic Riccati equation (7.60) is a scalar. Define

$$
(8.25) \qquad \begin{aligned} \rho(t) &\stackrel{\Delta}{=} B(t)\Big(\sum_{j=1}^m D_j(t)^\top D_j(t)\Big)^{-1} B(t)^\top \\ &\equiv B(t)\left(\sigma(t)\sigma(t)^\top\right)^{-1} B(t)^\top. \end{aligned}
$$

Then (7.60) reduces to

$$
(8.26) \qquad \begin{cases} \dot{P}(t) = (\rho(t) - 2r(t))\, P(t), \\ P(T) = \mu, \\ P(t)\left(\sigma(t)\sigma(t)^\top\right) > 0, \quad \text{a.e. } t \in [0, T]. \end{cases}
$$

Clearly, the solution of (8.26) is given by

$$
(8.27) \qquad P(t) = \mu e^{-\int_t^T (\rho(s) - 2r(s))ds}.
$$

Note that the third constraint in (8.26) is satisfied automatically due to the assumption (8.3). Moreover, equation (7.61) becomes

$$
(8.28) \qquad \begin{cases} \dot{\varphi}(t) = (\rho(t) - r(t))\, \varphi(t) - \gamma r(t) P(t), \\ \varphi(T) = 0, \end{cases}
$$

which evidently admits a unique solution $\varphi \in C([0, T]; \mathbb{R}^1)$. The optimal feedback control (7.64) gives

$$
(8.29) \qquad \begin{aligned} \overline{u}(t, y) &\equiv (\overline{u}_1(t, y), \ldots, \overline{u}_m(t, y)) \\ &= -(\sigma(t)\sigma(t)^\top)^{-1} B(t)^\top \Big(y + \frac{\varphi(t)}{P(t)}\Big). \end{aligned}
$$

Set $h(t) \stackrel{\Delta}{=} \frac{\varphi(t)}{P(t)}$. Then noting (8.26) and (8.28), one has

$$\begin{aligned}\dot{h}(t) &= \frac{P(t)\dot{\varphi}(t) - \dot{P}(t)\varphi(t)}{P(t)^2} \\ &= \frac{r(t)P(t)\varphi(t) - \gamma r(t)P(t)^2}{P(t)^2} \\ &= r(t)h(t) - \gamma r(t).\end{aligned}$$

Since $h(T) = 0$, we can solve $h(\cdot)$ to get

$$\frac{\varphi(t)}{P(t)} \equiv h(t) = \gamma\Big[1 - e^{-\int_t^T r(s)ds}\Big]. \tag{8.30}$$

Substituting (8.30) into (8.29), and noting (8.20), we arrive at

$$\begin{aligned}\overline{u}(t,x) &\equiv (\overline{u}_1(t,x), \ldots, \overline{u}_m(t,x)) \\ &= -(\sigma(t)\sigma(t)^\top)^{-1}B(t)^\top\Big[x - \gamma + \gamma(1 - e^{-\int_t^T r(s)ds})\Big] \\ &= (\sigma(t)\sigma(t)^\top)^{-1}B(t)^\top\Big[\gamma e^{-\int_t^T r(s)ds} - x\Big].\end{aligned} \tag{8.31}$$

The factor $(\sigma(t)\sigma(t)^\top)^{-1}B(t)^\top$ is called the *risk premium*. Under the above optimal feedback control, the wealth equation (8.6) evolves as

$$\begin{cases} dx(t) = \Big\{(r(t) - \rho(t))x(t) + \gamma e^{-\int_t^T r(s)ds}\rho(t)\Big\}dt \\ \qquad\qquad + B(t)(\sigma(t)\sigma(t)^\top)^{-1}\sigma(t)\Big[\gamma e^{-\int_t^T r(s)ds} - x(t)\Big]dW(t) \\ x(0) = x_0. \end{cases} \tag{8.32}$$

Moreover, applying Itô's formula to $x(t)^2$, we obtain

$$\begin{cases} dx(t)^2 = \Big\{(2r(t) - \rho(t))x(t)^2 + \gamma^2 e^{-2\int_t^T r(s)ds}\rho(t)\Big\}dt \\ \qquad\qquad +2x(t)B(t)\big[\sigma(t)\sigma(t)^\top\big]^{-1}\sigma(t)\Big[\gamma e^{-\int_t^T r(s)ds} - x(t)\Big]dW(t), \\ x(0)^2 = x_0^2. \end{cases} \tag{8.33}$$

Taking expectations on both sides of (8.32) and (8.33), we conclude that $Ex(t)$ and $Ex(t)^2$ satisfy the following two ordinary differential equations:

$$\begin{cases} dEx(t) = \{(r(t) - \rho(t))Ex(t) + \gamma e^{-\int_t^T r(s)ds}\rho(t)\}dt, \\ Ex(0) = x_0, \end{cases} \tag{8.34}$$

and

$$\begin{cases} dEx(t)^2 = \{(2r(t) - \rho(t))Ex(t)^2 + \gamma^2 e^{-2\int_t^T r(s)ds}\rho(t)\}dt, \\ Ex(0)^2 = x_0^2. \end{cases} \tag{8.35}$$

Solving (8.34) and (8.35), we can express $Ex(T)$ and $Ex(T)^2$ as explicit functions of γ,

$$
\begin{cases} Ex(T) = \alpha x_0 + \beta\gamma, \\ Ex(T)^2 = \delta x_0^2 + \beta\gamma^2, \end{cases} \tag{8.36}
$$

where

$$
\begin{cases} \alpha \stackrel{\Delta}{=} e^{\int_0^T (r(t)-\rho(t))dt}, \\ \beta \stackrel{\Delta}{=} 1 - e^{-\int_0^T \rho(t)dt}, \\ \delta \stackrel{\Delta}{=} e^{\int_0^T (2r(t)-\rho(t))dt}. \end{cases} \tag{8.37}
$$

By Theorem 8.2, an optimal solution of the problem $P(\mu)$, if it exists, can be found by selecting $\overline{\lambda}$ such that (noting (8.20))

$$
\overline{\lambda} = 1 + 2\mu E\overline{x}(T) = 1 + 2\mu\Big(\alpha x_0 + \beta\frac{\overline{\lambda}}{2\mu}\Big).
$$

This yields

$$
\overline{\lambda} = \frac{1 + 2\mu\alpha x_0}{1-\beta} = e^{\int_0^T \rho(t)dt} + 2\mu x_0 e^{\int_0^T r(t)dt}. \tag{8.38}
$$

Hence the optimal control for the problem $P(\mu)$ (if it exists) must be given by (8.31) with $\gamma = \overline{\gamma} \stackrel{\Delta}{=} \frac{\overline{\lambda}}{2\mu}$ and $\overline{\lambda}$ given by (8.38). In this case the corresponding variance of the terminal wealth is

$$
\begin{aligned}
\operatorname{Var}\overline{x}(T) &= E\overline{x}(T)^2 - [E\overline{x}(T)]^2 \\
&= \beta(1-\beta)\overline{\gamma}^2 - 2\alpha\beta x_0\overline{\gamma} + (\delta - \alpha^2)x_0^2 \\
&= \frac{1-\beta}{\beta}\Big[\beta^2\overline{\gamma}^2 - 2\frac{\alpha\beta^2 x_0\overline{\gamma}}{1-\beta} + \frac{\beta(\delta-\alpha^2)}{1-\beta}x_0^2\Big] \\
&= \frac{1-\beta}{\beta}\Big[(\beta\overline{\gamma} + \alpha x_0)^2 - 2\frac{\alpha\beta x_0\overline{\gamma}}{1-\beta} + \frac{\beta\delta - \alpha^2}{1-\beta}x_0^2\Big].
\end{aligned} \tag{8.39}
$$

Substituting $\beta\overline{\gamma} = E\overline{x}(T) - \alpha x_0$ in the above and noting (8.37), we obtain

$$
\begin{aligned}
\operatorname{Var}\overline{x}(T) &= \frac{1-\beta}{\beta}\Big\{[E\overline{x}(T)]^2 - 2\frac{\alpha}{1-\beta}x_0 E\overline{x}(T) + \frac{\beta\delta + \alpha^2}{1-\beta}x_0^2\Big\} \\
&= \frac{1-\beta}{\beta}\Big(E\overline{x}(T) - x_0 e^{\int_0^T r(t)dt}\Big)^2 \\
&= \frac{e^{-\int_0^T \rho(t)dt}}{1 - e^{-\int_0^T \rho(t)dt}}\Big(E\overline{x}(T) - x_0 e^{\int_0^T r(t)dt}\Big)^2.
\end{aligned} \tag{8.40}
$$

To summarize the above discussion, we have the following result.

Theorem 8.3. *Under the assumption (8.3), the efficient frontier of the bicriteria optimal portfolio selection problem (8.9), if it ever exists, is given by (8.40).*

The set (8.40) reveals explicitly the trade-off between the mean (return) and variance (risk). For example, if one has set an expected return level, then the above can tell the risk he/she has to take; and vice versa. In particular, if one cannot take any risk, namely, $\operatorname{Var}(\overline{x}(T)) = 0$, then $E\overline{x}(T)$ has to be $x_0 e^{\int_0^T r(t)dt}$, meaning that he/she can only put his/her money in the bond. Another interesting phenomenon is that the efficient frontier (8.40) involves a perfect square. This is due to the possible inclusion of the bond in a portfolio. In the case where the riskless bond is excluded from consideration, then the efficient frontier is no longer a perfect square, which means one cannot have a risk-free portfolio.

We were able to solve explicitly the LQ problem (8.21)–(8.22) associated with the mean–variance portfolio selection because of its very special structure. To be specific, in this problem the state variable $y(t)$ is scalar-valued, and the control cost R is equal to 0. These properties enable us to solve analytically the corresponding stochastic Riccati equation (8.26), which turns out to be a *linear* equation.

The mean–variance model studied above gives a good example in which the *uncertainty cost* (i.e., the additional cost resulting from the uncertainty in the system; see Section 1) is the sole cost that concerns us. To be specific, observe that there is no *direct* cost associated with a control (portfolio) in the cost functional (8.13) or (8.21), namely, $R \equiv 0$ in this case. However, the control, does influence the diffusion term of the system dynamics (see (8.22)), which gives rise to a "hidden" running cost $\sum_{j=1}^m D_j(t)^\top P(t) D_j(t) \equiv P(t)\sigma(t)\sigma(t)^\top$ (with $P(\cdot)$ given by (8.27)) that must be taken into consideration, and it may be regarded as a "cost equivalence" of the risk. To balance between this uncertainty/risk cost and the potential return is exactly the goal of the mean–variance portfolio selection problem.

9. Historical Remarks

While the LQ problem has its roots in least-squares estimation, which goes back to the very beginning of the calculus of variations (see Chapter 2, Section 8), Newton–Gould–Kaiser [1] should be credited for systematically applying the least-squares techniques to design feedback control systems. The deterministic LQ problem was first studied by Bellman–Glicksberg–Gross [2, Chapter 4] in 1958. R. E. Kalman [1] and A. M. Letov [1] elegantly solved the problem in a linear state feedback control form in 1960. A classical book on the theory is Anderson–Moore [1]. See also Lee–Markus [1], Wonham [3], and the survey paper of Willems [1]. In the classical deterministic case, one typically assumes that the weighting matrices satisfy $Q(t) \geq 0$, $R(t) > 0$, $G \geq 0$ (and $S(t) = 0$). Such a theory was extended to infinite dimensions as well (see J. L. Lions [1] and Lasiecka–Triggiani [1] for details). Since $R(t) \geq 0$ is necessary for deterministic LQ problems (both in finite- and infinite-dimensions) to be finite, one cannot do much about $R(t)$. In 1977, Molinari [1] studied finite-dimensional LQ problems allowing $Q(t)$ to be indefinite, and a similar theory involving the Riccati equation was established. This result was extended to infinite dimensions by You [1] and Chen [1] in the 1980s. See Li–Yong [2] and Wu–Li [1] for a detailed discussion and further extensions.

Stochastic LQ models, or as they are sometimes called, the *linear-quadratic-Gaussian (LQG) models*, including both completely and partially observed ones, are the most important examples of the stochastic control problems, especially in their applications in engineering design, where the physical uncertainty is modeled as white noise. The model involving Itô's equation was first studied by Kushner [1] in 1962 by the dynamic programming method. The first work for stochastic LQ problems using stochastic Riccati equations is due to Wonham [1,2] in 1968. The form of stochastic Riccati equations presented in this chapter is the same as that which appeared in Wonham [2]. There were some other later works, see McLane [1], Willems–Willems [1], Davis [2], Bensoussan [1], and de Souza–Fragoso [1]. There was a special issue of *IEEE Transactions on Automatic Control* on the LQG problem and related topics edited by Athans [1] in 1971, where good surveys and tutorials along with extensive references (up to that time) can be found. However, in those works, the assumption that the control weighting matrix $R(t)$ was positive definite, inherited from the deterministic case, was taken for granted.

In the mid-1990s Chen–Li–Zhou [1] investigated the stochastic LQ problems with random coefficients. They derived a stochastic Riccati equation that is a *nonlinear* backward stochastic differential of the Pardoux–Peng [1] type (see Chapter 7, Section 3; the equation reduces to (6.6) or (7.60) when all the coefficients are deterministic). Based on this new Riccati equation, they found that the LQ problem might still be solvable even if the matrix-valued function $R(t)$ is *indefinite* (and in particular, *negative definite*). They also gave conditions that ensure the solvability of the stochastic Riccati equation as well as the LQ problem for the case $C(t) \equiv 0$. Later, Chen–Zhou [1] discussed the case $C(t) \not\equiv 0$ (with indefinite $R(t)$) and derived an optimal feedback control by a decomposition technique. Ait Rami–Moore–Zhou [1] introduced a generalized stochastic Riccati equation involving the pseudo-inverse of a matrix and an additional algebraic constraint to treat the case where $R + D^\top PD$ is possibly *singular*. Moreover, the problem with integral constraints was studied by Lim–Zhou [1], and the discrete-time case was treated by Moore–Zhou–Lim [1]. Along another line, Chen–Yong [1] extensively used the methods of functional analysis and FBSDEs to study the LQ problems with general random coefficients. Among other things, they proved the local existence of adapted solutions to the stochastic Riccati equations with random coefficients, which led to the local solvability of the LQ problems with random coefficients. Notice that even the local existence is not obvious in this situation due to the high nonlinearity, the singularity, and the backward nature of the stochastic Riccati equations. Finally, (indefinite) stochastic LQ problems in an infinite time horizon and the associated stochastic (algebraic) Riccati equations are studied (both analytically and numerically) by Ait Rami–Zhou [1] via the linear matrix inequality (LMI) and semidefinite programming (SDP) techniques. This approach has been further developed extensively by Yao–Zhang–Zhou [1, 2], in the absence of the nonsingularity of $R + D^\top PD$, by employing a duality analysis of SDP. It turns out that one can solve LQ problems computationally based on some powerful SDP solvers (Boyd–El Ghaoui–Feron–Balakrishnan [1] and Nesterov–Nemirovskii [1]) without any longer having to involve Riccati equations.

The results contained in Section 2 are more or less standard. But the way of presentation has some new features. We use general functional analysis to transform the LQ problem to a minimization problem of a quadratic functional over some Hilbert space, which leads to necessary and sufficient conditions for solvability. Then we derive the Riccati equation using the maximum principle and indicate the possibility of using dynamic programming as well as the technique of completing squares to obtain the same equation. This in turn gives a concrete example of the intrinsic relation (and the essential equivalence in the LQ case) between the MP and DP. The discussion for the stochastic case essentially follows the same line. Examples appearing in Section 3 are taken from Chen–Yong [1]. Section 4 is a modification of a part of Chen–Yong [1]. Section 5 is based on an idea of Chen–Yong [1], with a new proof. Section 6 is based on Chen–Li–Zhou [1] with some new material arising from the additional nonhomogeneous terms in the dynamics and the cross term in the running cost. Section 7.1 is based on Chen–Zhou [1], and Section 7.2 on Chen–Li–Zhou [1], while Section 7.3 is taken from Chen–Yong [1].

The mean–variance model was formulated as a quadratic programming problem by Harry Markowitz in his Nobel-Prize-winning work (Markowitz [1]) in 1952 (see also Markowitz [2]) for a single-period investment. The most important contribution of this model is that it quantifies the risk using the variance, which enables investors to seek the highest return after evaluating their acceptable risk level. This approach has become the foundation of modern finance theory and has inspired hundreds of extensions and applications. In particular, in the case where the covariance matrix is positive definite and short-selling is allowed, an analytic solution was obtained by Merton [3]. Perold [1] developed a more general technique to locate the efficient frontier when the covariance matrix is nonnegative definite.

After Markowitz's first work, the mean–variance model was soon extended to multi-period portfolio selection; see, for example, Mossin [1], Samuelson [1], Elton–Gruber [1], and Grauer–Hakansson [1]. The mean–variance model in a continuous-time setting was developed a bit later; see Föllmer–Sondermann [1], Duffie–Jackson [1] and Duffie–Richardson [1]. The basic approach in these works is dynamic programming. Zhou–Li [1] applied the embedding technique (introduced by Li–Ng [1] for the multiperiod problem) to the continuous-time case, which leads to a stochastic LQ problem solvable by the results of Chen–Li–Zhou [1]. Section 8 in this chapter is based on Zhou–Li [1].

Chapter 7

Backward Stochastic Differential Equations

1. Introduction

In Chapter 3, in order to derive the stochastic maximum principle as a set of necessary conditions for optimal controls, we encountered the problem of finding adapted solutions to the adjoint equations. Those are terminal value problems of (linear) stochastic differential equations involving the Itô stochastic integral. We call them *backward stochastic differential equations* (BSDEs, for short). For an ordinary differential equation (ODE, for short), under the usual Lipschitz condition, both the initial value and the terminal value problems are well-posed. As a matter of fact, for an ODE, the terminal value problem on $[0,T]$ is equivalent to an initial value problem on $[0,T]$ under the time-reversing transformation $t \mapsto T-t$. However, things are fundamentally different (and difficult) for BSDEs when we are looking for a solution that is adapted to the given filtration. Practically, one knows only about what has happened in the past, but cannot foretell what is going to happen in the future. Mathematically, it means that we would like to keep the context within the framework of the Itô-type stochastic calculus (and do not want to involve the so-called *anticipative integral*). As a result, one cannot simply reverse the time to get a solution for a terminal value problem of SDE, as it would destroy the adaptiveness. Therefore, the first issue one should address in the stochastic case is how to correctly *formulate* a terminal value problem for stochastic differential equations (SDEs, for short).

The (adapted) solution of a BSDE turns out to be a *pair* of adapted stochastic processes. It is the second component that corrects the possible "nonadaptiveness" caused by the backward nature of the equations, including the given terminal value of the first component. By and large, we may also think of the pair this way: The first component of the pair represents the "mean evolution" of the dynamics, while the second one signifies the uncertainty/risk between the current time and the terminal time. It is interesting to note that the uncertainty/risk must be a part of the solution in order to have an adapted solution. Let us take a very simple example to roughly illustrate this idea.

Consider an investor who is making a one-year investment with two choices of assets, namely a bond with a (deterministic) annual return rate of 10%, and a stock with the following anticipated return: a positive return rate of 20% if the following year is a "good" year and a negative return rate of −20% if the year is a "bad" one. Now, the investor has the following objective for his/her wealth: having wealth of the amount a if the year is good and of the amount b if the year is bad (of course, $a \geq b$). The question is how to invest in order to achieve this financial goal.

Note that this is a problem with a future wealth given as a random variable, and the investor is not sure about its exact amount at the time of investment. Nevertheless, the problem can be easily solved as follows: Suppose the investor puts totally \$$y$, among which \$$z$ is invested in the stock. Then \$$(y-z)$ is the amount in the bond, and we have the following system of equations to solve:

$$
\begin{cases} 1.1(y-z)+1.2z=a, \\ 1.1(y-z)+0.8z=b, \end{cases} \tag{1.1}
$$

which admits a *unique* solution

$$
\begin{cases} y=\dfrac{5}{22}(3a+b), \\ z=\dfrac{5}{2}(a-b). \end{cases} \tag{1.2}
$$

One clearly sees that if the investor would like to exactly meet the future goal of wealth, which is a random variable depending on the future situation (in the above case, it is a random variable taking two values, a and b), then his/her investment should consist of two parts: (y,z), of which the part z "adjusts" the *uncertainty* between now and the future (in this particular example, z is the amount to be invested in the *risky* asset).

A special case is that in which $a=b$, meaning that the investor would like to have a definite amount of wealth next year. Then (1.2) gives $z=0$, namely, the "adjustment" part of the solution is zero, or the investor puts all the money into the riskless asset.

After the foundation for linear BSDEs has been established, it is very natural to extend to *nonlinear* BSDEs and *forward-backward stochastic differential equations* (FBSDEs, for short), which include the stochastic Hamiltonian systems presented in Chapter 3 as a special case. It turns out that solutions to some *nonlinear* partial differential equations (PDEs) may be represented via the solutions to BSDEs, which extends the famous Feynman–Kac formula. On the other hand, some FBSDEs may be solved via PDEs in an approach called the *four-step scheme.* These relationships between the PDEs and BSDEs/FBSDEs are analogies of those between the HJB equations and Hamiltonian systems in stochastic optimal control established in Chapter 5.

Throughout this chapter, we let $(\Omega,\mathcal{F},\{\mathcal{F}_t\}_{t\geq 0},\mathbf{P})$ be a complete probability space on which an m-dimensional standard Brownian motion $W(t)$ is defined, such that $\{\mathcal{F}_t\}_{t\geq 0}$ is the natural filtration of $W(t)$, augmented by all the $\mathbf{P}$-null sets in $\mathcal{F}$ (therefore, $t\mapsto\mathcal{F}_t$ is continuous; see the last paragraph of Section 2.2 in Chapter 1).

The rest of this chapter is organized as follows. We start with the original Bismut's formulation for linear BSDEs in Section 2 by using the martingale representation theorem. Nonlinear extension is studied in Section 3 for both deterministic and random-time durations. Section 4 is devoted to several Feynman–Kac-type formulae. In Section 5, FBSDEs and the four-step scheme are presented. An application of the BSDEs/FBSDEs to option pricing problems and the associated Black–Scholes formula is demonstrated in Section 6. Finally, historical remarks are given in Section 7.

2. Linear Backward Stochastic Differential Equations

We begin with a simple but illustrative example. Let $m = 1$ (i.e., the Brownian motion is one-dimensional), $T > 0$, and $\xi \in L^2_{\mathcal{F}_T}(\Omega; \mathbb{R})$. Consider the following terminal value problem of SDE:

$$\begin{cases} dY(t) = 0, & t \in [0,T], \\ Y(T) = \xi. \end{cases} \tag{2.1}$$

We want to find an $\{\mathcal{F}_t\}_{t\geq 0}$-adapted solution $Y(\cdot)$. However, this is impossible, since the only solution of (2.1) is

$$Y(t) = \xi, \qquad \forall t \in [0,T], \tag{2.2}$$

which is not necessarily $\{\mathcal{F}_t\}_{t\geq 0}$-adapted (unless ξ is $\mathcal{F}_0$-measurable, i.e., ξ is a constant). Namely, equation (2.1) is not well formulated if one expects any $\{\mathcal{F}_t\}_{t\geq 0}$-adapted solution of it. Hence, in order to get $\{\mathcal{F}_t\}_{t\geq 0}$-adapted "solutions" of (2.1), we should modify or reformulate (2.1).

Let us now find an appropriate reformulation for (2.1). In doing so, we require that the new formulation should keep most features of (2.1), and in particular, it should coincide with (2.1) when ξ is a (nonrandom) constant, since in that case, (2.1) does have an $\{\mathcal{F}_t\}_{t\geq 0}$-adapted solution given by (2.2).

We start with (2.2). It has been seen that the process $Y(\cdot)$ given by (2.2) satisfies (2.1) but is not necessarily $\{\mathcal{F}_t\}_{t\geq 0}$-adapted. A natural way of making (2.2) $\{\mathcal{F}_t\}_{t\geq 0}$-adapted is to redefine $Y(\cdot)$ as follows:

$$Y(t) = E(\xi|\mathcal{F}_t), \qquad t \in [0,T]. \tag{2.3}$$

Then, $Y(\cdot)$ is $\{\mathcal{F}_t\}_{t\geq 0}$-adapted and satisfies the terminal condition $Y(T) = \xi$ (since ξ is $\mathcal{F}_T$-measurable). Also, in the case where ξ is nonrandom, it is the same as that given by (2.2). However, $Y(\cdot)$ given by (2.3) no longer satisfies (2.1). Therefore, we need to reformulate (2.1) so that it admits $Y(\cdot)$ given by (2.3) as one of its $\{\mathcal{F}_t\}_{t\geq 0}$-adapted "solutions." Thus, our next goal is to find the equation that the process (2.3) satisfies.

To this end, note that the process $Y(\cdot)$ defined by (2.3) is a square-integrable $\{\mathcal{F}_t\}_{t\geq 0}$-martingale. By the martingale representation theorem (Chapter 1, Theorem 5.7), we can find an $\{\mathcal{F}_t\}_{t\geq 0}$-adapted process $Z(\cdot) \in L^2_{\mathcal{F}}(0,T;\mathbb{R})$ such that

$$Y(t) = Y(0) + \int_0^t Z(s)dW(s), \qquad \forall t \in [0,T], \quad \mathbf{P}\text{-a.s.} \tag{2.4}$$

From (2.3)–(2.4), it follows that (noting that ξ is $\mathcal{F}_T$-measurable)

$$\xi = Y(T) = Y(0) + \int_0^T Z(s)dW(s). \tag{2.5}$$

Hence, combining (2.4)–(2.5), one has

$$Y(t) = \xi - \int_t^T Z(s)dW(s), \quad \forall t \in [0,T]. \tag{2.6}$$

The differential form of (2.6) reads

$$\begin{cases} dY(t) = Z(t)dW(t), & t \in [0,T], \\ Y(T) = \xi. \end{cases} \tag{2.7}$$

Therefore, (2.7) seems to be an appropriate reformulation of (2.1). By comparing (2.7) with (2.1), we see that the term $Z(t)dW(t)$ has been added. The process $Z(\cdot)$ is not a priori known, and it is a *part* of the solution. As a matter of fact, the presence of the term $Z(t)dW(t)$ "corrects" the "non-adaptiveness" of the original $Y(\cdot)$ in (2.2). Thus, by an $\{\mathcal{F}_t\}_{t\geq 0}$-adapted solution, we should mean a *pair* of $\{\mathcal{F}_t\}_{t\geq 0}$-adapted processes $(Y(\cdot), Z(\cdot))$ satisfying (2.7) or equivalently (2.6).

From the above, we conclude that (2.7) admits an $\{\mathcal{F}_t\}_{t\geq 0}$-adapted solution $(Y(\cdot), Z(\cdot))$, for which $Y(\cdot)$ is given by (2.3) and $Z(\cdot)$ is determined by (2.4) via the martingale representation theorem.

At first glance, one might feel that the (adapted) solutions may not be unique, since (2.7) is an equation with two unknowns $Y(\cdot)$ and $Z(\cdot)$. However, it turns out that the $\{\mathcal{F}_t\}_{t\geq 0}$-adaptiveness gives another restriction that makes the equation have a unique solution $(Y(\cdot), Z(\cdot))$. In fact, applying the Itô formula to $|Y(t)|^2$, one has the following:

$$E|\xi|^2 = E|Y(t)|^2 + \int_t^T E|Z(s)|^2 ds, \qquad \forall t \in [0,T]. \tag{2.8}$$

Since equation (2.7) is linear, relation (2.8) gives the uniqueness of the $\{\mathcal{F}_t\}_{t\geq 0}$-adapted solution to (2.7).

Finally, note that in the case where ξ is a nonrandom constant, one can take $Y(t) \equiv \xi$ and $Z(t) \equiv 0$. By the uniqueness, it is the only solution of (2.7), which is now the same as (2.1).

Now we turn to the general situation. Consider the terminal value problem of the following linear stochastic differential equation:

$$\begin{cases} dY(t) = \Big\{A(t)Y(t) + \sum_{j=1}^m B_j(t)Z_j(t) + f(t)\Big\}dt \\ \qquad\qquad + Z(t)dW(t), \; t \in [0,T], \\ Y(T) = \xi, \end{cases} \tag{2.9}$$

where $A(\cdot), B_1(\cdot), \dots, B_m(\cdot)$ are bounded $\mathbb{R}^{k\times k}$-valued $\{\mathcal{F}_t\}_{t\geq 0}$-adapted processes, $f(\cdot) \in L^2_{\mathcal{F}}(0,T;\mathbb{R}^k)$, and $\xi \in L^2_{\mathcal{F}_T}(\Omega;\mathbb{R}^k)$. Our goal is to find $\{\mathcal{F}_t\}_{t\geq 0}$-adapted processes $Y(\cdot)$ and $Z(\cdot) \equiv (Z_1(\cdot), \dots, Z_m(\cdot))$, valued in $\mathbb{R}^k$ and $\mathbb{R}^{k\times m}$, respectively, such that (2.9) is satisfied. We emphasize the key issues here: The process $(Y(\cdot), Z(\cdot))$ that we are looking for is required to be *forward* adapted, whereas the process $Y(\cdot)$ is specified at the terminal time $t = T$ (thus the equation should be solved *backward*). Due to this, we call (2.9) a linear *backward stochastic differential equation* (BSDE). Clearly, (2.7) is the simplest case of this type of equation. Also, the first-order and second-order adjoint equations (3.8) and (3.9) in Chapter 3, Section 3.1, are specializations of (2.9).

Let us recall that $L^2_{\mathcal{F}}(\Omega; C([0,T]; \mathbb{R}^k))$ is the set of all $\{\mathcal{F}_t\}_{t\geq 0}$-adapted, $\mathbb{R}^k$-valued continuous processes $Y(\cdot)$ such that $E\sup_{t\in[0,T]}|Y(t)|^2 < \infty$. We now introduce the following definition.

Definition 2.1. A pair $(Y(\cdot), Z(\cdot)) \in L^2_{\mathcal{F}}(\Omega; C([0,T]; \mathbb{R}^k)) \times L^{2,loc}_{\mathcal{F}}(0,T; \mathbb{R}^{k\times m})$ is called an *adapted solution* of (2.9) if the following holds:

$$
(2.10)\qquad \begin{aligned} Y(t) = \xi - \int_t^T \{A(s)Y(s) + \sum_{j=1}^m B_j(s)Z_j(s) + f(s)\}ds \\ - \int_t^T Z(s)dW(s), \quad \forall t \in [0,T], \ \mathbf{P}\text{-a.s.} \end{aligned}
$$

Equation (2.9) is said to have a *unique* adapted solution if for any two adapted solutions $(Y(\cdot), Z(\cdot))$ and $(\tilde{Y}(\cdot), \tilde{Z}(\cdot))$, it must hold that

$$
\mathbf{P}\{Y(t) = \tilde{Y}(t), \ \forall t \in [0,T] \text{ and } Z(t) = \tilde{Z}(t), \ \text{a.e.}\, t \in [0,T]\} = 1.
$$

We have the following well-posedness result for the linear BSDE (2.9).

Theorem 2.2. *Let $A(\cdot), B_1(\cdot), \ldots, B_m(\cdot) \in L^\infty_{\mathcal{F}}(0,T; \mathbb{R}^{k\times k})$. Then, for any $f \in L^2_{\mathcal{F}}(0,T; \mathbb{R}^k)$ and $\xi \in L^2_{\mathcal{F}_T}(\Omega; \mathbb{R}^k)$, BSDE (2.9) admits a unique adapted solution $(Y(\cdot), Z(\cdot)) \in L^2_{\mathcal{F}}(\Omega; C([0,T]; \mathbb{R}^k)) \times L^2_{\mathcal{F}}(0,T; \mathbb{R}^{k\times m})$ and there exists a constant $K > 0$ such that*

$$
(2.11)\qquad E \sup_{t\in[0,T]} |Y(t)|^2 + \sum_{j=1}^m E\int_0^T |Z_j(t)|^2 dt \leq K\Big\{E|\xi|^2 + E\int_0^T |f(t)|^2 dt\Big\}.
$$

The proof presented below is based on the martingale representation theorem and heavily relies on the linearity of the equation. It is suggestive and constructive, which will help the reader understand the method of finding adapted solutions.

Proof. We first consider two SDEs for $\mathbb{R}^{k\times k}$-valued processes:

$$
(2.12)\qquad \begin{cases} d\Phi(t) = \{A(t)\Phi(t) + \sum_{j=1}^m B_j(t)B_j(t)\Phi(t)\}dt + \sum_{j=1}^m B_j(t)\Phi(t)dW^j(t), \\ \Phi(0) = I, \end{cases}
$$

$$
(2.13)\qquad \begin{cases} d\Psi(t) = -\Psi(t)A(t)dt - \sum_{j=1}^m \Psi(t)B_j(t)dW^j(t), \\ \Psi(0) = I. \end{cases}
$$

Since (2.12) and (2.13) are usual linear (forward) SDEs with bounded coefficients, they admit unique (strong) solutions $\Phi(\cdot)$ and $\Psi(\cdot)$, which are $\{\mathcal{F}_t\}_{t\geq 0}$-adapted. Now, by Itô's formula, we have

$$
\begin{aligned}
d[\Psi(t)\Phi(t)] &= -\Psi(t)A(t)\Phi(t)dt - \sum_{j=1}^{m} \Psi(t)B_j(t)\Phi(t)dW^j(t) \\
&\quad + \Psi(t)[A(t)\Phi(t) + \sum_{j=1}^{m} B_j(t)B_j(t)\Phi(t)]dt \\
&\quad + \sum_{j=1}^{m} \Psi(t)B_j(t)\Phi(t)dW^j(t) - \sum_{j=1}^{m} \Psi(t)B_j(t)B_j(t)\Phi(t)dt \\
&= 0.
\end{aligned} \tag{2.14}
$$

Hence, noting that $\Phi(0) = \Psi(0) = I$, we must have

$$
\Psi(t)^{-1} = \Phi(t), \qquad \forall t \in [0,T], \ \mathbf{P}\text{-a.s.} \tag{2.15}
$$

Next, we suppose $(Y(\cdot), Z(\cdot))$ is an adapted solution of (2.9). Applying Itô's formula to $\Psi(t)Y(t)$, we have

$$
\begin{aligned}
d[\Psi(t)Y(t)] &= -\Psi(t)A(t)Y(t)dt - \sum_{j=1}^{m} \Psi(t)B_j(t)Y(t)dW^j(t) \\
&\quad + \Psi(t)[A(t)Y(t) + \sum_{j=1}^{m} B_j(t)Z_j(t) + f(t)]dt \\
&\quad + \sum_{j=1}^{m} \Psi(t)Z_j(t)dW^j(t) - \sum_{j=1}^{m} \Psi(t)B_j(t)Z_j(t)dt \\
&= \Psi(t)f(t)dt + \sum_{j=1}^{m} \Psi(t)[Z_j(t) - B_j(t)Y(t)]dW^j(t).
\end{aligned} \tag{2.16}
$$

Thus,

$$
\begin{aligned}
\Psi(t)Y(t) &= \Psi(T)\xi - \int_t^T \Psi(s)f(s)ds \\
&\quad - \sum_{j=1}^{m} \int_t^T \Psi(s)[Z_j(s) - B_j(s)Y(s)]dW^j(s) \\
&= \theta + \int_0^t \Psi(s)f(s)ds \\
&\quad - \sum_{j=1}^{m} \int_t^T \Psi(s)[Z_j(s) - B_j(s)Y(s)]dW^j(s),
\end{aligned} \tag{2.17}
$$

where

$$
\theta \stackrel{\Delta}{=} \Psi(T)\xi - \int_0^T \Psi(s)f(s)ds. \tag{2.18}
$$

Taking $E(\cdot \mid \mathcal{F}_t)$ on both sides of (2.17), we get

$$\Psi(t)Y(t) = \int_0^t \Psi(s)f(s)ds + E(\theta|\mathcal{F}_t), \qquad t \in [0,T]. \tag{2.19}$$

Note that the right-hand side of (2.19) depends only on ξ and $f(\cdot)$. Also, we have seen that in achieving (2.19), the linearity of the equation has been used.

From (2.19), we see that one should define (noting (2.15))

$$Y(t) = \Phi(t)\Big\{\int_0^t \Psi(s)f(s)ds + E(\theta|\mathcal{F}_t)\Big\}, \qquad t \in [0,T], \tag{2.20}$$

with $\Phi(\cdot)$ and $\Psi(\cdot)$ being the solutions of (2.12) and (2.13), respectively, and θ being defined by (2.18). We now prove that the $Y(\cdot)$ defined by (2.20) together with some $Z(\cdot)$ will be an adapted solution of (2.9).

First of all, by (2.20) and (2.18), we have

$$Y(T) = \Phi(T)\Big\{\int_0^T \Psi(s)f(s)ds + \theta\Big\} = \Phi(T)\Psi(T)\xi = \xi. \tag{2.21}$$

Next, since $E(\theta|\mathcal{F}_t)$ is a square-integrable martingale, by the martingale representation theorem we can find a unique $\eta \equiv (\eta_1, \ldots, \eta_m) \in L^2_{\mathcal{F}}(0,T;\mathbb{R}^{k\times m})$ such that

$$E(\theta|\mathcal{F}_t) = E\theta + \sum_{j=1}^m \int_0^t \eta_j(s)dW^j(s), \qquad \forall t \in [0,T], \ \mathbf{P}\text{-a.s.} \tag{2.22}$$

Hence, it follows from (2.20) and (2.22) that

$$\begin{aligned} Y(t) &= \Phi(t)\Big\{\int_0^t \Psi(s)f(s)ds + \sum_{j=1}^m \int_0^t \eta_j(s)dW^j(s) + E\theta\Big\}, \\ &\stackrel{\Delta}{=} \Phi(t)r(t), \qquad \forall t \in [0,T], \ \mathbf{P}\text{-a.s.} \end{aligned} \tag{2.23}$$

Applying Itô's formula, we obtain

$$\begin{aligned} dY(t) &= \Big\{A(t)\Phi(t)r(t) + \sum_{j=1}^m B_j(t)B_j(t)\Phi(t)r(t)\Big\}dt \\ &\quad + \sum_{j=1}^m B_j(t)\Phi(t)r(t)dW^j(t) + \Phi(t)\Psi(t)f(t)dt \\ &\quad + \sum_{j=1}^m \Phi(t)\eta_j(t)dW^j(t) + \sum_{j=1}^m B_j(t)\Phi(t)\eta_j(t)dt \\ &= \Big\{A(t)Y(t) + \sum_{j=1}^m B_j(t)[B_j(t)Y(t) + \Phi(t)\eta_j(t)] + f(t)\Big\}dt \\ &\quad + \sum_{j=1}^m [B_j(t)Y(t) + \Phi(t)\eta_j(t)]dW^j(t). \end{aligned} \tag{2.24}$$

Therefore, by setting $Z(\cdot)=(Z_1(\cdot),\ldots,Z_m(\cdot))$ with

$$Z_j(t)\stackrel{\Delta}{=}B_j(t)Y(t)+\Phi(t)\eta_j(t),\quad t\in[0,T],\ \mathbf{P}\text{-a.s.},\ 1\le j\le m, \tag{2.25}$$

and using (2.21) and (2.24), we conclude that $(Y(\cdot),Z(\cdot))\in L^{2,loc}_{\mathcal{F}}(0,T;\mathbb{R}^k)\times L^{2,loc}_{\mathcal{F}}(0,T;\mathbb{R}^{k\times m})$ satisfy (2.9). We note that at this stage we have obtained only the *local* integrability of $(Y(\cdot),Z(\cdot))$. Next we show that in fact $(Y(\cdot),Z(\cdot))\in L^2_{\mathcal{F}}(\Omega;C(0,T;\mathbb{R}^k))\times L^2_{\mathcal{F}}(0,T;\mathbb{R}^{k\times m})$. To this end, for each n define an $\{\mathcal{F}\}_{t\ge0}$-stopping time $\tau_n\stackrel{\Delta}{=}\inf\{t\ge0:\int_0^t|Z(s)|^2ds\ge n\}\wedge T$. It is clear that τ_n increases to T ($\mathbf{P}$-a.s.) as $n\to\infty$. Applying Itô's formula to $|Y(t\wedge\tau_n)|^2$, we obtain

$$\begin{aligned}
&E|Y(0)|^2+E\int_0^{T\wedge\tau_n}\sum_{j=1}^m|Z_j(s)|^2ds\\
&=E|Y(T\wedge\tau_n)|^2\\
&\qquad-2E\int_0^{T\wedge\tau_n}\langle Y(s),A(s)Y(s)+\sum_{j=1}^m B_j(s)Z_j(s)+f(s)\rangle\,ds\\
&\le E|Y(T\wedge\tau_n)|^2+KE\int_0^{T\wedge\tau_n}\{|Y(s)|^2+|f(s)|^2\}ds\\
&\qquad+\frac12E\int_0^{T\wedge\tau_n}\sum_{j=1}^m|Z_j(s)|^2ds.
\end{aligned}\tag{2.26}$$

Hence,

$$\sum_{j=1}^m E\int_0^{T\wedge\tau_n}|Z_j(s)|^2ds\le K\Big\{E|Y(T\wedge\tau_n)|^2+E\int_0^{T\wedge\tau_n}|f(s)|^2ds\Big\}. \tag{2.27}$$

Letting $n\to\infty$ and using Fatou's lemma, we conclude that

$$\sum_{j=1}^m E\int_0^T|Z_j(s)|^2ds\le K\Big\{E|\xi|^2+E\int_0^T|f(s)|^2ds\Big\}. \tag{2.28}$$

This shows that $Z(\cdot)\in L^2_{\mathcal{F}}(0,T;\mathbb{R}^{k\times m})$.

We now prove $Y(\cdot)\in L^2_{\mathcal{F}}(\Omega;C([0,T];\mathbb{R}^k))$. From (2.24) and (2.21), we have

$$Y(t)=\xi-\int_t^T h(s)ds-\int_t^T Z(s)dW(s),\qquad t\in[0,T], \tag{2.29}$$

for some $h(\cdot)\in L^2_{\mathcal{F}}(0,T;\mathbb{R}^k)$. Thus, by Burkholder–Davis–Gundy's inequality and Doob's inequality, we have

$$
\begin{aligned}
&\Big(E\{\sup_{t\in[0,T]}|Y(t)|^2\}\Big)^{\frac{1}{2}}\\
&\le (E|\xi|^2)^{\frac{1}{2}}+\Big(E\{\int_0^T|h(s)|ds\}^2\Big)^{\frac{1}{2}}\\
&\quad+\Big(E|\int_0^T Z(s)dW(s)|^2\Big)^{\frac{1}{2}}+\Big(E\sup_{t\in[0,T]}|\int_0^t Z(s)dW(s)|^2\Big)^{\frac{1}{2}}\\
&\le (E|\xi|^2)^{\frac{1}{2}}+\sqrt{T}\Big(E\int_0^T|h(s)|^2ds\Big)^{\frac{1}{2}}+3\Big(E\int_0^T|Z(s)|^2ds\Big)^{\frac{1}{2}}<\infty.
\end{aligned}
\tag{2.30}
$$

This implies $Y(\cdot)\in L^2_{\mathcal{F}}(\Omega;C([0,T];\mathbb{R}^k)$, and therefore $(Y(\cdot),Z(\cdot))$ is an adapted solution of (2.9). Combining (2.28) and (2.30), we easily obtain the estimate (2.11). The uniqueness then follows, as the equation is linear. □

We notice that the above proof relies heavily on the linearity of the equation. The steps (2.16)–(2.19) are crucial for the proof. On the other hand, if the unknown process $(Y(\cdot),Z(\cdot))$ does not appear in the drift term in (2.9) (i.e., $A(t)=B_i(t)=0$), then steps (2.16)–(2.17) are not necessary, since in that case, $\Phi(t)\equiv\Psi(t)\equiv I$. Then the proof will be substantially simplified. We encourage the reader to carry out a proof for such a special case directly, which would be helpful for better understanding the idea of the above proof.

The solution pair $(Y(\cdot),Z(\cdot))$ actually has an interesting interpretation in functional analysis and duality analysis. More precisely, it is nothing but the (unique) *Riesz representation* of a certain functional on the Hilbert space $L^2_{\mathcal{F}}(0,T;\mathbb{R}^k)\times L^2_{\mathcal{F}}(0,T;\mathbb{R}^{k\times m})$ defined via a forward SDE that is *dual* to the BSDE (2.9). Indeed, define

$$
\begin{aligned}
&\pi(\varphi,\psi)\stackrel{\Delta}{=}E\Big\{\int_0^T\langle x(t),-f(t)\rangle\,dt+\langle x(T),\xi\rangle\Big\},\\
&\forall(\varphi,\psi)\equiv(\varphi,\psi_1,\dots,\psi_m)\in L^2_{\mathcal{F}}(0,T;\mathbb{R}^k)\times L^2_{\mathcal{F}}(0,T;\mathbb{R}^{k\times m}),
\end{aligned}
\tag{2.31}
$$

where $x(\cdot)$ is the solution of the following linear SDE:

$$
\begin{cases}
dx(t)=(-A(t)^\top x(t)+\varphi(t))dt\\
\qquad\qquad+\displaystyle\sum_{j=1}^m(-B_j(t)^\top x(t)+\psi_j(t))dW^j(t),\\
x(0)=0.
\end{cases}
\tag{2.32}
$$

Clearly, π is a linear bounded functional on the Hilbert space $L^2_{\mathcal{F}}(0,T;\mathbb{R}^k)\times L^2_{\mathcal{F}}(0,T;\mathbb{R}^{k\times m})$. Hence by the Riesz representation theorem (Yosida [1, Chapter III, Section 6]), there is a *unique* $(\overline{Y},\overline{Z})\in L^2_{\mathcal{F}}(0,T;\mathbb{R}^k)\times$

$L^2_{\mathcal{F}}(0,T;\mathbb{R}^{k\times m})$ such that

$$\pi(\varphi,\psi)=E\int_0^T\{\langle\varphi(t),\overline{Y}(t)\rangle+\langle\psi(t),\overline{Z}(t)\rangle\}dt. \tag{2.33}$$

On the other hand, if (Y,Z) is the solution of the BSDE (2.9), then applying Itô's formula to $\langle x(t),Y(t)\rangle$, we can easily derive that (2.33) holds with $(\overline{Y},\overline{Z})$ replaced by (Y,Z). Due to the uniqueness of the Riesz representation, $(\overline{Y},\overline{Z})\equiv(Y,Z)$.

We can also observe the duality between the forward equation (2.32) and the backward equation (2.9) in a way analogous to the deterministic case. In addition, the necessity of a *pair* solution of (2.9) is clear by the fact that its dual equation (2.32) has two parts: drift and diffusion. Notice that this kind of duality actually played an important role in deriving the stochastic maximum principle in Chapter 3.

3. Nonlinear Backward Stochastic Differential Equations

In this section we are going to study nonlinear backward stochastic differential equations. These equations have interesting applications in mathematical finance, nonlinear partial differential equations, and so on.

3.1. BSDEs in finite deterministic durations: Method of contraction mapping

In this subsection we consider the following BSDE in a fixed duration $[0,T]$:

$$\begin{cases} dY(t)=h(t,Y(t),Z(t))dt+Z(t)dW(t), & t\in[0,T),\ \text{a.s.}\\ Y(T)=\xi,\end{cases} \tag{3.1}$$

where $h:[0,T]\times\mathbb{R}^k\times\mathbb{R}^{k\times m}\times\Omega\to\mathbb{R}^k$ and $\xi\in L^2_{\mathcal{F}_T}(\Omega;\mathbb{R}^k)$. Note that h can be random. We have suppressed the argument $\omega\in\Omega$ and will do so hereafter. Our goal is to find a pair of $\{\mathcal{F}_t\}_{t\geq0}$-adapted processes $Y:[0,T]\times\Omega\to\mathbb{R}^k$ and $Z:[0,T]\times\Omega\to\mathbb{R}^{k\times m}$ satisfying the above BSDE.

Clearly, (3.1) is a nonlinear extension of (2.9). We note that the method used in the previous section does not apply to (3.1) any more. To establish the well-posedness of the nonlinear equation (3.1), we will use a different method based on the contraction mapping theorem. Thus, such a method is referred to as the *method of contraction mapping*.

Let us first introduce some notation. Define

$$\langle A,B\rangle\stackrel{\Delta}{=}\mathrm{tr}\,\{AB^\top\},\qquad \forall A,B\in\mathbb{R}^{k\times m}. \tag{3.2}$$

Clearly, (3.2) defines an inner product under which $\mathbb{R}^{k\times m}$ is a Hilbert space. Let $|\cdot|$ be the norm induced by (3.2) and $\|\cdot\|$ the usual matrix norm in $\mathbb{R}^{k\times m}$. Then

$$\begin{aligned}\|A\|&\stackrel{\Delta}{=}\sqrt{\max\sigma(AA^\top)}\leq\sqrt{\mathrm{tr}\,\{AA^\top\}}\equiv|A|\\ &\leq\sqrt{k\wedge m}\sqrt{\max\sigma(AA^\top)}\equiv\sqrt{k\wedge m}\,\|A\|,\qquad\forall A\in\mathbb{R}^{k\times m},\end{aligned} \tag{3.3}$$

where $\sigma(AA^\top)$ is the set of all eigenvalues of $AA^\top$. Next, for any $\beta \in \mathbb{R}$, we define $\mathcal{M}_\beta[0,T]$ to be the Banach space

$$\mathcal{M}_\beta[0,T] = L^2_{\mathcal{F}}(\Omega; C([0,T];\mathbb{R}^k)) \times L^2_{\mathcal{F}}(0,T;\mathbb{R}^{k\times m})\Big(\stackrel{\Delta}{=}\mathcal{M}[0,T]\Big), \tag{3.4}$$

equipped with the norm

$$\begin{aligned}&\|(Y(\cdot),Z(\cdot))\|_{\mathcal{M}_\beta[0,T]}\\ &\quad\stackrel{\Delta}{=}\Big\{E\Big(\sup_{t\in[0,T]}|Y(t)|^2e^{2\beta t}\Big)+E\int_0^T|Z(t)|^2e^{2\beta t}dt\Big\}^{1/2}.\end{aligned}\tag{3.5}$$

Since $0<T<\infty$, all the norms $\|\cdot\|_{\mathcal{M}_\beta[0,T]}$ with different $\beta\in\mathbb{R}$ are equivalent. However, it will be seen below that the introduction of different norms is useful.

Let us now make the following assumption on the function $h:[0,T]\times\mathbb{R}^k\times\mathbb{R}^{k\times m}\times\Omega\to\mathbb{R}^k$.

(B) For any $(y,z)\in\mathbb{R}^k\times\mathbb{R}^{k\times m}$, $h(t,y,z)$ is $\{\mathcal{F}_t\}_{t\ge0}$-adapted with $h(\cdot\,,0,0)\in L^2_{\mathcal{F}}(0,T;\mathbb{R}^k)$. Moreover, there exists an $L>0$ such that

$$\begin{aligned}&|h(t,y,z)-h(t,\bar y,\bar z)|\le L\{|y-\bar y|+|z-\bar z|\},\\ &\qquad\qquad\forall t\in[0,T],\ y,\bar y\in\mathbb{R}^k,\ z,\bar z\in\mathbb{R}^{k\times m},\ \mathbf{P}\text{-a.s.}\end{aligned}\tag{3.6}$$

Definition 3.1. A pair of processes $(Y(\cdot),Z(\cdot))\in\mathcal{M}[0,T]$ is called an *(adapted) solution* of (3.1) if the following holds:

$$Y(t)=\xi-\int_t^T h(s,Y(s),Z(s))ds-\int_t^T Z(s)ds,\quad \forall t\in[0,T],\ \mathbf{P}\text{-a.s.}\tag{3.7}$$

Moreover, the *uniqueness* of solutions can be defined in the same way as in Definition 2.1.

The main result of this subsection is the following.

Theorem 3.2. *Let (B) hold. Then for any given $\xi\in L^2_{\mathcal{F}_T}(\Omega;\mathbb{R}^k)$, the BSDE (3.1) admits a unique adapted solution $(Y(\cdot),Z(\cdot))\in\mathcal{M}[0,T]$.*

Proof. For any fixed $(y(\cdot),z(\cdot))\in\mathcal{M}[0,T]$, it follows from (B) that

$$h(\cdot)\equiv h(\cdot\,,y(\cdot),z(\cdot))\in L^2_{\mathcal{F}}(0,T;\mathbb{R}^k).\tag{3.8}$$

Consider the following BSDE:

$$\begin{cases}dY(t)=h(t,y(t),z(t))dt+Z(t)dW(t),\\ Y(T)=\xi.\end{cases}\tag{3.9}$$

This is a linear BSDE. By Theorem 2.2, it admits a unique adapted solution $(Y(\cdot),Z(\cdot))\in\mathcal{M}_\beta[0,T]$. Hence, we can define an operator $\mathcal{T}:\mathcal{M}_\beta[0,T]\to$

$\mathcal{M}_\beta[0,T]$ by $(y,z)\mapsto(Y,Z)$ via the BSDE (3.9). We are going to prove that for some $\beta>0$,

$$\|\mathcal{T}(y,z)-\mathcal{T}(\overline{y},\overline{z})\|_{\mathcal{M}_\beta[0,T]}\le\frac{1}{2}\|(y,z)-(\overline{y},\overline{z})\|_{\mathcal{M}_\beta[0,T]},\qquad \forall (y,z),\ (\overline{y},\overline{z})\in\mathcal{M}_\beta[0,T]. \tag{3.10}$$

This means that $\mathcal{T}$ is a contraction mapping on the Banach space $\mathcal{M}_\beta[0,T]$. Then we can use the contraction mapping theorem (Zeidler [1, p. 17]) to claim the existence and uniqueness of the fixed point of $\mathcal{T}$, which is the unique adapted solution of (3.1).

To prove (3.10), take any $(y(\cdot),z(\cdot)),(\overline{y}(\cdot),\overline{z}(\cdot))\in\mathcal{M}[0,T]$, and let

$$(Y(\cdot),Z(\cdot))=\mathcal{T}(y,z),\quad (\overline{Y}(\cdot),\overline{Z}(\cdot))=\mathcal{T}(\overline{y},\overline{z}).$$

Define

$$\begin{cases}\widehat{Y}(t)\stackrel{\Delta}{=}Y(t)-\overline{Y}(t), & \widehat{Z}(t)\stackrel{\Delta}{=}Z(t)-\overline{Z}(t),\\ \widehat{y}(t)\stackrel{\Delta}{=}y(t)-\overline{y}(t), & \widehat{z}(t)\stackrel{\Delta}{=}z(t)-\overline{z}(t),\\ \widehat{h}(t)\stackrel{\Delta}{=}h(t,y(t),z(t))-h(t,\overline{y}(t),\overline{z}(t)).\end{cases} \tag{3.11}$$

Let $\beta>0$ be undetermined. Applying Itô's formula to $|\widehat{Y}(t)|^2e^{2\beta t}$, we have (noting (3.6))

$$\begin{aligned}&|\widehat{Y}(t)|^2e^{2\beta t}+\int_t^T|\widehat{Z}(s)|^2e^{2\beta s}ds\\
&=-\int_t^T\{2\beta|\widehat{Y}(s)|^2+2\langle\widehat{Y}(s),\widehat{h}(s)\rangle\}e^{2\beta s}ds\\
&\quad-\int_t^T2e^{2\beta s}\langle\widehat{Y}(s),\widehat{Z}(s)dW(s)\rangle\\
&\le\int_t^T\{-2\beta|\widehat{Y}(s)|^2+2L|\widehat{Y}(s)|(|\widehat{y}(s)|+|\widehat{z}(s)|)\}e^{2\beta s}ds\\
&\quad-2\int_t^Te^{2\beta s}\langle\widehat{Y}(s),\widehat{Z}(s)dW(s)\rangle\\
&\le\int_t^T\{(-2\beta+\frac{2L^2}{\lambda})|\widehat{Y}(s)|^2+\lambda(|\widehat{y}(s)|^2+|\widehat{z}(s)|^2)\}e^{2\beta s}ds\\
&\quad-\int_t^T2e^{2\beta s}\langle\widehat{Y}(s),\widehat{Z}(s)dW(s)\rangle,\end{aligned} \tag{3.12}$$

where we take $\lambda \stackrel{\Delta}{=} \frac{2L^2}{\beta} > 0$. Then the above implies

$$
\begin{aligned}
&|\widehat{Y}(t)|^2 e^{2\beta t} + \int_t^T |\widehat{Z}(s)|^2 e^{2\beta s} ds \\
&\quad \le \lambda(T+1)\Big\{ \sup_{t\in[0,T]} \big(|\widehat{y}(t)|^2 e^{2\beta t}\big) + \int_0^T |\widehat{z}(s)|^2 e^{2\beta s} ds \Big\} \\
&\qquad - \int_t^T 2e^{2\beta s} \langle \widehat{Y}(s), \widehat{Z}(s) dW(s) \rangle .
\end{aligned} \tag{3.13}
$$

By taking the expectation, one obtains

$$
E\Big\{|\widehat{Y}(t)|^2 e^{2\beta t} + \int_t^T |\widehat{Z}(s)|^2 e^{2\beta s} ds\Big\} \le \lambda(T+1)\|(\widehat{y},\widehat{z})\|^2_{\mathcal{M}_\beta[0,T]}. \tag{3.14}
$$

On the other hand, by Burkholder–Davis–Gundy's inequality, we have (noting (3.14))

$$
\begin{aligned}
&E\Big\{ \sup_{t\in[0,T]} \Big| \int_t^T e^{2\beta s} \langle \widehat{Y}(s), \widehat{Z}(s) dW(s) \rangle \Big| \Big\} \\
&\le E\Big| \int_0^T e^{2\beta s} \langle \widehat{Y}(s), \widehat{Z}(s) dW(s) \rangle \Big| \\
&\quad + E\Big\{ \sup_{t\in[0,T]} \Big| \int_0^t e^{2\beta s} \langle \widehat{Y}(s), \widehat{Z}(s) dW(s) \rangle \Big| \Big\} \\
&\le 2E\Big\{ \sup_{t\in[0,T]} \Big| \int_0^t e^{2\beta s} \langle \widehat{Y}(s), \widehat{Z}(s) dW(s) \rangle \Big| \Big\} \\
&\le KE\Big\{ \int_0^T |\widehat{Y}(s)|^2 |\widehat{Z}(s)|^2 e^{4\beta s} ds \Big\}^{1/2} \\
&\le KE\Big\{ \Big(\sup_{t\in[0,T]} \big(|\widehat{Y}(t)|^2 e^{2\beta t}\big)\Big)^{1/2} \Big(\int_0^T |\widehat{Z}(s)|^2 e^{2\beta s} ds \Big)^{1/2} \Big\} \\
&\le \frac{1}{4} E\Big(\sup_{t\in[0,T]} \big(|\widehat{Y}(t)|^2 e^{2\beta t}\big)\Big) + K^2 E \int_0^T |\widehat{Z}(s)|^2 e^{2\beta s} ds \\
&\le \frac{1}{4} E\Big(\sup_{t\in[0,T]} \big(|\widehat{Y}(t)|^2 e^{2\beta t}\big)\Big) + K^2 \lambda(T+1)\|(\widehat{y},\widehat{z})\|^2_{\mathcal{M}_\beta[0,T]}.
\end{aligned} \tag{3.15}
$$

Consequently, from (3.13), we obtain (using (3.15))

$$
\begin{aligned}
&E\Big(\sup_{t\in[0,T]} \big(|\widehat{Y}(t)|^2 e^{2\beta t}\big)\Big) \\
&\le \lambda(T+1)\|(\widehat{y},\widehat{z})\|^2_{\mathcal{M}_\beta[0,T]} \\
&\quad + 2E\Big\{ \sup_{t\in[0,T]} \Big| \int_t^T e^{2\beta s} \langle \widehat{Z}(s)^T \widehat{Y}(s), dW(s) \rangle \Big| \Big\} \\
&\le (1+2K^2)\lambda(T+1)\|(\widehat{y},\widehat{z})\|^2_{\mathcal{M}_\beta[0,T]} + \frac{1}{2} E\Big(\sup_{t\in[0,T]} \big(|\widehat{Y}(t)|^2 e^{2\beta t}\big)\Big).
\end{aligned} \tag{3.16}
$$

Combining (3.14) and (3.16) yields (noting $\lambda = \frac{2L^2}{\beta}$)

$$\|(\widehat{Y},\widehat{Z})\|^2_{\mathcal{M}_\beta[0,T]} \le \frac{2(3+4K^2)(T+1)L^2}{\beta}\|(\widehat{y},\widehat{z})\|^2_{\mathcal{M}_\beta[0,T]}. \tag{3.17}$$

Then we can choose $\beta > 0$ large enough to get the contractivity of the operator $\mathcal{T}$ on $\mathcal{M}_\beta[0,T]$, which in turn implies the existence and the uniqueness of the adapted solution to (3.1). □

We now prove the continuous dependence of the solutions on the terminal value ξ as well as the function h.

Theorem 3.3. *Let $h, \overline{h} : [0,T]\times\mathbb{R}^k\times\mathbb{R}^{k\times m}\times\Omega \to \mathbb{R}^k$ satisfy (B) and let $\xi, \overline{\xi} \in L^2_{\mathcal{F}}(\Omega;\mathbb{R}^k)$. Let $(Y(\cdot),Z(\cdot)), (\overline{Y}(\cdot),\overline{Z}(\cdot)) \in \mathcal{M}[0,T]$ be the solutions of (3.1) corresponding to (h,ξ) and $(\overline{h},\overline{\xi})$, respectively. Then*

$$\begin{aligned}&\|(Y(\cdot)-\overline{Y}(\cdot), Z(\cdot)-\overline{Z}(\cdot))\|^2_{\mathcal{M}[0,T]}\\ &\le K\Big\{E|\xi-\overline{\xi}|^2 + E\int_0^T |h(s,Y(s),Z(s)) - \overline{h}(s,Y(s),Z(s))|^2 ds\Big\}.\end{aligned} \tag{3.18}$$

Proof. Define

$$\begin{cases}\widehat{Y}(\cdot) \stackrel{\Delta}{=} Y(\cdot)-\overline{Y}(\cdot), \quad \widehat{Z}(\cdot)\stackrel{\Delta}{=} Z(\cdot)-\overline{Z}(\cdot),\\ \widehat{\xi}\stackrel{\Delta}{=}\xi-\overline{\xi}, \quad \widehat{h}(\cdot)\stackrel{\Delta}{=} h(\cdot\,,Y(\cdot),Z(\cdot)) - \overline{h}(\cdot\,,Y(\cdot),Z(\cdot)).\end{cases} \tag{3.19}$$

Applying Itô's formula to $|\widehat{Y}(\cdot)|^2$, we obtain

$$\begin{aligned}&|\widehat{Y}(t)|^2 + \int_t^T |\widehat{Z}(s)|^2 ds\\ &= |\widehat{\xi}|^2 - 2\int_t^T \langle \widehat{Y}(s), h(s,Y(s),Z(s)) - \overline{h}(s,\overline{Y}(s),\overline{Z}(s))\rangle\, ds\\ &\quad -2\int_t^T \langle \widehat{Y}(s), \widehat{Z}(s)dW(s)\rangle\\ &\le |\widehat{\xi}|^2 + 2\int_t^T \{|\widehat{Y}(s)||\widehat{h}(s)| + L|\widehat{Y}(s)|(|\widehat{Y}(s)| + |\widehat{Z}(s)|)\}ds\\ &\quad -2\int_t^T \langle \widehat{Y}(s), \widehat{Z}(s)dW(s)\rangle\\ &\le |\widehat{\xi}|^2 + \int_t^T \{(1+2L+2L^2)|\widehat{Y}(s)|^2 + \frac{1}{2}|\widehat{Z}(s)|^2 + |\widehat{h}(s)|^2\}ds\\ &\quad -2\int_t^T \langle \widehat{Y}(s), \widehat{Z}(s)dW(s)\rangle.\end{aligned} \tag{3.20}$$

Taking the expectation in the above, we have

$$
\begin{aligned}
&E|\widehat{Y}(t)|^2 + \frac{1}{2}E\int_t^T |\widehat{Z}(s)|^2 ds \\
&\leq E|\widehat{\xi}|^2 + E\int_0^T |\widehat{h}(s)|^2 ds + (1+2L+2L^2)\int_t^T E|\widehat{Y}(s)|^2 ds, \quad t \in [0,T].
\end{aligned} \tag{3.21}
$$

Thus, by Gronwall's inequality,

$$
E|\widehat{Y}(t)|^2 + E\int_t^T |\widehat{Z}(s)|^2 ds \leq K\Big\{E|\widehat{\xi}|^2 + E\int_0^T |\widehat{h}(s)|^2 ds\Big\}, \qquad \forall t \in [0,T]. \tag{3.22}
$$

On the other hand, by Burkholder-Davis-Gundy's inequality, it follows from (3.20) that (see (3.15) for a similar argument)

$$
\begin{aligned}
E\{\sup_{t\in[0,T]} |\widehat{Y}(t)|^2\} &\leq K\Big\{E|\widehat{\xi}|^2 + E\int_0^T |\widehat{h}(s)|^2 ds\Big\} \\
&\quad + 2E\sup_{t\in[0,T]} \Big|\int_0^t \langle \widehat{Y}(s), \widehat{Z}(s)dW(s)\rangle\Big| \\
&\leq K\Big\{E|\widehat{\xi}|^2 + E\int_0^T |\widehat{h}(s)|^2 ds\Big\} \\
&\quad + K\Big(E\sup_{t\in[0,T]} |\widehat{Y}(t)|^2\Big)^{1/2}\Big(E\int_0^T |\widehat{Z}(s)|^2 ds\Big)^{1/2}.
\end{aligned} \tag{3.23}
$$

Then (3.18) follows easily from (3.23) and (3.22). □

We have established the well-posedness of BSDE (3.1) by the method of contraction mapping. From the numerical point of view, however, one might like to have some iterative scheme of computing the solution along with its convergence speed. In the theory of ordinary differential equations, one has the so-called *Picard iteration*. We now present a similar algorithm for BSDEs. Due to the nature of the BSDEs, the iteration is slightly different.

We assume that (B) holds and consider BSDE (3.1) with $\xi \in L^2_{\mathcal{F}_T}(\Omega;\mathbb{R}^k)$. Set

$$
Y^0(t) \equiv 0, \quad Z^0(t) \equiv 0, \qquad \forall t \in [0,T]. \tag{3.24}
$$

For any $i \geq 0$, we solve the following linear BSDE:

$$
\begin{cases} dY^{i+1}(t) = h(t, Y^i(t), Z^i(t))dt + Z^{i+1}(t)dW(t), & t \in [0,T), \\ Y^{i+1}(T) = \xi. \end{cases} \tag{3.25}
$$

We should note here that unlike the situation in the ordinary differential equation case, the above iteration is still *implicit*. The reason is that $Z^{i+1}(\cdot)$ is obtained via the martingale representation theorem at each iteration. By Theorem 2.2, (3.25) admits a unique adapted solution $(Y^{i+1}(\cdot), Z^{i+1}(\cdot)) \in$

$\mathcal{M}[0,T]$ for each $i \geq 0$. Let us now state the convergence result along with an error estimate.

Theorem 3.4. *Let assumption (B) hold and $\xi \in L^2_{\mathcal{F}_T}(\Omega;\mathbb{R}^k)$. Let $(Y(\cdot),Z(\cdot)) \in \mathcal{M}[0,T]$ be the unique adapted solution of (3.1) and let the sequence $(Y^i(\cdot),Z^i(\cdot)) \in \mathcal{M}[0,T]$ be constructed by (3.24)–(3.25). Then there exists a constant $K > 0$ such that*

$$\|(Y^i(\cdot),Z^i(\cdot)) - (Y(\cdot),Z(\cdot))\|_{\mathcal{M}[0,T]} \leq Ke^{-i}, \quad \forall i \geq 1. \tag{3.26}$$

Proof. By the definition of $\mathcal{T}$ (see the proof of Theorem 3.2), we have

$$(Y^{i+1},Z^{i+1}) = \mathcal{T}(Y^i,Z^i), \qquad i \geq 0. \tag{3.27}$$

Let us define

$$\begin{cases} \widehat{Y}^{i+1}(t) \stackrel{\Delta}{=} Y^{i+1}(t) - Y^i(t), \\ \widehat{Z}^{i+1}(t) \stackrel{\Delta}{=} Z^{i+1}(t) - Z^i(t), \end{cases} \qquad i \geq 0. \tag{3.28}$$

Then, by choosing $\beta > 0$ large enough (see (3.17)), we can have

$$\|(\widehat{Y}^{i+1},\widehat{Z}^{i+1})\|_{\mathcal{M}_\beta[0,T]} \leq e^{-1}\|(\widehat{Y}^i,\widehat{Z}^i)\|_{\mathcal{M}_\beta[0,T]}, \qquad i \geq 1. \tag{3.29}$$

Consequently, by induction and noting that $\beta > 0$, we obtain

$$\begin{aligned} \|(\widehat{Y}^{i+1},\widehat{Z}^{i+1})\|_{\mathcal{M}[0,T]} &\leq \|(\widehat{Y}^{i+1},\widehat{Z}^{i+1})\|_{\mathcal{M}_\beta[0,T]} \\ &\leq e^{-i}\|(Y^1,Z^1)\|_{\mathcal{M}_\beta[0,T]} \leq Ke^{-i}, \quad i \geq 1. \end{aligned} \tag{3.30}$$

Note that the constant $K > 0$ may depend on β, but it is uniform in $i \geq 1$. Hence, if $(Y,Z) \in \mathcal{M}[0,T]$ is the adapted solution of (3.1), then

$$\begin{aligned} \|(Y^i,Z^i) - (Y,Z)\|_{\mathcal{M}[0,T]} &\leq \sum_{\ell=i}^{\infty} \|(\widehat{Y}^{\ell+1},\widehat{Z}^{\ell+1})\|_{\mathcal{M}_\beta[0,T]} \\ &\leq Ke^{-i}, \qquad \forall i \geq 1, \end{aligned} \tag{3.31}$$

with the constant K possibly different from the one in (3.30). This proves (3.26). □

3.2. BSDEs in random durations: Method of continuation

In this subsection we are going to consider BSDEs in random time durations. By this we mean that the time durations under consideration are determined by some stopping times. Such a consideration is not simply a routine extension of the results presented in the previous subsection. Rather, it is motivated by the study of *elliptic* PDEs, which we will demonstrate in Section 4. Also, for optimal control problems in a random duration determined by a stopping time (see Chapter 2, Section 7.1), the first- and second-order adjoint equations along the optimal pair are BSDEs with a random duration. Another interesting application is in mathematical finance (see Section 6).

Let τ be an $\{\mathcal{F}_t\}_{t\geq 0}$-stopping time taking values in $[0,\infty]$. Consider the following BSDE (compare with (3.1)):

$$
\begin{cases} dY(t) = h(t, Y(t), Z(t))dt + Z(t)dW(t), & t \in [0, \tau), \\ Y(\tau) = \xi. \end{cases} \tag{3.32}
$$

The only difference between (3.1) and (3.32) lies in that they have different types of time durations. Of course, (3.1) is a special case of (3.32) in which the stopping time $\tau \equiv T$.

Note that the method of contraction mapping used in the previous subsection heavily relies on the assumption that the final time T is finite and deterministic. This seems necessary for finding a $\beta > 0$ such that the map $\mathcal{T}$ defined in the previous subsection (right before (3.10)) is a contraction on $\mathcal{M}_\beta[0,T]$ (see (3.17)). In the present case, the final time τ is an $\{\mathcal{F}_t\}_{t\geq 0}$-stopping time that could even be infinite for some $\omega \in \Omega$. Thus, the method used in Section 3.1 does not apply here. To overcome this difficulty and obtain the well-posedness for (3.32), we will introduce another method, called the *method of continuation.* This is essentially adopted from the theory of partial differential equations, where such a method has been used to establish the well-posedness of general second-order elliptic PDEs. The key idea is to prove some uniform a priori estimate for a class of BSDEs parametrized by, say, $\alpha \in [0,1]$ such that the BSDE corresponding to $\alpha = 0$ is a simple one whose well-posedness can be easily proved, and the BSDE corresponding to $\alpha = 1$ is (3.32). After obtaining such a uniform a priori estimate, by the continuation argument we can then establish the well-posedness of (3.32).

To carry out the method of continuation, let us first introduce some spaces. In what follows, we fix the stopping time τ. For any $\beta \in \mathbb{R}$, let $L^{2,\beta}_{\mathcal{F}}(0,\tau;\mathbb{R}^k)$ be the set of all $\mathbb{R}^k$-valued $\{\mathcal{F}_t\}_{t\geq 0}$-adapted processes $Y(\cdot)$ such that

$$
E\int_0^\tau |Y(t)|^2 e^{2\beta t}dt \stackrel{\Delta}{=} \lim_{T\to\infty} E\int_0^{T\wedge\tau} |Y(t)|^2 e^{2\beta t}dt < \infty. \tag{3.33}
$$

We will simply set $L^2_{\mathcal{F}}(0,\tau;\mathbb{R}^k) \stackrel{\Delta}{=} L^{2,0}_{\mathcal{F}}(0,\tau;\mathbb{R}^k)$. It is not hard to see that $L^{2,\beta}_{\mathcal{F}}(0,\tau;\mathbb{R}^k)$ is increasing as $\beta \in \mathbb{R}$ decreases. In particular, when $\beta < 0$, $L^{2,\beta}_{\mathcal{F}}(0,\tau;\mathbb{R}^k)$ contains $L^2_{\mathcal{F}}(0,\infty;\mathbb{R}^k)$, the set of all square-integrable $\{\mathcal{F}_t\}_{t\geq 0}$-adapted processes defined on $[0,\infty)$, valued in $\mathbb{R}^k$. Similarly, we can define $L^{2,\beta}_{\mathcal{F}}(0,\tau;\mathbb{R}^{k\times m})$. Next, we let $L^{2,\beta}_{\mathcal{F}}(\Omega;C([0,\tau];\mathbb{R}^k))$ be the set of all $\mathbb{R}^k$-valued $\{\mathcal{F}_t\}_{t\geq 0}$-adapted continuous processes $Y(\cdot)$ such that

$$
E\Big\{\sup_{t\in[0,\tau)} \big[|Y(t)|^2 e^{2\beta t}\big]\Big\} < \infty. \tag{3.34}
$$

We should note that in general, unlike the case $\tau \equiv T$, it is not necessarily true that

$$
L^{2,\beta}_{\mathcal{F}}(\Omega;C([0,\tau];\mathbb{R}^k)) \subseteq L^{2,\beta}_{\mathcal{F}}(0,\tau;\mathbb{R}^k), \tag{3.35}
$$

since τ could be ∞ on some subset of Ω, with a positive probability. Even if $\tau < \infty$ almost surely, (3.35) could still be false. Here is a simple example. Suppose τ is an $\{\mathcal{F}_t\}_{t\geq 0}$-stopping time such that $E\tau^2 < \infty$ and $E\tau^3 = \infty$. Then, for $Y(t) \equiv t$, we see that (with $\beta = 0$)

$$E \sup_{t\in[0,\tau)} |Y(t)|^2 = E\tau^2 < \infty; \tag{3.36}$$

but

$$E\int_0^\tau |Y(t)|^2 dt = \frac{1}{3}E\tau^3 = \infty. \tag{3.37}$$

Thus, unlike the case of deterministic finite time duration, we need to define the space (compare with (3.4))

$$\begin{aligned}\mathcal{M}_\beta[0,\tau] \triangleq \{L_{\mathcal{F}}^{2,\beta}(\Omega; C([0,\tau];\mathbb{R}^k)) \bigcap L_{\mathcal{F}}^{2,\beta}(0,\tau;\mathbb{R}^k)\} \\ \times L_{\mathcal{F}}^{2,\beta}(0,\tau;\mathbb{R}^{k\times m}),\end{aligned} \tag{3.38}$$

with the norm (compare with (3.5))

$$\begin{aligned}\|(Y(\cdot),Z(\cdot))\|_{\mathcal{M}_\beta[0,\tau]} = \Big\{E\Big[\sup_{t\in[0,\tau)}\{|Y(t)|^2 e^{2\beta t}\} \\ + \int_0^\tau |Y(s)|^2 e^{2\beta s} ds + \int_0^\tau |Z(s)|^2 e^{2\beta s} ds\Big]\Big\}^{1/2}, \\ \forall (Y(\cdot),Z(\cdot)) \in \mathcal{M}_\beta[0,\tau].\end{aligned} \tag{3.39}$$

One can verify that $\mathcal{M}_\beta[0,\tau]$ is a Banach space under the norm (3.39). Finally, we define $L_{\mathcal{F}_\tau}^{2,\beta}(\Omega;\mathbb{R}^k)$ to be the set of all $\mathbb{R}^k$-valued $\mathcal{F}_\tau$-measurable random variables ξ such that $E\big(|\xi|^2 e^{2\beta\tau}\big) < \infty$. Again, if $\beta < 0$, then $L_{\mathcal{F}_\tau}^{2,\beta}(\Omega;\mathbb{R}^k)$ contains $L_{\mathcal{F}_\tau}^2(\Omega;\mathbb{R}^k)$, the set of all square integrable $\mathcal{F}_\tau$-measurable $\mathbb{R}^k$-valued random variables.

Definition 3.5. A pair of processes $(Y(\cdot),Z(\cdot)) \in \mathcal{M}_\beta[0,\tau]$ is called an *(adapted) solution* of (3.32) *in the space* $\mathcal{M}_\beta[0,\tau]$ if the following hold:

$$\begin{aligned}Y(t\wedge\tau) = Y(T\wedge\tau) - \int_{t\wedge\tau}^{T\wedge\tau} h(s,Y(s),Z(s))ds \\ - \int_{t\wedge\tau}^{T\wedge\tau} Z(s)dW(s), \qquad \forall 0\leq t < T < \infty, \ \mathbf{P}\text{-a.s.},\end{aligned} \tag{3.40}$$

$$\lim_{T\to\infty} E\{|Y(T\wedge\tau) - \xi|^2 e^{2\beta(T\wedge\tau)}\} = 0. \tag{3.41}$$

Equation (3.32) is said to have a *unique* solution in $\mathcal{M}_\beta[0,\tau]$ if for any two solutions $\big(Y(\cdot),Z(\cdot)\big)$, $\big(\tilde{Y}(\cdot),\tilde{Z}(\cdot)\big) \in \mathcal{M}_\beta[0,\tau]$, it must hold that

$$\mathbf{P}\{Y(t) = \tilde{Y}(t), \ \forall t\in[0,\tau] \text{ and } Z(t) = \tilde{Z}(t), \ \text{a.e. } t\in[0,\tau]\} = 1.$$

Note that if there is a deterministic $T > 0$ such that $\tau \le T < \infty$ almost surely, then (3.40)–(3.41) is equivalent to the following:

$$Y(t) = \xi - \int_t^\tau h(s, Y(s), Z(s))ds - \int_t^\tau Z(s)dW(s), \qquad \forall t \in [0, \tau), \ \mathbf{P}\text{-a.s.} \tag{3.42}$$

However, in general we know only that if $(Y(\cdot), Z(\cdot)) \in \mathcal{M}_\beta[0,\tau]$ with $\beta > 0$, then the two integrals in (3.42) are convergent (we will see this below). In the case that $\beta < 0$, the integrals in (3.42) are not necessarily convergent. Thus, we have to define an adapted solution $(Y(\cdot), Z(\cdot))$ of (3.32) by (3.40)–(3.41) for the general case.

Now we make the following assumption on the map $h : [0,\infty) \times \mathbb{R}^k \times \mathbb{R}^{k\times m} \times \Omega \to \mathbb{R}^k$ (compare with Assumption (B) stated after (3.5)).

(B)′ For any $(y,z) \in \mathbb{R}^k \times \mathbb{R}^{k\times m}$, $(t,\omega) \mapsto h(t,y,z;\omega)$ is an $\{\mathcal{F}_t\}_{t\ge 0}$-adapted process. Moreover, there exist constants L, $L_0 > 0$ and $\mu, \beta \in \mathbb{R}$, such that $h(\cdot\,,0,0) \in L^{2,\beta}_{\mathcal{F}}(0,\tau;\mathbb{R}^k)$,

$$|h(t,y,z) - h(t,\bar y,\bar z)| \le L|y-\bar y| + L_0|z-\bar z|, \qquad \forall y,\bar y \in \mathbb{R}^k, \ z,\bar z \in \mathbb{R}^{k\times m}, \ \text{a.e. } t \in [0,\infty), \ \mathbf{P}\text{-a.s.}, \tag{3.43}$$

$$\langle h(t,y,z) - h(t,\bar y,z), y - \bar y\rangle \ge \mu|y-\bar y|^2, \qquad \forall y,\bar y \in \mathbb{R}^k, \ z \in \mathbb{R}^{k\times m}, \ \text{a.e. } t \in [0,\infty), \ \mathbf{P}\text{-a.s.}, \tag{3.44}$$

and

$$\beta > -\mu + \frac{L_0^2}{2}. \tag{3.45}$$

We point out here that in (3.44), the constant μ can be either positive or negative. From (3.43), one sees that (3.44) always holds if $\mu \le -L$. Thus, (3.44) gives some additional restriction in case $\mu > -L$. By and large, Assumption (B)′ gives some compatibility among τ, h, and β (note that τ appears in the definition of $L^{2,\beta}_{\mathcal{F}}(0,\tau;\mathbb{R}^k)$). In what follows, we will show that for such a compatible (τ, h, β), there is a class of terminal values ξ that makes the BSDE (3.32) admit adapted solutions in the space $\mathcal{M}_\beta[0,\tau]$.

Theorem 3.6. *Let (B)′ hold. Then, for any $\xi \in L^{2,\beta}_{\mathcal{F}_\tau}(\Omega;\mathbb{R}^k)$, BSDE (3.32) admits a unique adapted solution $(Y(\cdot), Z(\cdot)) \in \mathcal{M}_\beta[0,\tau]$. Moreover, there exists a constant $K > 0$ such that if $(\overline{Y}(\cdot), \overline{Z}(\cdot)) \in \mathcal{M}_\beta[0,\tau]$ is the adapted solution of (3.32) with h and ξ replaced by $\bar h$ and $\bar\xi$, respectively, where $\bar h$ satisfies (B)′ and $\bar\xi \in L^{2,\beta}_{\mathcal{F}_\tau}(\Omega;\mathbb{R}^k)$, then*

$$\|(Y(\cdot),Z(\cdot)) - (\overline{Y}(\cdot),\overline{Z}(\cdot))\|_{\mathcal{M}_\beta[0,\tau]} K E\Big\{|\xi-\bar\xi|e^{2\beta\tau} + \int_0^\tau |h(s,Y(s),Z(s)) - \bar h(s,Y(s),Z(s))|^2 e^{2\beta s}ds\Big\}^{1/2}. \tag{3.46}$$

Note that in the above theorem, if $\mu > L_0^2/2$, then β could be negative (see (3.45)). In such a case, the terminal state ξ can be any square-integrable $\mathcal{F}_\tau$-measurable random variable, regardless of the behavior of the stopping time τ. On the other hand, if $\tau \le T$ for some deterministic $T > 0$, then all the spaces $\mathcal{M}_\beta[0,\tau]$ and $L^{2,\beta}_{\mathcal{F}_\tau}(\Omega;\mathbb{R}^k)$ for different $\beta \in \mathbb{R}$ coincide, respectively. In such a case, we can always find a $\beta > 0$ satisfying (3.45). Thus, Theorem 3.6 holds for such a β, and hence for any $\beta \in \mathbb{R}$. Therefore, for the case $\tau \le T$, (3.44)–(3.45) are not necessary.

The proof of Theorem 3.6 is technical and lengthy. Readers may skip it at first reading.

We first prove the theorem for the case

$$\beta > -\mu + \frac{L_0^2}{2} = 0. \tag{3.47}$$

The general case will be reduced to this special one by some a transformation.

We first present the solvability of a simple BSDE (with a random time duration). Consider the following:

$$\begin{cases} dY(t) = h_0(t)dt + Z(t)dW(t), & t \in [0,\tau), \\ Y(\tau) = \xi. \end{cases} \tag{3.48}$$

The above equation is similar to (3.9) if we regard $h_0(t) = h(t, y(t), z(t))$. The difference is that (3.9) is in a finite deterministic time duration and (3.48) is in a random time duration. We use Definition 3.5 to define adapted solutions to (3.48). The following result is concerned with the solvability of (3.48).

Lemma 3.7. *Let $\beta > 0$, $h_0 \in L^{2,\beta}_{\mathcal{F}}(0,\tau;\mathbb{R}^k)$, and $\xi \in L^{2,\beta}_{\mathcal{F}_\tau}(\Omega;\mathbb{R}^k)$. Then (3.48) admits a unique adapted solution $(Y(\cdot), Z(\cdot)) \in \mathcal{M}_\beta[0,\tau]$. Moreover, there exists a constant $K > 0$ such that*

$$\|(Y(\cdot),Z(\cdot))\|^2_{\mathcal{M}_\beta[0,\tau]} \le K\Big\{E\big(|\xi|^2 e^{2\beta\tau}\big) + E\int_0^\tau |h_0(s)|^2 e^{2\beta s} ds\Big\}. \tag{3.49}$$

The idea of the proof is basically the same as that for (3.9) (i.e., that of Theorem 2.2). However, since now the time duration is determined by a stopping time, we need to make some interesting modifications. The reader is encouraged to provide a proof without looking at the following.

Proof. First of all, since $\beta > 0$, we have $\xi \in L^{2,\beta}_{\mathcal{F}_\tau}(\Omega;\mathbb{R}^k) \subseteq L^2_{\mathcal{F}_\tau}(\Omega;\mathbb{R}^n)$ and $h_0 \in L^{2,\beta}_{\mathcal{F}}(0,\tau;\mathbb{R}^k) \subseteq L^2_{\mathcal{F}}(0,\tau;\mathbb{R}^k)$. Thus, we can define

$$\begin{cases} M(t) \stackrel{\Delta}{=} E\Big\{\xi - \int_0^\tau h_0(s)ds \big| \mathcal{F}_t\Big\}, \\ Y(t) \stackrel{\Delta}{=} M(t\wedge\tau) + \int_0^{t\wedge\tau} h_0(s)ds \equiv Y(t\wedge\tau), \end{cases} \quad t \in [0,\tau), \ \mathbf{P}\text{-a.s.} \tag{3.50}$$

By Jensen's inequality, $E|M(t)|^2 < \infty$. This implies that $M(t)$ is a square-integrable $\{\mathcal{F}_t\}_{t\geq 0}$-martingale. By the martingale representation theorem, we have

$$M(t) = M(0) + \int_0^t Z(s)dW(s), \qquad \forall 0 \leq t \leq \tau, \ \mathbf{P}\text{-a.s.}, \tag{3.51}$$

for some $Z(\cdot) \in L_{\mathcal{F}}^{2,loc}(0,\infty;\mathbb{R}^{k\times m})$. Then (noting (3.50))

$$\begin{aligned} M(0) &= M(T\wedge\tau) - \int_0^{T\wedge\tau} Z(s)dW(s) \\ &= Y(T\wedge\tau) - \int_0^{T\wedge\tau} h_0(s)ds - \int_0^{T\wedge\tau} Z(s)dW(s), \quad T>0. \end{aligned} \tag{3.52}$$

Hence, combining (3.50)–(3.52), we obtain (for $0 \leq t < T < \infty$)

$$\begin{aligned} Y(t\wedge\tau) &\equiv M(t\wedge\tau) + \int_0^{t\wedge\tau} h_0(s)ds \\ &= M(0) + \int_0^{t\wedge\tau} Z(s)dW(s) + \int_0^{t\wedge\tau} h_0(s)ds \\ &= Y(T\wedge\tau) - \int_{t\wedge\tau}^{T\wedge\tau} h_0(s)ds - \int_{t\wedge\tau}^{T\wedge\tau} Z(s)dW(s). \end{aligned} \tag{3.53}$$

This implies the corresponding (3.40) for equation (3.48). We need to prove in addition that $(Y(\cdot),Z(\cdot)) \in \mathcal{M}_\beta[0,\tau]$, and (3.49) and (3.41) hold. To this end, observe that by (3.50) and Jensen's inequality, one has

$$\begin{aligned} &E\Big(|Y(T\wedge\tau)|^2 e^{2\beta(T\wedge\tau)}\Big) \\ &= E\Big\{\Big|E\Big(\xi - \int_{T\wedge\tau}^{\tau} h_0(s)ds\Big|\mathcal{F}_{T\wedge\tau}\Big)\Big|^2 e^{2\beta(T\wedge\tau)}\Big\} \\ &\leq E\Big\{\Big|\xi - \int_{T\wedge\tau}^{\tau} h_0(s)ds\Big|^2 e^{2\beta(T\wedge\tau)}\Big\} \\ &\leq 2E\big(|\xi|^2 e^{2\beta(T\wedge\tau)}\big) + 2E\Big(\Big|\int_{T\wedge\tau}^{\tau} h_0(s)ds\Big|^2 e^{2\beta(T\wedge\tau)}\Big) \\ &\leq 2E\big(|\xi|^2 e^{2\beta\tau}\big) \\ &\quad + 2E\Big\{\Big(\int_{T\wedge\tau}^{\tau} e^{-2\beta s}ds\Big)\Big(\int_{T\wedge\tau}^{\tau} e^{2\beta s}|h_0(s)|^2 ds\Big)e^{2\beta(T\wedge\tau)}\Big\} \\ &\leq 2E\big(|\xi|^2 e^{2\beta\tau}\big) + \frac{1}{\beta}E\int_0^{\tau} |h_0(s)|^2 e^{2\beta s}ds, \quad \forall T>0. \end{aligned} \tag{3.54}$$

Then, applying the Itô formula to $|Y(t)|^2e^{2\beta t}$, one has $(0 \le t < T < \infty)$

$$
\begin{aligned}
&|Y(T\wedge\tau)|^2 e^{2\beta(T\wedge\tau)} \\
&= |Y(t\wedge\tau)|^2 e^{2\beta(t\wedge\tau)} + 2\int_{t\wedge\tau}^{T\wedge\tau} e^{2\beta s}\langle Y(s), Z(s)dW(s)\rangle \\
&\quad + \int_{t\wedge\tau}^{T\wedge\tau} \{2\beta|Y(s)|^2 + 2\langle Y(s), h_0(s)\rangle + |Z(s)|^2\} e^{2\beta s} ds \\
&\ge |Y(t\wedge\tau)|^2 e^{2\beta(t\wedge\tau)} + \int_{t\wedge\tau}^{T\wedge\tau} \{(2\beta-\varepsilon)|Y(s)|^2 + |Z(s)|^2\} e^{2\beta s} ds \\
&\quad - \frac{1}{\varepsilon}\int_{t\wedge\tau}^{T\wedge\tau} |h_0(s)|^2 e^{2\beta s} ds + 2\int_{t\wedge\tau}^{T\wedge\tau} e^{2\beta s}\langle Y(s), Z(s)dW(s)\rangle,
\end{aligned}
\tag{3.55}
$$

where $\varepsilon \in (0,\beta)$. Taking the expectation in (3.55) and using (3.54), we obtain the following:

$$
\begin{aligned}
&E\Big\{|Y(t\wedge\tau)|^2 e^{2\beta(t\wedge\tau)} + \int_{t\wedge\tau}^{T\wedge\tau} (|Y(s)|^2 + |Z(s)|^2) e^{2\beta s} ds\Big\} \\
&\le KE\Big\{|Y(T\wedge\tau)|^2 e^{2\beta(T\wedge\tau)} + \int_{t\wedge\tau}^{T\wedge\tau} |h_0(s)|^2 e^{2\beta s} ds\Big\} \\
&\le KE\Big\{|\xi|^2 e^{2\beta\tau} + \int_0^\tau |h_0(s)|^2 e^{2\beta s} ds\Big\}, \quad \forall 0 \le t < T < \infty.
\end{aligned}
\tag{3.56}
$$

Consequently,

$$
\begin{aligned}
&E\Big\{|Y(t\wedge\tau)|^2 e^{2\beta(t\wedge\tau)} + \int_0^\tau (|Y(s)|^2 + |Z(s)|^2) e^{2\beta s} ds\Big\} \\
&\le KE\Big\{|\xi|^2 e^{2\beta\tau} + \int_0^\tau |h_0(s)|^2 e^{2\beta s} ds\Big\}, \quad \forall t \ge 0.
\end{aligned}
\tag{3.57}
$$

On the other hand, using Burkholder–Davis–Gundy's inequality and noting (3.55) and (3.57), we have

$$
\begin{aligned}
&E\Big(\sup_{t\in[0,T\wedge\tau)} (|Y(t)|^2 e^{2\beta t})\Big) \\
&\le E(|Y(T\wedge\tau)|^2 e^{2\beta(T\wedge\tau)}) + \frac{1}{\varepsilon}E\int_0^\tau |h_0(s)|^2 e^{2\beta s} ds \\
&\quad + 2E\sup_{t\in[0,T\wedge\tau)}\Big|\int_t^{T\wedge\tau} e^{2\beta s}\langle Y(s), Z(s)dW(s)\rangle\Big| \\
&\le K\Big(E(|\xi|^2 e^{2\beta\tau}) + E\int_0^\tau |h_0(s)|^2 e^{2\beta s} ds\Big) \\
&\quad + KE\Big(\int_0^{T\wedge\tau} |Y(s)|^2 |Z(s)|^2 e^{4\beta s} ds\Big)^{1/2}.
\end{aligned}
\tag{3.58}
$$

Then, as in (3.15)–(3.16), we can obtain

$$
\begin{aligned}
(3.59)\qquad & E\Big\{\sup_{t\in[0,T\wedge\tau)}\big(|Y(t)|^2e^{2\beta t}\big)\Big\}\\
&\le KE\Big\{|\xi|^2e^{2\beta\tau}+\int_0^\tau|h_0(s)|^2e^{2\beta s}ds+\int_0^\tau|Z(s)|^2e^{2\beta s}ds\Big\}.
\end{aligned}
$$

By Fatou's lemma, sending $T\to\infty$, we see that (3.59) holds with $T\wedge\tau$ on the left-hand side replaced by τ. Hence, combining (3.57) and (3.59) yields (3.49). This in particular shows that $(Y(\cdot),Z(\cdot))\in\mathcal{M}_\beta[0,\tau]$. Finally, let us prove (3.41). From (3.50), it follows that

$$
\begin{aligned}
& E\Big\{\big|Y(T\wedge\tau)-\xi\big|^2e^{2\beta(T\wedge\tau)}\Big\}\\
&=E\Big\{\Big|E\big(\xi-\int_0^\tau h_0(s)ds\big|\mathcal{F}_{T\wedge\tau}\big)-\xi+\int_0^{T\wedge\tau}h_0(s)ds\Big|^2e^{2\beta(T\wedge\tau)}\Big\}\\
(3.60)\qquad &=E\Big\{\Big|E(\xi|\mathcal{F}_{T\wedge\tau})-\xi-E\big(\int_{T\wedge\tau}^\tau h_0(s)ds\big|\mathcal{F}_{T\wedge\tau}\big)\Big|^2e^{2\beta(T\wedge\tau)}\Big\}\\
&\le 2E\Big\{\Big|E(\xi|\mathcal{F}_{T\wedge\tau})-\xi\Big|^2e^{2\beta(T\wedge\tau)}\Big\}+2E\Big|e^{\beta(T\wedge\tau)}\int_{T\wedge\tau}^\tau h_0(s)ds\Big|^2\\
&\le 2E\Big\{\Big|E(\xi|\mathcal{F}_{T\wedge\tau})-\xi\Big|^2e^{2\beta(T\wedge\tau)}\Big\}+\frac{1}{\beta}E\int_{T\wedge\tau}^\tau|h_0(s)|^2e^{2\beta s}ds.
\end{aligned}
$$

The second term on the right hand side of (3.60) goes to zero as $T\to\infty$. Moreover, by Lemma 3.8 below, the first term tends to zero also (as $T\to\infty$). This completes the proof. □

We now present the following lemma, which has been used in the above.

Lemma 3.8. *Let $\beta>0$ and $\xi\in L^{2,\beta}_{\mathcal{F}_\tau}(\Omega;\mathbb{R}^k)$. Then*

$$
(3.61)\qquad \lim_{T\to\infty}E\Big\{\Big|E(\xi|\mathcal{F}_{T\wedge\tau})-\xi\Big|^2e^{2\beta(T\wedge\tau)}\Big\}=0.
$$

Proof. Since $E(|\xi|^2e^{2\beta\tau})<\infty$, we have

$$
(3.62)\qquad \xi=0,\qquad \mathbf{P}\text{-a.s. on } (\tau=\infty).
$$

For any $T>0$, consider

$$
\begin{aligned}
& E\Big\{\Big|E(\xi|\mathcal{F}_{T\wedge\tau})-\xi\Big|^2e^{2\beta(T\wedge\tau)}\Big\}\\
(3.63)\qquad &=E\Big\{\Big|E(\xi|\mathcal{F}_{T\wedge\tau})-\xi\Big|^2e^{2\beta(T\wedge\tau)}I_{(\tau\le T)}\Big\}\\
&\quad+E\Big\{\Big|E(\xi|\mathcal{F}_{T\wedge\tau})-\xi\Big|^2e^{2\beta(T\wedge\tau)}I_{(\tau>T)}\Big\}\equiv I_1(T)+I_2(T).
\end{aligned}
$$

Note that (see Chapter 1, Proposition 3.7)

$$
(3.64)\qquad E(\xi|\mathcal{F}_{T\wedge\tau})I_{(\tau\le T)}=E(\xi|\mathcal{F}_\tau)I_{(\tau\le T)}=\xi I_{(\tau\le T)}.
$$

Consequently, for any $T > 0$,

$$I_1(T) = E\Big\{\Big|E(\xi|\mathcal{F}_{T\wedge\tau}) - \xi\Big|^2 e^{2\beta(T\wedge\tau)} I_{(\tau\le T)}\Big\} = 0. \tag{3.65}$$

On the other hand, (noting $\beta > 0$)

$$\begin{aligned} I_2(T) &\le 2E\Big\{\big[E(|\xi|^2|\mathcal{F}_{T\wedge\tau}) + |\xi|^2\big] e^{2\beta(T\wedge\tau)} I_{(\tau>T)}\Big\} \\ &\le 2E\Big\{E\big[|\xi|^2 e^{2\beta(T\wedge\tau)} I_{(\tau>T)}\big|\mathcal{F}_{T\wedge\tau}\big] + E\big(|\xi|^2 e^{2\beta\tau} I_{(\tau>T)}\big)\Big\} \\ &\le 4E\big(|\xi|^2 e^{2\beta\tau} I_{(T<\tau<\infty)}\big) \to 0, \quad (T\to\infty), \end{aligned} \tag{3.66}$$

since

$$\lim_{T\to\infty} P(T<\tau<\infty) = P\Big(\bigcap_{T>0}(T<\tau<\infty)\Big) = 0. \tag{3.67}$$

This proves the lemma. □

Next, we consider the following family of BSDEs:

$$\begin{cases} dY(t) = \big\{\alpha h(t, Y(t), Z(t)) + h_0(t)\big\}dt + Z(t)dW(t), & t\in[0,\tau), \\ Y(\tau) = \xi, \end{cases} \tag{3.68}$$

where $\alpha \in [0,1]$. Clearly, if $\alpha = 0$, (3.68) coincides with (3.48), which admits a unique adapted solution (if $\xi \in L^{2,\beta}_{\mathcal{F}_\tau}(\Omega;\mathbb{R}^k)$ and $h_0 \in L^{2,\beta}_{\mathcal{F}}(0,\tau;\mathbb{R}^k)$ with $\beta > 0$). On the other hand, if $\alpha = 1$, (3.68) reduces to an equation that is a little more general than (3.32), for which we want to establish the well-posedness. The following a priori estimate is crucial.

Lemma 3.9. *Let (B)′ and (3.47) hold. Let $\xi, \bar\xi \in L^{2,\beta}_{\mathcal{F}_\tau}(\Omega;\mathbb{R}^k)$ and $h_0, \bar h_0 \in L^{2,\beta}_{\mathcal{F}}(0,\tau;\mathbb{R}^k)$. Let $(Y(\cdot),Z(\cdot)), (\overline{Y}(\cdot),\overline{Z}(\cdot)) \in \mathcal{M}_\beta[0,\tau]$ be adapted solutions of (3.68) corresponding to the data (ξ,h_0) and $(\bar\xi,\bar h_0)$, respectively. Then there exists an absolute constant $K > 0$, independent of $\alpha\in[0,1]$, such that*

$$\begin{aligned} &\|(Y(\cdot),Z(\cdot)) - (\overline{Y}(\cdot),\overline{Z}(\cdot))\|^2_{\mathcal{M}_\beta[0,\tau]} \\ &\qquad \le KE\Big\{|\xi-\bar\xi|^2 e^{2\beta\tau} + \int_0^\tau |h_0(s) - \bar h_0(s)|^2 e^{2\beta s}ds\Big\}. \end{aligned} \tag{3.69}$$

Proof. We apply Itô's formula to $|Y(t)-\overline{Y}(t)|^2 e^{2\beta t}$ to get the following

$(0 \leq t < T < \infty)$:

$$
\begin{aligned}
&|Y(T\wedge\tau)-\overline{Y}(T\wedge\tau)|^2 e^{2\beta(T\wedge\tau)}\\
&\quad=|Y(t\wedge\tau)-\overline{Y}(t\wedge\tau)|^2 e^{2\beta(t\wedge\tau)}\\
&\qquad+\int_{t\wedge\tau}^{T\wedge\tau} e^{2\beta s}\big\{2\beta|Y(s)-\overline{Y}(s)|^2\\
&\qquad\quad+2\alpha\,\langle\, Y(s)-\overline{Y}(s), h(s,Y(s),Z(s))-h(s,\overline{Y}(s),Z(s))\,\rangle\\
&\qquad\quad+2\alpha\,\langle\, Y(s)-\overline{Y}(s), h(s,\overline{Y}(s),Z(s))-h(s,\overline{Y}(s),\overline{Z}(s))\,\rangle\\
&\qquad\quad+2\,\langle\, Y(s)-\overline{Y}(s), h_0(s)-\overline{h}_0(s)\,\rangle+|Z(s)-\overline{Z}(s)|^2\big\}ds\\
&\qquad+2\int_{t\wedge\tau}^{T\wedge\tau} e^{2\beta s}\,\langle\, Y(s)-\overline{Y}(s),(Z(s)-\overline{Z}(s))dW(s)\,\rangle\\
&\quad\geq|Y(t\wedge\tau)-\overline{Y}(t\wedge\tau)|^2 e^{2\beta(t\wedge\tau)}\\
&\qquad+\int_{t\wedge\tau}^{T\wedge\tau} e^{2\beta s}\Big\{\Big(2\beta+2\alpha\mu-\frac{\alpha^2L_0^2}{\lambda}-\varepsilon\Big)|Y(s)-\overline{Y}(s)|^2\\
&\qquad\quad+(1-\lambda)|Z(s)-\overline{Z}(s)|^2-\frac{1}{\varepsilon}|h_0(s)-\overline{h}_0(s)|^2\Big\}ds\\
&\qquad+2\int_{t\wedge\tau}^{T\wedge\tau} e^{2\beta s}\,\langle\, Y(s)-\overline{Y}(s),(Z(s)-\overline{Z}(s))dW(s)\,\rangle\,.
\end{aligned}
\tag{3.70}
$$

Since $\beta > 0$ and $\mu = L_0^2/2$ (see (3.47)), we can always find $\varepsilon > 0$ small enough and $\lambda < 1$ with $1-\lambda$ small enough such that

$$
2\beta+2\alpha\mu-\frac{\alpha^2L_0^2}{\lambda}-\varepsilon>0,\qquad \forall\alpha\in[0,1].
\tag{3.71}
$$

In fact, note that the left-hand side of (3.71) is concave in α. Thus, it being positive for $\alpha = 0, 1$ implies that it is positive for all $\alpha \in [0,1]$. It follows then from (3.70) that (by sending $T \to \infty$)

$$
\begin{aligned}
&E\Big\{|Y(t\wedge\tau)-\overline{Y}(t\wedge\tau)|^2 e^{2\beta(t\wedge\tau)}+\int_0^\tau |Z(s)-\overline{Z}(s)|^2 e^{2\beta s}ds\Big\}\\
&\qquad\leq KE\Big\{|\xi-\overline{\xi}|^2 e^{2\beta\tau}+\int_0^\tau |h_0(s)-\overline{h}_0(s)|^2 e^{2\beta s}ds\Big\}.
\end{aligned}
\tag{3.72}
$$

Again by the Burkholder–Davis–Gundy inequality, we obtain (3.69) in a way similar to (3.58)–(3.59). □

Lemma 3.9 gives the continuous dependence (uniformly in α) of the adapted solutions of (3.68) on the data (ξ, h_0). In particular, (3.69) implies that (3.68) admits at most one adapted solution. The following result is the continuation lemma, due to which we call our approach the method of continuation.

Lemma 3.10. *Let (B)′ and (3.47) hold. Then there exists a constant $\varepsilon_0 > 0$ having the following property: If for some $\alpha \in [0,1)$, (3.68) admits a unique adapted solution $(Y(\cdot), Z(\cdot)) \in \mathcal{M}_\beta[0,\tau]$ for any $\xi \in L^{2,\beta}_{\mathcal{F}_\tau}(\Omega;\mathbb{R}^k)$*

and $h_0 \in L^{2,\beta}_{\mathcal{F}}(0,\tau;\mathbb{R}^k)$, *then the same is true for (3.68) with* α *being replaced by any* $\alpha+\varepsilon$, *where* $\varepsilon \in [0,\varepsilon_0]$ *and* $\alpha+\varepsilon \le 1$.

Proof. Suppose there is an $\alpha \in [0,1)$ with the property specified in the statement of the lemma. Then for any ξ and h_0, define

$$Y^0(t)=0, \quad Z^0(t)=0, \tag{3.73}$$

and solve the following BSDEs successively:

$$\begin{cases} dY^{i+1}(t)=\{\alpha h(t,Y^{i+1}(t),Z^{i+1}(t))+\varepsilon h(t,Y^i(t),Z^i(t)) \\ \qquad\qquad +h_0(t)\}dt+Z^{i+1}(t)dW(t), \quad t\in[0,\tau), \\ Y^{i+1}(\tau)=\xi, \end{cases} \tag{3.74}$$

where $\varepsilon > 0$. Now, applying Lemma 3.9 to $(Y^{i+1}(\cdot),Z^{i+1}(\cdot))$ and $(Y^i(\cdot),Z^i(\cdot))$, we obtain the following:

$$\begin{aligned} &\|(Y^{i+1}(\cdot),Z^{i+1}(\cdot))-(Y^i(\cdot),Z^i(\cdot))\|^2_{\mathcal{M}_\beta[0,\tau]} \\ &\le KE\Big\{\int_0^\tau \varepsilon^2|h(s,Y^i(s),Z^i(s))-h(s,Y^{i-1}(s),Z^{i-1}(s))|^2e^{2\beta s}ds\Big\} \\ &\le \varepsilon^2K_0E\Big\{\int_0^\tau\{|Y^i(s)-Y^{i-1}(s)|^2+|Z^i(s)-Z^{i-1}(s)|^2\}e^{2\beta s}ds\Big\} \\ &\le \varepsilon^2K_0\|(Y^i(\cdot),Z^i(\cdot))-(Y^{i-1}(\cdot),Z^{i-1}(\cdot))\|^2_{\mathcal{M}_\beta[0,\tau]}, \qquad \forall i\ge 1. \end{aligned} \tag{3.75}$$

Note that the constant $K_0>0$ in (3.75) is independent of α and $\varepsilon>0$. Thus, we may choose $\varepsilon_0=(2K_0)^{-\frac{1}{2}}$. With this choice, for any $\varepsilon\in(0,\varepsilon_0]$ (with $\alpha+\varepsilon\le 1$), the corresponding sequence $\{Y^i(\cdot),Z^i(\cdot))\}$ is Cauchy in the space $\mathcal{M}_\beta[0,\tau]$, and hence there is a limit $(Y(\cdot),Z(\cdot))$ in this space. Clearly, this limit is an adapted solution to (3.68) with α being replaced by $\alpha+\varepsilon$. Finally, Lemma 3.9 gives the uniqueness. □

Now, we are ready to prove Theorem 3.6.

Proof of Theorem 3.6. First, we assume (3.47). Let us consider (3.68) with $\alpha=0$. This admits a unique adapted solution in $\mathcal{M}_\beta[0,\tau]$ (by Lemma 3.7) for any ξ, h_0. By Lemma 3.10, we can uniquely solve (3.68) with $\alpha=\varepsilon_0$, and then for $\alpha=2\varepsilon_0$, etc., and eventually, we can solve (3.68) with $\alpha=1$. Thus, we can solve (3.32). The stability estimate (3.46) can be proved in the same way as that for (3.69), and we leave the details to the reader.

Now, for the general case without (3.47), we set

$$\gamma \stackrel{\Delta}{=} -\mu+\frac{L_0^2}{2}, \tag{3.76}$$

and

$$\begin{cases} \widetilde{\mu}\stackrel{\Delta}{=}\gamma+\mu\equiv\dfrac{L_0^2}{2}, \\ \widetilde{\beta}\stackrel{\Delta}{=}\beta-\gamma>0. \end{cases} \tag{3.77}$$

Then,

$$-\widetilde{\mu}+\frac{L_0^2}{2}=0<\widetilde{\beta}. \tag{3.78}$$

This means that (3.47) holds if β and μ are replaced by $\widetilde{\beta}$ and $\widetilde{\mu}$. Next, define ($\omega\in\Omega$ is suppressed)

$$\begin{aligned}\widetilde{h}(t,y,z)&\stackrel{\Delta}{=}\gamma y+e^{\gamma t}h(t,e^{-\gamma t}y,e^{-\gamma t}z),\\&\qquad\qquad\forall(t,y,z)\in[0,\infty)\times\mathbb{R}^k\times\mathbb{R}^{k\times m}.\end{aligned} \tag{3.79}$$

From (B)$'$, we have (noting (3.76))

$$\widetilde{h}(\cdot\,,0,0)=e^{\gamma\cdot}h(\cdot\,,0,0)\in L_{\mathcal{F}}^{2,\widetilde{\beta}}(0,\tau;\mathbb{R}^k), \tag{3.80}$$

$$\begin{aligned}|\widetilde{h}(t,y,z)-\widetilde{h}(t,\overline{y},\overline{z})|&\le(|\gamma|+L)|y-\overline{y}|+L_0|z-\overline{z}|,\\&\qquad\forall t\in[0,\infty),\ y,\overline{y}\in\mathbb{R}^k,\ z,\overline{z}\in\mathbb{R}^{k\times m},\end{aligned} \tag{3.81}$$

$$\begin{aligned}\langle\widetilde{h}(t,y,z)-\widetilde{h}(t,\overline{y},z),y-\overline{y}\rangle&\ge\widetilde{\mu}|y-\overline{y}|^2,\\&\qquad\forall t\in[0,\infty),\ y,\overline{y}\in\mathbb{R}^k,\ z\in\mathbb{R}^{k\times m}.\end{aligned} \tag{3.82}$$

Thus, by the results we have just proved, for any $\widetilde{\xi}\in L_{\mathcal{F}_\tau}^{2,\widetilde{\beta}}(\Omega;\mathbb{R}^k)$, the following BSDE admits a unique adapted solution $(\widetilde{Y}(\cdot),\widetilde{Z}(\cdot))\in\mathcal{M}_{\widetilde{\beta}}[0,\tau]$:

$$\begin{cases}d\widetilde{Y}(t)=\widetilde{h}(t,\widetilde{Y}(t),\widetilde{Z}(t))dt+\widetilde{Z}(t)dW(t), & t\in[0,\tau),\\ \widetilde{Y}(\tau)=\widetilde{\xi},\end{cases} \tag{3.83}$$

with a stability estimate (3.46) where β is replaced by $\widetilde{\beta}$. Now, set

$$Y(t)\stackrel{\Delta}{=}e^{-\gamma t}\widetilde{Y}(t),\qquad Z(t)\stackrel{\Delta}{=}e^{-\gamma t}\widetilde{Z}(t),\qquad t\in[0,\tau). \tag{3.84}$$

One can easily check that $(Y(\cdot),Z(\cdot))\in\mathcal{M}_\beta[0,\tau]$, and that it is the unique adapted solution of (3.32). The estimate (3.46) also follows easily. □

To conclude this section, let us briefly discuss the following BSDE on $[0,\infty)$:

$$\begin{cases}dY(t)=h(t,Y(t),Z(t))dt+Z(t)dW(t), & t\in[0,\infty),\\ Y(\infty)=\xi.\end{cases} \tag{3.85}$$

This is a special case of (3.32) with $\tau\equiv\infty$. By Theorem 3.6, we have the following result.

Corollary 3.11. *Let (B)$'$ hold with $\tau\equiv\infty$. Then, for any $\xi\in L_{\mathcal{F}_\infty}^{2,\beta}(\Omega;\mathbb{R}^k)$, (3.85) admits a unique adapted solution $(Y(\cdot),Z(\cdot))\in\mathcal{M}_\beta[0,\infty]$.*

Note that for equation (3.85), the terminal condition $Y(\infty) = \xi$ is understood in the following sense:

$$\lim_{T\to\infty} E\{|Y(T) - \xi|^2 e^{2\beta T}\} = 0. \tag{3.86}$$

It is interesting that in the case $\beta < 0$, (3.85) is uniquely solvable for any ξ in $L^{2,\beta}_{\mathcal{F}_\infty}(\Omega;\mathbb{R}^k)$, which is a larger space than $L^2_{\mathcal{F}_\infty}(\Omega;\mathbb{R}^k)$. In other words, (3.85) might be uniquely solvable even if ξ is *not* square integrable. From (3.45), it is seen that this will be the case if $\mu > L_0^2/2$, which is valid when h is independent of z (i.e., $L_0 = 0$) and is uniformly monotone in y (i.e., $\mu > 0$).

4. Feynman–Kac-Type Formulae

There are intrinsic relations between stochastic differential equations and second-order partial differential equations of parabolic or elliptic type. This is very natural, since from the physics point of view, SDEs and these two kinds of PDEs describe one way or another "diffusion"-type phenomena in the real world. Such relations have been extensively studied in the context of optimal stochastic controls in Chapter 5. In this section we are going to use the solution of SDEs and BSDEs to represent the solutions of some second-order parabolic and elliptic PDEs. Such results are referred to as *Feynman–Kac-type formulae*, as the earliest result of this type was due to Feynman [1] and Kac [1]. To avoid too much technicality, we will omit some proofs.

4.1. Representation via SDEs

Let us consider the following Cauchy problem for a linear parabolic PDE:

$$\begin{cases} v_t + \sum_{i,j=1}^n a_{ij}(t,x)v_{x_ix_j} + \sum_{i=1}^n b_i(t,x)v_{x_i} + c(t,x)v + h(t,x) = 0, \\ \qquad\qquad (t,x) \in [0,T) \times \mathbb{R}^n, \\ v\big|_{t=T} = g(x), \end{cases} \tag{4.1}$$

where $a_{ij}, b_i, c, h : [0,T] \times \mathbb{R}^n \to \mathbb{R}$ and $g : \mathbb{R}^n \to \mathbb{R}$. We set $b(t,x) = (b_1(t,x),\dots,b_n(t,x))^\top$. Also, assume that for some m there exists a $\sigma : [0,T] \times \mathbb{R}^n \to \mathbb{R}^{n\times m}$ such that

$$a(t,x) \equiv (a_{ij}(t,x)) = \frac{1}{2}\sigma(t,x)\sigma(t,x)^\top, \qquad \forall (t,x) \in [0,T] \times \mathbb{R}^n. \tag{4.2}$$

Then (4.1) can be written in the following way:

$$\begin{cases} v_t + \frac{1}{2}\operatorname{tr}\{\sigma(t,x)\sigma(t,x)^\top v_{xx}\} + \langle b(t,x), v_x \rangle \\ \qquad + c(t,x)v + h(t,x) = 0, \qquad (t,x) \in [0,T) \times \mathbb{R}^n, \\ v\big|_{t=T} = g(x), \end{cases} \tag{4.3}$$

where $v_{xx} = (v_{x_i x_j})$ is the Hessian of v, and v_x is the gradient of v. In what follows, we will not distinguish (4.1) and (4.3).

Let us make the following assumption:

(F) The maps $b : [0,T] \times \mathbb{R}^n \to \mathbb{R}^n$, $\sigma : [0,T] \times \mathbb{R}^n \to \mathbb{R}^{n\times m}$, $c, f : [0,T] \times \mathbb{R}^n \to \mathbb{R}$, and $h : \mathbb{R}^n \to \mathbb{R}$ are uniformly continuous, c is bounded, and there exists a constant $L > 0$ such that for $\varphi(t,x) = b(t,x), \sigma(t,x), f(t,x), h(x)$,

$$\text{(4.4)}\qquad \begin{cases} |\varphi(t,x) - \varphi(t,\widehat{x})| \le L|x - \widehat{x}|, & \forall t \in [0,T], x, \widehat{x} \in \mathbb{R}^n, \\ |\varphi(t,0)| \le L, & \forall t \in [0,T]. \end{cases}$$

It is seen that (F) is basically the same as (S2)′ in Chapter 4, Section 3. The differences are that all the functions here are independent of the control variable u and there is one more function c (or the function c in (S2)′ is identically zero; this point of view will be made clear below).

One should note that $(a_{ij}(t,x))$ could be degenerate. Thus, in general, (4.3) may have no classical solution. However, we have the following result, which is a generalized version of the classical *Feynman–Kac formula.*

Theorem 4.1. *Let (F) hold. Then (4.3) admits a unique viscosity solution $v(\cdot\,,\cdot)$, and it has the following representation:*

$$\text{(4.5)}\qquad \begin{aligned} v(t,x) = E\Big\{ \int_t^T & h(s, X(s;t,x))e^{-\int_t^s c(r,X(r;t,x))dr} ds \\ & + g(X(T;t,x))e^{-\int_t^T c(r,X(r;t,x))dr} \Big\}, \quad (t,x) \in [0,T]\times\mathbb{R}^n, \end{aligned}$$

where $X(\cdot) \equiv X(\cdot\,;t,x)$ is the (unique) strong solution of the following SDE:

$$\text{(4.6)}\qquad \begin{cases} dX(s) = b(s,X(s))ds + \sigma(s,X(s))dW(s), & s \in [t,T], \\ X(t) = x, \end{cases}$$

with $(t,x) \in [0,T) \times \mathbb{R}^n$ and $W(\cdot)$ an m-dimensional standard Brownian motion starting from t (with $W(t) = 0$). In addition, if (4.3) admits a classical solution, then (4.5) gives that classical solution.

Proof. We consider the following (trivial) optimal stochastic control problem: The state equation is (4.6) with the state being $X(\cdot)$ and the control variable u being absent, or the control set U being a singleton. The cost functional is given by

$$\text{(4.7)}\qquad \begin{aligned} J(t,x;u(\cdot)) = E\Big\{ \int_t^T & h(X(s;t,x)))e^{-\int_t^s c(r,X(r;t,x))dr} ds \\ & + g(X(T;t,x))e^{-\int_t^T c(r,X(r;t,x))dr} \Big\}, \end{aligned}$$

with $(t,x) \in [0,T] \times \mathbb{R}^n$. Note that in Chapter 4 we have discussed only the case $c(t,x) \equiv 0$. For the above case, the theory is the same. Thus, the function $v(\cdot\,,\cdot)$ defined by (4.5) is the value function of the above (trivial) optimal control problem, and it turns out that (4.3) is exactly the corresponding HJB equation. Hence, v is the unique viscosity solution of (4.3) by Chapter 4, Theorems 5.2 and 6.1. In other words, the unique viscosity solution v of (4.3) is represented by (4.5). The rest of the theorem is obvious by the property of viscosity solutions (see Chapter 4, Proposition 5.8). $\square$

Notice that in the classical Feynman–Kac formula, equation (4.3) is nondegenerate, and the function h need only satisfy a weaker condition (see Karatzas–Shreve [3]). Also, a deterministic version of Theorem 4.1 has been given by Chapter 5, Corollary 3.3.

Next, we turn to the following terminal–boundary value problem for a parabolic equation:

$$
\begin{cases} v_t + \frac{1}{2}\mathrm{tr}\,\{\sigma(t,x)\sigma(t,x)^\top v_{xx}\} + \langle\, b(t,x), v_x\,\rangle \\ \qquad\qquad + c(t,x)v + h(t,x) = 0, \qquad (t,x)\in[0,T)\times G, \\ v\big|_{\partial G} = \psi(t,x), \\ v\big|_{t=T} = g(x), \end{cases} \tag{4.8}
$$

where $G\subseteq\mathbb{R}^n$ is a bounded domain with a C^1 boundary ∂G. We define

$$
\Psi(t,x) = \begin{cases} g(x), & (t,x)\in\{T\}\times\overline{G}, \\ \psi(t,x), & (t,x)\in[0,T)\times\partial G. \end{cases} \tag{4.9}
$$

Theorem 4.2. *Let (F) hold with all the functions defined on $[0,T]\times\overline{G}$, and let $\Psi(t,x)$ defined by (4.9) be continuous on $\big(\{T\}\times\overline{G}\big)\bigcup\big([0,T)\times\partial G\big)$. Then (4.8) admits a unique viscosity solution $v(\cdot,\cdot)$, which can be represented by the following:*

$$
\begin{aligned} v(t,x) = E\Big\{ &\int_t^\tau h(s,X(s;t,x))e^{-\int_t^s c(r,X(r;t,x))dr}ds \\ &+\Psi(\tau,X(\tau;t,x))e^{-\int_t^\tau c(r,X(r;t,x))dr}\Big\},\quad (t,x)\in[0,T]\times G, \end{aligned} \tag{4.10}
$$

where $X(\cdot\,;t,x)$ is the (unique) strong solution of SDE (4.6) and

$$
\tau\equiv\tau(t,x)\stackrel{\Delta}{=}\inf\{s\in[t,T]\mid X(s;t,x)\notin G\}. \tag{4.11}
$$

In addition, if (4.8) admits a classical solution, then (4.10) gives that classical solution.

We have not discussed the viscosity solutions in bounded domains. However, the theory is pretty much parallel. Thus, the proof of the above result can be carried out similarly to that of Theorem 4.1.

Now, let us state the result for elliptic PDEs. We consider the following boundary value problem:

$$(4.12)\qquad \begin{cases} \frac{1}{2}\text{tr}\,\{\sigma(x)\sigma(x)^\top v_{xx}\} + \langle\, b(x), v_x\,\rangle + c(x)v + h(x) = 0, & x \in G, \\ v\big|_{\partial G} = \psi(x), \end{cases}$$

where $G \subseteq \mathbb{R}^n$ is a bounded domain with a C^1 boundary ∂G. For this problem we have the following result, whose proof is similar to that of Theorems 4.1.

Theorem 4.3. *Let (F) hold where all the functions are defined on* $[0, T]\times\overline{G}$ *and are time-invariant. Moreover, either*

$$(4.13)\qquad \sigma(x)\sigma(x)^\top \geq \delta I, \qquad \forall x \in \overline{G},$$

or

$$(4.14)\qquad c(x) \leq -\delta, \qquad \forall x \in \overline{G},$$

holds for some $\delta > 0$. *Then (4.12) admits a unique viscosity solution* $v(\cdot)$ *having the following representation:*

$$(4.15)\qquad \begin{aligned} v(x) =E\Big\{ \int_0^\tau h(X(s;x))e^{-\int_t^s c(X(r;x))dr} ds \\ + g(X(\tau;x))e^{-\int_t^\tau c(X(r;t,x))dr}\Big\}, \qquad \forall x \in G, \end{aligned}$$

where $X(\cdot\,;x)$ *is the strong solution of the SDE*

$$(4.16)\qquad \begin{cases} dX(s) = b(X(s))ds + \sigma(X(s))dW(s), & s \in [0,\infty), \\ X(0) = x, \end{cases}$$

with $W(\cdot)$ *being an* m*-dimensional standard Brownian motion, and*

$$(4.17)\qquad \tau \equiv \tau(x) \stackrel{\Delta}{=} \inf\{s \geq 0 \mid X(s;x) \notin G\}.$$

In addition, if (4.12) admits a classical solution, then (4.15) gives that classical solution.

The proofs of the above results are based on the optimal stochastic control theory. Hence, in view of the nonlinearity of the HJB equation, we can actually obtain representation formulae for some more general *nonlinear* second-order parabolic and elliptic PDEs. This reveals some intrinsic relationship between the optimal stochastic control theory and the nonlinear PDEs in a different way. Let us now present one such result. The reader is encouraged to present other similar results.

Consider the following nonlinear parabolic PDE:

$$(4.18)\qquad \begin{cases} v_t + F(t,x,v,v_x,v_{xx}) = 0, & (t,x) \in [0,T) \times G, \\ v\big|_{\partial G} = \psi(t,x), \\ v\big|_{t=T} = g(x), \end{cases}$$

where $G \subseteq \mathbb{R}^n$ is a bounded domain with a C^1 boundary ∂G, $F : [0,T] \times G \times \mathbb{R} \times \mathbb{R}^n \times \mathcal{S}^n \to \mathbb{R}$, and ψ and g are as before. We assume that F can be represented by the following:

$$
(4.19)\quad \begin{aligned} F(t,x,v,p,P) = \inf_{u\in U} \Big\{ & \frac{1}{2}\mathrm{tr}\,\{\sigma(t,x,u)\sigma(t,x,u)^\top P\} \\ & + \langle\, b(t,x,u), p\,\rangle + c(t,x,u)v + h(t,x,u)\Big\}, \\ & (t,x,v,p,P) \in [0,T] \times G \times \mathbb{R} \times \mathbb{R}^n \times \mathcal{S}^n, \end{aligned}
$$

for some functions σ, b, c, and h along with some metric space U. It is known that (4.19) gives a very big class of nonlinear functions F in the totality of functions that are concave in (v,p,P). We assume the following.

(F)$'$ U is a Polish space and $T > 0$. The maps $b : [0,T]\times\mathbb{R}^n\times U \to \mathbb{R}^n$, $\sigma : [0,T]\times\mathbb{R}^n\times U \to \mathbb{R}^{n\times m}$, $c, f : [0,T]\times\mathbb{R}^n\times U \to \mathbb{R}$, and $h : \mathbb{R}^n \to \mathbb{R}$ are uniformly continuous, and c is bounded. Moreover, there exists a constant $L > 0$ such that for $\varphi(t,x,u) = b(t,x,u), \sigma(t,x,u), f(t,x,u), h(x)$,

$$
(4.20)\quad \begin{cases} |\varphi(t,x,u) - \varphi(t,\widehat{x},u)| \le L|x-\widehat{x}|, & \forall t\in[0,T], x,\widehat{x}\in\mathbb{R}^n, u\in U, \\ |\varphi(t,0,u)| \le L, & \forall (t,u)\in[0,T]\times U. \end{cases}
$$

We see that (F)$'$ is a combination of (S1)$'$–(S2)$'$ of Chapter 4, Section 3, plus a condition for the function c. We let $\mathcal{U}^w[t,T]$ be the set of all admissible controls under the weak formulation (see Chapter 4, Section 3). The following result is a further generalization of the Feynman–Kac formula.

Theorem 4.4. *Let (F)$'$ hold. Then (4.18) admits a unique viscosity solution $v(\cdot\,,\cdot)$, which can be represented by*

$$
(4.21)\quad \begin{aligned} v(t,x) = \inf_{u(\cdot)\in\mathcal{U}^w[t,T]} E\Big(& \int_t^\tau h(s,X(s),u(s))e^{-\int_t^s c(r,X(r),u(r))dr}ds \\ & + \Psi(\tau,X(\tau))e^{-\int_t^\tau c(r,X(r),u(r))dr}\Big), \quad (t,x)\in[0,T]\times G, \end{aligned}
$$

where $X(\cdot) \equiv X(\cdot\,;t,x,u(\cdot))$ is the (unique) solution (in the sense of Definition 6.5 in Chapter 1) of

$$
(4.22)\quad \begin{cases} dX(s) = b(s,X(s),u(s))ds + \sigma(s,X(s),u(s))dW(s), & s\in[t,T], \\ X(t) = x, \end{cases}
$$

with $(t,x)\in[0,T)\times G$, $u(\cdot)\in\mathcal{U}^w[t,T]$ and

$$
(4.23)\quad \tau(t,x,u(\cdot)) = \inf\{s\in[t,T] \mid X(s;t,x,u(\cdot)) \notin G\}.
$$

In addition, if (4.18) admits a classical solution, then (4.21) gives that classical solution.

The above can be further generalized. If F is not necessarily concave in (v,p,P), in some cases it may be represented as a sup inf or inf sup of some functions that are affine in (v,p,P). Then equation (4.18) will be the corresponding *lower* or *upper Hamilton–Jacobi–Isaacs (HJI) equation* for some two-person zero-sum *stochastic differential games.* In such cases, we can represent the (viscosity) solution of (4.18) as the *lower* or *upper value function* of such a differential game.

However, we realize immediately that these representations become more and more complicated. Note that representation (4.21) is an infimum of a complicated functional over a complicated space $\mathcal{U}^w[t,T]$, which is an infinite-dimensional space in general. Thus, we will not pursue further in this direction. Instead, in the following subsection, with the aid of BSDEs we will derive some simpler representation formulae for certain classes of nonlinear PDEs.

4.2. Representation via BSDEs

In this subsection we are going to represent solutions of some nonlinear PDEs through those of BSDEs.

We start with the following Cauchy problem:

$$\text{(4.24)}\quad \begin{cases} v_t + \frac{1}{2}\mathrm{tr}\,\{\sigma(t,x)\sigma(t,x)^\top v_{xx}\} + \langle\, b(t,x), v_x\,\rangle \\ \qquad\qquad + h(t,x,v,\sigma(t,x)^\top v_x) = 0, \qquad (t,x)\in[0,T)\times\mathbb{R}^n, \\ v\big|_{t=T} = g(x). \end{cases}$$

Notice that the equation above is nonlinear in v and v_x, but the nonlinearity in v_x takes a special form (it becomes general if σ is an invertible square matrix). This is necessary in order for its solution to be represented by that of a BSDE. We make the following assumption (compare with (F) stated after (4.3)).

(B)″ All the functions $\sigma:[0,T]\times\mathbb{R}^n\to\mathbb{R}^{n\times m}$, $b:[0,T]\times\mathbb{R}^n\to\mathbb{R}^n$, $h:[0,T]\times\mathbb{R}^n\times\mathbb{R}\times\mathbb{R}^m\to\mathbb{R}$, and $g:\mathbb{R}^n\to\mathbb{R}$ are continuous. Moreover, σ and b are uniformly Lipschitz continuous in $x\in\mathbb{R}^n$ and h is uniformly Lipschitz continuous in $(y,z)\in\mathbb{R}\times\mathbb{R}^m$ (uniform with respect to $(t,x)\in[0,T]\times\mathbb{R}^n$).

Let us now state the following representation theorem.

Theorem 4.5. *Let (B)″ hold. Then (4.24) admits a unique viscosity solution $v(\cdot\,,\cdot)$, and it has the following representation:*

$$\text{(4.25)}\qquad v(t,x) = EY(t;t,x)\equiv Y(t;t,x), \qquad \forall(t,x)\in[0,T]\times\mathbb{R}^n,$$

where $(Y(\cdot),Z(\cdot))\equiv(Y(\cdot\,;t,x),Z(\cdot\,;t,x))$ is the unique adapted solution of the following BSDE on a probability space $(\Omega,\mathcal{F},\mathbf{P})$:

$$\text{(4.26)}\quad \begin{cases} dY(s) = -h(s,X(s),Y(s),Z(s))ds + \langle\, Z(s), dW(s)\,\rangle \quad s\in[t,T], \\ Y(T) = g(X(T)), \end{cases}$$

where $W(\cdot)$ is an m-dimensional standard Brownian motion on $[t,T]$ (with $W(t)=0$), and $X(\cdot\,;t,x)$ is the strong solution of (4.6). In the case where (4.24) admits a classical solution, (4.25) gives that classical solution.

Proof. First of all, suppose (4.24) admits a classical solution $v(\cdot\,,\cdot)$. Note that under assumption (B)$''$, it follows from Theorem 3.2 that (4.26) admits a unique adapted solution $(Y(\cdot),Z(\cdot))$, which depends on (t,x). Now, applying Itô's formula to $v(s,X(s))$, we obtain the following:

$$\begin{aligned}
&g(X(T))-v(s,X(s))\\
&\quad=\int_s^T\Big\{v_t(r,X(r))+\frac{1}{2}\mathrm{tr}\,\big[\sigma(r,X(r))\sigma(r,X(r))^\top v_{xx}(r,X(r)\big]\\
&\qquad\qquad+\langle\, b(r,X(r)),v_x(r,X(r))\,\rangle\Big\}dr\\
&\qquad+\int_s^T\langle\, v_x(r,X(r)),\sigma(r,X(r))dW(r)\,\rangle\\
&\quad=-\int_s^T h\big(r,X(r),v(r,X(r)),\sigma(r,X(r))^\top v_x(r,X(r))\big)dr\\
&\qquad+\int_s^T\langle\,\sigma(r,X(r))^\top v_x(x),dW(r)\,\rangle.
\end{aligned}\tag{4.27}$$

This implies

$$\begin{aligned}
v(s,X(s))&=g(X(T))\\
&\quad+\int_s^T h\big(r,X(r),v(r,X(r)),\sigma(r,X(r))^\top v_x(r,X(r))\big)dr\\
&\quad-\int_s^T\langle\,\sigma(r,X(r))^\top v_x(x),dW(r)\,\rangle,\qquad s\in[t,T].
\end{aligned}\tag{4.28}$$

Hence, $\big(v(\cdot\,,X(\cdot)),\sigma(\cdot\,,X(\cdot))^\top v_x(\cdot\,,X(\cdot))\big)$ is an adapted solution of (4.26). By uniqueness of solutions to (4.26), we must have

$$\begin{cases}Y(s)=v(s,X(s)), & \forall s\in[t,T],\\ Z(s)=\sigma(s,X(s))^\top v_x(s,X(s)), & \text{a.e. } s\in[t,T],\end{cases}\qquad \mathbf{P}\text{-a.s.}\tag{4.29}$$

In particular, we have

$$v(t,x)=Y(t)\equiv Y(t;t,x)\equiv EY(t;t,x),\quad\forall(t,x)\in[0,T]\times\mathbb{R}^n.\tag{4.30}$$

This proves the second conclusion. Now, in the general case when (4.24) does not necessarily admit a classical solution, it is known that under assumption (B)$''$, (4.24) admits a unique viscosity solution $v(\cdot\,,\cdot)$, and more importantly, one has (see Chapter 4, Propositions 5.9 and 5.10)

$$v(t,x)=\lim_{\varepsilon\to0}v^\varepsilon(t,x),\tag{4.31}$$

uniformly in (t,x) over any compact sets, where v^ε is the classical solution of the following nondegenerate parabolic PDE:

$$
(4.32)\qquad \begin{cases} v_t + \frac{1}{2}\text{tr}\,\{\sigma_\varepsilon(t,x)\sigma_\varepsilon(t,x)^\top v_{xx}\} + \langle\, b_\varepsilon(t,x), v_x\,\rangle \\ \qquad + h_\varepsilon(t,x,v,\sigma_\varepsilon(t,x)^\top v_x) = 0, \qquad (t,x)\in[0,T)\times\mathbb{R}^n, \\ v\big|_{t=T} = g_\varepsilon(x). \end{cases}
$$

Here σ_ε, b_ε, h_ε, and g_ε are smooth functions that converge to σ, b, h, and g uniformly over compact sets, respectively, and

$$
(4.33)\qquad \sigma_\varepsilon(t,x)\sigma_\varepsilon(t,x)^\top \ge \varepsilon I + \sigma(t,x)\sigma(t,x)^\top, \qquad \forall (t,x)\in[0,T]\times\mathbb{R}^n.
$$

Then, from the above proof, we conclude that for any $(t,x)\in[0,T]\times\mathbb{R}^n$,

$$
(4.34)\qquad \begin{cases} Y^\varepsilon(s) = v^\varepsilon(s,X^\varepsilon(s)), \quad \forall s\in[t,T], \\ Z^\varepsilon(s) = \sigma_\varepsilon(s,X^\varepsilon(s))^\top v_x^\varepsilon(s,X^\varepsilon(s)), \quad \text{a.e. } s\in[t,T], \end{cases} \qquad \mathbf{P}\text{-a.s.},
$$

where $X^\varepsilon(\cdot)$ is the unique strong solution of the SDE

$$
(4.35)\qquad \begin{cases} dX^\varepsilon(s) = b_\varepsilon(s,X^\varepsilon(s))ds + \sigma_\varepsilon(s,X^\varepsilon(s))dW(s), \quad s\in[t,T], \\ X^\varepsilon(t) = x, \end{cases}
$$

and $(Y^\varepsilon(\cdot),Z^\varepsilon(\cdot))$ is the unique adapted solution of the BSDE

$$
(4.36)\qquad \begin{cases} dY^\varepsilon(s) = -h_\varepsilon(s,X^\varepsilon(s),Y^\varepsilon(s),Z^\varepsilon(s))ds + Z^\varepsilon(s)dW(s), \quad s\in[0,T], \\ Y^\varepsilon(T) = g_\varepsilon(X^\varepsilon(T)). \end{cases}
$$

Now let $X(\cdot)$ be the solution of (4.6) and let $(Y(\cdot),Z(\cdot))$ be the adapted solution of (4.26). Then, by the continuous dependence of the solutions to SDEs on the data (which can be proved by a routine argument using the Burkholder–Davis–Gundy and Gronwall inequalities), we have

$$
(4.37)\qquad E\sup_{s\in[t,T]} |X^\varepsilon(s) - X(s)|^2 = o(1), \quad \text{as } \varepsilon\to 0.
$$

By (3.18), we obtain (as $\varepsilon\to 0$)

$$
(4.38)\qquad E\sup_{s\in[t,T]} |Y^\varepsilon(s) - Y(s)|^2 + E\int_s^T |Z^\varepsilon(s) - Z(s)|^2 ds = o(1).
$$

This implies (noting the convergence (4.31) and representation (4.34))

$$
(4.39)\qquad \begin{aligned} &|v(t,x) - Y(t;t,x)| \\ &\quad\le |v(t,x) - v^\varepsilon(t,x)| + |Y^\varepsilon(t;t,x) - Y(t;t,x)| \\ &\quad\le |v(t,x) - v^\varepsilon(t,x)| + E\sup_{s\in[t,T]} |Y^\varepsilon(s) - Y(s)| = o(1), \quad \text{as } \varepsilon\to 0. \end{aligned}
$$

Hence (4.25) holds. $\square$

Next, we consider the following terminal–boundary value problem:

$$(4.40)\qquad \begin{cases} v_t + \frac{1}{2}\operatorname{tr}\{\sigma(t,x)\sigma(t,x)^\top v_{xx}\} + \langle b(t,x), v_x \rangle \\ \qquad\qquad + h(t,x,v,\sigma(t,x)^\top v_x) = 0, \qquad (t,x)\in[0,T)\times G, \\ v\big|_{\partial G} = \psi(t,x), \\ v\big|_{t=T} = g(x), \end{cases}$$

where $G \subseteq \mathbb{R}^n$ is a bounded domain with ∂G being C^1. To represent the solution of (4.40), we need to employ Theorem 3.6 instead. The result can be stated as follows.

Theorem 4.6. *Let (B)'' hold and let $\Psi(t,x)$ defined by (4.9) be continuous on $(\{T\}\times\overline{G})\times([0,T]\times\partial G)$. For any $(t,x)\in[0,T)\times G$, let $X(\cdot\,;t,x)$ be the unique strong solution of (4.6), and let $\tau(t,x)$ be defined by (4.11). Let $(Y(\cdot),Z(\cdot))$ be the unique adapted solution of the following BSDE (in a random duration):*

$$(4.41)\qquad \begin{cases} dY(s) = -h(s,X(s),Y(s),Z(s))ds + Z(s)dW(s), \quad s\in[t,\tau), \\ Y(\tau) = \Psi(\tau, X(\tau)). \end{cases}$$

Then the viscosity solution $v(t,x)$ of (4.40) is given by

$$(4.42)\qquad v(t,x) = EY(t;t,x) \equiv Y(t;t,x), \qquad (t,x)\in[0,T]\times G.$$

The proof is parallel to that for Theorem 4.5. The details are left to the reader. We point out that in the above representation theorem, since $\tau(t,x)\le T$, Theorem 3.6 applies in view of the remark made after Theorem 3.6.

To conclude this section let us discuss a more subtle case, namely, the nonlinear elliptic equation in a domain G. We consider the following problem:

$$(4.43)\qquad \begin{cases} \frac{1}{2}\operatorname{tr}\left[\sigma(x)\sigma(x)^\top v_{xx}\right] + \langle b(x), v_x \rangle + h(x,v,\sigma(x)^\top v_x) = 0, \quad x\in G, \\ v\big|_{\partial G} = \psi(x). \end{cases}$$

We make the following assumption (compare with (B)' stated after (3.42)).

(B)''' The functions $\sigma:\mathbb{R}^n\to\mathbb{R}^{n\times m}$, $b:\mathbb{R}^n\to\mathbb{R}^n$, $h:\mathbb{R}^n\times\mathbb{R}\times\mathbb{R}^m\to\mathbb{R}$, and $\psi:\mathbb{R}^n\to\mathbb{R}$ are continuous, and there exist $L, L_0>0$ and $\beta,\mu\in\mathbb{R}$ such that

$$(4.44)\qquad \begin{aligned} &|h(x,y,z)-h(x,\overline{y},\overline{z})| \le L|y-\overline{y}| + L_0|z-\overline{z}|, \\ &\qquad\qquad \forall x\in\overline{G},\ y,\overline{y}\in\mathbb{R},\ z,\overline{z}\in\mathbb{R}^m, \end{aligned}$$

$$(4.45)\qquad \begin{aligned} &\langle h(x,y,z)-h(x,\overline{y},z), y-\overline{y} \rangle \le -\mu|y-\overline{y}|^2, \\ &\qquad\qquad \forall x\in\mathbb{R}^n,\ y,\overline{y}\in\mathbb{R},\ z\in\mathbb{R}^m. \end{aligned}$$

$$\beta > -\mu + \frac{L_0^2}{2}. \tag{4.46}$$

Note that since the function h here and the one in Section 3 differ by a minus sign (compare (3.32) and (4.48) below), the condition (3.44) is changed properly to (4.45). The result now follows.

Theorem 4.7. *Let (B)''' hold. For any $x \in G$, let $X(\cdot\,;x)$ be the unique strong solution of (4.16) and let $\tau(x)$ be defined by (4.17). Further, suppose that*

$$E\{e^{2\beta\tau(x)}|g(X(\tau(x);x))|^2\} < \infty, \qquad \forall x \in G. \tag{4.47}$$

Then the BSDE

$$\begin{cases} dY(s) = -h(X(s),Y(s),Z(s))ds + Z(s)dW(s), & s \in [0,\tau), \\ Y(\tau) = g(X(\tau)) \end{cases} \tag{4.48}$$

admits a unique adapted solution $(Y(\cdot),Z(\cdot)) \equiv (Y(\cdot\,;x),Z(\cdot\,;x))$, and the viscosity solution $v(\cdot)$ of (4.43) can be represented by

$$v(x) = EY(0;x) = Y(0;x), \qquad \forall x \in \mathbb{R}^n. \tag{4.49}$$

Note that condition (4.47) gives some compatibility among the domain G, the SDE (4.16), and the BSDE (4.48). If G is bounded and $\sigma(t,x)$ is nondegenerate, then (4.47) automatically holds. Also, it is interesting that if $\beta < 0$ and g is bounded, then (4.47) holds. In particular, if g is bounded, h is independent of z (i.e., $L_0 = 0$), and $-h$ is uniformly monotone (i.e., $\mu > 0$), then (4.47) holds for some $\beta < 0$, and Theorem 4.7 is valid.

The proof of the above theorem is similar to that of Theorem 4.5. The nontrivial part is the unique solvability of the BSDE (4.48) (note here that τ may not be bounded by any deterministic constant), which has been shown in the previous section.

5. Forward–Backward Stochastic Differential Equations

In this section we are going to study the solvability for systems of generally *coupled* forward–backward stochastic differential equations (FBSDEs, for short). More precisely, let $(\Omega,\mathcal{F},\{\mathcal{F}_t\}_{t\ge0},\mathbf{P})$ be a complete filtered probability space on which an m-dimensional standard Brownian motion $W(t)$ is defined, such that $\{\mathcal{F}_t\}_{t\ge0}$ is the natural filtration of $W(t)$. We consider the following system:

$$\begin{cases} dX(t) = b(t,X(t),Y(t),Z(t))dt + \sigma(t,X(t),Y(t),Z(t))dW(t), \\ dY(t) = h(t,X(t),Y(t),Z(t))dt + Z(t)dW(t), \\ X(0) = x, \qquad Y(T) = g(X(T)). \end{cases} \tag{5.1}$$

In the above, $X(\cdot)$, $Y(\cdot)$, and $Z(\cdot)$ are the unknown processes, and they are required to be $\{\mathcal{F}_t\}_{t\ge0}$-adapted. Functions b, σ, h, and g are given, which are *not* random. The main feature of the above equation is that process $X(\cdot)$ satisfies a forward SDE, process $Y(\cdot)$ satisfies a BSDE, and they are *coupled.* We already know from the previous sections that the presence of process $Z(\cdot)$ is necessary for us possibly to find $\{\mathcal{F}_t\}_{t\geq 0}$-adapted processes $(X(\cdot), Y(\cdot))$ satisfying (5.1).

It should be noted that in stochastic optimal control problems, the stochastic Hamiltonian system introduced in Chapter 3 is an FBSDE, where the forward component $x(\cdot)$ and the backward components $(p(\cdot), q(\cdot))$ and $(P(\cdot), Q(\cdot))$ are coupled through the maximum condition. On the other hand, (5.1) can be regarded as a generalization of the two-point boundary value problem for ordinary differential equations.

5.1. General formulation and nonsolvability

We let

$$\widehat{\mathcal{M}}[0,T] = L^2_{\mathcal{F}}(\Omega; C([0,T];\mathbb{R}^n)) \times L^2_{\mathcal{F}}(\Omega; C([0,T];\mathbb{R}^k)) \times L^2_{\mathcal{F}}(0,T;\mathbb{R}^{k\times m}). \tag{5.2}$$

Similar to one in the previous section, we have the following definition.

Definition 5.1. A triple of processes $(X, Y, Z) \in \widehat{\mathcal{M}}[0,T]$ is called an *(adapted) solution* of the FBSDEs (5.1) if the following holds:

$$\begin{cases} X(t) = x + \displaystyle\int_0^t b(s, X(s), Y(s), Z(s))ds \\ \qquad\qquad + \displaystyle\int_0^t \sigma(s, X(s), Y(s), Z(s))dW(s), \\ Y(t) = g(X(T)) - \displaystyle\int_t^T h(s, X(s), Y(s), Z(s))ds - \int_t^T Z(s)dW(s), \\ \qquad\qquad\qquad \forall t \in [0,T], \quad \mathbf{P}\text{-a.s.} \end{cases} \tag{5.3}$$

Equation (5.1) is said to have a *unique* adapted solution if for any two adapted solutions (X, Y, Z) and $(\tilde{X}, \tilde{Y}, \tilde{Z})$,

$$\mathbf{P}\Big\{(X(t), Y(t)) = (\tilde{X}(t), \tilde{Y}(t)), \ \forall t \in [0,T] \\ \text{and } Z(t) = \tilde{Z}(t), \ \text{a.e. } t \in [0,T]\Big\} = 1.$$

We point out that coupled FBSDEs are not necessarily solvable. To see this, let us present the following simple result.

Proposition 5.2. *Let the following two-point boundary value problem for a system of linear ordinary differential equations admit no solutions:*

$$\begin{cases} \begin{pmatrix} \dot{X}(t) \\ \dot{Y}(t) \end{pmatrix} = \mathcal{A} \begin{pmatrix} X(t) \\ Y(t) \end{pmatrix}, \\ X(0) = x, \qquad Y(T) = GX(T), \end{cases} \tag{5.4}$$

where $\mathcal{A}$ and G are certain matrices. Then, for any bounded $\sigma : [0,T] \times \mathbb{R}^n \times \mathbb{R}^k \times \mathbb{R}^{k\times m} \to \mathbb{R}^{n\times m}$, the FBSDE

$$
\left\{\begin{aligned} & d\begin{pmatrix} X(t) \\ Y(t) \end{pmatrix} = \mathcal{A}\begin{pmatrix} X(t) \\ Y(t) \end{pmatrix} dt + \begin{pmatrix} \sigma(t, X(t), Y(t), Z(t)) \\ Z(t) \end{pmatrix} dW(t), \\ & X(0) = x, \qquad Y(T) = GX(T), \end{aligned}\right. \tag{5.5}
$$

admits no adapted solution.

Proof. Suppose (5.5) admits an adapted solution $(X(\cdot), Y(\cdot), Z(\cdot))$. Then $(EX(\cdot), EY(\cdot))$ is a solution of (5.4), leading to a contradiction. This proves the assertion. □

There are many examples where systems like (5.4) do not admit a solution. Here is a very simple one ($n = m = 1$):

$$
\left\{\begin{aligned} & \dot{X} = Y, \\ & \dot{Y} = -X, \\ & X(0) = x, \quad Y(T) = X(T). \end{aligned}\right. \tag{5.6}
$$

We can easily show that for $T = \ell\pi + \frac{\pi}{4}$ (ℓ is a nonnegative integer), the above two-point boundary value problem does not admit any solution for any $x \in \mathbb{R} \setminus \{0\}$ and it admits infinitely many solutions for $x = 0$.

By rescaling the time and embedding a one-dimensional ODE into a higher-dimensional space, one can easily construct nonsolvable linear systems of ordinary differential equations over any given time duration and in any (finite) dimensional spaces. Thus, by Proposition 5.2, we find a large class of nonsolvable FBSDEs.

5.2. The four-step scheme, a heuristic derivation

Now we are going to introduce a method for solving the FBSDE (5.1) over any time duration $[0,T]$.

Let us give a heuristic derivation first. Suppose that (X, Y, Z) is an adapted solution to (5.1) or equivalently (5.3). We assume that Y and X are related by

$$
Y(t) \equiv (Y^1(t), \dots, Y^k(t)) = \theta(t, X(t)), \quad \forall t \in [0,T], \quad \mathbf{P}\text{-a.s.}, \tag{5.7}
$$

where θ is some function to be determined. We suppose that $\theta(t,x)$ is C^1 in t and C^2 in x. Then by Itô's formula, we have, for $1 \le \ell \le k$,

$$
\begin{aligned} dY^\ell(t) &= d\theta^\ell(t, X(t)) \\ &= \Big\{\theta_t^\ell(t, X(t)) + \langle\, \theta_x^\ell(t, X(t)), b(t, X(t), \theta(t, X(t)), Z(t))\,\rangle \\ &\qquad + \frac{1}{2}\text{tr}\,\big[\theta_{xx}^\ell(t, X(t))(\sigma\sigma^\top)(t, X(t), \theta(t, X(t)), Z(t))\big]\Big\}dt \\ &\quad + \langle\, \theta_x^\ell(t, X(t)), \sigma(t, X(t), \theta(t, X(t)), Z(t))dW(t)\,\rangle\,. \end{aligned} \tag{5.8}
$$

Comparing (5.8) with (5.1), we see that for θ to be the right choice, it suffices that the following holds for $\ell = 1, \ldots, k$:

$$(5.9)\qquad \begin{cases} h^\ell(t, X(t), \theta(t, X(t), Z(t)) \\ \qquad = \theta^\ell_t(t, X(t)) + \langle \theta^\ell_x(t, X(t)), b(t, X(t), \theta(t, X(t)), Z(t)) \rangle \\ \qquad\quad + \frac{1}{2}\mathrm{tr}\,\big[\theta^\ell_{xx}(t, X(t))(\sigma\sigma^\top)(t, X(t), \theta(t, X(t)), Z(t))\big], \\ \theta(T, X(T)) = g(X(T)), \end{cases}$$

and

$$(5.10)\qquad \theta_x(t, X(t))\sigma(t, X(t), \theta(t, X(t)), Z(t)) = Z(t).$$

The above argument suggests that we design the following *four-step scheme* to solve the FBSDE (5.1).

The four-step scheme:

Step 1. Find $z(t, x, y, p)$ satisfying the following:

$$(5.11)\qquad \begin{aligned} &z(t, x, y, p) = p\sigma(t, x, y, z(t, x, y, p)), \\ &\qquad\qquad \forall (t, x, y, p) \in [0, T] \times \mathbb{R}^n \times \mathbb{R}^k \times \mathbb{R}^{k\times n}. \end{aligned}$$

Step 2. Use the function z obtained above to solve the following parabolic system for $\theta(t, x)$:

$$(5.12)\qquad \begin{cases} \theta^\ell_t + \frac{1}{2}\mathrm{tr}[\theta^\ell_{xx}(\sigma\sigma^\top)(t,x,\theta,z(t,x,\theta,\theta_x))] + \langle b(t,x,\theta,z(t,x,\theta,\theta_x)), \theta^\ell_x\rangle \\ \quad -h^\ell(t,x,\theta,z(t,x,\theta,\theta_x)) = 0, \quad (t,x)\in(0,T)\times\mathbb{R}^n, \ 1\le\ell\le k, \\ \theta(T,x) = g(x), \qquad x\in\mathbb{R}^n. \end{cases}$$

Step 3. Use θ and z obtained in Steps 1–2 to solve the following forward SDE:

$$(5.13)\qquad X(t) = x + \int_0^t \tilde b(s, X(s))ds + \int_0^t \tilde\sigma(s, X(s))dW(s),$$

where

$$(5.14)\qquad \begin{cases} \tilde b(t,x) \stackrel{\Delta}{=} b(t, x, \theta(t,x), z(t, x, \theta(t,x), \theta_x(t,x))), \\ \tilde\sigma(t,x) \stackrel{\Delta}{=} \sigma(t, x, \theta(t,x), z(t, x, \theta(t,x), \theta_x(t,x))). \end{cases}$$

Step 4. Set

$$(5.15)\qquad \begin{cases} Y(t) \stackrel{\Delta}{=} \theta(t, X(t)), \\ Z(t) \stackrel{\Delta}{=} z(t, X(t), \theta(t, X(t)), \theta_x(t, X(t))). \end{cases}$$

Should this scheme be realizable, (X, Y, Z) would give an adapted solution of (5.1). As a matter of fact, we have the following result.

Theorem 5.3. *Assume that (5.11) admits a unique solution $z(t,x,y,p)$ that is uniformly Lipschitz continuous in (x,y,p) with $z(t,0,0,0)$ being bounded, and that (5.12) admits a classical solution $\theta(t,x)$ with bounded θ_x and θ_{xx}. Assume further that the functions $b(t,x,y,z)$ and $\sigma(t,x,y,z)$ are uniformly Lipschitz continuous in (x,y,z) with $b(t,0,0,0)$ and $\sigma(t,0,0,0)$ being bounded. Then the process $(X(\cdot),Y(\cdot),Z(\cdot))$ determined by (5.13) and (5.15) is an adapted solution to (5.1). Moreover, if h is also uniformly Lipschitz continuous in (x,y,z) and the (uniform) Lipschitz constant of $z\mapsto\sigma(t,x,y,z)$, denoted by $L_\sigma\ge 0$, satisfies*

$$\sup_{(t,x)\in[0,T]\times\mathbb{R}^n}|\theta_x^\ell(t,x)|L_\sigma<1,\qquad 1\le\ell\le k,\tag{5.16}$$

then the adapted solution is unique and is determined by (5.13) and (5.15).

Proof. Under our conditions both $\widetilde{b}(t,x)$ and $\widetilde{\sigma}(t,x)$ defined by (5.14) are uniformly Lipschitz continuous in x. Therefore, for any $x\in\mathbb{R}^n$, (5.13) has a unique strong solution. Then, by defining $Y(t)$ and $Z(t)$ via (5.15) and applying Itô's formula, we can easily check that (5.1) is satisfied. Hence, (X,Y,Z) is a solution of (5.1).

It remains to show the uniqueness. We claim that any solution (X,Y,Z) of (5.1) must be of the form we constructed using the four-step scheme. To show this, let (X,Y,Z) be any solution of (5.1). Define

$$\widetilde{Y}(t)\stackrel{\Delta}{=}\theta(t,X(t)),\qquad \widetilde{Z}(t)\stackrel{\Delta}{=}z(t,X(t),\theta(t,X(t)),\theta_x(t,X(t))).\tag{5.17}$$

By our assumption, (5.11) admits a unique solution. Thus, it follows from (5.17) that

$$\widetilde{Z}(t)=\theta_x(t,X(t))\sigma(t,X(t),\widetilde{Y}(t),\widetilde{Z}(t)),\qquad\forall t\in[0,T].\tag{5.18}$$

Now, applying Itô's formula to $\theta(t,X(t))$ and noting (5.12) and (5.15), we have the following (for notational simplicity, we suppress t in $X(t)$, etc.):

$$\begin{aligned}
d\widetilde{Y}^\ell(t)=&d\theta^\ell(t,X(t))\\
=&\Big\{\theta_t^\ell(t,X)+\langle\,\theta_x^\ell(t,X),b(t,X,Y,Z)\,\rangle\\
&+\frac{1}{2}\mathrm{tr}\,[\theta_{xx}^\ell(t,X)(\sigma\sigma^\top)(t,X,Y,Z)]\Big\}dt\\
&+\langle\,\theta_x^\ell(t,X),\sigma(t,X,Y,Z)dW(t)\,\rangle\\
=&\Big\{\langle\,\theta_x^\ell(t,X),b(t,X,Y,Z)-b(t,X,\widetilde{Y},\widetilde{Z})\,\rangle\\
&+\frac{1}{2}\mathrm{tr}\Big[\theta_{xx}^\ell(t,X)\big\{(\sigma\sigma^\top)(t,X,Y,Z)\\
&\qquad-(\sigma\sigma^\top)(t,X,\widetilde{Y},\widetilde{Z})\big\}\Big]+h^\ell(t,X,\widetilde{Y},\widetilde{Z})\Big\}dt\\
&+\langle\,\theta_x^\ell(t,X),\sigma(t,X,Y,Z)dW(t)\,\rangle.
\end{aligned}\tag{5.19}$$

Then it follows from (5.1) and (5.18) that

$$(5.20)\qquad \begin{aligned} &E|\widetilde{Y}(t) - Y(t)|^2 \\ &= -E\int_t^T \sum_{\ell=1}^k \Big\{ 2(\widetilde{Y}^\ell - Y^\ell)\Big[\langle \theta_x^\ell(s,X), b(s,X,Y,Z) - b(s,X,\widetilde{Y},\widetilde{Z})\rangle \\ &\quad + \frac{1}{2}\mathrm{tr}\,\{\theta_{xx}^\ell(s,X)[(\sigma\sigma^\top)(s,X,Y,Z) - (\sigma\sigma^\top)(s,X,\widetilde{Y},\widetilde{Z})]\} \\ &\quad + h^\ell(s,X,\widetilde{Y},\widetilde{Z}) - h^k(s,X,Y,Z)\Big] \\ &\quad + \mathrm{tr}\Big[\{\sigma(s,X,Y,Z) - \sigma(s,X,\widetilde{Y},\widetilde{Z})\}^\top \theta_x^\ell(s,X) + \widetilde{Z} - Z\Big] \\ &\quad \cdot \Big[\{\sigma(s,X,Y,Z) - \sigma(s,X,\widetilde{Y},\widetilde{Z})\}^\top \theta_x^\ell(s,X) + \widetilde{Z} - Z\Big]^\top \Big\} ds. \end{aligned}$$

By (5.16), we see that (recall the norm of matrices, see (3.2)–(3.3))

$$(5.21)\qquad \begin{aligned} &\Big|\big[\sigma(s,X,Y,Z) - \sigma(s,X,Y,\widetilde{Z})\big]^\top \theta_x^\ell(s,X) + \widetilde{Z} - Z\Big|^2 \\ &\geq \big(1 - |\theta_x^\ell(s,X)|L_\sigma\big)^2 |\widetilde{Z} - Z|^2 \geq \delta|\widetilde{Z} - Z|^2, \end{aligned}$$

for some $\delta > 0$. Hence, by the boundedness of θ_x, θ_{xx} and the uniform Lipschitz continuity of b, σ, h, we have

$$(5.22)\qquad \begin{aligned} &E|\widetilde{Y}(t) - Y(t)|^2 + \int_t^T E|Z(s) - \widetilde{Z}(s)|^2 ds \\ &\leq K\int_t^T E|\widetilde{Y}(s) - Y(s)|(|\widetilde{Y}(s) - Y(s)| + |\widetilde{Z}(s) - Z(s)|)ds. \end{aligned}$$

Using Gronwall's inequality, we conclude that

$$(5.23)\qquad Y(t) = \widetilde{Y}(t), \quad Z(t) = \widetilde{Z}(t), \quad \text{a.e. } t \in [0,T], \ \mathbf{P}\text{-a.s.}$$

Thus any solution of (5.1) must have the form that is constructed via the four-step scheme, proving our claim.

Finally, let (X, Y, Z) and $(\widetilde{X}, \widetilde{Y}, \widetilde{Z})$ be any two solutions of (5.1). By the previous argument we have

$$(5.24)\qquad \begin{cases} Y(t) = \theta(t, X(t)), & Z(t) = z(t, X(t), \theta(t, X(t)), \theta_x(t, X(t))), \\ \widetilde{Y}(t) = \theta(t, \widetilde{X}(t)), & \widetilde{Z}(t) = z(t, \widetilde{X}(t), \theta(t, \widetilde{X}(t)), \theta_x(t, \widetilde{X}(t))). \end{cases}$$

Hence $X(t)$ and $\widetilde{X}(t)$ satisfy exactly the same forward SDE (5.13) with the same initial state x. Thus by strong uniqueness we must have $X(t) = \widetilde{X}(t)$, $\forall t \in [0,T]$, $\mathbf{P}$-a.s., which in turn shows that $Y(t) = \widetilde{Y}(t)$, $Z(t) = \widetilde{Z}(t)$, $\forall t \in [0,T]$, $\mathbf{P}$-a.s. by (5.24). The proof is now complete. □

Note that when σ is independent of z, we may take $L_\sigma = 0$, and then (5.16) holds automatically.

One can clearly see the resemblance between (5.8) and (5.10) above and equations (4.13) and (4.23) in Chapter 5. Therefore, part of the relationship between the stochastic maximum principle and dynamic programming can also be explained by the four-step scheme. More specifically, to solve the stochastic Hamiltonian system (which is an FBSDE), solving the HJB equation is Step 2 in the four-step scheme via the key relation $p(t) = -V_x(t, \bar{x}(t))$ (namely, $\theta = -V_x$ here; see Chapter 5, (4.13)). On the other hand, the approach of using stochastic Riccati equations in the LQ problems can also be regarded as an example of the four-step scheme (see Chapter 6, (6.2), in which case $\theta(t, x) = -P(t)x - \varphi(t)$).

5.3. Several solvable classes of FBSDEs

From the previous subsection we see that to solve FBSDE (5.1), it suffices to check when the four-step scheme can be realized. In this subsection we are going to find several such classes of FBSDEs.

1. A general case.

Let us make the following assumptions for this case.

(FB1) $m = n$. In addition, the functions b, σ, h, and g are smooth functions taking values in $\mathbb{R}^n$, $\mathbb{R}^{n\times n}$, $\mathbb{R}^k$, and $\mathbb{R}^k$, respectively, and their first-order derivatives in x, y, z are all bounded uniformly by some constant $L > 0$.

(FB2) The function σ is independent of z, and there exists a positive continuous function $\nu(\cdot)$ and a constant $\mu > 0$ such that for all $(t, x, y, z) \in [0, T] \times \mathbb{R}^n \times \mathbb{R}^k \times \mathbb{R}^{n\times n}$,

$$\nu(|y|)I \leq \sigma(t, x, y)\sigma(t, x, y)^T \leq \mu I, \tag{5.25}$$

$$|b(t, x, 0, 0)| + |h(t, x, 0, z)| \leq \mu. \tag{5.26}$$

(FB3) The function g is bounded in $C^{2+\alpha}(\mathbb{R}^n; \mathbb{R}^k)$ for some $\alpha \in (0, 1)$.

Throughout this subsection, by "smooth" we mean that the involved functions possess partial derivatives of all necessary orders. We prefer not to indicate the exact order of smoothness for the sake of simplicity of presentation. Since σ is independent of z, equation (5.11) is (trivially) uniquely solvable for z. In the present case, FBSDE (5.1) reads as follows:

$$\begin{cases} dX(t) = b(t, X(t), Y(t), Z(t))dt + \sigma(t, X(t), Y(t))dW(t), \\ dY(t) = h(t, X(t), Y(t), Z(t))dt + Z(t)dW(t), \\ X(0) = x, \qquad Y(T) = g(X(T)); \end{cases} \tag{5.27}$$

and (5.12) takes the following form:

$$\begin{cases} \theta^\ell_t + \frac{1}{2}\text{tr}\left[\theta^\ell_{xx}(\sigma\sigma^\top)(t, x, \theta)\right] + \langle\, b(t, x, \theta, z(t, x, \theta, \theta_x)), \theta^\ell_x \,\rangle \\ \qquad - h^\ell(t, x, \theta, z(t, x, \theta, \theta_x)) = 0, \ (t, x) \in (0, T) \times \mathbb{R}^n, \ 1 \leq \ell \leq k, \\ \theta(T, x) = g(x), \qquad x \in \mathbb{R}^n. \end{cases} \tag{5.28}$$

Let us first apply the result of Ladyzhenskaya–Solonnikov–Ural'tseva [1] to solve (5.28). Consider the following terminal–boundary value problem:

$$\begin{cases} \theta^\ell_t + \sum_{i,j=1}^n a_{ij}(t,x,\theta)\theta^\ell_{x_i x_j} + \sum_{i=1}^n b_i(t,x,\theta,z(t,x,\theta,\theta_x))\theta^\ell_{x_i} \\ \qquad - h^\ell(t,x,\theta,z(t,x,\theta,\theta_x)) = 0, \ (t,x)\in[0,T]\times B_R, \ 1\le \ell\le k, \\ \theta\mid_{\partial B_R} = g(x), \qquad |x| = R, \\ \theta(T,x) = g(x), \qquad x\in B_R, \end{cases} \tag{5.29}$$

where B_R is the ball centered at the origin with radius $R>0$ and

$$\begin{cases} \big(a_{ij}(t,x,y)\big) = \frac{1}{2}\sigma(t,x,y)\sigma(t,x,y)^\top, \\ (b_1(t,x,y,z),\dots,b_n(t,x,y,z))^\top = b(t,x,y,z), \\ (h^1(t,x,y,z),\dots,h^k(t,x,y,z))^\top = h(t,x,y,z). \end{cases}$$

Clearly, under the present situation, the function $z(t,x,y,p)$ determined by (5.11) is smooth. The following lemma is a variant of Ladyzhenskaya–Solonnikov–Ural'tseva [1, Chapter VII, Theorem 7.1].

Lemma 5.4. *Suppose that all the functions a_{ij}, b_i, h^ℓ, and g are smooth, and g is bounded in $C^{2+\alpha}(\mathbb{R}^n;\mathbb{R}^k)$ for some $\alpha\in(0,1)$. Suppose in addition that for all $(t,x,y)\in[0,T]\times\mathbb{R}^n\times\mathbb{R}^k$ and $p\in\mathbb{R}^{k\times n}$, it holds that*

$$\nu(|y|)I \le \big(a_{ij}(t,x,y)\big) \le \mu(|y|)I, \tag{5.30}$$

$$|b(t,x,y,z(t,x,y,p))| \le \mu(|y|)(1+|p|), \tag{5.31}$$

$$\left|\frac{\partial}{\partial x_\alpha}a_{ij}(t,x,y)\right| + \left|\frac{\partial}{\partial y^\beta}a_{ij}(t,x,y)\right| \le \mu(|y|), \tag{5.32}$$

for some continuous functions $\mu(\cdot)$ and $\nu(\cdot)$ with $\nu(r)>0$,

$$|h(t,x,y,z(t,x,y,p))| \le [\varepsilon(|y|) + P(|p|,|y|)](1+|p|^2), \tag{5.33}$$

where $P(|p|,|y|)\to 0$ as $|p|\to\infty$ and $\varepsilon(|y|)$ is small enough, and

$$\sum_{\ell=1}^k h^\ell(t,x,y,z(t,x,y,p))y^\ell \ge -L(1+|y|^2), \tag{5.34}$$

for some constant $L>0$. Then (5.29) admits a unique classical solution.

Using Lemma 5.4, we can now prove the solvability of (5.28) under our assumptions.

Theorem 5.5. *Let (FB1)–(FB3) hold. Then (5.28) admits a unique classical solution $\theta(t,x)$ with $\theta(t,x)$, $\theta_t(t,x)$, $\theta_x(t,x)$, and $\theta_{xx}(t,x)$ bounded. Consequently, FBSDE (5.27) admits a unique adapted solution.*

Proof. We first show that all the required conditions in Lemma 5.4 are satisfied. Since σ is independent of z, the function $z(t,x,y,p)$ determined by (5.11) satisfies

$$|z(t,x,y,p)| \le K|p|, \quad \forall (t,x,y,p) \in [0,T] \times \mathbb{R}^n \times \mathbb{R}^k \times \mathbb{R}^{k\times n}. \tag{5.35}$$

Now, (5.30) and (5.32) follow from (FB1) and (FB2); (5.31) follows from (FB1), (5.26), and (5.35); and (5.33)–(5.34) follow from (FB1) and (5.26). Therefore, by Lemma 5.4 there exists a unique bounded solution $\theta(t,x;R)$ of (5.29) for which $\theta_t(t,x;R)$, $\theta_x(t,x;R)$, and $\theta_{xx}(t,x;R)$ together with $\theta(t,x)$ are bounded uniformly in $R > 0$. Using a diagonalization argument one further shows that there exists a subsequence $\theta(t,x,R)$ that converges to $\theta(t,x)$ as $R \to \infty$, and $\theta(t,x)$ is a classical solution of (5.28), with $\theta_t(t,x)$, $\theta_x(t,x)$, and $\theta_{xx}(t,x)$ all being bounded.

Since all the coefficients as well as all the possible solutions are smooth with required bounded partial derivatives, the uniqueness of solutions to (5.28) follows from a standard argument using Gronwall's inequality.

Finally, by Theorem 5.3, FBSDE (5.27) is uniquely solvable. □

2. Case where h has linear growth in z.

Although Theorem 5.5 gives a general solvability result of the FBSDE (5.27), condition (5.26) in (FB2) is rather restrictive. For instance, the case where the coefficient $h(t,x,y,z)$ grows linearly in z is excluded. This case, however, is very important for applications in optimal stochastic control theory. Specifically, in the Pontryagin maximum principle (see Chapter 3) for optimal stochastic control, the adjoint equations are of the form where the corresponding h is affine in z (see Chapter 3, (3.8) and (3.9)). Thus we would like to discuss this case separately.

In order to relax the condition (5.26), we consider the following special FBSDE:

$$\begin{cases} dX(t) = b(t,X(t),Y(t),Z(t))dt + \sigma(t,X(t))dW(t), \\ dY(t) = h(t,X(t),Y(t),Z(t))dt + Z(t)dW(t), \\ X(0) = x, \qquad Y(T) = g(X(T)). \end{cases} \tag{5.36}$$

We assume that σ is independent of y and z, but allow h to have linear growth in z. In this case, the corresponding parabolic system (5.12) becomes (compare with (5.28))

$$\begin{cases} \theta^\ell_t + \frac{1}{2}\mathrm{tr}\left[\theta^\ell_{xx}\sigma(t,x)\sigma(t,x)^\top\right] + \langle b(t,x,\theta,z(t,x,\theta,\theta_x)),\theta^\ell_x\rangle \\ \qquad -h^\ell(t,x,\theta,z(t,x,\theta,\theta_x))=0, \ (t,x)\in[0,T]\times\mathbb{R}^n, \ 1\le \ell\le k, \\ \theta(T,x) = g(x), \qquad x\in\mathbb{R}^n. \end{cases} \tag{5.37}$$

Since now h has linear growth in z, the result of Ladyzhenskaya–Solonnikov–Ural'tseva [1] does not apply. We use a result of Wiegner [1] instead. To this end, let us rewrite (5.37) in divergence form:

$$
(5.38)\qquad \begin{cases} \theta_t^\ell + \displaystyle\sum_{i,j=1}^n (a_{ij}(t,x)\theta_{x_i}^\ell)_{x_j} = f^\ell(t,x,\theta,\theta_x), \\ \qquad\qquad\qquad\qquad (t,x)\in[0,T]\times\mathbb{R}^n, \quad 1\le\ell\le k, \\ \theta(T,x)=g(x), \qquad x\in\mathbb{R}^n, \end{cases}
$$

where

$$
(5.39)\qquad \begin{cases} (a_{ij}(t,x)) \triangleq \dfrac{1}{2}\sigma(t,x)\sigma(t,x)^\top, \\ f^\ell(t,x,y,p) \triangleq \displaystyle\sum_{i,j=1}^n (a_{ij})_{x_j}(t,x)p_i^\ell - \sum_{i=1}^n b_i(t,x,y,z(t,x,y,p))p_i^\ell \\ \qquad\qquad - h^\ell(t,x,y,z(t,x,y,p)). \end{cases}
$$

By Wiegner [1], for any $T>0$, (5.38) has a unique classical solution, global in time, provided that the following conditions hold:

$$
(5.40)\qquad \nu I \le (a_{ij}(t,x)) \le \mu I, \qquad \forall (t,x)\in[0,T]\times\mathbb{R}^n,
$$

$$
(5.41)\qquad \begin{aligned} &\sum_{\ell=1}^k y^\ell f^\ell(t,x,y,p) \le \varepsilon_0|p|^2 + K(1+|y|^2), \\ &\qquad\qquad \forall (t,x,y,p)\in[0,T]\times\mathbb{R}^n\times\mathbb{R}^k\times\mathbb{R}^{n\times k}, \end{aligned}
$$

where ν,μ,K,ε_0 are constants with ε_0 being small enough. Therefore, we need the following assumption:

(FB2)$'$ There exist positive constants ν,μ such that

$$
(5.42)\qquad \nu I \le \sigma(t,x)\sigma(t,x)^\top \le \mu I, \qquad \forall (t,x)\in[0,T]\times\mathbb{R}^n,
$$

$$
(5.43)\qquad \begin{aligned} &|b(t,x,y,z)|,\ |h(t,x,0,0)| \le \mu, \\ &\qquad\qquad \forall (t,x,y,z)\in[0,T]\times\mathbb{R}^n\times\mathbb{R}^k\times\mathbb{R}^{k\times n}. \end{aligned}
$$

Theorem 5.6. *Suppose that (FB1), (FB2)$'$, and (FB3) hold. Then (5.36) admits a unique adapted solution (X,Y,Z).*

Proof. In the present case, for the function $z(t,x,y,p)$ determined by (5.11), we still have (5.35). Also, conditions (5.40) and (5.41) hold, which leads to the existence and uniqueness of classical solutions of (5.38) or (5.37). Next, applying Theorem 5.3, we can show that there exists a unique adapted solution (X,Y,Z) of (5.36). □

Since $h(t,x,y,z)$ is only assumed to be uniformly Lipschitz continuous in (y,z) (see (FB1)), we have

$$
(5.44)\qquad \begin{aligned} &|h(t,x,y,z)| \le K(1+|y|+|z|), \\ &\qquad\qquad \forall (t,x,y,z)\in[0,T]\times\mathbb{R}^n\times\mathbb{R}^k\times\mathbb{R}^{k\times n}. \end{aligned}
$$

In other words, the function h is allowed to have linear growth in (y, z).

3. Case where $k = 1$.

Unlike the previous ones, this is the case in which the existence and uniqueness result can be derived for a more general system than (5.28) and (5.37). The main reason is that in this case, the function $\theta(t, x)$ is scalar valued, and the theory for (single) quasilinear parabolic PDEs is more satisfactory than that for *systems* of quasilinear parabolic PDEs. Consequently, the corresponding results for the FBSDEs will allow more complicated nonlinearities. Remember that in the present case, the backward component is one-dimensional, but the forward part is still n-dimensional. This is exactly the case for some option pricing problems (see Section 6).

We now consider (5.1) with $k = 1$. Under (FB1), W is an n-dimensional standard Brownian motion, and b, σ, h, and g take values in $\mathbb{R}^n$, $\mathbb{R}^{n\times n}$, $\mathbb{R}$, and $\mathbb{R}$, respectively. Also, X, Y, and Z take values in $\mathbb{R}^n$, $\mathbb{R}$, and $\mathbb{R}^n$, respectively. In what follows we will use our four-step scheme to solve (5.1). To this end, we first need to solve (5.11) for z. In the present case, using the convention that all the vectors are column vectors, we should rewrite (5.11) as follows:

$$z = \sigma(t, x, y, z)^\top p. \tag{5.45}$$

Let us introduce the following assumption.

(FB2)$''$ There exists a positive continuous function $\nu(\cdot)$ and constants $C, \beta > 0$ such that for all $(t, x, y, z) \in [0, t] \times \mathbb{R}^n \times \mathbb{R} \times \mathbb{R}^n$,

$$\nu(|y|)I \leq \sigma(t, x, y, z)\sigma(t, x, y, z)^\top \leq KI, \tag{5.46}$$

$$\langle [\sigma(t, x, y, z)^\top]^{-1} z - [\sigma(t, x, y, \widehat{z})^\top]^{-1}\widehat{z}, z - \widehat{z}\rangle \geq \beta|z - \widehat{z}|^2, \tag{5.47}$$

$$|b(t, x, 0, 0)| + |h(t, x, 0, 0)| \leq K. \tag{5.48}$$

Note that condition (5.47) amounts to saying that the map $z \mapsto [\sigma(t, x, y, z)^\top]^{-1} z$ is uniformly monotone. This is a sufficient condition for (5.45) to be uniquely solvable for z. Some other conditions are also possible, for example, the map $z \mapsto -[\sigma(t, x, y, z)^\top]^{-1} z$ is uniformly monotone. We have the following result for the unique solvability of FBSDE (5.1) with $k = 1$.

Theorem 5.7. *Let (FB1) and (FB2)$''$ hold with $k = 1$. Then there exists a unique smooth function $z(t, x, y, p)$ that solves (5.45) and satisfies (5.35). If in addition (FB3) holds, then FBSDE (5.1) admits a unique adapted solution determined by the four-step scheme.*

We should note that the existence and uniqueness of solutions to (5.12) in the present case ($k = 1$) follow from Ladyzhenskaya–Solonnikov–Ural'tseva [1, Chapter V, Theorem 8.1]. Therefore, Theorem 5.7 can be proved similarly. Details are left to the reader. We see that the condition (5.48) together with (FB1) implies that both functions b and h are allowed to have linear growth in y and z.

6. Option Pricing Problems

In this section we are going to present an application of the theory developed in the previous sections to so-called *option pricing problems.* Note, however, that the theory of BSDEs and/or FBSDEs represents only one possible approach to handling option pricing problems.

6.1. European call options and the Black–Scholes formula

Let us first recall the market model introduced in Chapter 2, Section 3.2. Our notation will be a bit different from that in Chapter 2, in order to match the notation that we have been using in this chapter. For simplicity, we consider a market where only 2 assets (or securities) are traded continuously. One of the assets is called a bond, whose price process $X_0(t)$ is subject to the following ordinary differential equation:

$$\begin{cases} dX_0(t) = r(t)X_0(t)dt, & t \in [0,T], \\ X_0(0) = x_0. \end{cases} \tag{6.1}$$

The other asset is called a stock, whose price process $X(t)$ satisfies the following stochastic differential equation:

$$\begin{cases} dX(t) = b(t)X(t)dt + \sigma(t)X(t)dW(t), & t \in [0,T], \\ X(0) = x. \end{cases} \tag{6.2}$$

Here, $W(t)$ is a standard one-dimensional Brownian motion defined on some fixed filtered complete probability space $(\Omega, \mathcal{F}, \{\mathcal{F}_t\}_{t\geq 0}, \mathbf{P})$ with $\{\mathcal{F}_t\}_{t\geq 0}$ being the natural filtration of $W(\cdot)$ augmented by all the $\mathbf{P}$-null sets. We refer to the processes $r(\cdot)$, $b(\cdot)$, and $\sigma(\cdot)$ as the interest rate (of the bond), the appreciation rate, and the volatility (of the stock), respectively. All these processes are assumed to be $\{\mathcal{F}_t\}_{t\geq 0}$-adapted. Sometimes, we also call $b(\cdot)$ and $\sigma(\cdot)$ the average return rate and dispersion, respectively. Here we suppose

$$\sigma(t) \geq \sigma_0 > 0, \qquad \forall t \in [0,T], \tag{6.3}$$

for some constant σ_0. This assumption means that the "noise" that affects the stock price persistently exists. Usually, people expect that $Eb(t) > Er(t)$. This is very natural, since otherwise nobody is willing to invest in the stock. However, the following discussion does not depend on this assumption.

Now, a *European call option* is a contract that is signed between the *seller* (also called *issuer* or *writer*) and the *buyer* (or *holder*) of the option at time $t = 0$. By holding it, the holder has the right to buy one share of the stock from the writer at a prespecified price, say q, at a specified time $t = T$. The price q is called the *exercise price* (or *strike price*), and the time T is called the *maturity* (or *expiration*) *date*. Note that a European call option can be exercised only on the maturity date, as opposed to an *American call option*, which can be exercised at any time before the maturity date. At time $t = T$, if the market price $X(T)$ of the stock is lower than q, then the contract is worthless to the holder, since he/she can buy the stock at a lower price $X(T)$ than q in the market if he/she wishes. However, if it happens that the price $X(T)$ is higher than q, then the holder will exercise the contract, buy the stock at price q from the writer, and then sell it immediately at the price $X(T)$ in the market to get a profit of $X(T) - q$. This implies that the contract (option) is worth $(X(T) - q)^+$ at maturity $t = T$. Symmetrically, the loss to the *writer* of the option is exactly the same amount $(X(T) - q)^+$. The question is, What is the (fair) price of this option at time $t = 0$? This is the *option pricing* problem.

Let us make some observation on this problem. As mentioned, the price of the option at $t = T$ is the amount that the holder of the option would obtain as well as the amount that the writer would lose at that time. Now, suppose this option has a price y at $t = 0$. Then the writer of the option has y as the *initial endowment* at $t = 0$. He/she has to invest this amount of money in some way (called the *replication* of the option) in the market (where there are one bond and one stock available) so that at time $t = T$, his/her total wealth, denoted by $Y(T)$, resulting from the investment of y, should at least compensate his/her loss $(X(T) - q)^+$, namely,

$$Y(T) \geq (X(T) - q)^+. \tag{6.4}$$

It is clear that for the same investment strategy, the larger the initial endowment y, the larger the final wealth $Y(T)$. Hence the writer of the option would like to set y large enough so that (6.4) can be guaranteed. On the other hand, if it happens that for some y the resulting final wealth $Y(T)$ is strictly larger than the loss $(X(T) - q)^+$, then the price y of this option at $t = 0$ is considered to be too high. In this case, the buyer of the option instead of buying the option, would make his/her own investment to make more profit $Y(T)$ than just $(X(T) - q)^+$. From the above observation, we see that the *fair price* for the option at time $t = 0$ should be such a y that the corresponding optimal investment would result in a wealth process $Y(\cdot)$ satisfying

$$Y(T) = (X(T) - q)^+. \tag{6.5}$$

Next, let us find the equation that the wealth process $Y(\cdot)$ should satisfy. To this end, denote by $\pi(t)$ the amount that the investor (the seller of the option) invested in the stock. The remaining amount $Y(t) - \pi(t)$ is invested in the bond. Clearly, $\pi(\cdot)$ determines a strategy of the investment, which is called a *portfolio*. Different portfolios would result in different values of final wealth $Y(T)$. Thus, we can regard the portfolio as the control. Next, we assume that $N_0(t)$ and $N(t)$ are the numbers of shares that the investor holds for the bond and the stock, respectively. Then the total wealth is

$$Y(t) = N_0(t)X_0(t) + N(t)X(t), \qquad t \in [0,T]. \tag{6.6}$$

If the trade is made at some discrete time moments, say at t and $t+h$, and there is no infusion or withdrawal of funds, then

$$Y(t+h)-Y(t) = N_0(t)\{X_0(t+h)-X_0(t)\}+N(t)\{X(t+h)-X(t)\}. \tag{6.7}$$

Thus, the continuous-time analogue of the above should be the following (noting (6.1)–(6.2)):

$$\begin{aligned} dY(t) &= N_0(t)dX_0(t) + N(t)dX(t) \\ &= N_0(t)X_0(t)r(t)dt + N(t)X(t)\{b(t)dt + \sigma(t)dW(t)\} \\ &= \{r(t)Y(t) + [b(t) - r(t)]\pi(t)\}dt + \sigma(t)\pi(t)dW(t). \end{aligned} \tag{6.8}$$

By setting $Z(t) \overset{\Delta}{=} \sigma(t)\pi(t)$, we obtain the following FBSDEs:

$$\begin{cases} dX(t) = b(t)X(t)dt + \sigma(t)X(t)dW(t), \\ dY(t) = \Big\{r(t)Y(t) + [b(t) - r(t)]\dfrac{Z(t)}{\sigma(t)}\Big\}dt + Z(t)dW(t), \\ X(0) = x, \quad Y(T) = (X(T) - q)^+. \end{cases} \tag{6.9}$$

We see that this is a decoupled linear FBSDE. Hence, one has the following result, which is a corollary of Theorem 2.2.

Proposition 6.1. *There exists a unique adapted solution* $(X(\cdot), Y(\cdot), Z(\cdot))$ *to (6.9). The option price is given by* $Y(0)$, *and the portfolio* $\pi(\cdot)$ *is given by*

$$\pi(t) = \frac{Z(t)}{\sigma(t)}, \qquad \forall t \in [0,T]. \tag{6.10}$$

Notice that the above result gives the existence of a fair price for the European call option, with all the coefficients $r(\cdot)$, $b(\cdot)$, and $\sigma(\cdot)$ allowed to be random. As a matter of fact, the exercise price q could also be random ($\mathcal{F}_T$-measurable) as far as the existence is concerned. However, Proposition 6.1 does not tell how to calculate the price. Let us now use the four-step scheme to achieve such a goal. To this end, however, we need to assume that $r(t)$, $b(t)$, and $\sigma(t)$ are all deterministic.

Step 1. Set

$$z(t,x,y,p) = \sigma(t)xp, \quad (t,x,y,p) \in [0,T] \times \mathbb{R}^3. \tag{6.11}$$

Step 2. Solve the following PDE:

$$\begin{cases} \theta_t + \dfrac{\sigma(t)^2x^2}{2}\theta_{xx} + r(t)x\theta_x - r(t)\theta = 0, \\ \theta\big|_{t=T} = (x-q)^+. \end{cases} \tag{6.12}$$

Step 3. Solve the SDE (6.2).

Step 4. Set

$$\begin{cases} Y(t) = \theta(t, X(t)), \\ Z(t) = \sigma(t)X(t)\theta_x(t, X(t)). \end{cases} \tag{6.13}$$

Then the option price (at $t = 0$) is given by

$$y = Y(0) = \theta(0, x). \tag{6.14}$$

It should be noted that since equation (6.12) is *independent* of $b(t)$ (the average return rate of the stock), the option price depends only on the interest rate $r(\cdot)$ of the bond and the volatility $\sigma(\cdot)$ of the stock. We refer to (6.13)–(6.14) as a *generalized Black–Scholes formula.*

When $r(t) \equiv r$, $b(t) \equiv b$, and $\sigma(t) \equiv \sigma > 0$ are all constants, (6.12) is called the *Black–Scholes equation* and (6.14) is called the *Black–Scholes formula.* In this case, one can derive the explicit form of $\theta(t, x)$. For the reader's convenience, let us now carry out a derivation of the explicit form of $\theta(t, x)$.

We need to solve the following equation:

$$\begin{cases} \theta_t + \dfrac{\sigma^2 x^2}{2}\theta_{xx} + rx\theta_x - r\theta = 0, \quad (t, x) \in [0, T) \times \mathbb{R}, \\ \theta\big|_{t=T} = (x - q)^+. \end{cases} \tag{6.15}$$

First of all, let us note that if θ is a classical solution of (6.15), then at $x = 0$, θ as a function of t satisfies a linear ODE with terminal value 0 (assuming $q \geq 0$). Hence $\theta\big|_{x=0} = 0, \quad t \in [0, T]$. Therefore, θ solves

$$\begin{cases} \theta_t + \dfrac{\sigma^2 x^2}{2}\theta_{xx} + rx\theta_x - r\theta = 0, \quad (t, x) \in [0, T) \times (0, \infty), \\ \theta\big|_{x=0} = 0, \quad t \in [0, T], \\ \theta\big|_{t=T} = (x - q)^+, \quad x \in (0, \infty). \end{cases} \tag{6.16}$$

Here we are interested in $\theta(t, x)$ only for $x > 0$, as the stock price is never negative. Now, we solve (6.16) by the following steps:

(i) Let $\xi \stackrel{\Delta}{=} \ln x$ and $\varphi(t, \xi) \stackrel{\Delta}{=} \theta(t, e^\xi)$. Then $\varphi(t, \xi)$ satisfies

$$\begin{cases} \varphi_t + \dfrac{\sigma^2}{2}\varphi_{\xi\xi} + (r - \dfrac{\sigma^2}{2})\varphi_\xi - r\varphi = 0, \quad (t, \xi) \in [0, T) \times \mathbb{R}, \\ \varphi\big|_{t=T} = (e^\xi - q)^+, \quad \xi \in \mathbb{R}. \end{cases} \tag{6.17}$$

(ii) Let $s \stackrel{\Delta}{=} \gamma t$ and $\psi(s, \xi) \stackrel{\Delta}{=} e^{-\frac{\alpha s}{\gamma} - \beta\xi}\varphi(\frac{s}{\gamma}, \xi)$, with

$$\alpha \stackrel{\Delta}{=} r + \frac{1}{2\sigma^2}\Big(r - \frac{\sigma^2}{2}\Big)^2, \ \beta \stackrel{\Delta}{=} -\frac{1}{\sigma^2}\Big(r - \frac{\sigma^2}{2}\Big), \ \gamma \stackrel{\Delta}{=} \frac{\sigma^2}{2}.$$

Then $\psi(s,\xi)$ satisfies the following:

$$\begin{cases}\psi_s+\psi_{\xi\xi}=0, & (s,\xi)\in[0,\gamma T)\times\mathbb{R},\\ \psi\big|_{s=\gamma T}=e^{-\alpha T-\beta\xi}(e^\xi-q)^+, & \xi\in\mathbb{R}.\end{cases}\tag{6.18}$$

We can verify directly that the solution of the above terminal value problem for the heat equation is given by

$$\begin{aligned}\psi(s,\xi)&=\frac{1}{\sqrt{4\pi(\gamma T-s)}}\int_{\mathbb{R}}e^{-\frac{(\xi-\eta)^2}{4(\gamma T-s)}}e^{-\alpha T-\beta\eta}(e^\eta-q)^+d\eta\\ &=\frac{e^{-\alpha T}}{\sqrt{4\pi(\gamma T-s)}}\int_{\ln q}^{\infty}e^{-\frac{(\xi-\eta)^2}{4(\gamma T-s)}-(\beta-1)\eta}d\eta\\ &\quad-\frac{qe^{-\alpha T}}{\sqrt{4\pi(\gamma T-s)}}\int_{\ln q}^{\infty}e^{-\frac{(\xi-\eta)^2}{4(\gamma T-s)}-\beta\eta}d\eta.\end{aligned}\tag{6.19}$$

Consequently, returning to $\varphi(t,\xi)$ with some change of variables, we obtain

$$\begin{aligned}\varphi(t,\xi)&=\frac{e^\xi}{\sqrt{2\pi}}\int_{-\infty}^{\frac{\xi-\ln q+(r+\frac{\sigma^2}{2})(T-t)}{\sigma\sqrt{T-t}}}e^{-\frac{z^2}{2}}dz\\ &\quad-\frac{qe^{-r(T-t)}}{\sqrt{2\pi}}\int_{-\infty}^{\frac{\xi-\ln q+(r-\frac{\sigma^2}{2})(T-t)}{\sigma\sqrt{T-t}}}e^{-\frac{z^2}{2}}dz.\end{aligned}\tag{6.20}$$

This leads to the final form

$$\begin{aligned}\theta(t,x)&=\frac{x}{\sqrt{2\pi}}\int_{-\infty}^{\frac{\ln(\frac{x}{q})+(r+\frac{\sigma^2}{2})(T-t)}{\sigma\sqrt{T-t}}}e^{-\frac{z^2}{2}}dz\\ &\quad-\frac{qe^{-r(T-t)}}{\sqrt{2\pi}}\int_{-\infty}^{\frac{\ln(\frac{x}{q})+(r-\frac{\sigma^2}{2})(T-t)}{\sigma\sqrt{T-t}}}e^{-\frac{z^2}{2}}dz.\end{aligned}\tag{6.21}$$

If we set

$$\begin{cases}N(x)\triangleq\dfrac{1}{\sqrt{2\pi}}\displaystyle\int_{-\infty}^{x}e^{-\frac{z^2}{2}}dz,\\[2ex] d_\pm(t,x)\triangleq\dfrac{\ln(\frac{x}{q})+(r\pm\frac{\sigma^2}{2})(T-t)}{\sigma\sqrt{T-t}},\end{cases}\tag{6.22}$$

then

$$\theta(t,x)=xN(d_+(t,x))-qe^{-r(T-t)}N(d_-(t,x)),\tag{6.23}$$

which is the familiar (classical) form of the Black–Scholes formula for European call options.

6.2. Other options

In this subsection we briefly discuss some other option pricing problems. Throughout this subsection we assume that the market is described by (6.1)–(6.2).

1. European put options

A *European put option* is a contract signed at $t = 0$, and by holding it, the holder has the right to sell one share of the stock to the writer at a prespecified price q and a specified time $t = T$. Thus, for such a case, the loss to the writer at time $t = T$ is $(q - X(T))^+$. Hence, instead of the FBSDE (6.9), we need to solve the following:

$$
\begin{cases} dX(t) = b(t)X(t)dt + \sigma(t)X(t)dW(t), \\ dY(t) = \Big\{r(t)Y(t) + [b(t) - r(t)]\dfrac{Z(t)}{\sigma(t)}\Big\}dt + Z(t)dW(t), \\ X(0) = x, \quad Y(T) = (q - X(T))^+. \end{cases} \tag{6.24}
$$

Comparing (6.9) and (6.24), we see that the only difference is the terminal condition for $Y(\cdot)$. Hence, by Theorem 2.2, we have the existence of the fair price for such an option. It is also possible to obtain a generalized Black–Scholes formula using the four-step scheme when all the coefficients are deterministic. Further, if the coefficients are constants, we can recover the classical form of the Black–Scholes formula for European put options. We leave the details to the reader.

2. Asian options

Consider a contract that is signed at $t = 0$, and by holding it, the holder has the option to buy one share of the stock at time $t = T$ at the average price of this stock during the time interval $[T_0, T]$, for some $T_0 \in [0, T)$. Thus, the loss to the writer at time $t = T$ is $(\frac{1}{T-T_0}\int_{T_0}^T X(t)dt - q)^+$. This is called an *Asian call option*. For pricing such an option, we introduce

$$
\overline{X}(t) \triangleq \frac{1}{T - T_0}\int_0^t X(s)I_{[T_0,T]}(s)ds, \qquad t \in [0, T]. \tag{6.25}
$$

Then we need to solve the following FBSDE:

$$
\begin{cases} dX(t) = b(t)X(t)dt + \sigma(t)X(t)dW(t), \\ d\overline{X}(t) = \dfrac{1}{T - T_0}X(t)I_{[T_0,T]}(t)dt, \\ dY(t) = \Big\{r(t)Y(t) + [b(t) - r(t)]\dfrac{Z(t)}{\sigma(t)}\Big\}dt + Z(t)dW(t), \\ X(0) = x, \ \overline{X}(0) = 0, \ Y(T) = (\overline{X}(T) - q)^+. \end{cases} \tag{6.26}
$$

Again, this is a decoupled FBSDE. Thus, Theorem 2.2 applies to obtain the existence of the fair price $y = Y(0)$ for this option, where all the coefficients are allowed to be $\{\mathcal{F}_t\}_{t\geq 0}$-adapted. One can also deal with the problem using the four-step scheme. Finally, the *Asian put options* can be discussed in a similar manner. We leave the details to the reader.

3. Barrier options

There are various forms of barrier options. The most common ones are *knock-out* and *knock-in* options. Since these options can be treated similarly, here let us discuss only one type of knock-out option called the *down-and-out call option*. Consider a contract signed at time $t = 0$ with a specified time period $[0, T]$ and a specified stock price q (called a *barrier*) that is lower than the current market price $X(0)$. If the price $X(t)$ of the stock ever hits the barrier q during $[0, T]$, the contract becomes void (*out*). If the price $X(t)$ stays above q all the time during $[0, T]$, then the holder has the right to buy the stock at price q on the maturity day $t = T$ from the writer of the option. It is seen that by holding such an option, the holder is guaranteed to buy the stock (if he/she so wishes) at a price no higher than q during $[0, T]$.

In order to find the price of such an option, we introduce the following stopping time

$$\tau = \inf\{t \in [0, T] \mid X(t) = q\}, \quad \text{where } \inf \phi \stackrel{\Delta}{=} T. \tag{6.27}$$

Then the loss to the writer at time $t = \tau$ is $(X(\tau) - q)^+$. Hence, we need to solve the following decoupled FBSDE on a random time duration $[0, \tau]$:

$$\begin{cases} dX(t) = b(t)X(t)dt + \sigma(t)X(t)dW(t), \\ dY(t) = \Big\{r(t)Y(t) + [b(t) - r(t)]\dfrac{Z(t)}{\sigma(t)}\Big\}dt + Z(t)dW(t), \\ X(0) = x, \ Y(\tau) = (X(\tau) - q)^+. \end{cases} \tag{6.28}$$

Now, by Theorem 3.6, we have the existence of an adapted solution $(X(\cdot), Y(\cdot), Z(\cdot))$ that gives the existence (and uniqueness) of the price for this option.

7. Historical Remarks

There have been two types of backward stochastic differential equations studied in the literature. One of them allows nonadapted (anticipative) solutions typically by Kunita [2] using time-reversal, and by Ocone–Pardoux [1] using an anticipative integral initiated by Skorohod [2] and developed by Nualart–Pardoux [1] and others. The other one requires adapted (nonanticipative) solutions. Bismut [1,2,3] first introduced a linear BSDE with adapted solutions when he was studying adjoint equations of the stochastic optimal control problem. It should be noted that a BSDE having adapted solutions is required very naturally in deriving a stochastic maximum principle (see Chapter 3). This approach was later developed by Bensoussan [1,3]. The discussion in Section 2 follows closely that of Bismut [3] and Bensoussan [1]. The research on this type of BSDE was reactivated in the weekly seminar of the Fudan group during 1987–1989, in studying the necessary conditions for stochastic optimal controls with the control variable entering into the diffusion coefficients of the state equation and the control domain being not necessarily convex. Making use of the adjoint equations, Peng [1] proved the general maximum principle (of the form presented in Chapter 3), and Zhou [2,4] established the relationship between the maximum principle and dynamic programming (see Chapter 5), both in the late

1988. Motivated by this, and combining the works by Bismut [1,2,3] and Bensoussan [1], in 1989, Peng and Pardoux (who was at that time visiting Fudan University for a short period) found that they could prove the well-posedness of the nonlinear BSDE directly. This led to the fundamental paper of Pardoux–Peng [1]. A little later, Duffie–Epstein [1] independently introduced the nonlinear BSDE when studying the so-called *recursive utility* in finance. Section 3.1 is a modification of Pardoux–Peng [1] while Section 3.2 is an improvement of Peng [2], with some ideas adopted from Yong [8]. A sound application of the nonlinear BSDEs is the derivation of a *nonlinear* Feynman–Kac formula that was obtained by Peng [2]. More research on representing quasilinear PDEs via BSDEs was carried out by Pardoux–Peng [3] and Peng [7]. See Yong [9] for a survey on the relationship among various types of differential equations. The material in Section 4.2 is a modification of Peng [2].

Antonelli [1] first studied forward–backward stochastic differential equations (FBSDEs, for short) in the early 1990s. Since the contraction mapping technique was used, he had to assume that either the Lipschitz constant of the coefficients or the time duration is small enough. In 1992, Ma and Yong started their investigation of FBSDEs. They first introduced the method of stochastic optimal control to approach FBSDEs in an arbitrary finite time duration, without constraint on the size of the Lipschitz constants (Ma–Yong [1]). Inspired by this work, as well as some ideas of invariant embedding of Bellman–Kalaba–Wing [1] and Scott [1], Ma–Protter–Yong [1] proposed the so-called *four-step scheme* to solve an FBSDE via a parabolic PDE, which can be regarded as a sort of reverse of the Feynman–Kac formula. More research on FBSDEs can be found in Hu–Peng [3], Yong [8,10], Pardoux–Tang [1], Peng–Wu [1], and Ma–Yong [4]. A summary of updated works on FBSDEs can be found in the book by Ma–Yong [6]. Section 5 is mainly based on Ma–Protter–Yong [1].

While BSDEs stem from stochastic controls, Kohlmann–Zhou [1] recently interpreted BSDEs as some *forward* optimal stochastic control problems starting from *given* initial values. They regard the first component of the solution pair of a BSDE as a state variable while regarding the second component as a *control* variable and solve the problem via the stochastic LQ method. This gives a new perspective on BSDEs.

It is now widely accepted that the work by Bachelier [1] in 1900 announced the birth of mathematical finance (as well as Brownian motion). See Merton [6] for a survey on the history of this subject. The modern foundation of option pricing theory was established by the Nobel-prize-winning work of Black–Scholes [1] and Merton [4]. Standard references on options are Cox–Rubinstein [1] and Hull [1]. Musiela–Rutkowski [1] is a good book on this subject.

One of the most appealing features of BSDEs and FBSDEs is that they can be applied to finance problems and give deep insights into them. Section 6, which is partly based on El Karoui–Peng–Quenez [1], gives such an example of option pricing problems. Our way of deriving the Black–

Scholes formula using the four-step scheme is inspired by Ma–Protter–Yong [1]. The method we used in Sectino 6 in treating Asian options and barrier options by BSDEs seems to be new. Other applications can be found in Cvitanić–Ma [1], Duffie–Ma–Yong [1], El Karoui–Peng–Quenez [1], and Buckdahn–Hu [1,2,3].

Motivated by pricing European contingent claims in incomplete markets, Yong [11,12] recently studied some BSDEs and FBSDEs with possibly degenerate correction terms (namely, the term $Z(t)dW(t)$ is replaced by $\sigma(t)Z(t)dW(t)$ for some matrix-valued process $\sigma(t)$, which could be degenerate). Solvability of such BSDEs heavily depends on the choice of terminal conditions. This is also closely related to the controllability of stochastic systems.

One key assumption in the BSDE theory is that the underlying filtration must be generated by the Brownian motion (due to the martingale representation). Recently, Kohlmann–Zhou [2] considered a case where the filtration is larger than the Brownian filtration and established an existence result for the BSDE. The result is then applied to a contingent claim problem where different agents have different information.

Using BSDEs, Peng [6] introduced a so-called g-expectation, which is a generalization of the conventional mathematical expectation. See Peng [8] for recent developments in this direction.

Backward stochastic partial differential equations (BSPDEs) were introduced by Bensoussan [4] in the linear case in order to derive a maximum principle for SPDEs. He employed the so-called *reduction to Robust form* to treat the problem. However, this approach applied only to the case where the drift operator of the original state equation was bounded. Zhou [7] used a different approach, namely, finite-dimensional approximation, to prove the existence and uniqueness of the solutions to linear BSDEs even when their forward adjoints have unbounded drift. Moreover, Zhou [5] obtained the Sobolev norm estimates for the solutions to the linear BSPDEs. Later, Ma–Yong [3,5] established a general theory for linear BSPDEs. Nonlinear BSPDEs were first studied by Peng [3] as the dynamic programming equation for stochastic optimal control with random coefficients. More work on the problem was carried out by Peng [4], Hu–Peng [2], and Pardoux–Peng [2].

BSDE theory remains a new and largely unexplored area. This chapter represents the first systematic and updated account of the theory.

References

Ait Rami, M. and Zhou, X.Y.

[1] Linear matrix inequalities, Riccati equations, and indefinite stochastic linear–quadratic controls, preprint.

Ait Rami, M., Moore, J. B., and Zhou, X.Y.

[1] Indefinite stochastic linear quadratic control problem and generalized differential Riccati equation, preprint.

Akella, R. and Kumar, P.R.

[1] Optimal control of production rate in a failure-prone manufacturing system, *IEEE Trans. Auto. Control.*, AC-31 (1986), 116–126.

Alexandrov, A.D.

[1] Almost everywhere existence of the second differential of a convex function and some properties of convex functions, *Leningrad Univ. Ann. (Math. Ser.),* 37 (1939), 3–35 (in Russian).

Anderson, B.D.O. and Moore, J.B.

[1] *Linear Optimal Control,* Prentice-Hall, Englewood Cliffs, New Jersey, 1971.

Antonelli, A.

[1] Backward-forward stochastic differential equations, *Ann. Appl. Prob.,* 3 (1993), 777–793.

Arkin, V. and Saksonov, M.

[1] Necessary optimality conditions in problems of the control of stochastic differential equations, *Dokl. SSSR,* 244 (1979), 11–15.

Arrow, K.

[1] *Essays in the Theory of Risk-Bearing,* North-Holland, Amsterdam, 1970.

Arrow, K.J. and Kurz, M.

[1] *Public Investment, the Rate of Return, and Optimal Fiscal Policy,* John Hopkins Press, Baltimore, 1970.

Asmussen, S. and Taksar, M.

[1] Controlled diffusion models for optimal dividend pay-out, *Insurance: Math. & Econom.*, 20 (1997), 1–15.

Athens, M.

[1] Special issues on linear-quadratic-Gaussian problem, *IEEE Trans. Auto. Control.,* AC-16 (1971), 527–869.

Bachelier, L.

[1] Théorie de la spéculation, *Ann Sci.École Norm. Sup.*, 17 (1900), 21–86; English Transl. in *The Random Character of Stock Market Prices,* P.H. Coonter (ed.), MIT Press, Cambridge, 1964, 17–78.

Balakrishnan, A.V.

[1] *Applied Functional Analysis,* Springer-Verlag, New York, 1976.

Baldursson, F.M. and Karatzas, I.

[1] Irreversible investment and industry equilibrium, *Finance & Stoch.*, 1 (1997), 69–89.

Baras, J.S., Elliott, R.J., and Kohlmann, M.

[1] The partially observed stochastic minimum principle, *SIAM J. Control & Optim.*, 27 (1989), 1279–1292.

Barbu, V.

[1] Hamilton–Jacobi equations and nonlinear control problems, *J. Math. Anal. Appl.*, 120 (1986), 494–509.

[2] The dynamic programming equation for the time-optimal control problem in infinite dimensions, *SIAM J. Control & Optim.*, 29 (1991), 445–456.

Barbu, V. and Da Prato, G.

[1] *Hamilton Jacobi Equations in Hilbert Spaces*, Pitman, London, 1983.

[2] Hamilton–Jacobi equation in Hilbert spaces: variational and semigroup approach, *Ann. Math. Pura. Appl.*, 142 (1985), 303–349.

Barbu, V., Da Prato, G., and Popa, C.

[1] Existence and uniqueness of the dynamic programming equation in Hilbert space, *Nonlinear Anal.*, 7 (1983), 283–299.

Bardi, M. and Capuzzo-Dolcetta, I.

[1] *Optimal Control and Viscosity Solutions of Hamilton–Jacobi–Bellman Equations*, Birkhäuser, Boston, 1997.

Barles, G.

[1] Deterministic impulse control problems, *SIAM J. Control & Optim.*, 23 (1985), 419–432.

Barron, E.N. and Jensen, R.

[1] The Pontryagin maximum principle from dynamic programming and viscosity solutions to first-order partial differential equations, *Trans. Amer. Math. Soc.*, 298 (1986), 635–641.

Başar, T. and Bernhard, P.

[1] H_∞ *Optimal Control and Related Minimax Design Problems*, Birkhäuser, Boston, 1991.

Bass, F.M.

[1] A new product growth model for consumer durables, Manag. Sci., 15 (1969), 215–227.

Bather, J.A. and Chernoff, H.

[1] Sequential decisions in the control of a spaceship (finite fuel), *J. Appl. Prob.*, 49 (1967), 584–604.

Bellman, R.

[1] On the theory of dynamic programming, *Proc. Nat. Acad. Sci. USA*, 38 (1952), 716–719.

[2] *An Introduction to the Theory of Dynamic Programming*, Rand Corporation Report R-245, 1953.

[3] Dynamic programming and a new formalism in the calculus of variations, *Proc. Nat. Acad. Sci. USA*, 40 (1954), 231–235.

[4] Monotone approximation in dynamic programming and the calculus of variations, *Proc. Nat. Acad. Sci. USA*, 40 (1954), 1073–1075.

[5] *Dynamic Programming*, Princeton Univ. Press, Princeton, New Jersey, 1957.

[6] A Markovian decision process, *J. Math. Mech.*, 6 (1957), 679–684.

[7] Dynamic programming and stochastic control process, *Inform. & Control*, 1 (1958), 228–239.

Bellman, R. and Blackwell, D.

[1] On a particular non-zero sum game, *Rand Corporation*, RM-250, 1949.

Bellman, R., Glicksberg, I., and Gross, O.

[1] On some variational problems occurring in the theory of dynamic programming, *Rend. Circ. Mat. Palermo* (2), 3 (1954), 1–35.

[2] *Some Aspects of the Mathematical Theory of Control Processes*, Rand Corporation, Santa Monica, California, 1958.

Bellman, R. and Kalaba, R.

[1] *Selected Papers on Mathematical Trends in Control Theory*, Dover Publications, New York, 1964.

Bellman, R., Kalaba, R.E., and Wing, G. M.

[1] Invariant embedding and the reduction of two-point boundary-value problems to initial value problems, *Proc. Nat'l Acad. Sci. USA*, 46 (1960), 1646–1649.

Bellman, R. and LaSalle, J.P.

[1] On non-zero sum games and stochastic processes, *Rand Corporation*, RM-212, 1949.

Beneš, V.E.

[1] Existence of optimal strategies based in specified information, for a class of stochastic decision problems, *SIAM J. Control*, 8 (1970), 179–188.

[2] Existence of optimal stochastic control laws, *SIAM J. Control*, 9 (1971), 446–472.

[3] Exact finite dimensional filters for certain diffusions with nonlinear drift, *Stochastics*, 5 (1985), 65–92.

Beneš, V.E. and Karatzas, I.

[1] Estimation and control for linear partially observable systems with non-Gaussian initial distribution, *Stoch. Proc. & Their Appl.*, 14 (1983), 233–248.

Beneš, V.E., Shepp, L.A., and Witsenhaussen, H.S.

[1] Some solvable stochastic control problems, *Stochastics*, 4 (1980), 39–83.

Bensoussan, A.

[1] Lecture on stochastic control, *Lecture Notes in Math.*, Vol.972, Springer-Verlag, Berlin, 1981.

[2] *Stochastic Control by Functional Analysis Methods*, North-Holland, 1982.

[3] Stochastic maximum principle for distributed parameter system, *J. Franklin Inst.*, 315 (1983), 387–406.

[4] Maximum principle and dynamic programming approaches of the optimal control of partially observed diffusions, *Stochastics*, 9 (1983), 169-222.

[5] *Perturbation Methods in Optimal Control*, John-Wiley & Sons, New York, 1988.

[6] *Stochastic Control of Partially Observable Systems*, Cambridge University Press, Cambridge, 1992.

Bensoussan, A., Crouhy, M., and Proth, J.-M.

[1] *Mathematical Theory of Production Planning*, North-Holland, New York, NY, 1983.

Bensoussan, A. and Frehse, J.

[1] On Bellman equations of ergodic controls in $\mathbb{R}^n$, *J. reine U. angew. Math.*, 429 (1992), 125–160.

Bensoussan, A., Frehse, J., and Nagai, H.

[1] Some results on risk-sensitive control with full observation, *Appl. Math. Optim.*, 37 (1998), 1–41.

Bensoussan, A., and Lions, J.L.

[1] Problèmes de temps d'arrêt optimal et inéquations variationnelles paraboliques, *Appl. Anal.*, 3 (1973), 267–294.

[2] Nouvelle formulation de problèmes de contrôle impulsionnel et applications, *C. R. Acad. Sci. Paris*, 277 (1973), 1189–1192.

[3] *Impulse Control and Quasi-Variational Inequalities*, Bordes, Paris, 1984.

Bensoussan, A. and Nagai, H.

[1] An ergodic control problem arising from the principal eigenfunction of an elliptic operator, *J. Math. Soc. Japan*, 43 (1991), 49–65.

[2] Min-max characterization of a small noise limit on risk sensitive control, *SIAM J. Control & Optim.*, 35 (1997), 1093–1115.

Bensoussan, A. and Nisio, M.

[1] Nonlinear semigroup arising in the control of diffusions with partial observation, *Stochastics*, 30 (1990), 1–45.

Bensoussan, A., Sethi, S.P., Vickson, R., and Derzko, N.

[1] Stochastic production planning with production constraints, *SIAM J. Control & Optim.*, 22 (1984), 920-935.

Bensoussan, A. and Van Shuppen, J.H.

[1] Optimal control of partially observable stochastic systems with an exponential-of-integral performance index, *SIAM J. Control Optim.*, 23 (1985), 599–613.

Benton, S.

[1] *The Hamilton–Jacobi Equation: A Global Approach*, Academic Press, New York, 1977.

Berkovitz, L.D.

[1] *Optimal Control Theory*, Springer-Verlag, New York, 1974.

[2] Optimal feedback controls, *SIAM J. Control & Optim.*, 27 (1989), 367–377.

Berkovitz, L.D. and Fleming, W.H.

[1] A tribute to E. J. McShane, *SIAM J. Control & Optim.*, 27 (1989), 909–915.

Billingsley, P.

[1] *Probability and Measure*, Wiley, New York, 1985.

Bismut, J.M.

[1] Analyse convexe et probabilités, Thèse, Faculté des Sciences de Paris, Paris, 1973.

[2] Théorie probabiliste du contrôle des diffusions, *Mem. Amer. Math. Soc.*, 4 (1976), 1–130.

[3] An introductory approach to duality in optimal stochastic control, *SIAM Rev.*, 20 (1978), 62–78.

Black, F. and Scholes, M.

[1] The pricing of options and corporate liability, *J. Polit. Economy*, 81 (1973), 637–659.

Boetius, F. and Kohlmann, M.

[1] Connection between optimal stopping and singular stochastic control, *Stoch. Proc. & Their Appl.*, 77 (1998), 253–281.

Boerner, H.,

[1] Carathéodory's Eingang zur Variationsrechnung, *Jahresber. Deutsche Math.-Ver.* 56 (1953), 31–58.

[2] Variationsrechnung à la Carathéodory und das Zermelo'sche Navigationsproblem, *Selecta Mathematica V*, Heidelberger Taschenbücher Nr. 201, Springer-Verlag, Berlin, 1979, 23–67.

Boltyanski, V.G.

[1] The maximum principle in the theory of optimal processes, *Dokl. Akad. Nauk SSSR*, 119 (1958), 1070–1073 (in Russian).

[2] The maximum principle—How it came to be? Report No. 526, Mathematisches Institut, Technische Universität München, 1994.

Boltyanski, V.G., Gamkrelidze, R.V., and Pontryagin, L.S.

[1] On the theory of optimal processes, *Dokl. Akad. Nauk SSSR*, 10 (1956), 7–10 (in Russian).

[2] The theory of optimal processes, I: The maximum principle, *Izv. Akad. Nauk SSSR Ser. Mat.*, 24 (1960), 3–42 (in Russian); English transl. in *Amer. Math. Soc. Transl.* 18 (1961), no.2, 341–382.

Borkar, V.S.

[1] *Optimal Control of Diffusion Processes*, Pitman Research Notes in Math., No.203, Longman Sci. & Tech., Harlow, UK, 1989.

Borovkov, A.

[1] On limit laws for service processes in multi-channel systems, *Siberian Math. J.*, 8 (1967), 746–763.

Boyd, S, El Ghaoui, L., Feron, E., and Balakrishnan, V.

[1] *Linear Matrix Inequality in Systems and Control Theory*, SIAM, Philadelphia, 1994.

Branicky, M.S., Borkar, V.S., and Mitter, S.K.
[1] A unified framework for hybrid control: Model and optimal control theory, *IEEE Trans. Auto. Control.*, AC-43 (1998), 31–45.

Breeden, D.T.
[1] An intertemporal asset pricing model with stochastic consumption and investment opportunities, *J. Finan. Econ.*, 7 (1979), 265–296.

Brockett, R.W.
[1] Nonlinear systems and nonlinear estimation theory, in *Mathematics of Filtering and Identification and Applications,* M. Hazewinkel and J.C. Willems (eds.), Reidel, Dordrecht, 1981.

Brockett, R.W. and Clark, J.M.C.
[1] The geometry of the conditional density functions, in *Analysis and Optimization of Stochastic Systems,* O.L.R. Jacobs (ed.), Academic Press, New York, 1980, 299–309.

Buckdahn, R. and Hu, Y.
[1] Pricing of American contingent claims with jump stock price and constrained portfolios, *Math. Oper. Res.*, 23 (1998), 177–203.
[2] Hedging contingent claims for a large investor in an incomplete market, *Adv. Appl. Probab.*, 30 (1998), 239–255.
[3] Hedging contingent claims with Wiener-Poisson driven stock price, preprint.

Cadenillas, A. and Haussmann, U.G.
[1] The stochastic maximum principle for a singular control problem, *Stoch. & Stoch. Rep.*, 49 (1994), 211–237.

Cadenillas, A. and Karatzas, I.
[1] The stochastic maximum principle for linear convex optimal control with random coefficients, *SIAM J. Control & Optim.*, 33 (1995), 590–624.

Cannarsa, P.
[1] Regularity properties of solutions to Hamilton Jacobi equations in infinite dimensions and nonlinear optimal control, *Diff. Int. Eqs.*, 2 (1989), 479–493.

Cannarsa, P. and Da Prato, G.
[1] Nonlinear optimal control with infinite horizon for distributed parameter systems and stationary Hamilton–Jacobi equations, *SIAM J. Control & Optim.*, 27 (1989), 861–875.
[2] Some results on nonlinear optimal control problems and Hamilton–Jacobi equations in infinite dimensions, *J. Funct. Anal.*, 90 (1990), 27–47.

Capuzzo-Dolcetta, I. and Evans, L.C.
[1] Optimal switching for ordinary differential equations, *SIAM J. Control & Optim.*, 22 (1984), 143–161.

Carathéodory, C.
[1] Die methode der geodätischen Äquisidtanten und das Problem von Lagrange, *Acta Math.*, 47 (1926), 199–236.

[2] *Calculus of Variations and Partial Differential Equations of the First Order, I*, B.G.Teubner, Berlin, 1935 (in German); Holden-Day, San Francisco, 1965 (in English).

Cesari, L.

[1] *Optimization: Theory and Applications*, Springer-Verlag, New York, 1982.

Chaleyat-Maurel, M. and Michel, D.

[1] Des résultats de non existence de filtre de dimension finie, *Stochastics*, 13 (1984), 83–102.

Chaney, R.W.

[1] Second order directional derivatives for non smooth functions, *J. Math. Anal. Appl.*, 128 (1987), 495–511.

Charalambous, C.D. and Elliott, R.J.

[1] Certain nonlinear partially observable stochastic optimal control problems with explicit control laws equivalent to LEQG/LQG problems, *IEEE Trans. Auto. Control*, AC-42 (1997), 482–497.

Chen, H. and Mandelbaum, A.

[1] Stochastic discrete flow networks: Diffusion approximations and bottlenecks, *Ann. Probab.*, 19 (1991), 1463–1519.

Chen, S.

[1] Riccati equations arising in infinite-dimensional optimal control problems, *Control Theory & Appl.*, 2 (1985), no.4, 64–72 (in Chinese).

Chen, S., Li, X., and Zhou, X.Y.

[1] Stochastic linear quadratic regulators with indefinite control weight costs, *SIAM J. Control & Optim.*, 36 (1998), 1685–1702.

Chen, S., and Yong, J.,

[1] Stochastic linear quadratic optimal control problems, preprint.

Chen, S. and Zhou, X.Y.

[1] Stochastic linear quadratic regulators with indefinite control weight costs, II, preprint.

Chernoff, H.

[1] Optimal stochastic controls, *Sankhya, Ser. A*, 30 (1968), 221–252.

Chi, T., Liu, J.J., and Chen, H.

[1] Optimal stopping rule for a project with uncertain completion time and partial salvageability, *IEEE Trans. Eng. Manag.*, 44 (1997), 54–66.

Chow, P.L., Menaldi, J.L., and Robin, M.

[1] Additive control of stochastic linear systems with finite horizon, *SIAM J. Control & Optim.*, 23 (1985), 858–899.

Citron, S.J.

[1] *Elements of Optimal Control*, Holt, Rinehart and Winston, New York, 1969.

Clarke, F.H.

[1] The maximum principle under minimal hypotheses, *SIAM J. Control & Optim.*, 14 (1976), 1078–1091.

[2] *Optimization and Nonsmooth Analysis*, Wiley, New York, 1983.

Clarke, F.H. and Vinter, R.B.

[1] Local optimality conditions and Lipschitzian solutions to the Hamilton–Jacobi equations, *SIAM J. Control & Optim.*, 21 (1983), 856–870.

[2] The relationship between the maximum principle and dynamic programming, *SIAM J. Control & Optim.*, 25 (1987), 1291–1311.

Cominetti, R. and Correa, R.

[1] A generalized second-order derivative in nonsmooth optimization, *SIAM J. Control & Optim.*, 28 (1990), 789–809.

Cook, I. and Mayes, P.

[1] *Introduction to Innovation and Technology Transfer*, Artech House, Inc., Boston, 1996.

Courant, R. and Hilbert, D.

[1] *Methods of Mathematical Physics, I & II*, John Wiley & Sons, New York, 1962.

Cox, J.C.

[1] Optimal consumption and portfolio rules when assets follow a diffusion process, working paper #658, Graduate School of Business, Stanford University, 1983.

Cox, J.C. and Huang, C.F.

[1] Optimal consumption and portfolio policies when asset prices follow a diffusion process, *J. Econ. Theory*, 49 (1989), 33–83.

Cox, J.C., Ingersoll, J.E., and Ross, S.A.

[1] An intertemporal general equilibrium model of asset prices, *Econometrica*, 53 (1985), 363–384.

Cox, R.M. and Karatzas, I.,

[1] Stationary control of Brownian motion in several dimensions, *Adv. Appl. Probab.*, 17 (1985), 531–561.

Cox, J.C. and Rubinstein, M.

[1] *Options Markets*, Prentice-Hall, Englewood Cliffs, New Jersey, 1985.

Crandall, M.G. and Lions, P.L.

[1] Conditions d'unicité pour les solutions generalisés des équations d'Hamilton–Jacobi du premier ordre, *C. R. Acad. Sci. Paris Sér. I, Math.*, 292 (1981), 487–502.

[2] Viscosity solutions of Hamilton–Jacobi equations, *Trans. Amer. Math. Soc.*, 277 (1983), 1–42.

[3] Hamilton Jacobi equation in infinite dimensions I: uniqueness of viscosity solutions; II: existence of viscosity solutions; III; IV: Hamiltonians with unbounded linear terms; V: unbounded linear terms and *B*-continuous solutions, *J. Funct. Anal.*, 62 (1985), 379–396; 65 (1986), 368–425; 68 (1986), 214–247; 90 (1990), 237–283; 97 (1991), 417–465.

Crandall, M.G. Evans, L.C., and Lions, P.L.

[1] Some properties of viscosity solutions of Hamilton–Jacobi equations, *Trans. Amer. Math. Soc.*, 282 (1984), 487–502.

Crandall, M.G., Ishii, H., and Lions, P.L.
[1] User's guide to viscosity solutions of second order partial differential equations, *Bull. Amer. Math. Soc. (NS)*, 27 (1992), 1–67.

Cvitanić, J. and Ma, J.
[1] Hedging options for a large investor and forward–backward SDEs, *Ann. Appl. Probab.*, 6 (1996), 370–398.

Davis, M.H.A.
[1] On the existence of optimal policies in stochastic control, *SIAM J. Control*, 11 (1973), 587–594.
[2] *Linear Estimation and Stochastic Control*, Chapman & Hall, London, 1977.
[3] A deterministic approach to stochastic optimal control with application to anticipative control, *Stoch. & Stoch. Rep.*, 40 (1992), 203–256.
[4] *Markov Models and Optimization*, Chapman & Hall, London, 1993.

Davis, M.H.A., and Varaiya, P.P.
[1] Dynamic programming conditions for partially observable stochastic systems, *SIAM J. Control*, 11 (1973), 226–261.

de Souza, C.E. and Fragoso, M.D.
[1] On the existence of maximal solution for generalized algebraic Riccati equations arising in stochastic control, *Systems & Control Lett.*, 14 (1990), 233–239.

Di Blasio, G.
[1] Global solutions for a class of Hamilton–Jacobi equations in Hilbert spaces, *Numer. Funct. Anal. Optim.*, 8 (1985–86), 261–300.

Diestel, J. and Uhl, J.J.,Jr.
[1] *Vector Measures*, AMS Providence, R.I., 1977.

Dixit, A.K. and Pindyck, R.S.
[1] *Investment under Uncertainty*, Princeton University Press, Princeton, 1994.

Dockner, E. and Jorgensen, S.
[1] Optimal advertising policies for diffusion models of new product innovation in monopolistic situations, *Manag. Sci.*, 34 (1988), 119–130.

Dokuchaev, N. and Zhou, X,Y.
[1] Stochastic control problems with endpoint contingent conditions, preprint.

Dong, G. and Bian, B.
[1] Uniqueness of viscosity solutions of fully nonlinear second order parabolic PDEs, *Chin. Ann. Math. Ser. B*, 11 (1990), 156–170.

Donsker, M.D.
[1] An invariant principle for certain probability limit theorems, *Mem. Amer. Math. Soc.*, 6 (1951).

Dorfman, R.
[1] An economic interpretation of optimal control theory, *Amer. Econ. Rev.*, 59 (1969), 817–831.

Duffie, D. and Epstein, L.

[1] Stochastic differential utility, *Econometrica*, 60 (1992), 353–394.

Duffie, D. and Jackson, M.

[1] Optimal hedging and equilibrium in a dynamic futures market, *J. Econ. Dynam. Control.*, 14 (1990), 21–33.

Duffie, D. and Lions, P.L.

[1] PDE solutions of stochastic differential utility, *J. Math. Econom.*, 21 (1993), 577–606.

Duffie, D., Ma, J., and Yong, J.

[1] Black's consol rate conjecture, *Ann. Appl. Prob.*, 5 (1995), 356–382.

Duffie, D. and Richardson, H.

[1] Mean–variance hedging in continuous time, *Ann. Appl. Probab.*, 1 (1991), 1–15.

Duncan, T.

[1] Doctoral dissertation, Dept. of Electrical Engineering, Stanford University, 1967.

[2] Stochastic systems in Riemannian manifolds, *J. Optim. Theory & Appl.*, 27 (1979), 399–421.

Duncan, T. and Upmeier, H.

[1] Stochastic control problems and spherical functions on symmetric spaces, *Trans. Amer. Math. Soc.*, 347 (1995), 1083–1130.

Duncan, T. and Varaiya, P.P.

[1] On the solutions of a stochastic control system, *SIAM J. Control*, 9 (1971), 354–371.

Dupuis, P. and Ishii, H.

[1] On Lipschitz continuity of the solution mapping to the Skorohod problem, with applications, *Stoch. & Stoch. Rep.*, 35 (1991), 31–62.

Dynkin, E.B.

[1] The optimal choice of the instant for stopping a Markov process, *Dolk. Acad. Nauk USSR*, 150 (1963), 238–240.

Ekeland, I.

[1] On the variational principle, *J. Math. Anal. Appl.*, 47 (1974), 324–353.

[2] Nonconvex minimization problems, *Bull Amer. Math. Soc. (New Series)*, 1 (1979), 443–474.

[3] *Convexity Methods in Hamiltonian Mechanics*, Springer-Verlag, 1990.

El Karoui, N. and Chaleyat-Maurel, M.

[1] Un problème de réflexion et ses applications au temps local et aux équation différentielles stochastique sur $\mathbb{R}$ cas continu, *Temps Locaux. Asterisque*, 52-53 (1978), 117–144.

El Karoui, N., Huu Nguyen, D., and Jeanblanc-Picqué, M.

[1] Compactification methods in the control of degenerate diffusions: existence of an optimal control, *Stochastics*, 20 (1987), 169–219.

El Karoui, N., Peng, S., and Quenez, M.C.

[1] Backward stochastic differential equations in finance, *Math. Finance*, 7 (1997), 1–71.

Elliott, R.J.

[1] The optimal control of a stochastic system, *SIAM J. Control & Optim.*, 15 (1977), 756–778.

[3] *Stochastic Calculus and Applications,* Springer-Verlag, New York, 1982.

Elliott, R.J. and Kohlmann, M.

[1] The variational principle and stochastic optimal control, *Stochastics,* 3 (1980), 229–241.

[2] The second order minimum principle and adjoint process, *Stoch. & Stoch. Rep.*, 46 (1994), 25–39.

Elton, E.J. and Gruber, M.J.

[1] *Finance as a Dynamic Process,* Englewood Cliffs: Prentice Hall, 1975.

Evans, L.C.

[1] A second order elliptic equation with a gradient constraint, *Comm. Part. Diff. Eqs.*, 4 (1979), 552–572.

Evans, L.C. and Friedman, A.

[1] Optimal stochastic switching and the Dirichlet problem for the Bellman equation, *Trans. Amer. Math. Soc.*, 253 (1979), 365–389.

Evans, L.C. and Gariepy, R.F.

[1] *Measure Theory and Fine Properties of Functions,* CRC Press, Boca Raton, FL, 1992.

Feynman, R.P.

[1] Space-time approach to nonrelativistic quantum mechanics, *Rev. Mod. Phys.*, 20 (1948), 367–387.

Feynman, R.P. and Hibbs, A.R.

[1] *Quantum Mechanics and Path Integrals,* McGraw Hill, New York, 1965.

Fitzpatrick, B. and Fleming, W.H.

[1] Numerical methods for optimal investment-consumption models, *Math. Oper. Res.*, 16 (1991), 823–841.

Fleming, W.H.

[1] The Cauchy problem for degenerate parabolic equations, *J. Math. Mech.*, 13 (1964), 987–1008.

[2] Optimal control of partially observable diffusions, *SIAM J. & Control,* 6 (1968), 194–214.

[3] The Cauchy problem for a nonlinear first order differential equation, *J. Diff. Eqs.*, 5 (1969), 530–555.

[4] Stochastic control for small noise intensities, *SIAM J. & Control,* 9 (1971), 473-517.

[5] Generalized solutions in optimal stochastic control, in *Differential Games and Control Theory III,* E. Roxin, P.-T. Liu, and R.L. Sternberg (eds.), Marel Dekker, New York, 1977, 147–165.

[6] Optimal control of Markov processes, *Proc. Int. Congress of Math., Warzawa*, 1983, 71–84.

[7] *Future Directions in Control Theory—A Mathematical Perspective*, SIAM Philadelphia, 1988.

Fleming, W.H. and Hernández-Hernández, D.

[1] Risk-sensitive control of finite state machines on an infinite horizon I, *SIAM J. Control & Optim.*, 35 (1997), 1790–1810.

[2] Risk-sensitive control of finite state machines on an infinite horizon II, *SIAM J. Control & Optim.*, to appear.

Fleming, W.H. and James, M.R.

[1] The risk-sensitive index and the H_2 and H_∞ norms for nonlinear systems, *Math. Control, Sign. & Systems*, 8 (1995), 199–221.

Fleming, W.H. and McEneaney, W.M.

[1] Risk-sensitive optimal control and differential games, *Lect. Notes in Control and Inform. Sci.*, 184 (1992), 185–197.

[2] Risk-sensitive control on an infinite horizon, *SIAM J. Control & Optim.*, 33 (1995), 1881–1915.

Fleming, W.H. and Nisio, M.

[1] On the existence of optimal stochastic controls, *J. Math. Mech.*, 15 (1966), 777–794.

Fleming, W.H. and Pardoux, E.

[1] Existence of optimal controls for partially observed diffusions, *SIAM J. Control*, 20 (1982), 261–283.

Fleming, W.H. and Rishel, R.W.

[1] *Deterministic and Stochastic Optimal Control*, Springer-Verlag, New York, 1975.

Fleming, W.H., Sethi, S.P., and Soner, H.M.

[1] An optimal stochastic production planning problem with randomly fluctuating demand, *SIAM J. Control & Optim.*, 25 (1987), 1494–1502.

Fleming, W.H. and Sheu, S.J.

[1] Optimal long term growth rate of expected utility of wealth, preprint.

Fleming, W.H. and Soner, H.M.

[1] *Controlled Markov Processes and Viscosity Solutions*, Springer-Verlag, New York, 1992.

Fleming, W.H. and Zariphopoulou, T.

[1] An optimal investment/consumption model with borrowing constraints, *Math. Oper. Res.*, 16 (1991), 802–822.

Florentin, J.J.

[1] Optimal control of continuous time, Markov, stochastic systems, *J. Electronics Control*, 10 (1961), 473–488.

Föllmer, H. and Sondermann, D.

[1] Hedging of non-redundant contingent claims, in *Contributions to Mathematical Economics*, A. Mas-Colell and W. Hildenbrand (eds.), North-Holland, Amsterdam, 1986, 205–223.

Frankowska, H.

[1] Hamilton–Jacobi equations: viscosity solutions and generalized gradients, *J. Math. Anal. Appl.*, 141 (1989), 21–26.

Friedman, A.

[1] *Stochastic Differential Equations and Applications, Vol. I-II,* Academic Press, New York, 1975.

Fujisaki, M., Kallianpur, G., and Kunita, H.

[1] Stochastic differential equations for the nonlinear filtering problem, *Osaka J. Math.*, 9 (1972), 19–40.

Gamkrelidze, R.V.

[1] On the theory of optimal processes in linear systems, *Dokl. Akad. Nauk SSSR,* 116 (1957), 9–11 (in Russian).

Giaquinta, M. and Hildebrandt, S.

[1] *Calculus of Variations, I, II,* Springer-Verlag, Berlin, 1996.

Gihman, I.I. and Skorohod, A.V.

[1] *The Theory of Stochastic Processes,* Springer-Verlag, Berlin, 1972.

Gilbarg, D. and Trudinger, N.S.

[1] *Elliptic Partial Differential Equations of Second Order, 2nd Edition,* Springer-Verlag, 1983.

Glynn, P.W.

[1] Diffusion approximations, in *Stochastic Models, Handbooks in Operations Research and Management Science, Vol.2,* D.P. Heyman and M.J. Sobel (eds.), North-Holland, Amsterdam, 145–198.

Goldstein, H.

[1] *Classical Mechanics,* Addison-Wesley, Massachusetts, 1959.

Goldstine, H.H.

[1] *A History of the Calculus of Variations from the 17th through the 19th Century,* Springer-Verlag, New York, 1980.

Grauer, R.R. and Hakansson, N.H.

[1] On the use of mean–variance and quadratic approximations in implementing dynamic investment strategies: a comparison of returns and investment policies, *Manag. Sci.*, 39 (1993), 856–871.

Grisvard, P.

[1] *Elliptic Problems in Nonsmooth Domains,* Pitman, Boston, 1985.

Harrison, J.M. and Nguyen, V.

[1] The QNET method for two-moment analysis of open queueing networks, *Queueing Syst. Theory Appl.*, 6 (1990), 1–32.

Harrison, J.M. and Reiman, M.I.

[1] Reflected Brownian motion on an orthant, *Ann. Probab.*, 9 (1981), 302–308.

Harrison, J.M. and Taksar, M.

[1] Instantaneous control of a Brownian motion, *Math. Oper. Res.*, 8 (1983), 439–453.

Harrison, J.M. and Wein, L.M.

[1] Scheduling networks of queues: Heavy traffic analysis of a simple open network, *Queueing Syst. Theory Appl.*, 5 (1989), 265–280.

[2] Scheduling networks of queues: Heavy traffic analysis of a two-station closed network, *Oper. Res.*, 38 (1990), 1051–1064.

Harrison, J.M. and Williams, R.J.

[1] Brownian models of open queueing networks with homogeneous customer populations, *Stochastics*, 22 (1987), 77–115.

Harrison, J.M., Williams, R.J., and Chen, H.

[1] Brownian models of closed queueing networks with homogeneous customer populations, *Stochastics*, 29 (1990), 37–74.

Haussmann, U.G.

[1] General necessary conditions for optimal control of stochastic systems, *Math. Prog. Study*, 6 (1976), 34–48.

[2] On the existence of optimal controls for partially observed diffusions, *SIAM J. Control & Optim.*, 20 (1982), 385–407.

[3] Existence of optimal Markovian controls for degenerate diffusions, *Lect. Notes Control & Inform. Sci.*, 78 (1985), 171–186.

[4] *A Stochastic Maximum Principle for Optimal Control of Diffusions*, Pitman Research Notes in Math., No.151, Longman Sci. & Tech., Harlow, UK, 1986.

[5] The maximum principle for optimal control of diffusions with partial information, *SIAM J. Control & Optim.*, 25 (1987), 341–361.

[6] A probabilistic approach to the generalized Hessian, *Math. Oper. Res.*, 17 (1992), 411–443.

[7] Generalized solutions of the Hamilton–Jacobi equations of stochastic control, *SIAM J. Control & Optim.*, 32 (1994), 728–743.

Haussmann, U.G. and Pardoux, E.

[1] A conditionally almost linear filtering problem with non-Gaussian initial conditions, *Stochastics*, 23 (1988), 241–275.

Haussmann, U.G. and Suo, W.

[1] Singular control of stochastic differential equations I: existence of optimal controls, *SIAM J. Control & Optim.*, 33 (1995), 916–936.

[2] Singular control of stochastic differential equations II: dynamic programming, *SIAM J. Control & Optim.*, 33 (1995), 937–959.

Hazewinkel, M., Marcus, S.I., and Sussmann, H.J.

[1] Nonexistence of finite dimensional filters for conditional statistics of the cubic sensor problem, *Syst. Control Lett.*, 3 (1983), 331–340.

Hermes, H. and LaSalle, J.P.

[1] *Functional Analysis and Time Optimal Control*, Academic Press, New York, 1969.

Hestenes, M.R.

[1] A general problem in the calculus of variations with applications to paths of least time, *Research Mem.* RM-100, Rand Corporation, Santa Monica, California, 1950.

Hojgaard, B. and Taksar, M.

[1] Controlling risk exposure and dividends pay-out schemes: Insurance company example, *Math. Finance*, to appear.

Horsky, D. and Simon, L.S.

[1] Advertising and the diffusion of new products, *Marketing Sci.*, 2 (1983), 1–17.

Hu, Y.

[1] Maximum principle for optimal control of Markov processes, *Acta Math. Sinica*, 33(1) (1990), 43–56 (in Chinese).

Hu, Y. and Peng, S.

[1] Maximum principle for semilinear stochastic evolution control systems, *Stoch. & Stoch. Rep.*, 33 (1990), 159–180.

[2] Adapted solution of backward semilinear stochastic evolution equations, *Stoch. Anal. & Appl.*, 9 (1991), 445–459.

[3] Solution of forward–backward stochastic differential equations, *Prob. Theory & Rel. Fields*, 103 (1995), 273–283.

Hu, Y. and Yong, J.

[1] Maximum principle for stochastic optimal impulse controls, *Chin. Ann. Math. Ser. A, Suppl.*, 12 (1991), 109–114 (in Chinese).

Hull, J.

[1] *Options, Futures, and Other Derivative Securities*, 2nd ed., Prentice-Hall, Englewood Cliffs, 1993.

Iglehart, D.L. and Whitt, W.

[1] Multiple channel queues in heavy traffic, I, *Adv. Appl. Probab.*, 2 (1970), 150-177.

[2] Multiple channel queues in heavy traffic, II: Sequences, networks, and batches, *Adv. Appl. Probab.*, 2 (1970), 355–369.

Ikeda, N. and Watanabe, S.

[1] *Stochastic Differential Equations and Diffusion Processes*, 2nd Edition, North Holland-Kodansha, Amsterdam-Tokyo, 1989.

Isaacs, R.

[1] Games of pursuit, *Rand Corporation Report*, P-257, 17 Nov. 1951.

[2] *Differential Games*, John Wiley & Sons, Inc., New York, 1965.

[3] Some fundamentals of differential games, *Topics in Differential Games*, A. Blaquière, North-Holland Publishing, Amsterdam, Holland, 1973, 1–42.

Ishii, H.

[1] Uniqueness of unbounded viscosity solution of Hamilton–Jacobi equations, *Indiana Univ. Math. J.*, 33 (1984), 721–748.

[2] On uniqueness and existence of viscosity solutions of fully nonlinear second order elliptic PDE's, *Comm. Pure Appl. Math.*, 42 (1989), 15–45.

Ishii, H. and Koike, S.

[1] Remarks on elliptic singular perturbation problems, *Appl. Math. Optim.*, 23 (1991), 1–15.

Itô, K.,

[1] Differential equations determining Markov processes, *Zenkoku Shijo Sugaku Danwakai*, 1077 (1942), 1352–1400 (in Japanese).

[2] Stochastic integral, *Proc. Imperial Acad. Tokyo*, 20 (1944), 519–524.

Jacobson, D.H.

[1] Optimal stochastic linear systems with exponential criteria and their relation to deterministic differential games, *IEEE Trans. Auto. Control*, AC-18 (1973), 124–131.

James, M.R.

[1] Asymptotic analysis of nonlinear stochastic risk-sensitive control and differential games, *Math. Contr. Sign. Syst.*, 5 (1992), 401–417.

James, M.R., Baras, J.S., and Elliott, R.J.

[1] Risk-sensitive control and dynamic games for partially observed discrete time nonlinear systems, *IEEE Trans. Auto. Control*, AC-39 (1994), 780–792.

Jeanblanc-Pique, M. and Shiryaev, A.

[1] Optimization of the stream of dividends (In Russian), *Russian Math. Surveys*, 50 (1995), 25–46.

Jensen, R.

[1] The maximum principle for viscosity solutions of second order fully nonlinear partial differential equations, *Arch. Rat. Mech. Anal.*, 101 (1988), 1–27.

Kabanov, Yu.M.

[1] On the Pontryagin maximum principle for SDEs with a Poisson-type driving noise, preprint.

Kac, M.

[1] On distribution of certain Wiener functionals, *Trans. Amer. Math. Soc.*, 65 (1949), 1–13.

[2] On some connection between probability theory and differential and integral equations, *Proc. 2nd Berkeley Sympos. Math. Stat. and Probab.*, (1951), 189–215.

Kailath, T.

[1] An innovation approach to least-square estimation. Part I: Linear filtering in additive white noise, *IEEE Trans. Auto. Control*, AC-13 (1968), 646–655.

Kailath, T. and Frost, P.

[1] An innovation approach to least-square estimation. Part II: Linear smoothing in additive white noise, *IEEE Trans. Auto. Control*, AC-13 (1968), 656–660.

Kallianpur, G.

[1] *Stochastic Filtering Theory*, Springer-Verlag, New York, 1980.

Kalman, R.E.

[1] Contributions to the theory of optimal control, *Bol. Soc. Mat. Mexicana*, 5 (1960), 102–119.

[2] On the general theory of control systems, *Proc. 1st IFAC Congress. Moscow,* Butterworths, London, 1960.
[3] The theory of optimal control and the calculus of variations, *Proc. of Conf. on Mathematical Optimization Techniques (Santa Monica, 1960), R. Bellman (ed.),* Univ. of California Press, CA 1963, 309–331.

Kalman, R.E. and Bucy, R.S.
[1] New results in linear filtering and prediction theory, *Trans. ASME Ser. D (J. Basic Eng.),* 83 (1961), 95–108.

Kalman, R.E., Falb, P.L., and Arbib, M.A.
[1] *Topics in Mathematical System Theory,* McGraw-Hill Book Company, New York, 1969.

Karatzas, I.
[1] On a stochastic representation for the principal eigenvalue of a second order differential equation, *Stochastics,* 3 (1980), 305–321.
[2] A class of singular stochastic control problems, *Adv. Appl. Prob.,* 15 (1983), 225–254.
[3] Probabilistic aspects of finite-fuel stochastic control, *Proc. Natl. Acad. Sci. USA,* 82 (1985), 5579–5581.
[4] Adaptive control of a diffusion to a goal and a parabolic Monge-Ampère-type equation, *Asian J. Math.,* 1 (1997), 324–341.

Karatzas, I., Lehoczky, J., and Shreve, S.E.
[1] Optimal portfolio and consumption decisions for a "small investor" on a finite horizon, *SIAM J. Control & Optim.,* 25 (1987), 1157–1186.

Karatzas, I. and Ocone, D.L.
[1] Feynman–Kac formula for a degenerate planar diffusion and an application in stochastic control, *Stoch. Anal.,* 273–296. Academic Press, Boston, 1991.
[2] The resolvent of a degenerate diffusion on the plane, with application to partially observed stochastic control, *Ann. Appl. Probab.,* 2(1992), 629–668.
[3] The finite-horizon version for a partially-observed stochastic control problem of Benes and Rishel, *Stoch. Anal. Appl.,* 11 (1993), 569–605.

Karatzas, I. and Shreve, S.E.
[1] Connections between optimal stopping and singular stochastic control I, Monotone follower problems, *SIAM J. Control & Optim.,* 22 (1984), 856–877.
[2] Connections between optimal stopping and singular stochastic control II, Reflected follower problems, *SIAM J. Control & Optim.,* 23 (1985), 433–451.
[3] *Brownian Motion and Stochastic Calculus,* Springer-Verlag, New York, 1988.

Klepper, S.
[1] Entry, exit, growth, and innovation over the product life cycle, *Amer. Econ. Rev.,* 86 (1996), 562–583.

Kohlmann, M.

[1] Concepts for the derivation of optimal partially observed controls, *Res. Notes Math.*, 150 (1986), 223–272.

Kohlmann, M. and Zhou, X.Y.

[1] Backward stochastic differential equations and stochastic controls: A new perspective, preprint.

[2] The informed and uninformed agent prices of a contingent claim, preprint.

Krichagina, E.V., Lou, S., Sethi, S.P., and Taksar, M.

[1] Production control in a failure-prone manufacturing system: Diffusion approximation and asymptotic optimality, *Ann. Appl. Probab.*, 3 (1993), 421–453.

Kružkov, S.N.

[1] The Cauchy problem in the large for certain nonlinear first order differential equations, *Soviet Math. Dokl.*, 1 (1960), 474–477.

[2] The Cauchy problem in the large for nonlinear equations and for certain quasilinear systems of the first order with several variables, *Soviet Math. Dokl.*, 5 (1964), 493–496.

[3] On solutions of first-order nonlinear equations, *Soviet Math. Dokl.*, 7 (1966), 376–379.

[4] Generalized solutions of nonlinear first order equations with several independent variables II, *Math. USSR Sbornik*, 1 (1967), 93–116.

[5] First order quasilinear equations in several independent variables, *Math. USSR Sbornik*, 10 (1970), 217–243.

Krylov, N.V.

[1] Sequences of convex functions and estimates of the maximum of the solution of a parabolic equation, *Siberian Math. J.*, 17 (1976), 290–303.

[2] *Controlled Diffusion Processes*, Springer-Verlag, New York, 1980.

[3] *Nonlinear Elliptic and Parabolic Equations of the Second Order*, Reidel, Dordrecht, 1987.

Krylov, N.V., and Rozovskii, B.L.

[1] On the Cauchy problem for linear stochastic partial differential equations, *Izv. Akad. Nauk SSSR Ser. Mat.*, 41 (1977), 1329–1347; *Math. USSR-Izv.*, 11 (1977), 1267–1284.

[2] Stochastic evolution equations, *J. Soviet Math.*, 16 (1981), 1233–1277.

[3] On characteristics of the degenerate parabolic Ito equations of the second order, *Proc. Petrovskii Sem.*, 8 (1982), 153–168 (in Russian).

[4] Stochastic partial differentials and diffusion processes, *Uspekhi Mat. Nauk.*, 37:6 (1982), 75–95; *Russian Math. Surveys*, 37:6 (1982), 81–105.

Kulatilaka, N.

[1] Valuing the flexibility of flexible manufacturing systems, *IEEE Trans. Eng. Manag.*, 35 (1988), 250–257.

Kulldorff, M.

[1] Optimal control of a favorable game with a time-limit, *SIAM J. Control & Optim.*, 31 (1993), 52–69.

Kumar, P.R. and Van Schuppen, J.H.

[1] On the optimal control of stochastic systems with an exponential-of-integral performance index, *J. Math. Anal. Appl.*, (1981), 312–332.

Kunita, H.

[1] Cauchy problem for stochastic partial differential equations arising in non linear filtering theory, *Syst. Control Lett.*, 1 (1981), 37–41.

[2] On backward stochastic differential equations, *Stochastics*, 6 (1981/82), 293–313.

[3] *Stochastic Flows and Stochastic Differential Equations*, Cambridge Univ. Press, New York, 1990.

Kurtz, T.G.

[1] Martingale problem for constrained Markov processes, in *Recent Advances in Stochastic Calculus*, J.S. Baras and V. Mirelli (eds.), Springer-Verlag, New York, 1990.

Kushner, H.J.

[1] Optimal stochastic control, *IRE Trans. Auto. Control*, AC-7 (1962), 120–122.

[2] On the stochastic maximum principle: Fixed time of control, *J. Math. Anal. Appl.*, 11 (1965), 78–92.

[3] On the stochastic maximum principle with "average" constraints, *J. Math. Anal. Appl.*, 12 (1965), 13–26.

[4] On the existence of optimal stochastic controls, *SIAM J. Control*, 3 (1965), 463–474.

[5] Sufficient conditions for the optimality of a stochastic control, *SIAM J. Control*, 3 (1966), 499–508.

[6] *Stochastic Stability and Control*, Academic Press, New York, 1967.

[7] Necessary conditions for continuous parameter stochastic optimization problems, *SIAM J. Control*, 10 (1972), 550–565.

[8] Existence results for optimal stochastic controls, *J. Optim. Theory & Appl.*, 15 (1975), 347–359.

Kushner, H.J. and Dupuis, P.

[1] *Numerical Methods for Stochastic Control Problems in Continuous Time*, Springer-Verlag, New York, 1993.

Kushner, H.J., and Martins, L.F.

[1] Routing and singular control for queueing networks in heavy traffic, *SIAM J. Control & Optim.*, 28 (1990), 1209–1233.

Kushner, H.J. and Ramachandran, K.M.

[1] Nearly optimal singular controls for wideband noise driven system, *SIAM J. Control & Optim.*, 26 (1988), 569–591.

[2] Optimal and approximately optimal control policies for queues in heavy traffic, *SIAM J. Control & Optim.*, 27 (1989), 1293–1318.

Kushner, H.J. and Schweppe, F.C.

[1] A maximum principle for stochastic control systems, *J. Math. Anal. Appl.*, 8 (1964), 287–302.

Kushner, H.J. and Yin, G.

[1] *Stochastic Approximation Algorithms and Applications,* Springer-Verlag, New York, 1997.

Ladyzhenskaya, O.A., Solonnikov, V.A., and Ural'tseva, N.N.,

[1] *Linear and Quasilinear Equations of Parabolic Type,* AMS Providence, RI, 1968.

Lasiecka, I. and Triggiani, R.

[1] *Differential and Algebraic Riccati Equations with Applications to Boundary/Point Control Problems: Continuous Theory and Approximation Theory, Lecture Notes in Control & Inform. Sci.*, Vol. 164, Springer-Verlag, Berlin, 1991.

Lee, E.B. and Markus, L.

[1] *Foundations of Optimal Control Theory,* John Wiley, New York, 1967.

Lanczos, C.

[1] *The Variational Principles of Mechanics, 4th ed.,* Dover, New York, 1970.

Lasry, J.M.

[1] Doctoral dissertation, Université Paris IX-Dauphine, 1974.

Lehoczky, J., Sethi, S.P, Soner, H.M., and Taksar, M.

[1] An asymptotic analysis of hierarchical control of manufacturing systems under uncertainty, *Math. Oper. Res.*, 16 (1991), 596–608.

Leland, H.E. and Toft, T.B.

[1] Optimal capital structure, endogenous bankruptcy and the term structure of credit spread, *J. of Finance*, 3 (1996), 987–1019.

Lenhart, S.M.

[1] Viscosity solutions associated with impulse control problems for piecewise deterministic processes, *Int. J. Math. & Math. Sci.*, 12 (1989), 145–158.

Lenhart, S.M. and Yamada, N.

[1] Viscosity solutions associated with switching games for piecewise-deterministic processes, *Stoch. & Stoch. Rep.*, 38 (1992), 27–47.

Letov, A.M.

[1] Analytic design of regulators, *Avtomat. i Telemekh.*, (1960), no. 4, 436–446; no. 5, 561–571; no. 6, 661–669 (in Russian). English transl. in *Automat. Remote Control,* 21 (1960).

Li, D. and Ng, W.L.

[1] Optimal dynamic portfolio selection : Multi-period mean–variance formulation, preprint.

Li, X. and Tang, S.

[1] General necessary conditions for partially observed optimal stochastic controls, *J. Appl. Prob.*, 32 (1995), 1118–1137.

Li, X. and Yong, J.

[1] Necessary conditions of optimal control for distributed parameter systems, *SIAM J. Control & Optim.*, 29 (1991), 895–908.

[2] *Optimal Control Theory for Infinite Dimensional Systems*, Birkhäuser, Boston, 1995.

Lim, A.E.B. and Zhou, X.Y.

[1] Optimal stochastic LQR control with integral quadratic constraints and indefinite control weights, *IEEE Trans. Auto. Control.*, to appear.

Lions, J.L.

[1] *Optimal Control of Systems Governed by Partial Differential Equations*, Springer-Verlag, New York, 1971.

Lions, P.L.

[1] *Generalized Solutions of Hamilton–Jacobi Equations*, Pitman Research Notes in Math., No.69, Pitman, Boston, 1982.

[2] Optimal control of diffusion processes and Hamilton–Jacobi-B equations, Part I: The dynamic programming principle and applications; Part II: Viscosity solutions and uniqueness; Part III: Regularity of the optimal cost function, *Comm. PDEs*, 8 (1983), 1101–1134; 1229–1276; *Nonlinear PDEs and Appl., College de France Seminar, Vol. V*, Pitman, Boston, 1983.

[3] Viscosity solutions and optimal stochastic control in infinite dimensions, Parts I, II, III, *Acta Math.*, 161 (1988), 243–278; *Lecture Notes in Math., Springer-Verlag*, No.1390 (1988), 147–179; *J. Funct. Anal.*, 86 (1989), 1–18.

Lions, P.L. and Sznitman, A.S.

[1] Stochastic differential equations with reflecting boundary conditions, *Comm. Pure Appl. Math.*, 3 (1984), 511–537.

Liptser, R.Sh. and Shiryayev, A.N.

[1] *Statistics of Random Processes I-II*, Springer-Verlag, New York, 1977.

Liu, J.J.

[1] On the dynamics of stochastic diffusion of manufacturing technology, *Oper. Res.*, to appear.

Liu, J.J. and Zhou, X.Y.

[1] Technology diffusion under control-imputed disturbances, preprint.

Ma, J., Protter, P., and Yong, J.

[1] Solving forward–backward stochastic differential equations explicitly — a four step scheme, *Prob. Th. & Rel. Fields*, 98 (1994), 339–359.

Ma, J. and Yong, J.

[1] Solvability of forward–backward SDEs and the nodal set of HJB equations, *Chin. Ann. Math. Ser. B*, 16 (1995), 279–298.

[2] Regular–singular stochastic controls for higher dimensional diffusions —dynamic programming approach, *IMA preprint*, #1170, 1993.

[3] Adapted solution of a degenerate backward SPDE, with applications, *Stoch. Proc. & Appl.* 70 (1997), 59–84.

[4] Approximate solvability of forward–backward differential equations, *Appl. Math. Optim.*, to appear.
[5] On linear, degenerate backward stochastic partial differential equations, *Prob. Theory & Rel. Fields,* to appear,
[6] *Forward–Backward Stochastic Differential Equations and Their Applications,* in preparation.

Mangasarian, O.L.
[1] Sufficient conditions for the optimal control of nonlinear systems, *SIAM J. Control,* 4 (1966), 139–152.

Marcus, S.I.
[1] Algebraic and geometric methods in nonlinear filtering, *SIAM J. Contr. Optim.*, 32 (1984), 817–844.

Markowitz, H.M.
[1] Portfolio selection, *J. of Fin.*, 7 (1952), 77–91.
[2] *Portfolio Selection: Efficient Diversification of Investments,* John Wiley & Sons, New York, 1959.

McLane, P.J.
[1] Optimal stochastic control of linear systems with state- and control-dependent disturbances, *IEEE Trans. Auto. Control,* 16 (1971), 793–798.

McShane, E.J.
[1] On multipliers for Lagrange problems, *Amer. J. Math.,* 61 (1939), 809-819.
[2] The calculus of variations from the beginning through optimal control theory, in *Optimal Control and Differential Equations,* A.B. Schwazkopf *et al.* (eds.), Academic Press, New York, 1978, 3–51; also *SIAM J. Control & Optim.*, 27 (1989), 916–939.

Menaldi, J.L. and Robin, M.
[1] On some cheap control problem for diffusion processes, *Trans. Amer. Math. Soc.*, 278 (1983), 771–802.

Menaldi, J.L. and Taksar, M.
[1] Optimal correction problem of a multi-dimensional stochastic system, *Automatica,* 25 (1989), 223–232.

Merton, R.C.
[1] Lifetime portfolio selection under uncertainty: The continuous time case, *Rev. Econ. Statist.*, 51 (1969), 239–265.
[2] Optimum consumption and portfolio rules in a continuous time model, *J. Econ. Theory,* 3 (1971), 373-413; Erratum 6 (1973), 213–214.
[3] An analytic derivation of the efficient portfolio frontier, *J. Fin. Quant. Anal.*, 7 (1972), 1851–1872.
[4] Theory of rational option pricing, *Bell J. Economics & Manag. Sci.*, 4 (1973), 141–183.
[5] *Continuous-Time Finance,* Oxford, Basil Blackwell, 1992.

[6] Influence of mathematical models in finance on practice: past, present and future, in *Mathematical Models in Finance,* S.D. Howison, F.P. Kelly, and P. Wilmott (eds.), Chapman & Hall, London, 1995, 1–14.

Meyer, P.A.

[1] *Probability and Potentials,* Blaisdell Publishing Company, Waltham, Mass., 1966.

Mitter, S.K.

[1] On the analogy between mathematical problems on nonlinear filtering and quantum physics, *Richerche di Automatica,* 10 (1979), 163–216.

Modigliani, F. and Hohn, F.

[1] Production planning over time and the nature of the expectation and planning horizon, *Econometrica,* 23 (1955), 46–66.

Molinari, B.P.

[1] The time-invariant linear-quadratic optimal control problem, *Automatica,* 13 (1977), 347–357.

Moore, J.B., Zhou, X.Y., and Lim, A.E.B.

[1] LQG control of linear stochastic systems with control dependent noise, *Syst. Control Lett.,* 36 (1999), 199–206.

Mortensen, R.E.

[1] Doctoral Dissertation, Dept. of Electrical Engineering, University of California at Berkeley, 1966.

Mossin, J.

[1] Optimal multiperiod portfolio policies, *J. of Bus.,* 41 (1968), 215–229.

Musiela, M. and Rutkowski, M.

[1] *Martingale Methods in Financial Modeling,* Springer-Verlag, 1997.

Nagai, H.

[1] Bellman equations of risk sensitive control, *SIAM J. Control & Optim.,* 34 (1996), 74–101.

Nesterov, Y and Nemirovskii, A

[1] *Interior-Point Polynomial Algorithms in Convex Programming,* SIAM, Philadelphia, 1994.

Neustadt, L.W.

[1] An abstract variational theory with applications to a broad class of optimization problems, *SIAM J. Control,* 4 (1966), 505–527.

Newton, G.C., Gould, L.A., and Kaiser, J.F.

[1] *Analytical Design of Linear Feedback Controls,* Wiley, New York, 1957.

Nisio, M.

[1] Some remarks on stochastic optimal controls, in *Proc. 3rd USSR-Japan Sympos. Prob. Theory,* Lecture Notes in Math. No. 550, Springer-Verlag, Berlin/Heidelberg/New York, 1976, 446–460.

[2] On a nonlinear semigroup attached to stochastic optimal control, *Publ. Res. Inst. Math. Sci.,* 12 (1976).

[3] *Lectures on Stochastic Control Theory*, ISI Lecture Notes No. 9, Kaigai Publ., Osaka, 1981.

[4] Optimal control of stochastic partial differential equations and viscosity solutions of Bellman equations, *Nagoya Math. J.*, 123 (1991), 13–37.

Nualart, D. and Pardoux, E.

[1] Stochastic calculus with anticipating integrands, *Probab. Th. Rel. Fields*, 78 (1988), 535–581.

Ocone, D.L.

[1] Finite dimensional estimation algebras in nonlinear filtering, in *Mathematics of Filtering and Identification and Applications*, M. Hazewinkel and J.C. Willems (eds.), Reidel, Dordrecht, 1981.

Ocone, D.L. and Pardoux, E.

[1] A generalized Itô-Ventzell formula. Application to a class of anticipating stochastic differential equations, *Ann. Inst. H. Poincaré Probab. Statist.*, 25 (1989), 39–71.

Pardoux, E.

[1] Stochastic partial differential equations and filtering of diffusion processes, *Stochastics*, 3 (1979), 127–167.

[2] Backward stochastic differential equations and applications, *Proc. ICM*, 1994.

Pardoux, E. and Peng, S.

[1] Adapted solution of backward stochastic equation, *Syst. & Control Lett.*, 14 (1990), 55–61.

[2] Backward stochastic differential equations and quasilinear parabolic partial differential equations, *Lecture Notes in Control & Inform. Sci., Vol. 176*, Springer-Verlag, New York, 1992, 200–217.

[3] Backward doubly stochastic differential equations and systems quasilinear SPDEs, *Prob. Theory & Rel. Fields*, 98 (1994), 209–227.

Pardoux, P. and Tang, S.

[1] The study of forward backward stochastic differential equation and its application in quasilinear PDEs, preprint.

Parthasarathy, K.R.

[1] *Probability Measures on Metric Spaces*, Academic Press, New York, 1967.

Peng, S.

[1] A general stochastic maximum principle for optimal control problems, *SIAM J. Control & Optim.*, 28 (1990), 966–979.

[2] Probabilistic interpretation for systems of quasilinear parabolic partial differential equations, *Stoch. & Stoch. Rep.*, 37 (1991), 61–74.

[3] Stochastic Hamilton–Jacobi–Bellman equations, *SIAM J. Control & Optim.*, 30 (1992), 284–304.

[4] A generalized dynamic programming principle and Hamilton–Jacobi–Bellman equation, *Stoch. & Stoch. Rep.*, 38 (1992), 119–134.

[5] Backward stochastic differential equation and its application in optimal control, *Appl. Math. & Optim.*, 27 (1993), 125–144.

[6] Backward SDE and related g-expectation, preprint.
[7] Adapted solution of backward stochastic differential equations and related partial differential equations, preprint.
[8] Monotonic limit theorem of BSDE and nonlinear decomposition theorem of Doob-Meyer's type, preprint.

Peng, S. and Wu, Z.
[1] Fully coupled forward-backward stochastic differential equations with arbitrarily large time duration, preprint.

Peng, S. and Yong, J.
[1] Determination of controllable set for a controlled dynamic system, *J. Austral. Math. Soc., Ser. B,* 33 (1991), 164–179.

Perold, A.F.
[1] Large-scale portfolio optimization, *Manag. Sci.*, 30 (1984), 1143–1160.

Pesch, H.J. and Bulirsch, R.
[1] The maximum principle, Bellman's equation, and Carathéodory work, *J. Optim. Theory & Appl.*, 80 (1994), 199–225.

Pindyck, R.S.
[1] Investment of uncertain costs, *J. Finance Econ.*, 34 (1993), 53–76.

Pliska, S.
[1] A stochastic calculus model of continuous trading: Optimal portfolios, *Math. Oper. Res.*, 11 (1986), 371–382.

Pogorelov, A.V.
[1] *Monge-Ampère Equations of Elliptic Type*, P. Noordhoff Ltd., Groningen, 1964.

Pontryagin, L.S.
[1] Optimal processes of regulation, *Uspekhi Mat. Nauk*, 14 (1959), no.1(85), 3–20 (in Russian); English transl. in *Amer. Math. Soc. Transl.* 18 (1961), no.2.
[2] Optimal processes of regulation, *Proc. Internat. Congr. Math. (Edinburgh, 1958)*, Cambridge Univ. Press, 1960, 182–202 (in Russian).
[3] The mathematical theory of optimal control processes and differential games, *Proc. Steklov Inst. Math.*, (1986), No.4, 123–159.

Pontryagin,L.S., Boltyanski,V.G., Gamkrelidze,R.V., and Mischenko,E.F.
[1] *Mathematical Theory of Optimal Processes*, Wiley, New York, 1962.

Prohorov, Y.
[1] Convergence of random processes and limit theorems in probability theory, *Theory Probab. Appl.*, 1 (1956), 157–214.

Radner, R. and Shepp, L.
[1] Risk vs. profit potential: A model for corporate strategy, *J. Econ. Dynamics & Contr.*, 20 (1996), 1373–1393.

Ray, D.B.
[1] On spectra of second order differential operators, *Trans. Amer. Math. Soc.*, 77 (1949), 299–321.

Reiman, M.I.

[1] Open queueing networks in heavy traffic, *Math. Oper. Res.*, 9 (1984), 441–458.

Reinganum, J.

[1] On the diffusion of new technology: A game theoretic approach, *Rev. Econ. Studies*, 48 (1981), 395–405.

Richard, S.

[1] Optimal consumption, portfolio, and life insurance rules for an uncertain lived individual in a continuous time model, *J. Finance Econ.*, 2 (1975), 187–203.

Rishel, R.W.

[1] Necessary and sufficient dynamic programming conditions for continuous time stochastic optimal control, *SIAM J. Control*, 8 (1970), 559–571.

Roberts, K. and Weitzman, M.L.

[1] Funding criteria for research, development, and exploration projects, *Econometrica*, 49 (1981), 1261–1287.

Robin, M.

[1] Long-term average cost control problems for continuous time Markov processes: A survey, *Acta Appl. Math.*, 1 (1983), 281–300.

Rockafellar, R.T.

[1] *Convex Analysis*, Princeton Univ. Press, Princeton, 1970.

[2] First- and second-order epi differentiability in nonlinear programming, *Trans. Amer. Math. Soc.*, 307 (1988), 75–108.

Rogers, E.M.

[1] *The Diffusion of Innovations*, The Free Press, New York, 1995.

Rosenberg, J.

[1] Factors affecting the diffusion of technology, *Explorations in Econ. History*, 10 (1972), 3–33.

Roxin, E.

[1] The existence of optimal controls, *Michigan Math. J.*, 9 (1962), 109–119.

Rozovskii, B.L.

[1] Nonnegative L^1-solutions of second order stochastic parabolic equations with random coefficients, *Trans. Math. Eng.*, (1985), 410–427.

Rund, H.

[1] *The Hamilton–Jacobi Theory in the Calculus of Variations*, Robert E. Krieger Publishing Co., Huntington, N.Y., 1973.

Runolfsson, T.

[1] The equivalence between infinite-horizon optimal control of stochastic systems with exponential-of-integral performance index and stochastic differential games, *IEEE Trans. Auto. Control*, (1994), 1551–1563.

Saisho, Y. and Tanaka, H.

[1] Stochastic differential equations for mutually reflected Brownian balls, *Osaka J. Math.*, 23 (1986), 725–740.

Samaratunga, C., Sethi, S.P., and Zhou, X.Y.

[1] Computational evaluation of hierarchical production control policies for stochastic manufacturing systems, *Oper. Res.*, 45 (1997), 258–274.

Samuelson, P.A.

[1] Lifetime portfolio selection by dynamic stochastic programming, *Rev. Econ. Stat.*, 51 (1969), 239–246.

Scott, M.R.

[1] *Invariant Embedding and its Applications to Ordinary Differential Equations—An Introduction,* Addison-Wesley, London, 1973.

Sethi, S.P., Zhang, Q., and Zhou, X.Y.

[1] Hierarchical controls in stochastic manufacturing systems with machines in tandem, *Stoch. & Stoch. Rep.*, 41 (1992), 89–118.

[2] Hierarchical controls in stochastic manufacturing systems with convex costs, *J. Optim. Theory & Appl.* 80 (1994), 299–318.

Sethi, S.P. and Zhou, X.Y.

[1] Dynamic stochastic job shops and hierarchical production planning, *IEEE Trans. Auto. Control,* AC-39 (1994), 2061–2076.

Shreve, S.E., Soner, H.M., and Xu, G.L.

[1] Optimal investment and consumption with two bonds and transaction costs, *Math. Finance,* 1 (1991), 53–84.

Shyriaev, A.N.

[1] *Optimal Stopping Rules,* Springer-Verlag, New York, 1978.

Skorohod, A.V.

[1] Stochastic equations for diffusion process in a bounded region, *Theory Prob. Appl.*, 6 (1961), 264–274.

[2] On a generalization of a stochastic integral, *Theor. Probab. Appl.*, 20 (1975), 219–233.

Soner, H.M.

[1] Optimal control with state space constraint I, II, *SIAM J. Control & Optim.*, 24 (1986), 552–562; 24 (1986), 1110–1122.

[2] Singular perturbations in manufacturing systems, *SIAM J. Control & Optim.*, 31 (1993), 132–146.

Soner, H.M., and Shreve, S.E.

[1] Regularity of the value function of a two-dimensional singular stochastic control problem, *SIAM J. Control & Optim.*, 27 (1989), 876–907.

[2] A free boundary problem related to singular stochastic control; the parabolic case, *Comm. Part. Diff. Eqs.*, 16 (1991), 373–424.

Speyer, J.L.

[1] An adaptive terminal guidance scheme based on an exponential cost criterion with application to homing missile guidance, *IEEE Trans. Auto. Control,* AC-21 (1976), 371–375.

Speyer, J.L., Deyst, J., and Jacobson, D.H.

[1] Optimization of stochastic linear systems with additive measurement and process noise using exponential performance criteria, *IEEE Trans. Auto. Control*, AC-19 (1974), 358–366.

Stampacchia, G.

[1] Hilbert's twenty-third problem extension of the calculus of variations, *Mathematical Developments Arising from Hilbert Problems*, AMS, Providence, RI, 1976, 611–628.

Stojanovic, S. and Yong, J.

[1] Optimal switching for systems governed by nonlinear evolution equations, *Numer. Funct. Anal. Optim.*, 9 (1987), 995–1030.

[2] Optimal switching for partial differential equations I, II, *J. Math. Anal. Appl.*, 138 (1989), 418–438; 439–460.

Strook, D.W. and Varadhan, S.R.S.

[1] Diffusion processes with boundary conditions, *Comm. Pure Appl. Math.*, 24 (1971), 147–225.

Subbotin, A.I.

[1] A generalization of the basic equation of the theory of differential games, *Soviet Math. Dokl.*, 22 (1980), 358–362.

[2] *Generalized Solutions of First Order PDEs: The Dynamic Optimization Perspective*, Birkhäuser, Boston, 1995.

Subbotina, N.N.

[1] The maximum principle and the superdifferential of the value function, *Problems Contr. Inform. Theory*, 18 (1989), 151–160.

Sussmann, H.J.

[1] Geometry and optimal control, preprint.

Taksar, M.

[1] Skorohod problems with nonsmooth boundary conditions, *J. Comput. & Appl. Math.*, 40 (1992), 233–251.

Taksar, M. and Zhou, X.Y.

[1] Optimal risk and dividend control for a company with a debt liability, *Insurance: Math. & Econom.*, 22 (1998), 105–122.

Tanaka, H.

[1] Stochastic differential equations with reflection boundary condition in convex regions, *Hiroshima Math. J.*, 9 (1979), 163–177.

[2] Stochastic differential equation corresponding to the spatially homogeneous Boltzmann equation of Maxwellian and non-cutoff type, *J. Fac. Sci. Univ. Tokyo*, 34 (1987), 351–369.

Tang, S.

[1] The maximum principle for partially observed optimal of stochastic differential equations, *SIAM J. Control & Optim.*, 36 (1998), 1596–1617.

Tang, S. and Li, X.

[1] Necessary conditions for optimality control of stochastic systems with random jumps, *SIAM J. Control & Optim.*, 32 (1994), 1447–1475.

Tang, S. and Yong, J.

[1] Finite horizon stochastic optimal switching and impulse controls with a viscosity solution approach, *Stoch. & Stoch. Rep.*, 45 (1993), 145–176.

Tarres, R.

[1] Comportement asymptotique d'un problème de controle stochastique, *Cahier de math. de la Decision*, No. 8215, Universite de Paris IX, 1982.

[2] Asymptotic evolution of a stochastic control problem, *SIAM J. Control & Optim.*, 23 (1985), 614–631.

Tataru, D.

[1] Viscosity solutions for the dynamic programming equations, *Appl. Math. Optim.*, 25 (1992), 109–126.

[2] Viscosity solutions of Hamilton–Jacobi equations with unbounded nonlinear terms, to appear.

Van Moerbecke, P.

[1] On optimal stopping and free boundary problems, *Arch. Rat. Mech.*, (1976).

Varadhan, S.R.S. and Williams, R.J.

[1] Brownian motion in a wedge with oblique reflection, *Comm. Pure Appl. Math.*, 38 (1985), 405–433.

Vinter, R.B.

[1] New results on the relationship between dynamic programming and the maximum principle, *Math. Control Sign. Syst.*, 1 (1988), 97–105.

Wald, A.

[1] *Sequential Analysis*, John Wiley & Sons, New York, 1947.

[2] *Statistical Decision Functions*, John Wiley & Sons, New York, 1950.

Walsh, J.B.

[1] An introduction to stochastic partial differential equations, *Lect. Notes in Math.*, 1180 (1986), 266–423.

Warga, J.

[1] *Optimal Control of Differential and Functional Equations*, Academic Press, New York, 1972.

Wein, L.M.

[1] Scheduling networks of queues: Heavy traffic analysis of a two-station network with controllable inputs, *Oper. Res.*, 38 (1990), 1065–1078.

Whitt, W.

[1] On the heavy-traffic limit theorem for $GI/G/\infty$ queues, *Adv. Appl. Probab.*, 14 (1982), 171–190.

[2] Departures from a queue with many busy servers, *Math. Oper. Res.*, 9 (1984), 534–544.

Whittle, P.

[1] Risk sensitive linear quadratic Gaussian control, *Adv. Appl. Probab.*, 13 (1974), 764–777.

[2] *Risk Sensitive Optimal Control*, John Wiley, New York, 1990.

Wiegner, M.

[1] Global solutions to a class of strongly coupled parabolic systems, *Math. Ann.* 292 (1992), 711–727.

Willems, J.L. and Willems, J.C.

[1] Feedback stabilizability for stochastic systems with state and control dependent noise, *Automatica,* 12 (1976), 277–283.

Willems, J.C.

[1] Least squares stationary optimal control and the algebraic Riccati equation, *IEEE Trans. Auto. Control,* 16 (1971), 621–634.

Wong, W.S.

[1] New classes of finite-dimensional nonlinear filters, *Syst. Control Lett.,* 3 (1983), 155–164.

Wong, W.S. and Yau, S.S.T.

[1] The estimation algebra of nonlinear filtering systems, preprint.

Wonham, W.M.

[1] On the separation theorem of stochastic control, *SIAM J Control,* 6 (1968), 312–326.

[2] On a matrix Riccati equation of stochastic control, *SIAM J. Control,* 6 (1968), 681–697.

[3] *Linear Multivariable Control: a Geometric Approach, 2nd Edition,* Springer-Verlag, 1979.

Wu, H. and Li, X.

[1] A linear quadratic problem with unbounded control in Hilbert spaces, preprint.

Yao, D.D., Zhang, S., and Zhou, X.Y.

[1] LQ control without Riccati equations: Deterministic systems, preprint.

[2] LQ control without Riccati equations: Stochastic systems, preprint.

Yau, S.T. and Yau, S.S.T.

[1] Finite-dimensional filters with nonlinear drift XI: Explicit solution of the generalized Komolgorov equation in Brockett-Mitter program, preprint.

Yong, J.

[1] Systems governed by ordinary differential equations with continuous, switching and impulse controls, *Appl. Math. Optim.,* 20 (1989), 223–235.

[2] Optimal switching and impulse controls for distributed parameter systems, *System Sci. & Math. Sci.,* 2 (1989), 137–160.

[3] Differential games with switching strategies, *J. Math. Anal. Appl.,* 145 (1990), 455–469.

[4] A zero-sum differential game in a finite duration with switching strategies, *SIAM J. Control & Optim.,* 28 (1990), 1234–1250.

[5] Existence of the value for a differential game with switching strategies in a Banach space, *System Sci. & Math. Sci.,* 4 (1991), 321–340.

[6] *Dynamic Programming Method and the Hamilton–Jacobi–Bellman Equations,* Shanghai Sci. Press, 1992 (in Chinese).

[7] Zero-sum differential games involving impulse controls, *Appl. Math. Optim.*, 29 (1994), 243–261.

[8] Finding adapted solutions of forward–backward stochastic differential equations—method of continuation, *Prob. Theory & Rel. Fields,* 107 (1997), 537–572.

[9] Relations among ODEs, PDEs, FSDEs, BSDEs, and FBSDEs, *Proc. 36th CDC,* 2779–2784.

[10] Linear forward–backward stochastic differential equations, *Appl. Math. Optim.*, 39 (1999), 93–119.

[11] Some results on the reachable sets of linear stochastic systems, *Proc. 37th CDC,* 1335–1340.

[12] European type contingent claims in an incomplete market with constrained wealth and portfolio, preprint.

Yosida, K.

[1] *Functional Analysis, 6th ed.*, Springer-Verlag, Berlin, 1980.

You, Y.

[1] Optimal control for linear systems with quadratic indefinite criterion on Hilbert spaces, *Chin. Ann. Math. Ser. B,* 4 (1983), 21–32.

Young, L.C.

[1] *Calculus of Variations and Control Theory,* W.B. Saunders, Philadelphia, 1969.

Zakai, M.

[1] On the optimal filtering of diffusion processes, *Z. Wahrsch. Geb.*, 11 (1969), 230–243.

Zeidler, E.

[1] *Nonlinear Functional Analysis and Its Applications, I: Fixed-Point Theorems,* Springer-Verlag, 1986.

Zeleny, M.

[1] *Multiple Criteria Decision Making,* McGraw-Hill, New York, 1981.

Zhang, Q.

[1] Controlled partially observed diffusions with correlated noise, *Appl. Math. Optim.*, 22 (1990), 265–285.

Zhou, X.Y.

[1] Maximum principle, dynamic programming and their connection in deterministic control, *J. Optim. Theory & Appl.*, 65 (1990), 363–373.

[2] The connection between the maximum principle and dynamic programming in stochastic control, *Stoch. & Stoch. Rep.*, 31 (1990), 1–13.

[3] Remarks on optimal controls of stochastic partial differential equations, *Syst. Control Lett.*, 16 (1991), 465–472.

[4] A unified treatment of maximum principle and dynamic programming in stochastic controls, *Stoch. & Stoch. Rep.*, 36 (1991), 137–161.

[5] A duality analysis on stochastic partial differential equations, *J. Funct. Anal.*, 103 (1992), 275–293.

[6] On the existence of optimal relaxed controls of stochastic partial differential equations, *SIAM J. Control & Optim.*, 30 (1992), 247–261.

[7] On the necessary conditions of optimal controls for stochastic partial differential equations, *SIAM J. Control & Optim.*, 31 (1993), 1462–1478.

[8] A class of semilinear stochastic partial differential equations and their controls: Existence results, *Stoch. Proc. & Their Appl.*, 44 (1993), 89–106.

[9] Verification theorems within the framework of viscosity solutions, *J. Math. Anal. Appl.*, 177 (1993), 208–225.

[10] Sufficient conditions of optimality for stochastic systems with controllable diffusions, *IEEE Trans. Auto. Control*, AC-41 (1996), 1176–1179.

[11] Deterministic near-optimal controls Part II: Dynamic programming and viscosity solution approach, *Math. Oper. Res.*, 21 (1996), 655–674.

[12] Stochastic near-optimal controls: Necessary and sufficient conditions for near-optimality, *SIAM J. Control & Optim.*, 36 (1998), 929–947.

Zhou, X.Y. and Li, D.

[1] Explicit efficient frontier of a continuous-time mean–variance portfolio selection problem, in *Control of Distributed Parameter and Stochastic Systems*, S. Chen *et al.* (eds.), Kluwer, Boston/Dordrecht/London, 1999, 323–330.

Zhou, X.Y., Yong, J., and Li, X.

[1] Stochastic verification theorems within the framework of viscosity solutions, *SIAM J. Control & Optim.*, 35 (1997), 243–253.

Zhu, H.

[1] Dynamic programming and variational inequalities in singular stochastic control, Ph.D. thesis, Brown University, 1991.

[2] Generalized solution in singular stochastic control: the nondegenerate problem, *Appl. Math. Optim.*, 25 (1992), 225–245.

Index

Applications of Mathematics

(continued from page ii)

33 Embrechts/Klüppelberg/Mikosch, **Modelling Extremal Events** (1997)
34 Duflo, **Random Iterative Models** (1997)
35 Kushner/Yin, **Stochastic Approximation Algorithms and Applications** (1997)
36 Musiela/Rutkowski, **Martingale Methods in Financial Modeling: Theory and Application** (1997)
37 Yin/Zhang, **Continuous-Time Markov Chains and Applications** (1998)
38 Dembo/Zeitouni, **Large Deviations Techniques and Applications**, Second Ed. (1998)
39 Karatzas/Shreve, **Methods of Mathematical Finance** (1998)
40 Fayolle/Iasnogorodski/Malyshev, **Random Walks in the Quarter Plane** (1999)
41 Aven/Jensen, **Stochastic Models in Reliability** (1999)
42 Hernández-Lerma/Lasserre, **Further Topics on Discrete-Time Markov Control Processes** (1999)
43 Yong/Zhou, **Stochastic Controls: Hamiltonian Systems and HJB Equations** (1999)

Made in the USA
Monee, IL
13 September 2026

62976977R00256